Proofs in Competition Math: Volume 1

Dennis Chen, Freya Edholm, Alex Toller

First edition: April 2019

Contents

Acknowledgments

This book is dedicated to everyone who has ever dreamt of studying math, in any way, shape, or form. To anyone who picks this book up, you have my thanks. May you have a truly legendary mathematical journey, just like the one all of us are currently undertaking!

In particular, I would like to thank several individuals and groups who have played monumental roles in shaping not only my mathematical aspirations, but also my entire life trajectory. My undying gratitude extends to the educators in my life, who have provided countless opportunities and opened countless doors. The teachers at my middle school, Coral Academy of Science, and my high school, North Star Online School, as well as the professors at the University of Nevada, Reno, where I took math courses throughout high school, have planted within me the seeds of desire to learn, and have introduced me to unique and exciting fields of math. In particular, I would like to thank Dr. Edward Keppelmann of the University of Nevada, Reno, as he had devoted his time and efforts to collaborating with me in creating and organizing the Mathematical Talent Exhibition - a series of contests I had written and co-organized with the Northern Nevada Math Club. His continued patience and passion, and heartfelt support for the cause of reaching out to elementary and high school students alike and fostering a love of math within them, is truly appreciated.

A special mention goes to Mr. Emre Gul, coach of my middle school's Math Club, who lit the passionate fire of genuine interest and kindled within me a true thirst for knowledge and intellectual exploration. From the moment he convinced me to join the first after-school practice session of my sixth grade spring semester and introduced me to the unforgettably magical experience that was my first ever MathCounts competition, I decided to pursue math as a lifelong dream, not because I had been forcefully behooved lest I feel a nagging guilt or suffer external consequence, but rather because I had been so wonderfully inspired to learn math for the sake of learning math by all those who supported me along the way, and continue to do so to this day.

In a similar vein, I am also deeply thankful to Mrs. Sherry Griffin, president of the Northern Nevada Math Club, which I joined in the middle of my seventh grade year. She had been instrumental in introducing me to not only the competitive aspect of mathematics, but also the collaborative teamwork aspect, as well as the pure, unadulterated joy of both learning from others, and passing down the invaluable knowledge I have learned from my fellow teammates to the future generations of bright-eyed young mathematicians. In addition, her tireless efforts have paved the way for a new generation of math students to gain the most out of the competition math sphere, through monthly contests, weekly practices, and events such as math debates and fun math-related gatherings. In entering my high school years, I began to associate with peers, teachers, and coaches from the Davidson Academy of Nevada, for whom I express my gratitude. Not only are the students of the Academy to this day some of my closest friends, the staff of the Academy themselves have stepped up several times to administer prominent math competitions such as the AMC 10/12 and the AIME, providing students such as myself with the opportunity to further improve their math skills and bond with fellow aspiring mathematicians.

In this regard, I am also thankful to the organizers of competitions such as the Stanford and Berkeley Math Tournaments, the Caltech-Harvey Mudd Math Competition, and the Harvard-MIT Math Tournament, for providing absolutely wonderful experiences that have left their footprints on me, both mathematically and sentimentally. These contests have served as perfect bonding experiences with my teammates, as well as invaluable learning experiences. Often during practices, we would find ourselves commiserating over our failed attempts at solving the toughest problems, going over the steps and scratching our heads, scribbling away on a blackboard and scratching our heads some more, and then either finally coming to an insight that inspired both celebration and awe, or a collective vow to destroy any other problems like it the next time around! These were the types of moments that truly made my competitive math career.

Finally, and arguably foremost, I am indebted to my wholly supportive and loving parents, who have always been by my side through the toughest of times. They were never the sort to give up, even under the mounting pressure of a tough situation. Whenever it seemed as if all hope was lost in the way of breaking new ground in my life, they would always find a way to make it happen regardless. They turned even the most remote of pipe dreams into reality; they were, and still are, my superheroes.

Preface

All too often, math[1] is presented formulaically and independently of exposition, with little to no contextualization or applicable real-world connection. Through common school mathematics, students often memorize and regurgitate formulas that seem meaningless in that they lack the necessary accompanying context to fully understand them beyond the surface level. Students may find themselves excelling in school math classes by memorizing formulas, but not their applications or the motivation behind them. As a consequence, understanding derived in this manner is tragically based on little or no proof.

This is why studying proofs is of paramount importance! Proofs help us understand the nature of mathematics and show us the key to appreciating its elegance. Many readers have likely had personal experiences with the regurgitative, computational side of mathematics: the school math teacher relentlessly drilling formulas into you for an upcoming test, books trying to help you prepare for said tests, those timed arithmetic drills that are the bane of every elementary schooler's existence... Though the formulas themselves are indeed versatile in modern day mathematics, without the rigor or motivation driving them, they lose all purpose. Simply memorizing a formula is akin to memorizing the spelling of a word but failing to assimilate its definition.

But even getting past the concern of "why should this be true?" students often face the equally-pressing question of "when will I ever need this in life?" This book aims to remedy both of these issues at a wide range of levels, from the fundamentals of competition math all the way to the Olympiad level and beyond. The book is constructed in hopes that the reader will gain something from not only a laundry list of formulas, but their real-world applications as well, along with their proofs and several accompanying exercises involving interdisciplinary, higher-order thought.

And don't worry if you don't know all of the mathematics in this book; there will be prerequisites for each skill level, giving you a better idea of your current position, your strengths and weaknesses, and, hopefully, allowing you to set realistic goals as a math student. So, mathematical minds, we set you off!

---

[1] Or maths, if you aren't from the US. If you insist!

0.1　Prerequisites

Beginner

The following are *required* for a thorough understanding of the content in the Beginner sections of this book:

1. A willingness to learn!

2. A minimal background in middle school math (namely an intuitive understanding of how operations, variables, equations, and functions work).

The following are *recommended, but not required:*

1. Notes and a writing utensil (or a note-taking application on your computer/tablet/phone).

2. At least some experience/exposure/familiarity with competition math (such as MathCounts, AMC 8).

3. Minimal exposure to proofs (being familiar with the different methods such as induction and proof by contradiction).

Intermediate

In addition to the Beginner prerequisites, the following are *required* for a thorough understanding of the content in the Intermediate sections of this book:

1. A solid understanding of the middle school math curriculum (A or A+ grades in middle school math courses), and some exposure to high school math.

2. At least minimal exposure to the idea behind a mathematical proof, and some understanding of what distinguishes a rigorous proof from intuition.

The following are *recommended, but not required:*

1. Sufficient experience/exposure/familiarity with middle school competition math to score at least 40/46 on a typical MathCounts Chapter Competition, 23/25 on the AMC 8, 100/150 on the AMC 10, or 80/150 on the AMC 12.

Advanced

In addition to the Beginner and Intermediate prerequisites, the following are *required* for a thorough understanding of the content in the Advanced sections of this book:

1. Familiarity and comfort with the various methods of proof.

2. A solid understanding of the high school math curriculum (A or A+ grades in high school math courses up to Pre-Calculus, or a 750+ score on the SAT II Math Level II exam (note: this is not a strict benchmark, only a guideline. However, if you are consistently scoring below 700, the Advanced sections may assume understanding of material that you have not fully mastered yet)).

3. A genuine interest in understanding rigorous mathematics stemming from a love of learning! These sections of the text will weed out those who are not highly devoted to passionately learning the subject matter. (Forewarning: if you're in it solely for the money or the glory, mathematics might not be the best fit for you...Mathematics is a *very* intense field!)

The following are *recommended, but not required:*

1. Sufficient experience with competition math to qualify for the AIME (120/150 on the AMC 10, or 100/150 on the AMC 12).

2. Successful completion of a post-secondary level "introduction to proofs" course or equivalent (such as a course in symbolic logic).

3. At least some exposure to the types of problems that are commonly asked in higher-level proof-based competitions, such as the USA(J)MO/IMO, the Team portions of HMMT February, and USAMTS.

0.2 Prerequisite Exams

Beginner Exam

Directions: Answer all of the following 7 questions, which you will have 60 minutes to complete. Unless otherwise specified, all answers must be complete, simplified, and fully justified with supporting work and mathematically accurate reasoning.

The test begins on the following page. It is recommended to either print this test out and write your answers on separate scratch paper, or type your responses out and score them using the scoring rubric after you are finished.

Note: Throughout this test, "compute", "find", or "evaluate" all mean that you should provide a numerical answer *along with justification.* A correct answer without supporting work will not receive credit.

1. (15 pts.)

   (a) State the definition of a permutation of a set.

   (b) Consider the set $S = \{1, 2, 3, 3\}$. How many distinct permutations does S have?

   (c) How many distinct permutations does the set $T = \{1, 1, 2, 2, 3, 3, 4, 4\}$ have?

2. (15 pts.)

   (a) Define each of the following symbols: $\forall, \exists, \in, \supset, \subset, \emptyset, \mathbb{R}, \mathbb{Z}, \mathbb{Q}, \mathbb{C}$.

   (b) Is the following statement true? Why or why not? "$\forall a \in \mathbb{Q}, \exists b, c \in \mathbb{Z}$ such that $a = \dfrac{b}{c}$."

   (c) Is the following statement true? Why or why not? "$\forall a \in \mathbb{Q}, \exists b, c \in \mathbb{N}$ such that $a = \dfrac{b}{c}$."

3. (20 pts.)

   (a) Let
   $$S = \sum_{k=1}^{n} k.$$
   Prove that $S = \dfrac{n^2 + n}{2}$.

   (b) Let
   $$S = \sum_{k=1}^{n} k^2.$$
   Prove that $S = \dfrac{n(n+1)(2n+1)}{6}$.

   (c) Let
   $$S = \sum_{k=1}^{n} k^3.$$
   Prove that
   $$S = \left(\sum_{k=1}^{n} k \right)^2$$
   for all $n \geq 1$.

   (d) Show that $(1 \cdot (1^2 - 1 + 1)) + (2 \cdot (2^2 - 2 + 1)) + (3 \cdot (3^2 - 3 + 1)) + (4 \cdot (4^2 - 4 + 1)) + \ldots + (n \cdot (n^2 - n + 1)) = \dfrac{n(n+1)(3n^2 - n + 4)}{12}$.

4. (15 pts.) In rectangle $ABCD$, let $AB = 2$ and $BC = 1$.

   (a) If $ABCD$ is a cyclic quadrilateral inscribed in circle O, what is the area of O?

   (b) Let P be a point on $\overline{AB}$ such that $\angle ACP = 15°$. Compute, with explanation, the area of trapezoid $APCD$.

   (c) Let d_0 be the length of the diagonal of $ABCD$. Let d_1 be the length of the diagonal of the rectangle with side lengths 2 and d_0, let d_2 be the length of the diagonal of the rectangle with side lengths d_0 and d_1, and so forth. Furthermore, let u_i be the remainder upon dividing d_i^2 by 5. Show that the u_i are periodic and determine their period.

   (d) Refer to part (c). Compute
   $$\sum_{i=0}^{2019} u_i.$$

x

5. (12 pts.)

   (a) Provide a careful and complete statement of the Binomial Theorem.

   (b) What is the coefficient of the $x^2 y^3$ term in the expansion of $(x + y)^5$?

   (c) What is the coefficient of the $x^{20} y^{18}$ term in the expansion of $(20x + 18y)^{38}$?

6. (13 pts.) Your brother set a four-digit passcode on your phone, but he forgot what it was! It is your challenge to figure out the code.

   (a) How many possible codes could he have set?

   (b) If your brother does remember that each digit of the passcode is distinct, how many codes could he have set?

   (c) What if the code was known to have the digits in non-decreasing order? In this part, the digits are assumed to be non-distinct.

7. (10 pts.) Let x and y be positive real numbers such that $(x^2 y^2 + 1)(x^2 y^2 - 1) = 48$. Find, with proof, the maximum value of $x^4 + y^4$.

End of exam.

Intentionally blank page.

Intermediate Exam

Directions: Answer all of the following 6 questions, which you will have 90 minutes to complete. Unless otherwise specified, all answers must be complete, simplified, and fully justified with supporting work and mathematically accurate reasoning.

The test begins on the following page. It is recommended to either print this test out and write your answers on separate scratch paper, or type your responses out and score them using the scoring rubric after you are finished.

Note: Throughout this test, "compute", "find", or "evaluate" all mean that you should provide a numerical answer *along with justification*. A correct answer without supporting work will not receive credit.

1. (12 pts.) Let $f(x) = ax^2 + bx + a$, where a and b are constants.

   (a) If one of the roots of the equation $f(x) = 0$ is $x = 17$, what is the other root? Explain your answer.

   (b) What if one of the roots of the equation $f(x) = 0$ were the general root $x = r$? Find, with proof, the other root in terms of r.

2. (20 pts.)

   (a) Define a *quadratic residue* and *quadratic nonresidue* modulo p, for p a prime.

   (b) What are the QRs and QNRs modulo 19?

   (c) Define the Legendre and Jacobi symbols.

   (d) State (but do not prove) the Law of Quadratic Reciprocity.

3. (15 pts.)

   (a) Let p be a prime number, and let $\tau(n)$ be the number of positive divisors of n. When is $\tau(p^p) > \tau((p+1)^{p+1})$?

   (b) Let p be any positive integer, and let $\tau(n)$ be the number of positive divisors of n. When is $\tau(p^p) > \tau((p+1)^{p+1})$?

4. (15 pts.) In $\triangle ABC$, let $AB = 25$, $BC = 39$, $CA = 40$. Let I be the incenter of $\triangle ABC$. In each of triangles $\triangle ABI$, $\triangle AIC$, and $\triangle BIC$, construct centroids O_1, O_2, O_3 respectively. Furthermore, let $\overline{IO_1}$, $\overline{IO_2}$, $\overline{IO_3}$ intersect $\overline{AB}$, $\overline{AC}$, $\overline{BC}$ at D, E, F respectively. Compute the area of $\triangle DEF$.

5. (20 pts.) Consider the 2×2 matrix
$$\begin{bmatrix} 4x & x^2 \\ 2 & 3 - x \end{bmatrix}$$

   (a) Compute $\det M$ in terms of x.

   (b) For what values of x is M singular? What about M^2?

   (c) In terms of x, compute M^{-1}.

   (d) If $x = 1$, what are the eigenvalues and eigenvectors of M?

6. (18 pts.)

   (a) An ant begins at $(0,0)$ in the Cartesian plane. Each minute, it moves one unit up, right, down, or left, each direction with equal probability. Compute the probability that, after four minutes, the ant ends back up at $(0,0)$.

   (b) Briefly describe the *gambler's ruin* problem and its outcome for two players Alpha and Beta (who begin with d and $N - d$ dollars respectively, for some positive integer N).

End of exam.

Intentionally blank page.

Advanced Exam

Directions: Answer all of the following 3 questions, which you will have 120 minutes to complete. Unless otherwise specified, all answers must be complete, simplified, and fully justified with supporting work and mathematically accurate reasoning.

The test begins on the following page. It is recommended to either print this test out and write your answers on separate scratch paper, or type your responses out and score them using the scoring rubric after you are finished.

Note: Throughout this test, "compute", "find", or "evaluate" all mean that you should provide a numerical answer *along with justification.* A correct answer without supporting work will not receive credit.

1. (50 pts.)

   (a) Let $f : \mathbb{R} \mapsto \mathbb{R}$ be a function from the real numbers to the real numbers. Rigorously define what it means for f to be convex.

   (b) Show that the intersection of any number of convex sets is convex.

   (c)   i. Provide a careful and complete statement of Jensen's Inequality.
         ii. Prove Jensen's Inequality.

   (d) Let $x_1, x_2, x_3, \ldots x_n \in \mathbb{R}^+$. Show that
   $$\left(\frac{x_1 + x_2 + x_3 + \ldots + x_n}{n} \right)^{x_1 + x_2 + x_3 + \ldots + x_n} \leq \prod_{i=1}^{n} x_i^{x_i}$$

   (e) Find the largest real value of c for which
   $$\sum_{i=1}^{5} x_i^2 \geq c \left(\sum_{i=1}^{4} x_i x_{i+1} \right)$$
   holds for any $x_i \in \mathbb{R}$ $(1 \leq i \leq 5)$.

   (f) Let $a, b, c > 0$. Prove that
   $$\sum_{cyc} 2a^3 + a^2 b \geq \sum_{cyc} 3ab^2$$

   (g)   i. Provide a careful and complete statement of the Cauchy-Schwarz inequality.
         ii. *Using Jensen's Inequality*, prove the Cauchy-Schwarz inequality.

2. (30 pts.)

   (a) Give a careful and complete statement of the Multinomial Theorem.

   (b) With explanation, compute each of the following:
       i. $\dbinom{-4}{2}$
       ii. $\dbinom{4.5}{3}$
       iii. $\dbinom{2\sqrt{2}}{2}$
       (You cannot use the gamma function!)

   (c) Let $G(a_n; x)$ denote the generating function of the sequence (a_n) in terms of x. Propose and prove a formula for $G(a_n^2; x)$.

   (d) What is the power series of the function $f(x) = \dfrac{\arctan(x)}{\frac{1}{2} + \frac{x^2}{2}}$ about $x = 0$?

   (e) Show that the Maclaurin series of any polynomial is equal to itself.

   (f) Give an example of an infinitely-differentiable function f whose Maclaurin series does not converge to $f(x)$ for any $x > 0$.

3. (20 pts.)

   (a) Give a complete and careful statement, as well as a proof, of the Chebyshev Inequality.

   (b) According to the Chebyshev Inequality, the probability of some random variable X with mean μ lying in the interval $[\mu - a\sigma, \mu + a\sigma]$ is at least $\dfrac{1}{n}$. What is a in terms of n?

   (c) State Burnside's Lemma, along with the axioms of a group. You do not need to prove the lemma.

   (d) Using only the colors red, white, and blue, in how many ways can we color the six faces of a cube? (Rotations are considered identical.)

End of exam.

Intentionally blank page.

Throughout this Book...

...you will encounter a variety of opportunities to enhance your understanding of the content, and to hone your ability to apply it to various contexts, in the form of *examples, exercises,* and *extension exercises.*

An example is a problem similar to one that you may encounter in an actual competition setting that demonstrates the concept, idea, or motivation at hand. The problem will be solved immediately after it is stated; take these as a learning opportunity, and as guides if you find yourself stuck on an exercise. It is not encouraged, however, to merely copy a procedure from an example for an exercise, as this leads to little or no actual learning in the long run.

An exercise is a problem within a set at the end of a section. Much like an example, an exercise is solvable with all of the knowledge gained from the section, but unlike an example, it will not have its solution immediately following it. Instead, the Solutions to Exercises part at the end contains the solutions to all exercises throughout the book (with exercises re-printed for your convenience).

An extension exercise is a type of exercise that tests understanding of concepts at a higher and deeper level. It may require you to integrate formulas, definitions, and/or theorems from several different areas studied, to analyze the motivation behind a proof rather than merely memorize it, or to convert intuition into mathematically rigorous proof. Since they are designed explicitly for the purpose of challenging those readers who have already mastered the fundamental concepts, extension exercises are especially proof-focused and rarely computational.

Part I

The Nature of Mathematical Arguments

Chapter 1

Defining "Proof" and Proof Terminology

In ordinary contexts, *proof* may refer to evidence that helps establish a truth or statement, a demonstration that shows something to have a certain characteristic, or perhaps anything that compels acceptance, whether it be grounded in truth or completely nonsensical. In the mathematical context, however, a *proof* refers to an unambiguously correct justification that goes beyond the mere shadow of a doubt" by showing that there is no other possible outcome that could exist, for otherwise it would contradict some basic axiom of mathematics or result in an absurdity.

Unlike in a court of law, merely convincing the other party that a statement is *likely* or *probably* true is not good enough - even to a 99.99% degree of certainty. Without a 100% conviction that the statement is accurate, based on mathematical truth, we cannot truly consider something to be a proof. In other words, we can't rely on popular opinion, intuition, or the gut feeling that something is right - it needs to actually be right, and furthermore obtained through completely rigorous means.

Moreover, mathematical proofs have an underlying grammatical structure, one which mathematicians use to communicate despite linguistic barriers, and one which solidifies the universal, standardized nature of proofs in the mathematical sphere. Each and every piece of vocabulary has a precise meaning, deviation from which in even the slightest would result in confusion, ambiguity, and the obfuscation of the logic within the proof itself. These include ordinarily harmless English conjunctions such as "and," "or," and "if," which, if used improperly in a proof, can completely alter its meaning. Hence, it is of the utmost importance that we study at least the most commonly-used proof terminology.

Consider the word "and," for instance. In common conversation, this particular particle usually serves to add on another point of interest to the speaker's arsenal; in mathematics, it is a conditional used in statements that affects whether they are true or false. Or works similarly to and in mathematics, but has almost the complete opposite meaning in ordinary conversation. We usually wish to make our list of options or choices in vernacular more or less limited through these conjunctions, but in mathematical proofs, they do not refer to choices or alternatives.

The word "if" is another point of contention that carries vastly different definitions between the non-mathematical and mathematical spheres. Ordinarily, "if" is a conditional, as well as in mathematics, but its mathematical meaning is more nuanced. Mathematicians usually use if as a gateway of sorts from one proposition to another, as in the example statement "If it rains tomorrow, then we cant have a picnic".

"Not" is simple enough: its meaning is quite similar between the common and mathematical spheres. In both contexts, it is a negation of a proposition or part of a statement. In fact, the standalone statement "not A" is called the *negation* of A. It may or may not change the truth/falsehood of a statement, depending on other conditionals.

1.1 Logical Notation

To write "or" in mathematical notation, we use the notation $p \lor q$. To write "and", we use $\land$, so $p \land q$ means "p and q". "Not" is written with the symbol $\neg$ or $\sim$ preceding a statement, but we will denote "not" by $\neg$ in this book. Thus, "not p" becomes $\neg p$. In addition, we can form new operators using a combination of these three fundamental ones, such as the NAND (not and) operator and the NOR (not or) operator.

The symbol $\implies$ is the implication symbol, and is used to denote that one statement p being true implies that another statement q is also true (in the form $p \implies q$). This is also known as the *material conditional*. More succinctly stated, $p \implies q$ is true if either p is false, q is true, or both. The only way for this statement to be false is if p is true, but q is false. (Well examine why in more detail in the next subsection.) Alternatively, the implication symbol may be written as $\rightarrow$ or $\supset$. The symbol $\iff$ is the equivalence symbol, which in the context of the statement $p \iff q$, says that the statement is true when both p and q are true or when they are both false. This is known as a *bidirectional conditional,* and is alternatively read "p is true iff q is true" (see the next subsection for further details).

Finally, we introduce the notion of *logical quantifiers.* Statements may be true under some condition, namely for only specific values of an input variable in its domain, or perhaps for all input values in the domain. In the former case, we use the quantifier $\exists$ (read: "there exists"), as in the example true statement

$$\exists a \in \mathbb{Z} \text{ s.t. } a^2 = 1. \tag{1.1}$$

Note: In the statement above, $\mathbb{Z}$ denotes the integers, that is, the set $\{\dots, -2, -1, 0, 1, 2, \dots\}$, and s.t. is the abbreviation for the phrase "such that".

Similarly, the quantifier representing "for all" is $\forall$, as in the true statement

$$\forall x > 1, x \in \mathbb{R}, x < x^2. \tag{1.2}$$

Exercise 1. *For further demonstrative purposes, both of the following statements are false. Why?*

$$\forall b \in \mathbb{R}, b^2 > 0.$$

$$\exists c \in \mathbb{R} \text{ s.t. } c^2 = -1.$$

1.2 What is a Statement?

A *statement,* in the mathematical sense, is a sentence that is either true or false. Statements cannot be subjective, nor can they be ambiguous or otherwise up to interpretation.

Example 1.2.0.1. *Is the sentence "I like the color red" a statement?*

Solution 1.2.0.1. *The sentence "I like the color red" would syntactically be a statement in English grammar, but would not be a mathematical statement. This is because it expresses a subjective opinion that does not follow from the axioms of mathematical logic. However, it does form a thesis, namely that red is a color liked by the speaker, so it does qualify as an English statement.*

Henceforth, consider the statement from the previous subsection: "If it rains tomorrow, then we can't have a picnic." For now, we will assume that this statement is true. We will dissect this, and many, many others like it, from a mathematical point of view rather than a common one. The phrase "it rains tomorrow" is the first part of the statement, and the phrase "we can't have a picnic" is the second part of the statement. *If-then statements,* such as this one, in mathematics set up conditions for the statement to be true or false. In fact, if-then statements are also called *conditional statements* for this reason. Here, we must keep in mind that rain equates to not having a picnic, and sunshine equates to being able to have a picnic.

Let's examine possible alterations of the statement to illustrate this point:

"If it's sunny tomorrow, then we can't have a picnic."

This statement would be false, as there is no such implication that sunshine alone leads to not being able to have a picnic. However, sunshine does not automatically mean that having a picnic is feasible, as we observe with this next statement:

"If it's sunny tomorrow, then we can have a picnic."

Is this statement true? Not necessarily. For all we know, it might not be raining. There could also be high winds, a parade cutting through the park, or maybe even a giant monster rampaging through the city! Anything other than rain could be a culprit that leads to the ultimate outcome of not being able to have a picnic, despite the lack of rain. That is, the statement that "If A, then B" does not automatically imply "If not A, then not B." This alternate form of the original statement is called the *inverse statement*.

There is also another piece of terminology that, unlike the other three, is not commonly used in regular speech: "iff," meaning "if and only if." When used in a statement of the form "B is true iff A is true," this means that if A is true, then so is B; if A is false, then so is B. It is not possible for one statement to be true and the other to be false.

As such, we can think of this type of statement as a two-way directional that needs to be satisfied in both cases. A must imply B, and B must imply A for the statement to be true. Observe that, if one of the statements, suppose A, is false, then $A \implies B$ (read: "A implies B") is automatically true, since an implication can only be false if the original statement is true (there is nothing said about when the statement is false). But if B is true, then B does not imply A; however, if B is also false, then B does imply A. We can apply the same argument to when we assume B is false; hence, we obtain the result that the statement "B is true iff A is true" holds whenever A and B are either both true or both false, proving the bi-directional condition for iff statements.

An example of a true statement using the "iff" conjunction is as follows:

$2^2 = 4$ iff the smallest prime number is even.

Here's another example:

$1 + 1 = 2$ iff we are using the base 10 number system.

And here are two examples of false "iff" statements:

5 is a perfect square iff 0 is a whole number.

$$a = b \text{ iff } a^2 - b^2 = 0.$$

Note that, in both false examples above, only one direction of implication is true (i.e. in the first statement, 0 is a whole number if 5 is a perfect square, since the proposition that 5 is a perfect square is false and so the second part of the statement is unconditionally true, but the opposite direction, that 5 is a perfect square if 0 is a whole number is untrue, since 0 is in fact a whole number, but 5 is not a perfect square. In the second example, $a = b$ does in fact imply $a^2 - b^2 = 0$, but not the other way around (a counterexample is any ordered pair of the form $(a, b) = (k, -k)$).

In summary, it is best to think of the conjunctions as not words that share their English meanings but instead as operators that have very specific, unique meanings that cannot be tampered with lest there be a distortion of logical sense and correctness (or at least as words that, in their mathematical contexts, share their standard conversational meanings only loosely and tangentially).

Exercise 2. *Which of the following are, or are not, statements? Justify your answers.*

1. *I like cheesecake but not ice cream.*

2. *59 is a prime number.*

3. *Is 6 a perfect number?*

4. *The expanded form of $(2k + 1)^2$ is $4k^2 + 4k + 1$.*

5. $15 + 5 = 30.$

1.3 Statements and Their Various Forms

Consider the generic statement "If p, then q," where p and q are the two parts of the statement. We will call this statement A for brevity. We can define three modifications of A: the *inverse*, the *converse*, and the *contrapositive*.

The inverse of A is the statement "If not p, then not q" ($\neg p \implies \neg q$).

The converse of A is "If q, then p" ($q \implies p$).

The contrapositive of A is "If not q, then not p" ($\neg q \implies \neg p$).

Example 1.3.0.1. *What are the inverse, converse, and contrapositive of the statement "If I meditate, then I won't be stressed?"*

Solution 1.3.0.1. *Here are the three statements:*

Inverse: "If I don't meditate, then I will be stressed."

Converse: "If I won't be stressed, then I'll meditate."

Contrapositive: "If I'm stressed, then I won't meditate."

Exercise 3. *Write the inverse, converse, and contrapositive of the following statements, then classify each of statements 1-5 as true or false with justification.*

1. *If the sky is blue, then 4 is a composite number.*

2. *1 is a composite number if a cube has 8 vertices.*

3. *If $x^2 + y^2 = 0$, and x and y are real numbers, then $x = y = 0$.*

4. *If x and y are positive integers less than or equal to 10 with $xy = 25$, then $x + y = 10$.*

5. *$2^4 = 4^2$ if 90 is less than 3^4.*

6. *If you have a good diet, you won't get sick.*

7. *If you study for the test, you will get an A.*

Extension Exercise 1. *Let A be a statement in the form "if p, then q." Let B be the contrapositive of the inverse of A, let C be the converse of the contrapositive of B, and let D be the inverse of the converse of C. What can we infer about B, C, and D?*

1.4 The Truth Table

Truth tables are tables that classify each of the outcomes of logical operators on input variables. They are used to demonstrate the truth or falsehood of a function given an input of either true or false.

1.4.1 Unary Operators

Logical operators can be either *unary* or *binary;* i.e. invoking either one or two arguments, as well as ternary, quaternary, quinary, etc., though we will only focus on unary and binary operators. The most rudimentary examples of unary operators are the *logical true* and *logical false* operators, which always return true and false, respectively: (Here p refers to the Boolean true/false value of a statement; i.e. the *input variable*.)

p	T
T	T
F	T

Table 1.1: The logical true operator.

p	F
T	F
F	F

Table 1.2: The logical false operator.

There are also the *logical identity* and *logical negation* operators, which return the input variable and the negation of the input variable, respectively. Below is the identity operator, represented by p itself:

p	p
T	T
F	F

Table 1.3: The logical identity operator.

Below is the negation operator, represented by $\neg p$:

p	$\neg p$
T	F
F	T

Table 1.4: The logical negation operator.

There exist several other unary operators, but these four are the most commonly used.

1.4.2 Binary Operators and The Extended Truth Table

$\oplus$ represents an exclusive disjunction, which is a logical operator that returns true when either of the two arguments is true, but not both.[2] For instance, if we let p be the statement "All humans are mammals," and q be the statement "All mammals are humans," then $p \oplus q$ is true, since p is true and q is false, and as such, one, but not both, of the statements is true. However, if we changed q to the true statement "Some mammals are humans," then $p \oplus q$ would no longer be true.

NOR is the "not or" operator, or the negation of the "or" operator. Similarly, NAND ("not and") is

---

[2]This symbol is also used in functional analysis, abstract/linear algebra, and other advanced disciplines of math as the direct sum symbol, but those topics are beyond the scope of this book.

the negation of the "and" operator.

"iff" (also denoted $p \iff q$) represents whether or not p is true if q is true (and vice versa). Id est, it returns true when p and q have the same Boolean truth value. In a similar vein, $p \to q$ and $q \to p$ return true when the statements "if p, then q" and "if q, then p" return true, respectively.

We henceforth list all of the aforementioned operators in one extended truth table:

p	q	T	F	$p \vee q$	$p \wedge q$	$\neg p$	$\neg q$	XOR	NOR	NAND	iff	$p \to q$	$q \to p$
T	T	T	F	T	T	F	F	F	F	F	T	T	T
T	F	T	F	T	F	F	T	T	F	T	F	F	T
F	T	T	F	T	F	T	F	T	F	T	F	T	F
F	F	T	F	F	F	T	T	F	T	T	T	T	T

Table 1.5: The extended truth table.

Exercise 4. *Suppose we define an operator* $+$ *such that* $p + q$ *is true when* $(p \text{NOR} q) \oplus (p \vee (p \iff q))$ *is false.*

(a) *Draw the truth table for* $+$ *given the four possibilities for* (p, q).

(b) *Draw the truth table for* $(p + q) + (p + q)$. *Your table should only have three rows, including the labels.*

(c) *Say an operator* $\circ$ *commutes if* $p \circ q = q \circ p$ *(i.e. returns the same truth values for all possible ordered Boolean pairs* (p, q)*). Does* $+$ *commute? Justify your answer.*

Extension Exercise 2. *Say an operator* $\circ$ *has a left identity* L *if* $L \circ q = q$ *whether* q *is true or false. Similarly, say* $\circ$ *has a right identity* R *if* $p \circ R = p$ *whether* p *is true or false.*

(a) *Under what conditions will an operator have one identity, but not the other?*

(b) *Are there operators that commute, but do not have either a left or right identity? If so, what are they?*

Extension Exercise 3. *Consider table 1.5 (the extended truth table) above. There are 14 columns in the table; however, there are* $2^4 = 16$ *possible operators from 4 possible Boolean pairs of input variables* (p, q).

(a) *Briefly describe the two missing operators. Do they commute, and do they have a left identity, right identity, or both?*

(b) *If we introduced a third Boolean input variable* r, *how many operators (columns) would be in a complete truth table?*

(c) *If we introduced more input variables* v_i *so that there are then* V *input variables, how many operators (columns) would be in a complete truth table?*

7

Chapter 2

Arguments and Their Forms

Now that we've gone through some basic logical notation and introduced conditionals and truth tables, let's talk about arguments and the forms they can take. In essence, arguments are the building blocks of proofs, which are, well, the main focus of this book. There are two types of arguments: *valid arguments* and *invalid arguments*. The former are what we will be seeking to create and use to prove what we want to prove!

Before we go on, though, we need to define a valid argument:

Valid Argument – a sequence of logical statements so that:

1. A statement in the sequence is true because of the previous statement.

2. The first statements in the sequence are known to be true.

3. The last statement is the conclusion, which will be true.

By contrast, an *invalid argument* is a conclusion does not necessarily logically follow from previous statements. These are not the arguments we wish to create!

Now, why do we bother differentiating valid and invalid arguments? Aren't valid arguments the only ones that matter? Well, yeah, valid arguments are the only ones we want to use to prove anything; after all, don't we want our proofs to be logically tight? However, it is necessary to know what makes an argument invalid to avoid falling into that pitfall ourselves. In addition, being able to differentiate valid and invalid arguments is a crucial skill in real life. People make arguments all the time. How do we know what to believe? By determining if their arguments are valid!

So yeah, that's the basic gist of arguments. In the rest of this chapter, we'll go through some examples of valid and invalid arguments as well as some techniques to show that arguments are valid.

2.1 Showing an Argument is Valid

Let's start off with a couple examples of real-life valid and invalid arguments to get some intuition.

Example 2.1.0.1.
Given 1: George majors in computer science.
Given 2: If you major in CS, then you must take discrete math.
Conclusion: George needs to take discrete math.

Solution 2.1.0.1. *Let's label the statements. Let statement A be "George majors in computer science," statement B be "you are a CS major," and statement C be "you need to take discrete math." The givens are A and $B \rightarrow C$. Then the conclusion is C – George needs to take discrete math. This is a* valid *argument because A is included in B, and the if-then promises C for B.*

8

We just saw an example of a *valid argument*. Now, let's look at an example of an *invalid argument* to get a feel for the difference.

Example 2.1.0.2.
Given 1: George is taking discrete math.
Given 2: If you're majoring in CS, then you must take discrete math.
Conclusion: George majors in CS.

Solution 2.1.0.2. *Again, let's label the statements. Let statement A be "George is taking discrete math," statement B be "you are a CS major," and statement C be "you need to take discrete math." The givens are A and $B \to C$. However, we can't conclude that George is majoring in CS just because he is taking discrete math (A, which is included in C) because we do not have that $C \to B$. Therefore, the argument in this example is* invalid.

Alright, that's great; now, we have some examples of valid and invalid arguments. However, what do we do when arguments get more complicated? We need a more general (and foolproof) procedure than writing paragraphs as we did above; that only consistently works for short arguments like the ones we had in the examples. That said, here is a procedure to determine whether an argument is valid without losing track of statements and premises:

1. Make a truth table.

2. Pick out the rows where the givens (premises) are true.

3. If the conclusion is true for those lines, then the argument is valid.

Let's see an example of how this works.

Example 2.1.0.3. *Consider the following argument. Is it valid?*

- $p \vee q$

- $p \to \neg q$

- $p \to r$

Therefore, r.

Solution 2.1.0.3. *We start by setting up our truth table:*

p	q	r	$p \vee q$	$\neg q$	$p \to \neg q$	$p \to r$	givens true?	conclusion $= r$

Notice that the first three columns are the values of p, q, and r. These are the three statements that we are considering the relationships between. In order to account for every possible truth value of the given premises, we must account for every possible array of truth values of p, q, and r.

Notice that r has a duplicate entry in the truth table. The reason is so its truth value, as the conclusion of the argument, can easily be compared with the the truth values of the premises. Our goal is to ensure that true premises lead to a true conclusion. (See the bold rows below.) Let's fill in the rest of the truth table:

p	q	r	$p \vee q$	$\neg q$	$p \to \neg q$	$p \to r$	***givens true?***	***conclusion $= r$***
T	T	T	T	F	F	T	***F***	***T***
T	T	F	T	F	F	F	***F***	***F***
T	F	T	T	T	T	T	***T***	***T***
T	F	F	T	T	T	F	***F***	***F***
F	T	T	T	F	T	T	***T***	***T***
F	T	F	T	F	T	T	***T***	***F***
F	F	T	F	F	T	F	***T***	***T***
F	F	F	F	F	T	T	***T***	***F***

We've gotten a false conclusion with true premises in rows 6 and 8 (when p and r are both false, so the argument is invalid.

2.2 Basic Valid Argument Forms

Now that we've gone through how to determine whether or not an argument is valid, let's look at some of the basic forms valid arguments take on. Many of these have names, which we'll provide, but there is no reason to memorize them. However, it is crucial to gain an understanding of how these basic valid argument forms work, as they can be employed in the context of proofs.

2.2.1 Modus Ponens

Here is the basic layout of a Modus Ponens argument form.

Given 1: If p, then q. (In logical notation: $p \rightarrow q$.)
Given 2: p
Conclusion: q

Example 2.2.1.1. *Let's reconsider example 2.1.0.1, except you now are George.*

Given 1: You major in computer science.
Given 2: If you major in computer science, then you need to take discrete math.
Conclusion: You need to take discrete math.

Solution 2.2.1.1. *This (valid) argument is an example of Modus Ponens, although it may be hard to see in its original form. However, let's see what happens if we swap the positions of Givens 1 and 2:*

> *Given 2: If you major in computer science, then you need to take discrete math.*

> *Given 1: You major in computer science.*

> *Conclusion: You need to take discrete math.*

Let p be the statement "you major in computer science," and let q be the statement "you need to take discrete math." That is, in logical notation, Given 2 says $p \rightarrow q$. Then Given 1, in logical notation, says p. By Modus Ponens, we can conclude q, that is, you need to take discrete math (as stated in the conclusion).

2.2.2 Modus Tollens

Here is the basic layout of a Modus Tollens argument form.

Premise 1: If p, then q. (In logical notation: $p \rightarrow q$.)
Premise 2: $\neg q$
Conclusion, $\neg p$.

Why is Modus Tollens a valid argument form? Recall that a conditional statement, "if p, then q," is logically equivalent to its contrapositive, "if not q, then not p." In logical notation, we say that $p \rightarrow q$ is equivalent to $\neg q \rightarrow \neg p$. So, we can rewrite Premise 1 as follows (the other premises are copied below for convenience):

Premise 1: $\neg q \rightarrow \neg p$
Premise 2: $\neg q$
Conclusion, $\neg p$.

Let's look at an example now.

Example 2.2.2.1.
Given 1: If you are a CS major, then you need to take Discrete Math.

Given 2: You don't need to take Discrete Math.

Conclusion: You're not a CS major.

Solution 2.2.2.1. *This (valid) argument is an example of Modus Tollens, which, once again, we can see by rewriting Given 1 in the form of the contrapositive:*

> *Given 1: If you do not need to take Discrete Math, then you are not a CS major.*

> *Given 2: You don't need to take Discrete Math.*

> *Conclusion: You're not a CS major.*

Let p be the statement "you do not need to take Discrete Math," and let q be the statement "you are not a CS major." That is, in logical notation, Given 1 says $p \to q$. Then Given 2, in logical notation, says p. By Modus Ponens, we can conclude q, that is, you're not a CS major (as stated in the conclusion).

Alternatively, we could have considered the argument in its original form: let r be the statement "you are a CS major," and let s be the statement "you need to take Discrete Math. Then Given 1 states (in logical notation) $r \to s$. Given 2 (in logical notations) is $\neg s$. Therefore, we can assert our conclusion, $\neg r$, via Modus Tollens.

2.2.3 Generalization (Disjunctive Addition)

This argument form may seem especially obvious, but it is still worth noting.

Premise: p
Conclusion: $p \lor q$

Remember that the logical operator $\lor$ is not the exclusive or operator; that is, $p \lor q$ is true regardless of the truth value of q so long as p is true.

Consider the following statement: "You are reading this book or you were born on Mars." You're here right now following along with the text of the book, so it's definitely true that you're reading this book. It doesn't matter whether or not you were born on Mars, as the premise that you're reading the book is true. Therefore, the given statement is also true.

2.2.4 Specialization (Conjunctive Simplification)

Premise: $p \land q$
Conclusion: p (or q, if you wish)

2.2.5 Disruptive Syllogism

Premise 1: $p \lor q$
Premise 2: $\neg p$
Conclusion: q

2.2.6 Transitivity (Hypothetical Syllogism)

Premise 1: $p \to q$
Premise 2: $q \to r$
Conclusion: $p \to r$

2.2.7 Division into Cases

Premise 1: $p \vee q$
Premise 2: $p \to r$
Premise 3: $q \to r$
Conclusion: r

2.2.8 Examples and Exercises

Now that we've gone through some basic valid argument forms, let's go through some examples that apply the forms.

Example 2.2.8.1. *Prove that the following argument is valid.*

1. $p \vee q$

2. $q \to r$

3. $p \wedge (s \to t)$

4. $\neg r$

5. $\neg q \to (u \wedge s)$

Therefore, t.

Solution 2.2.8.1.

Step	Statement	Reason
A	$\neg r$	4
B	$\neg q$	A, 2, Modus Tollens
C	$u \wedge s$	B, 5, Modus Ponens
D	$p \wedge (s \to t)$	3
E	$s \to t$	D, Specialization
F	s	C
G	t	F, E

Example 2.2.8.2. *Prove that the following argument is valid.*

1. $\neg p \vee q \to r$

2. $s \vee \neg q$

3. $\neg t$

4. $p \to t$

5. $\neg p \wedge r \to \neg s$

Therefore, $\neg q$.

Solution 2.2.8.2.

Step	Statement	Reason
A	$\neg t$	3
B	$\neg p$	A, 4, Modus Tollens
C	$\neg p \vee q$	B
D	r	C, 1, Modus Ponens
E	$\neg p \wedge r$	B, D
F	$\neg s$	E, 5, Modus Ponens
G	$\neg q$	F, 2, Disruptive Syllogism

12

...and we are done!

Now, you try!

Exercise 5. *Prove that the following argument is valid.*

1. $\neg p \to r \wedge \neg s$

2. $t \to s$

3. $u \to \neg p$

4. $\neg w$

5. $u \vee w$

Therefore, $\neg t$.

Chapter 3

Types of Mathematical Proof

3.1 Direct Proof (Proof by Construction)

A *direct proof* uses postulates, axioms, theorems and lemmas that have already been proven, and other givens directly to establish the truth or falsehood of a statement, without making any of its own assumptions.[3] Direct proofs of the general statement "if p, then q" prove that q is true using only the fact that p is true, along with the corresponding conclusions that can be drawn from the truth of statement p.

Here is a traditional example using direct proof, namely the definitions of even and odd integers:

Example 3.1.0.1. *Prove that the sum of an even and an odd integer is always an odd integer.*

Solution 3.1.0.1. *Note that any even integer m can be written in the form $m = 2k$, where k is an integer (by definition, all even integers are divisible by 2), and that any odd integer n can be written in the form $n = 2k + 1$, where k is an integer. Then $m + n = 2k + (2k + 1) = 4k + 1 = 2(2k) + 1$. If k is an integer, then $2k$ must also be an integer, so $m + n$ is an odd integer.*

Example 3.1.0.2. *Prove that if a perfect square of an integer is even, it is also divisible by 4.*

Solution 3.1.0.2. *Let n be an integer. Then n can be either even or odd. If it is even, then it can be written as $n = 2k$ for some integer k. Hence, $n^2 = 4k^2 = 4(k^2)$. Clearly, k^2 is an integer, so n^2 is a multiple of 4. However, if n is odd, then $n = 2k+1$ for some integer k and $n^2 = 4k^2+4k+1 = 4(k^2+k)+1 = 2(2k^2+2k)+1$, which is not even, let alone a multiple of 4.*

Exercise 6. *Prove or disprove that $x^2 + y^2$ is a multiple of 8 iff x and y are both even.*

3.2 Proof by Induction

A *proof by induction* is a form of proof that, like a direct proof, works directly from the original statement "if p, then q," but unlike a direct proof, makes an assumption that the statement itself holds for some n. A proof by induction begins by verifying that the statement is true for some *base case*, usually $n = 0$ or $n = 1$ (or the smallest applicable value of n). Then, it makes the assumption that the statement is true for some n, and proves that it is also true for $n + 1$. In this way, if the statement holds for $n = 1$, it will also hold for $n = 2$; if it holds for $n = 2$, it will hold for $n = 3$, and so forth. Thus, such a proof will eventually show that the statement holds for all n.

We denote by $S(n)$ the statement in terms of n; e.g. if S is the general statement "$1+2+3+\ldots+n = \dfrac{n(n+1)}{2}$," then $S(2)$ is the true statement "$1 + 2 = \dfrac{2 \cdot 3}{2}$." The goal of a proof by induction is to show $S(1)$ (or $S(0)$,

---

[3]Though the two-column proof is a form of direct proof commonly taught and practiced in the standard secondary school mathematics curriculum, it will not be covered here due to its similarity to the methods used to prove algebraic axioms in §3.

or whatever the base case happens to be), and by extension, all of $S(2), S(3), S(4), \ldots$ are true as well.

Usually this is accomplished by means of substitution; rather than n, we use $n + 1$ in the given expression that is part of the statement we are trying to prove, then simplify/rearrange it to obtain an equivalent expression solely in terms of $n + 1$ that parallels the original expression in terms of n.

Example 3.2.0.1. *Prove that $2^n > 2n + 1$ for $n \geq 3$.*

Solution 3.2.0.1. *Let $S(n)$ be the statement "$2^n > 2n + 1$." Note that here, the base case is not $n = 1$, but $n = 3$ (as specified in the problem statement). $S(3)$ is trivially true: $2^3 = 8 > 7 = 2 \cdot 3 + 1$. Henceforth, assume $S(n)$ is true, so that n is some positive integer satisfying $2^n > 2n + 1$. We show $S(n+1)$ holds as well.*

We can write $S(n+1)$ as the statement $2^{n+1} > 2(n+1) + 1 = 2n + 3$. Assuming $2^n > 2n + 1$, it follows that $2^{n+1} > 2(2n + 1) = 4n + 2$. In addition, $4n + 2 > 2n + 3$ for all $n \geq 3$, and we have $2^{n+1} > 4n + 2 > 2n + 3$; therefore, by the transitive property, $2^{n+1} > 2n + 3$ for all $n \geq 3$. This proves $S(n + 1)$ for all $n \geq 3$, and since $S(3)$ is true, $S(4), S(5), S(6), \ldots$ are all true by extension, completing the proof.

Example 3.2.0.2. *Show that $\sum_{k=1}^{n}(2k + 1) = (n + 1)^2 - 1$ for all $n \in \mathbb{N}$.*

Solution 3.2.0.2. *Let $S(n)$ be the statement for an upper bound $n \in \mathbb{N}$. Our base case is $S(1)$, which is trivially true as $\sum_{k=1}^{1}(2k + 1) = 3 = (1 + 1)^2 - 1$. We henceforth assume $S(n)$ holds for some $n \in \mathbb{N}$ and prove that $S(n + 1)$ holds as well.*

$S(n + 1)$ is the statement $\sum_{k=1}^{n+1}(2k + 1) = (n + 2)^2 - 1$. Under the assumption given by $S(n)$, $\sum_{k=1}^{n+1}(2k + 1) = ((n + 1)^2 - 1) + (2(n + 1) + 1) = n^2 + 4n + 3 = (n + 2)^2 - 1$, so $S(n + 1)$ is true whenever $S(n)$ is true and the statement is proven for all $n \in \mathbb{N}$.

Example 3.2.0.3. *Prove that $n! < n^{n-1}$ for all $n \geq 3$.*

Solution 3.2.0.3. *$n = 3$ is the base case: clearly, $3! = 6 < 9 = 3^{3-1}$, so $S(3)$ holds true. $S(n + 1)$ is the statement $(n + 1)! < (n + 1)^n$; it follows from $S(n)$ that $(n + 1)! < (n + 1) \cdot n^{n-1}$. Furthermore, $(n + 1) \cdot n^{n-1} = (n + 1) \cdot \dfrac{n^n}{n} = \dfrac{n + 1}{n} \cdot n^n > n^n$, and $(n + 1)! < n^n$, which itself can be proven by induction (exercise to the reader). Hence, $S(n + 1)$ holds and $S(n)$ is proven for all $n \geq 3$.*

3.3 Proof by Strong Induction

Whereas induction relies on proving that the statement $S(n + 1)$ is true when $S(n)$ is true for all natural numbers n, in strong induction, we want to show that $S(n + 1)$ is true when all of the statements $S(1), S(2), S(3), \ldots, S(n)$ are true. (As a consequence, regular induction follows from strong induction; however, the reverse is not true.)

Example 3.3.0.1. *Let $\{a_i\}_{i=1}^{\infty}$ be a sequence defined recursively by $a_1 = 1$, $a_2 = 3$, and for $n \geq 3$, $a_n = 2a_{n-1} + a_{n-2}$. Show that, for all positive integers n, a_n is odd.*

Solution 3.3.0.1. *Base case: Consider $n = 1$ and $n = 2$. $S(1), S(2)$ are trivially true (given in the problem).*

Now we proceed with the inductive step. Let $j \geq 2$ be an integer, and assume that, for all $1 \leq i \leq j$, $S(i)$ is true (i.e. that a_i is odd). This is our inductive hypothesis. Consider a_{j+1}, where $j + 1 \geq 3$. By the recursive definition of the sequence, we have $a_{j+1} = 2a_j + a_{j-1}$.[4] We have $a_j = 2k_1 + 1$ and $a_{j-1} = 2k_2 + 1$ for some integers k_1, k_2, so $2a_j + a_{j-1} = (4k_1 + 2) + (2k_2 + 1) = 4k_1 + 2k_2 + 3 = 2(2k_1 + k_2 + 1) + 1$, which is odd since $2k_1 + k_2 + 1 \in \mathbb{Z}$. This proves $S(j + 1)$, and by extension $S(j)$ for all $j \geq 1$.

---

[4]Observe that it is mathematically valid to invoke the inductive hypothesis here, as $1 \leq j \leq j$ and $1 \leq j - 1 \leq j$.

15

Example 3.3.0.2 (Extending the Chicken McNugget Theorem). *Show that it is always possible to buy 152 or more nuggets, if you can buy 9-nugget packs and 20-nugget packs.*

Solution 3.3.0.2. *Call a positive integer n constructible if it is possible to buy exactly n nuggets by buying 9-nugget packs and 20-nugget packs. Let $S(n)$ be the statement, "n is constructible." The base case here is not one value of n, but nine (since we must cover all of the values from 152 to 160, and then proceed to show how we can cover exactly n nuggets for $n \geq 161$). $S(152)$ is true (one could buy eight 9-nugget packs and four 20-nugget packs). Similarly, $S(153)$ through $S(160)$ are true (proving these statements is left as an exercise to the reader).*

Assuming we can "construct" all nugget totals from 152 to n inclusive (where $n \geq 160$), we show how to construct $n+1$ nuggets. We assume $n+1 \geq 161$. Since $n+1 \geq 161$, $(n+1) - 9 \geq 152$. By the inductive hypothesis, we can construct $(n+1) - 9$ nuggets by buying a 9-nugget packs and b 20-nugget packs. That is, $(n+1) - 9 = 9a + 20b$ for all $(n+1) - 9 \geq 152$.

We then have $n+1 = (9a + 20b) + 9 = 9(a+1) + 20b$, so it is always possible to construct $n+1$ nuggets for $n \geq 160$. Since we can construct all numbers of nuggets from 152 to 160, inclusive, as well as all numbers of nuggets that are at least 161, this shows that all numbers of nuggets that are at least 152 are constructible.

3.4 Proof by Contradiction

In a *proof by contradiction*, also known as an *indirect proof, apagogical argument*, or *reductio ad impossiblem*,[5] we assume the opposite of the statement is true, then show that this assumption leads to an absurdity or violation of an axiom. *Reductio ad absurdum*, the generalized form of a proof by contradiction, in the words of G. H. Hardy, "is one of a mathematician's finest weapons, [a] far finer gambit than any chess gambit: a chess player may offer the sacrifice of a pawn or even a piece, but a mathematician offers the game" (Hardy).

Example 3.4.0.1. *Show that 2 is the only even prime number.*

Solution 3.4.0.1. *We assume the contrary (there there exists a prime number greater than 2 that is also even). Let this number be p. However, p being even (and not equal to 2) implies that 2 is a factor of p, which contradicts the definition of a prime number (that its only positive factors are 1 and itself).*

Example 3.4.0.2. *Show that $\sqrt{2}$ is irrational; i.e. that it cannot be written in simplest form as $\frac{a}{b}$ where a and b are integers.*

Solution 3.4.0.2. *Assume the contrary. Then $\sqrt{2} = \frac{a}{b}$, where $\gcd(a, b) = 1$, and so squaring both sides, we get $2 = \frac{a^2}{b^2}$, or $a^2 = 2b^2$. It follows that a^2 must be even, and so a itself must also be even (for if it were odd, it would be of the form $2k+1$ for k an integer, and $a^2 = 4k^2 + 4k + 1$, which is odd. This, in itself, is a mini-proof by contradiction). However, it follows from a being even that a^2 is a multiple of 4 (as $a = 2k$ for some integer k, and consequently, $a^2 = 4k^2 = 4(k^2)$. Thus, b^2 would have to be even as well, which means b is even. (Were b^2 to be odd, $2b^2$ would be even, but it would not be a multiple of 4; consider the prime factorization for insight as to why this is true.) But if both a and b are even, then 2 is a factor of both, and the fraction $\frac{a}{b}$ would not be in simplest form. Hence $\sqrt{2}$ cannot be written in the form $\frac{a}{b}$ where $\gcd(a, b) = 1$, and so it is irrational.*

Exercise 7. *Show that $\sqrt{p}$ is irrational for any prime number p.*

3.5 Non-Constructive Proof

The proof by contradiction is in fact a subcategory of the more general *non-constructive proof*. In a non-constructive proof, we aim to prove existence, but without providing a concrete example of that which we have proved exists.

---

[5] A specific form of the more general *reductio ad absurdum*.

Example 3.5.0.1. *Prove that there are an infinite number of prime numbers.*

Solution 3.5.0.1. *Assume the contrary. Denote by $\{p_1, p_2, p_3, \ldots, p_n\}$ the sequence of prime numbers, where there is a "largest" prime number p_n. But $p_1 p_2 p_3 \cdots p_n + 1$ is not a multiple of any of $p_1, p_2, p_3, \ldots, p_n$, so it must have a prime factor larger than p_n. Hence there is no largest prime number, and so the number of prime numbers is infinite.*

Note that this proof is non-constructive since we have not established any specific example of a number of the form $p_1 p_2 p_3 \cdots p_n + 1$, but have nevertheless shown that it must exist.

Here is an example of a non-constructive proof that does not rely on proof by contradiction, to illustrate that the two types of proof are not one and the same:

Example 3.5.0.2. *Prove that there exist irrational numbers a and b such that a^b is rational.*

Solution 3.5.0.2. *Consider the number $\sqrt{2}^{\sqrt{2}}$, which may be either rational or irrational. If it is rational, then $a = b = \sqrt{2}$ immediately proves the statement. If it is irrational, take $a = \left(\sqrt{2}\right)^{\sqrt{2}}$ and $b = \sqrt{2}$. Then $a^b = \left(\sqrt{2}\right)^2 = 2$, which is rational, proving the existence of $a, b \in \mathbb{R} \backslash \mathbb{Q}$ satisfying the theorem. [Note: The notation $\mathbb{R} \backslash \mathbb{Q}$ refers to the set of numbers that are real but not rational; this set is also known as the set of irrational numbers.]* [6]

3.6 Proof by Exhaustion

Proof by exhaustion, also known as *complete induction* or *proof by cases*, revolves around the idea of case-work. Much like solving a combinatorics problem, proofs often involve several different cases, which can often be checked manually and exhausted one by one. This is in fact a subcategory of the direct proof, but it is less commonly used.

A proof by exhaustion must satisfy two checks: first, it must show the set of cases to be exhaustive (i.e. that no other cases are possible, which is often accomplished via a proof by contradiction), and second, it must individually prove each case within that exhaustive set.

Example 3.6.0.1. *Prove that all perfect squares leave a remainder of 0, 1, or 4 when divided by 8.*

Solution 3.6.0.1. *Let k be an integer. Note that it suffices to consider the remainder when k is divided by 4, as k^2 leaves the same remainder as $(8 - k)^2 = 64 - 16k + k^2$ when divided by 8. We now prove the statement for each of the following cases:*

- *$k \equiv 0 \pmod 4$: If k is divisible by 4, then k^2 will be divisible by 16 (as $k = 4n$ for some integer n and $k^2 = 16n^2$), and so it will also be divisible by 8. Thus it leaves a remainder of zero when divided by 8.*

- *$k \equiv 1 \pmod 4$: If $k = 4n + 1$ for some integer n, then $k^2 = 16n^2 + 8n + 1 = 8(2n^2 + n) + 1 \equiv 1 \pmod 8$.*

- *$k \equiv 2 \pmod 4$: If $k = 4n + 2$, then $k^2 = 16n^2 + 16n + 4 = 8(2n^2 + 2n) + 4 \equiv 4 \pmod 8$.*

- *$k \equiv 3 \pmod 4$: If $k = 4n + 3$, then $k^2 = 16n^2 + 24n + 9 = 8(2n^2 + 3n + 1) + 1 \equiv 1 \pmod 8$.*

(Note: $k \equiv a \pmod b$ means that k leaves a remainder of a when divided by b.)

For each of the four cases, the statement is true. In addition, these are the only four cases possible, so the statement must be true for all integers k.

---

[6]It turns out that the quantity $\left(\sqrt{2}\right)^{\sqrt{2}}$ is in fact irrational, by the Gelfond-Schneider Theorem (Warrens 2012).

3.6.1   Not a Proof, Archimedes, but Close Enough

Proof by exhaustion is not to be confused with the *method of exhaustion,* famously used by Archimedes to establish lower and upper bounds for π, the area of a unit circle. Archimedes' method, which does not exhaust a finite number of cases but rather merely approximates the common limiting value of two convergent sequences using a finite number of *steps,* proceeds by incrementally adding 1 to the value of n, starting from 3, and inscribing and circumscribing a regular n-gon about the unit circle, then bounding the area of the circle between the areas of the inscribed and circumscribed polygons. For $n = 3$, for instance, Archimedes' approximation resembled the following:

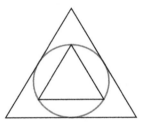

Figure 3.1: Trapping the unit circle between inscribing and circumscribing equilateral triangles.

The inscribed equilateral triangle has side length $\sqrt{3}$, so its area is $\dfrac{3\sqrt{3}}{4}$. The circumscribing equilateral triangle has side length $2\sqrt{3}$, so its area is $3\sqrt{3}$. Hence the area of the unit circle must lie between $\dfrac{3\sqrt{3}}{4} \approx 1.299$ and $3\sqrt{3} \approx 5.196$. So far not the best approximation - but Archimedes has only gotten started.

For $n = 4$, this was Archimedes' diagram:

Figure 3.2: Trapping the unit circle between inscribing and circumscribing squares.

Here the lower and upper bounds for π (the area of the unit circle) are 2 and 4 respectively. Still not even close, but improving by leaps and bounds already! Despite initial impressions, for larger values of n, Archimedes' approximations are actually admirably close. Here is the diagram for $n = 2019$:

Figure 3.3: The approximation diagram for $n = 2019$.

The inscribing and circumscribing polygons look virtually identical to the unit circle! And indeed, the lower and upper bounds, respectively, are only $5.0709 \cdot 10^{-6}$ and $2.5355 \cdot 10^{-6}$ off from the actual value of π. Clearly, Archimedes' method will never actually reach π, but for sufficiently large n the bounds are astonishingly close.[7]

---

[7]This demonstration also ties in with the idea of *limits,* which represent the values that quantities approach to within

3.7 Computer-Assisted Proof

Computer-assisted proofs (CAPs) are surprisingly common in professional mathematics, proving theorems that deal with exorbitantly large numbers or which involve optimization problems that are impossible to solve by hand (for instance, the traveling salesman problem for the entire United States). Many CAPs are proofs by exhaustion, working through many computationally-intensive cases to prove the final theorem. Much like any other proof by exhaustion/casework, a CAP must not only prove the cases themselves, but also show that the results of the computer-assisted computations imply the theorem.

It is in this type of proof that math connects especially with computer science. Artificial intelligence, machine reasoning techniques, and automated search methods (e.g. comparison-based algorithmic sorting, such as binary sort and merge sort) are all various subcategories of computer assistance.

One issue that arises with CAPs is the limitations of the computers that are used to carry out the computations in the proofs. Computer programs, especially less recent ones, are more prone to mathematical error or malfunction that leads to an error, such as a floating-point error, rounding error, or overflow/underflow. As such, some human intervention is required to ensure the accuracy of the proof, and to maximize the efficiency with which the computers carry out the necessary computations.

The following is the first theorem proven with assistance from a computer in 1976, the four-color theorem:

Theorem 3.1: Four-Color Theorem

Given any separation of a plane into contiguous regions (a *map*), no more than four colors are ever required to color the map such that any two adjacent regions have different colors.[a]

---

[a]For those interested, a human-verifiable proof can be found in the paper "A Human-Checkable Four Color Proof" at https://arxiv.org/pdf/1708.07442.pdf (Barbosa).

---

infinitesimally small margins of error, but never actually reach.

Chapters 1-3 Review Exercises

Exercise 8. *Classify each of the following as true, false, inconclusive, or not a statement:*

(a) $\forall a \in \mathbb{R}\backslash\mathbb{Q}$, $a^2 \in \mathbb{Q}$.

(b) *A real number has an imaginary part of 1.*

(c) π *is better than* e.

(d) *I live at* $(1,2)$ *on the Cartesian plane. The nearest gas station is at* $(2,3)$. *The gas station is not further than 1 mile from my house.*

(e) $A > B$, *and* $C > B$. *Then* $A > C$.

Exercise 9. *Given the following two premises:*

- *"All poisonous objects are bad-tasting."*

- *"Broccoli is bad-tasting."*

Why is the following an example of a faulty conclusion? "Broccoli is a poisonous object." How could we change the premises so that we can make a conclusion regarding broccoli?

Exercise 10. *Show that the sum of the squares of an odd integer and an integer divisible by 4 leaves a remainder of 1 when divided by 8.*

Exercise 11. *List the inverse, converse, and contrapositive of each of the following statements, and classify each as true or false with explanation:*

(a) *"If I am sick, I won't be able to go to school."*

(b) *"If a regular octahedron has an equal number of faces and vertices, then a cube with side length s has surface area given by* $6s^2$.*"*

(c) *"If the statement '$\neg(p \wedge (q \oplus \neg p)) \vee (q \iff p)$' is true, then either p and q are both true or both false."*

Exercise 12. *Prove by induction that* $2^n > n^3$ *for all* $n \geq 10$.

Exercise 13. *Prove that* $\displaystyle\sum_{k=1}^{n} 2^k = 2^{n+1} - 2$ *for all* $n \geq 1$ *by induction.*

Exercise 14. *Prove that* $1 \cdot 2 \cdot 3 + 2 \cdot 3 \cdot 4 + 3 \cdot 4 \cdot 5 + \ldots + n \cdot (n+1) \cdot (n+2) = \dfrac{n(n+1)(n+2)(n+3)}{4}$ *for all* $n \in \mathbb{N}$.

Extension Exercise 4. *Using* **strong** *induction, prove the following:*

Theorem 3.2: Factorial Representation Theorem

Every positive integer N can be written in the form $N = a_1 \cdot 1! + a_2 \cdot 2! + a_3 \cdot 3! + \ldots + a_k \cdot k!$, where k is a positive integer and $a_i \in \{0, 1, \ldots, i\}$ for all $i \in \mathbb{N}$.

Extension Exercise 5. *Let $f : \mathbb{N} \to \mathbb{N}$ be an increasing function with $f(1) = 1$ such that $f(n+1) - f(n) \geq f(n) - f(n-1)$ for all $n \in \mathbb{N}$. Show that*

$$\sum_{k=1}^{n} f(k) \leq f\left(\sum_{k=1}^{n} k\right)$$

Part II

Algebra

Chapter 4

Beginning Algebra

Why Study Algebra?

One of the cornerstones of both ancient and modern mathematics, *algebra,* alongside geometry, number theory and discrete mathematics, lays the foundation for mathematical problem solving and proof. It is, of these disciplines, the one that most directly and coherently unifies mathematics as a whole, serving as the primary means to not only solve equations derived from geometric, discrete, or number-theoretic means, but give context and meaning to them as well. Henceforth, we will study *elementary* or *beginning* algebra with a focus on proving the definitions and theorems presented within; in other words, we will explore the study of equations and their manipulation with a rigorous motivation for how the axioms governing algebraic operations come to be.

4.1   The Axioms of Algebraic Equations

The heart and soul of algebra lie in the study of *algebraic equations,* statements that relay a comparison:

Definition 4.1: Equation

An *equation,* in algebra, is a form of statement that establishes *equality* between two quantities, numerical or variable.

This definition is in contrast to an *expression* or *formula,* which is merely a finite combination of mathematical symbols that obeys the conditions of well-formedness.[8]

What qualifies as an expression or an equation? Here are some examples:

$$x + y + z \text{ and } a^2 - b^2 \text{ are both expressions, but } not \text{ equations, whereas}$$
$$2 = 2,\ 3 + 1 = 4, \text{ and } a^2 + b^2 = c^2 \text{ are all equations (and also expressions).}$$

Note that all equations are expressions (as equations are necessarily well-formed), but *not* all expressions are equations. Furthermore, equations can be true or false, much like logical statements. Despite being patently false, $1 + 1 = 3$ is still an equation.

All equations consist of two expressions; one to the left of the $=$ sign, and one to its right, which are referred to as the equation's *left-hand side* (LHS) and *right-hand side* (RHS) respectively. To *solve* an equation in one or more variables is to obtain a single value (or set of values) for each variable; this is also known

---

[8]Much like a combination of letters in written language that forms a word, a symbol combination is considered *well-formed* if it carries coherent meaning that can be universally used and understood (linguistic barriers notwithstanding, since math is a globally-understood language, with minor dialectical differences from country to country).

as *isolating* the variables. In an *identity equation,* the equation is true for all values of the variable(s), but in a *conditional equation,* the equation holds only for specific values of the variable(s).

Equations can also be classified based on the nature of their variables. If the highest exponent of a variable anywhere in the equation is 0, then all quantities are numerical and the equation is trivial to solve (as there are no variables; the LHS and RHS are either equal, or unequal, which yield infinite and zero solutions respectively). If the highest exponent, or *degree,* is 1, then the equation is a *linear equation,* since it can be represented geometrically in the Cartesian coordinate plane as a straight line. All linear equations are of the form $ax + by = c$, where a, b, and c are constants. If the degree is 2, then the equation is a *quadratic equation,* of the form $ax^2 + bx + c = 0$ (where c can be adjusted so that the equation is equal to zero if it isn't zero initially). If the degree is 3, the equation is *cubic,* if it is 4, the equation is *quartic,* and if it is 5 or higher, it is usually referred to as simply as "of the nth degree," or more briefly, just as "n^{th} degree" equations, although specialized terms do exist to describe such equations.

To solve an equation, we must perform operations onto both sides of the equation simultaneously, always abiding by the following axioms:

Theorem 4.1: Axioms of Algebraic Equations

- Addition property of equality: If $a = b$, then $a + c = b + c$. (This is also the subtraction property of equality.)

- Multiplication property of equality: If $a = b$, then $ac = bc$. (This is also the division property of equality.)

- Distributive property: For all constants a, b, and c, $a(b + c) = ab + ac$.

- Substitution property of equality: If $a = b$, then we can substitute b in for a within another equation.

- Reflexive property of equality: For all real numbers a, $a = a$.

- Symmetric property of equality: If $a = b$, then $b = a$.

- Transitive property of equality: If $a = b$ and $b = c$, then $a = c$.

These axioms are themselves based on associativity, commutativity, and the existence of the additive identity:

Theorem 4.2: Associativity, Commutativity, and the Identity

- Associative property: Addition and multiplication are both associative operations; that is, $(a + b) + c = a + (b + c)$ and $(ab) \times c = a \times (bc)$ for all real numbers a, b, c.[a]

- Commutative property: For all real numbers a and b, $a + b = b + a$ and $ab = ba$.

- 0 is the *additive identity;* i.e. it satisfies the equations $a + 0 = a$ and $0 + a = a$, being both a left and right identity.

---

[a]Indeed, we can generalize this to any number of variables. The proof of the generalized associative property is left as an exercise to the reader.

Proof. To prove associativity, commutativity, and the existence of the additive identity, we must first introduce the arithmetical axioms of Italian mathematician Giuseppe Peano. (Here $\mathbb{N}$ is the set of natural numbers, that is, the set $\{1, 2, 3, \dots\}$.)

Theorem 4.3: Peano's Axioms

1. $0 \in \mathbb{N}$.

2. For $n \in \mathbb{N}$, define $S(n)$ to be the *successor function* of n.

3. $\forall n \in \mathbb{N}, S(n) \in \mathbb{N}$.

4. $\forall m, n \in \mathbb{N}, m = n \iff S(m) = S(n)$.

5. $\forall n \in \mathbb{N}, S(n) \neq 0$.

The main argument in proving associativity involves induction on n. Let S be the set of all c such that, for all $a, b \in \mathbb{N}, (a + b) + c = a + (b + c)$. Then we have $a + (b + 1) = a + b' = (a + b)' = (a + b) + 1$. Hence, $1 \in S$. Next, assume that $c \in S$ for some $c \geq 2$. We show that $c + 1 = c' \in S$ as well: $a + (b + c') = a + (b + c)' = (a + (b + c))' = ((a + b) + c)' = (a + b) + c'$ and $c' \in S$ as desired. Thus, the associative property holds for all $a, b, c \in \mathbb{N}$.

To prove commutativity, we first prove the following lemma:

Lemma 4.3: Commutativity for $b = 1$

For all $a \in \mathbb{N}$, $a + 1 = 1 + a$.

Proof. Let S_1 be the set of all a such that $a + 1 = 1 + a$. Clearly, $1 \in S_1$. Assume that $a \in S_1$; we show $a' \in S_1$: $1 + a' = (1 + a)' = (a + 1)' = (a')' = a' + 1$ and thus $a' \in S_1$. We conclude that the associative property holds for $b = 1$. $\qquad \square$

We now have the necessary tools to prove the commutative property for all $b \in \mathbb{N}$. As before, let S be the set of all $b \in \mathbb{N}$ such that $a + b = b + a$ for all $a \in \mathbb{N}$. By Lemma 4.3, $1 \in S$. For the inductive step, we assume $b \in S$ and show that $b' \in S$. Let $T(b')$ be the set of all a such that $a + b' = b' + a$. $1 \in T(b')$ by Lemma 4.3. For $a \in \mathbb{N}, a' + b' = (a' + b)' = (b + a')' = ((b + a)')' = ((a + b)')' = (a + b')' = (b' + a)' = b' + a'$. Hence $a' \in T(b')$, and $b' \in S$, proving the theorem for all $a, b \in \mathbb{N}$. We can also prove the associative and commutative laws of multiplication similarly, using the fact that $a \cdot b' = (a \cdot b) + a$. (The proofs are left as an exercise to the reader.)

Finally, to prove that 0 is the additive identity, we can re-apply the proof of Lemma 4.3.

The distributive property can also be proven in a similar fashion. Once again, let S be the set of all c for which $(a + b) \cdot c = a \cdot c + b \cdot c$ for all $a, b \in \mathbb{N}$. For the base case, $c = 1$, we have $(a + b) \cdot 1 = a + b = a \cdot 1 + b \cdot 1$, proving that $1 \in S$. Next we assume $c \in S$ and show $c' \in S$:

$$(a + b) \cdot c' = ((a + b) \cdot c) + (a + b) \tag{4.1}$$

$$= ((a \cdot c) + (b \cdot c)) + (a + b) \tag{4.2}$$

$$= (((a \cdot c) + (b \cdot c)) + a) + b \tag{4.3}$$

$$= ((a \cdot c) + (a + (b \cdot c))) + b \tag{4.4}$$

$$= (((a \cdot c) + a) + (b \cdot c)) + b \tag{4.5}$$

$$= ((a \cdot c) + a) + ((b \cdot c + b) \tag{4.6}$$

$$= (a \cdot c') + (b \cdot c') \tag{4.7}$$

This proves the Distributive Property for all $a, b, c \in \mathbb{N}$. $\qquad \square$

These axioms can be used to prove the solution process for algebraic equations. Here are a couple of examples with one- and two-equation systems:

Example 4.1.0.1. *Prove that the solution to the equation $20x + 18 = -2$ is $x = -1$.*

Solution 4.1.0.1. *By the addition property of equality, let $c = -18$, such that $(20x + 18) + (-18) = (-2) - 18 \implies 20x = -20$. Then by the multiplication property, let $c = \dfrac{1}{20}$, so that $20x \cdot \dfrac{1}{20} = -20 \cdot \dfrac{1}{20} \implies x = -1$, as desired.*

Example 4.1.0.2. *Show that $(x, y) = (3, 2)$ is the only solution to the system of equations*

$$3x + 2y = 13, 2x - \frac{1}{2}y = 5$$

Solution 4.1.0.2. *We can think of the LHS and RHS of each equation as their own separate quantities. Let $a = 3x + 2y, b = 13, c = 2x - \dfrac{1}{2}y, d = 5$. Then we have $a = b, c = d$. By the additive property, we obtain $a + 4c = b + 4c$, and by the substitution property, $4c = 4d$, so $a + 4c = b + 4d \implies 11x = 13 + 4(5) = 33$. Applying the multiplication property ($c = \dfrac{1}{11}$) yields $x = 3$. Finally, by the substitution property, $a = b$, and $a = 3x + 2y$, so $a = 3 \cdot 3 + 2y = 13 \implies 2y = 4$ by the additive property. One more application of the multiplication property yields $y = 2$, hence the solution $(x, y) = (3, 2)$ as desired.*

We won't be using this method of proof extensively throughout the rest of the book, but this is a standard exercise in thinking in terms of rigorous proof and not just in terms of intuition or "guess-and-check". Indeed, this level of rigor with every step is often unnecessary even in professional mathematics, and with computational problems such as systems of linear equations often consumes more time and has a greater risk of error. The main purpose of studying this particular proof method is to hopefully ease the reader into the ideas and motivation behind mathematical proof in general.

Exercise 15. *For (a) through (c), use only the seven axioms in the subsection (listed in Theorem 4.1).*

(a) Show that $x = 1$ is the solution to $3x + 4 = 7$.

(b) Show that $x = \dfrac{c - b + d}{a}$ is the solution to $ax + b - d = c$.

(c) Show that $x = \pm a$ is the solution to $x^2 - a^2 = 0$.

(d) When is the Reflexive Property necessary to prove a solution to an equation? What about the Symmetric Property?

Extension Exercise 6. *Define a binary operator ξ such that $a\xi b = \left| a^2 - b^2 \right|$ for all real numbers a and b. What can we conclude about the associativity and commutativity of ξ?*

4.2   An Introduction to Proofs Involving Inequalities

An *inequality* is a comparison between two expressions, relating them in terms of equality, *strict inequality*, or *non-strict inequality*. That is, two expressions can either be equal, one can be strictly less ($<$) or strictly greater ($>$) than the other, or one can be less than or equal to ($\leq$) or greater than or equal to ($\geq$) the other.

Theorem 4.4: Axioms of Inequalities

Much like equations, inequalities are governed by axioms, but only two of the axioms for equations apply to inequalities:

- Addition property: If $a > b$, then $a + c > b + c$. If $a < b$, then $a + c < b + c$. (The same holds for non-strict inequality.)

- Multiplication property: If $a > b$, then for any **nonzero** real number c, if $c > 0$, then $ac > bc$, and if $c < 0$, then $ac < bc$.

The last part of the revised multiplication property for inequalities is especially important, as neglecting it often creates undue confusion. When we multiply two sides of an inequality by a negative number, we must reverse the direction of the inequality. Consider the true inequality $2 > 1$, for instance. If we multiply both sides by -1, we end up with the false inequality $-2 > -1$. However, $-2 < -1$ is indeed true.

Example 4.2.0.1. *Is it true that, for all real numbers a, b, c, d, $ab > cd$ if $a > c$ and $b > d$?*

Solution 4.2.0.1. *No. Consider the counterexample $(a, b, c, d) = (2, 1, -3, -1)$. In general, it does not follow from $a > c$ that $ab > cd$, by virtue of $b > d$, since two of the variables can be negative, which would reverse the direction of the inequality.*

To be able to manipulate inequalities so that false results such as that in Example 4.0.1 can become valid, we must place additional restrictions on a, b, c, and d, which operate under an *order relation* on $\mathbb{R}$:

Theorem 4.5: Axioms of Order Relations

For the following axioms, we define the order relation in terms of $\mathbb{R}^+$. (Here $\mathbb{R}^+$ denotes the subset of $\mathbb{R}$ that consists of the positive real numbers.)

- For all $x \in \mathbb{R}$, one of the following is true: $-x \in \mathbb{R}^+$, $x = 0$, $x \in \mathbb{R}^+$.

- For all $x_1, x_2 \in \mathbb{R}^+$, $x_1 + x_2 \in \mathbb{R}^+$.

- For all $x_1, x_2 \in \mathbb{R}^+$, $x_1 x_2 \in \mathbb{R}^+$.

Furthermore, we must define in terms of this order relation what it means for a number to be less than or greater than another:

Definition 4.2: Inequality Under the Order Relation

- $x > y$: For some $p \in \mathbb{R}^+$, $x = y + p$.

- $x < y$: For some $p \in \mathbb{R}^+$, $x = y - p$.

We can now prove the axioms in Theorem 4.1.

Addition property. Let $a = b + p$, where $p \in \mathbb{R}^+$. Thus, $a + c = (b + c) + p > b + c$. $\qquad\square$

Multiplication property. As before, let $a = b + p$, $p \in \mathbb{R}^+$. Then for $c > 0, c \in \mathbb{R}^+$, $ac = c(b + p) = bc + pc$. Since $pc \in \mathbb{R}^+$,[9] $ac > bc$, as desired. $\qquad\square$

Exercise 16. *Let $x \in \mathbb{R}$. Prove that $-1 \leq 2 - 3\cos\left(1 - x^2\right) \leq 5$.*

4.2.1 The Trivial Inequality

Theorem 4.6: Trivial Inequality

For all $x \in \mathbb{R}$, $x^2 \geq 0$.

Proof. We take a proof-by-exhaustion approach, using axiom 1 from Theorem 4.2: namely, the fact that if $x \in \mathbb{R}$, either $x = 0$, $x \in \mathbb{R}^+$, or $-x \in \mathbb{R}^+$. We proceed to check each case individually.

- $x = 0$: This yields $x^2 = 0 \geq 0$ immediately.

---

[9]This follows from $\mathbb{R}^+$ being *closed under multiplication*, meaning that the product ab of any two elements $a, b \in \mathbb{R}^+$ will also be an element of $\mathbb{R}^+$. The proof of this relies on axioms traditionally taught in a post-secondary group theory (abstract algebra) course.

- $x \in \mathbb{R}^+$: $x > 0$, so write $x = 0 + p$ for some $p \in \mathbb{R}^+$. Then $x^2 = 0 + 2p + p^2 = p(2+p)$, and $p(2+p) \in \mathbb{R}^+$ since $p, 2 + p \in \mathbb{R}^+$.

- $-x \in \mathbb{R}^+$. $x < 0$, so write $-x = 0 + p$, $p \in \mathbb{R}^+$. Then $x^2 = (-x)^2 = 0 + 2p + p^2 \geq 0$.

It follows that $x^2 \geq 0$ for all $x \in \mathbb{R}$. $\qquad \square$

Alternatively, we may proceed by contradiction and then casework:

Proof. Assume the contrary: that there exists a real number x such that $x^2 < 0$. Then there are three cases, as before: $x = 0$, $x > 0$, and $x < 0$.

- For $x = 0$, there is a clear contradiction: $0^2 = 0 \not< 0$.

- For $x > 0$, $x^2 < 0$ gives $x < \dfrac{0}{x}$ upon dividing both sides by x (allowable since x is positive, not changing the direction of the inequality), but this implies $x < 0$, which is a contradiction.

- For $x < 0$, $x^2 < 0$ gives $x > \dfrac{0}{x} = 0$, which is again a contradiction.

Hence $x^2 \geq 0$ for all $x \in \mathbb{R}$. $\qquad \square$

4.2.2 Examples

Example 4.2.2.1. *Let x and y be real numbers. Show that $(x + y)^2 \geq 4xy$.*

Solution 4.2.2.1. *We expand the LHS: $(x+y)^2 = x^2 + 2xy + y^2$, then use the Addition Property for $c = -4xy$ to obtain $x^2 - 2xy + y^2 \geq 0$, or $(x - y)^2 \geq 0$ after factoring the LHS. Since $x, y \in \mathbb{R}$, $x - y \in \mathbb{R}$,[10] and so Trivial Inequality proves the statement. (This is in fact a standard application of the AM-GM Inequality; see §10 for more information on this inequality.)*

4.2.3 Exercises

Exercise 17. *Let x, y, and z be real numbers. Show that $xy + xz + yz \leq x^2 + y^2 + z^2$.*

Exercise 18. *Let x, y, and z be real numbers. Show that $(0, 0, 0)$ is the only solution to the equation $10xy + 12yz + 2xz = 2x^2 + 29y^2 + 10z^2$.*

4.3 Introduction to Functions

4.3.1 Defining a Relation

To understand what a function is, we must first define a *relation* between two sets of values:

> **Definition 4.3: Relation**
>
> A *relation* $R : X \mapsto Y^a$ between two sets X and Y is a collection of ordered pairs (x, y) where $x \in X$ and $y \in Y$.
>
> ---
> $^a R$ maps elements in X to elements in Y.

If objects x and y are in the sets X and Y, respectively, then they are said to be *related* if $(x, y) \in R$.

---
[10]$\mathbb{R}$ is not only closed under multiplication, but under addition as well.

4.3.2 Defining the Function

Equations and inequalities are relations, but they are confined to merely relating two quantities to each other. A *function* takes the fundamental idea behind the formation of an equation or inequality and expands it to an infinitude of potential quantities.

At its very core, a function is a relationship that produces an *output* in terms of the *input*. The input and output must both originate from sets, namely the *domain* and *codomain* respectively.

First we must more carefully define what these terms mean:

> **Definition 4.4: Function Vocabulary Definitions**
>
> - The *domain* is the set of all valid inputs.
>
> - The *range* is the set of all valid outputs.
>
> - The *codomain* is the set of all outputs that are actually realized by the function.
>
> - The *input* is any element of the domain.
>
> - The *output* is any element of the range.

In more rigorous terms, we define a function as follows:

> **Definition 4.5: Definition of a Function**
>
> A function $f : X \mapsto Y$ from the set X to the set Y assigns to each input $x \in X$ an output $y \in Y$. A function must assign to each x a unique y, but a certain value of y may correspond to more than one value of x.

The familiar notation customarily used to denote this is $f(x) = y$.

Example 4.3.2.1. *Which of the following relations are functions? (All inputs are mapped to the outputs in the corresponding position. For instance, $f : \{1, 2, 3\} \mapsto \{1, 2, 3\}$ denotes the function given by $f(1) = 1$, $f(2) = 2$, $f(3) = 3$.)*

(a) $f : \{1, 3, 4, 5\} \mapsto \{1, 1, 2, 3\}$

(b) $f : \{0, 1, 2, 2\} \mapsto \{0, 2, 3, 4\}$

(c) $f : \{5, -1, 5, 59\} \mapsto \{100, 2018, 2018, 2019\}$

(d) $f : \mathbb{R}^+ \to \mathbb{R}^+$, $f^2(x) = x$

Solution 4.3.2.1.

(a) This is a function, since here, $f(1) = f(3) = 1$, $f(4) = 2$, and $f(5) = 3$. There are no two outputs being mapped to one input.

(b) This is not a function, as $f(2) = 3$ and $f(2) = 4$ at the same time.

(c) This is a function, although $f(-1) = f(-5) = 2018$.

(d) This is a function, since $f(x) = |\sqrt{x}|$, which when plotted in the Cartesian plane, satisfies the VLT ($|\sqrt{x}|$ has only one value for any given input $x \in \mathbb{R}^+$). Note that the answer would be different had the relation instead taken its input over $\mathbb{R}$ (since $\sqrt{x}$ would then be a complex number), or had it been defined by $f(x) = \pm\sqrt{x}$.

4.3.3 The Geometric Perspective: Horizontal and Vertical Line Tests

There is a key distinction that must be made between *relations* and *functions*. Whereas a relation is any set of ordered pairs that can be plotted in the Cartesian coordinate plane, a function must pass the *Vertical Line Test,* or VLT:

Theorem 4.7: Horizontal and Vertical Line Tests

- Horizontal Line Test (HLT): Determines whether or not a relation maps a y-value to a unique x-value and, therefore, whether the relation has an inverse. If there exists a horizontal line l that passes through two or more points that lie on the curve, the relation corresponding to that curve is said to fail the HLT. However, if a function passes the HLT, then it is said to be *injective* or *one-to-one*, and it has an inverse.

- Vertical Line Test (VLT): Determines whether or not a relation is a function. If there exists a vertical line l that passes through two or more points that lie on the curve, the relation corresponding to that curve is said to fail the VLT. Otherwise, the curve is a function! :)

Note that a relation can still be a function despite failing the HLT, though it will not be one-to-one. That is, there will be two or more inputs for a given output (the function does not uniquely map each input to an output).

Example 4.3.3.1. *For each of the relations in Example 5.1.1, list whether it passes or fails the HLT and VLT.*

Solution 4.3.3.1.

(a) *This relation fails the HLT, since $f(1) = f(3) = 3$ (there are two inputs mapping to the same output), but it passes the VLT by virtue of being a function at all (more specifically, no two outputs map to the same input).*

(b) *This passes the HLT but fails the VLT (since there are no two inputs mapping to the same output, but $f(2)$ has two distinct values).*

(c) *This passes the VLT but fails the HLT.*

(d) *This passes both the HLT and VLT.*

4.3.4 Properties of Continuous Functions

Functions abide by different properties depending on whether or not they obey the condition of *continuity*. To define continuity, we must first develop the framework of *limits* using the traditional δ-ϵ definition as a foundation:[11]

The Delta-Epsilon Definition of the Limit and Continuity

Definition 4.6: The Delta-Epsilon Limit Definition

Let $f(x)$ be a function whose domain consists of an interval that contains $x = a$; the domain may or may not contain $x = a$ itself. Then we write $\lim_{x \to a} f(x) = L$ if for every $\epsilon \in \mathbb{R}^+$ there exists a $\delta \in \mathbb{R}^+$ such that $|f(x) - L| < \epsilon$ whenever $0 < |x - a| < \delta$.

---

[11]This level of rigor in defining a limit is traditionally established in a post-secondary introductory real analysis course; usually, in a pre-calculus or calculus curriculum, the definition of the "limit" is confined to intuition, in the sense that we treat limits as "that which we approach, but never actually reach." However, this intuition alone does not suffice when proving theorems regarding the continuity of functions.

Shedding Light on the Concept

If we consider the function f as a curve in $\mathbb{R}^2$,[12] then we say that f is continuous at $x = a$ if for every $\epsilon > 0$ there exists a $\delta > 0$ such that, whenever a point x is within δ of a (i.e. $x \in [a - \delta, a + \delta]$), then $f(x)$ is within ϵ of L (i.e. $f(x) \in [L - \epsilon, L + \epsilon]$). Here is a graphical illustration of the δ-ϵ definition:

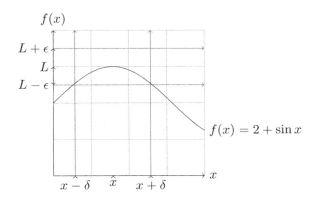

Figure 4.1: The graph of a continuous function according to the δ-ϵ definition.

Now we have all of the necessary tools to define continuity:

Definition 4.7: Definitions of Function Continuity

A function $f(x)$ is continuous at $x = a$ iff $f(a) = \lim_{x \to a} f(x)$.

We can also define continuity in the following manner. Say $f(x)$ is continuous at $x = a$ iff, for every real-valued sequence $\{a_i\}_{i=1}^{\infty}$ converging to a (meaning that $\lim_{n \to \infty} a_n = a$, more compactly written as $\{a_i\}_{i=1}^{\infty} \to a$), the sequence $\{f(a_i)\}_{i=1}^{\infty} \to f(a)$.

Equivalently, rephrased in terms of the δ-ϵ definition, we say that $f(x)$ is continuous at $x = a$ iff, for every $\epsilon > 0$, there exists a $\delta > 0$ such that $|f(y) - f(x)| < \epsilon$ for all y such that $|y - x| < \delta$.

Furthermore, f is continuous on an interval I iff it is continuous at $x = a$ for all $a \in I$.

Proving Function Limits and Continuity

To prove the existence and value of a limit, we use the values of a (the limiting x-value), L (the limit), and $f(x)$ (the literal definition of the function in terms of x), as follows:

Example 4.3.4.1. *Prove that*
$$\lim_{x \to 3} (4x - 2) = 10$$

Solution 4.3.4.1. *Since $a = 3$, we require a $\delta > 0$ for all $\epsilon > 0$ such that $|(4x - 2) - 10| < \epsilon$ whenever $0 < |x - 3| < \delta$. That is, $|4x - 12| < \epsilon$ whenever $|x - 3| < \delta$. The first inequality simplifies to $|x - 3| < \frac{1}{4}\epsilon$ after factoring out the 4 from the absolute value. Solving for δ, we find $\delta = \frac{1}{4}\epsilon$ satisfies the necessary condition. We have proven existence, as desired.*

Example 4.3.4.2. *Prove that*
$$\lim_{x \to 1} \frac{x^2 - 1}{x - 1} = 2$$

---

[12]The standard notation for the 2-dimensional Cartesian coordinate plane. Similarly, $\mathbb{R}^n$ denotes the n-dimensional coordinate plane.

Solution 4.3.4.2. *We are given that $a = 1$, $L = 2$, and $f(x) = \dfrac{x^2 - 1}{x - 1}$. For $\delta, \epsilon \in \mathbb{R}$, whenever $|x - 1| < \delta$, we have $|f(x) - L| = \left| \dfrac{x^2 - 1}{x - 1} - 2 \right|$. This expression simplifies to $|x - 1| < \epsilon = \delta$. This completes the proof.*

To prove continuity by the first definition (f is continuous at $x = a$ iff $f(a) = \lim\limits_{x \to a} f(x)$), it suffices to use the δ-ϵ method for limits:

Example 4.3.4.3. *Show that $f(x) = 2x^2 + 1$ is continuous at $x = 1$.*

Solution 4.3.4.3. *We show that $\lim\limits_{x \to 1} f(x) = f(1) = 3$. $a = 1$, $L = 3$, so we have $\left| 2x^2 + 1 \right| < \epsilon$ whenever $|x - 1| < \delta$, or $\delta = \sqrt{\dfrac{\epsilon - 1}{2}} - 1$, proving existence.*

For problems such as the following, we fix a and ϵ and manipulate the ϵ equality to derive a value for δ in terms of ϵ, thereby proving existence:

Example 4.3.4.4. *Show that $f(x) = \sqrt{x}$ is continuous for all $x \geq 0$.*

Solution 4.3.4.4. *Fix a and let $L = \sqrt{a}$. Then we require the existence of a $\delta > 0$ for all $\epsilon > 0$ such that $|\sqrt{x} - \sqrt{a}| < \epsilon$ whenever $|x - a| < \delta$. We have $|\sqrt{x} - \sqrt{a}| \leq |\sqrt{x} + \sqrt{a}|$ for all x, a. It follows that $|\sqrt{x} - \sqrt{a}|^2 \leq |\sqrt{x} - \sqrt{a}| \, |\sqrt{x} + \sqrt{a}| = |x - a| < \delta$, from which we obtain $|\sqrt{x} - \sqrt{a}| < \epsilon$ if $\delta = \epsilon^2$, as desired.[13]*

(Note: Proving uniform continuity on $[0, \infty)$ using sub-definitions 2 and 3 in Definition 5.4 is left as an exercise to the reader.)

Types of Discontinuities

A function f is *discontinuous* at a point $x = a$ whenever $f(a) \neq \lim\limits_{x \to a} f(x)$. This may occur when f is undefined at $x = a$ (see Example 5.3.2 for an example of such an f), when the limits from the left-hand side of $x = a$ and the right-hand side of $x = a$ differ,[14] or when those limits are infinite or non-existent.

> **Definition 4.8: Types of Function Discontinuities**
>
> Let f be a function defined on an interval $[a, b] \in \mathbb{R}$ with a discontinuity at $x_0 \in [a, b]$.
>
> - Removable (hole) discontinuity: f has a removable discontinuity at $x = a$ if $\lim\limits_{x \to a} f(a) = L < \infty$, but $f(a) \neq L$. Most often, f is undefined at $x = a$, and this discontinuity can be "removed" by dividing both the numerator and denominator of f by a factor of $x - a$.
>
> - Jump (step) discontinuity: f has a jump discontinuity at $x = a$ if $\lim\limits_{x \to a^-} f(a) = L_1$ and $\lim\limits_{x \to a^+} f(a) = L_2$ exist, but $L_1 \neq L_2$.
>
> - Endpoint (essential) discontinuity: f has an endpoint discontinuity if either $\lim\limits_{x \to a^-} f(a)$ or $\lim\limits_{x \to a^+} f(a)$ do not exist or are infinite. (As an example, consider $f(x) = \sqrt{x}$; then f has an endpoint discontinuity at $x = 0$, since $\lim\limits_{x \to 0} f(x)$ does not exist, in turn because $\lim\limits_{x \to 0^+} f(x) = 0$, but $\lim\limits_{x \to 0^-} f(x)$ does not exist as f is undefined for $x < 0$.) The *infinite discontinuity* is a special case of the endpoint discontinuity, wherein both one-sided limits are $\pm \infty$.

Example 4.3.4.5. *Is the function $f(x) = \dfrac{3x^2 - 9}{x - \sqrt{3}}$ continuous over $\mathbb{R}$? If not, what type of discontinuity(ies) does it have?*

---

[13] This type of continuity is known as *uniform continuity* on an interval $I \subset \mathbb{R}$.

[14] Known as one-sided limits, and are denoted by $\lim\limits_{x \to a^-} f(x)$ and $\lim\limits_{x \to a^+} f(x)$ respectively. The notation $\lim\limits_{x \to a} f(x)$ denotes the *two-sided limit* L of $f(x)$ at $x = a$, and is only defined when $\lim\limits_{x \to a^-} f(x) = \lim\limits_{x \to a^+} f(x) = L$.

Solution 4.3.4.5. *f has a removable discontinuity at $x = \sqrt{3}$, since $f(\sqrt{3})$ is undefined, but $x - \sqrt{3}$ is a factor of the numerator and denominator.*

4.3.5 Function Monotonicity

Another fundamental property of functions is *monotonicity;* i.e. the constant state of increase or decrease of the value of a function. Rigorously, we define monotonicity as follows:[15]

> ### Theorem 4.8: Function Monotonicity
>
> A function f is *monotonically increasing* on an interval $\mathbb{R} \supset I = [a, b]$ if, for all $x_1, x_2 \in I$, $x_1 \geq x_2 \implies f(x_1) \geq f(x_2)$. On the contrary, f is *monotonically decreasing* on I if $x_1 \geq x_2 \implies f(x_1) \leq f(x_2)$. (Note that, in this context, "increasing" and "decreasing" are synonymous with "non-decreasing" and "non-increasing.")

We also say that f is *order-preserving* if it is monotonically increasing on an interval I, and *order-reversing* if it is monontonically decreasing on I. In addition, we say that monotonicity is *strict* if the inequalities in the definition have the symbols $<$ or $>$ rather than $\leq$ or $\geq$, i.e. that the function is strictly increasing or strictly decreasing on I.

Example 4.3.5.1. *Prove that if f is strictly monotonic on an interval $I = [a, b]$, it must also pass the HLT on I.*

Solution 4.3.5.1. *WLOG[16] assume f is monotonically increasing. By the definition of strict monotonicity, for all $x_1, x_2 \in I$, if $x_1 > x_2$, then $f(x_1) > f(x_2)$. Thus, $f(x_1) \neq f(x_2)$ for $x_1 > x_2$, and so no horizontal line can pass through two or more points on the function. (In order for f to fail the HLT, there must exist $x_1, x_2 \in I$ such that $f(x_1) = f(x_2)$.) The proof for when f is monotonically decreasing is identical.*

Example 4.3.5.2. *Suppose f is monotonic on $\mathbb{R} \supset I = [a, b]$. Need f be continuous?*

Solution 4.3.5.2. *No. f may have jump discontinuities (but no other types of discontinuities). If f is monotonic on I,[17] then $\forall x_1, x_2 \in I$, $f(x_1) \geq f(x_2)$. For some $x = x_0 \in [a, b]$, it is possible to have $\lim_{x \to x_0^-} f(x) \neq \lim_{x \to x_0^+} f(x)$, provided they both exist.*

This is because the one-sided limit from the left is bounded above [18] by $f(x_0)$, while the one-sided limit from the right is bounded below by $f(x_0)$. Since f is increasing in (a, x_0), as well as in (x_0, b), both one-sided limits exist.

4.3.6 Injectivity, Surjectivity, and Bijectivity

For our final topic in this chapter, let's revisit the Horizontal and Vertical Line Tests and add more rigor to our current definitions by working the function classifications of *injectivity*, *surjectivity*, and *bijectivity*:

> ### Definition 4.9: Classifications of Functions
>
> The following are the classifications of continuous functions:
>
> - Injective (one-to-one): A function $f : X \mapsto Y$ is injective iff it passes the HLT; i.e. iff, $\forall x \in X$, $f(x) = y$ uniquely. That is, any horizontal line should pass through the graph of function no more than once.

---

[15]The general term "monotonic" may refer to a function that is either monotonically increasing or monotonically decreasing.
[16]WLOG stands for "without loss of generality" and denotes the assumption of a specific case, which does not alter the proof.
[17]Here we assume WLOG that f is monotonically increasing; the proof for when f is monotonically decreasing is identical.
[18]We say that f is *bounded above* by s if $\exists s \in I$ such that $s \geq f(x)$ for all $x \in I$. Similarly, we say f is *bounded below* by s if $\exists s \in I$ such that $s \leq f(x)$ for all $x \in I$.

- Surjective (onto): A function $f : X \mapsto Y$ is surjective iff, $\forall y \in Y$, $\exists x \in X$ such that $f(x) = y$. That is, any horizontal line should pass through the graph of the function at least once.

- Bijective: A function $f : X \mapsto Y$ is bijective iff it is both injective and surjective. That is, any horizontal line should pass through the graph of the function exactly once.

For non-continuous (discrete) functions, these definitions still apply, except we examine the inputs and outputs individually, not considering the HLT or VLT.

Example 4.3.6.1. *Study the injectivity, surjectivity, and bijectivity of the function $f : \mathbb{R} \to \mathbb{R}$, $f(x) = e^x$.*

Solution 4.3.6.1. *f is injective; we prove this by contradiction. Assume $\exists x_1, x_2 \in \mathbb{R}$ such that $x_1 \neq x_2$ but $f(x_1) = f(x_2)$. However, f is monotonically increasing on $\mathbb{R}$,[19] so there do not exist any such x_1, x_2 - contradiction.*

f is not surjective; $f(x) = e^x$ never attains any values in the interval $(-\infty, 0]$ for $x \in \mathbb{R}$.[20] It follows that f is not bijective, since bijectivity requires both injectivity and surjectivity.

---

[19]The proof involves taking the derivative f' and showing that $f'(x) > 0$ for all $x \in \mathbb{R}$.
In this case, $f' = e^x > 0$ for all $x \in \mathbb{R}$.
[20]However, if we defined $f : \mathbb{R} \to (0, \infty)$, $f(x) = e^x$, then f would be surjective, as $f(x) = e^x$ attains all values in the interval $(0, \infty)$. This example illustrates the importance of defining not only the relation part of the function, but also the domain and range.

4.3.7 Exercises

Note: All functions are from $\mathbb{R}$ to $\mathbb{R}$.

Exercise 19. *For each of the following, either provide a proof if the statement is true, or provide a counterexample if the statement is false.*

(a) *All bijective functions are strictly monotonic.*

(b) *All strictly monotonic functions are bijective.*

(c) *All bijective continuous functions are strictly monotonic.*

(d) *All strictly monotonic continuous functions are bijective.*

Exercise 20. *Prove that*

$$\lim_{x \to 1} \frac{x^n - 1}{x - 1} = n$$

for all $n \in \mathbb{N}$.

Extension Exercise 7. *Prove that all linear functions $f(x) = ax + b$ ($a \neq 0$) are continuous at all $x \in \mathbb{R}$.*

Extension Exercise 8. *Prove that*

$$\lim_{x \to \infty} x = \infty.$$

Extension Exercise 9. *Let $f : \mathbb{R} \to \mathbb{R}$ and $g : \mathbb{R} \to \mathbb{R}$ be continuous functions. Show that $f(g(x))$ and $g(f(x))$ are also continuous.*

4.4 Introduction to the Binomial Theorem

In dealing with polynomials, we can manipulate expressions with up to three or four terms with ease, deriving elegant and clean properties from them. However, polynomials are a vast, unlimited space; eventually, they become far too cumbersome and massive to handle. What we require beyond the three- or four-term polynomial is a way to generalize an expansion - beginning with the humble *binomial expansion.*

A *binomial* is any expression consisting of two terms (much like a *monomial* is a single-term expression, and a *trinomial* is a three-term expression). Ordinarily, these are quite straightforward to deal with. If we raise them to a power, however, they rapidly escalate in complexity and messiness.

This is what we would say were it not for the trusty *Binomial Theorem*! The Binomial Theorem allows us to compute the coefficients of any binomial expansion, without having to distribute by hand. We begin with the most common and fundamental form of the theorem:

Theorem 4.9: Binomial Theorem

For all non-negative integers n, $(a+b)^n = \binom{n}{0} \cdot a^n + \binom{n}{1} \cdot a^{n-1}b + \binom{n}{2} \cdot a^{n-2}b^2 + \ldots + \binom{n}{n-2} \cdot a^2 b^{n-2} + \binom{n}{n-1} \cdot ab^{n-1} + \binom{n}{n} \cdot b^n = \sum_{k=0}^{n} \binom{n}{k} \cdot a^k b^{n-k}.$

(Here $\binom{n}{k}$ denotes the *binomial coefficient*; i.e. the quantity $\dfrac{n!}{k!(n-k)!}$, representing the number of ways to choose k distinct objects from n total objects.)

Before proving the theorem, we should make note of some observations. Firstly, note that all of the coefficients are symmetric, as $\binom{n}{k} = \binom{n}{n-k}$ for all $n \geq k$; this can also be observed by expanding $(x+y)^n$ for small values of n. Also note that the coefficients; i.e. the values of $\binom{n}{k}$ for a given value of n, are the elements in the nth row of *Pascal's Triangle,* where row 0 is the topmost row, consisting of a single 1.

At last, it's time to prove this remarkable result. We can prove the base form of the theorem in at least two different ways:

Combinatorial proof. Consider the expansion of $(a+b)^n = (a+b)(a+b)\ldots(a+b)$ (multiplied out n times). Each term has degree n because n terms of degree 1 are multiplied together. In how many ways can we create a term with n a's multiplied together and 0 b's? There's only $\binom{n}{0} = 1$ way: choosing all of the a's. In how many ways can we create a term with $(n-1)$ a's and 1 b multiplied together? Will the first binomial be the b, the last, or which one? There are $\binom{n}{1} = n$ ways to choose which of the b terms to multiply to $(n-1)$ a's.

Now, let's look at this more generally. What is the coefficient of the $a^k b^{n-k}$ term (k is an integer such that $0 \leq k \leq n$)? There are $\binom{n}{k}$ ways to choose which k of the $(a+b)$ terms will be a's, leaving the remaining $(n-k)$ $(a+b)$ terms as b's. Thus, the coefficient of the $a^k b^{n-k}$ term is $\binom{n}{k}$. It follows that

$(a+b)^n = \binom{n}{0}a^n b^0 + \binom{n}{1}a^{n-1}b^1 + \cdots + \binom{n}{n-1}a^1 b^{n-1} + \binom{n}{n}a^0 b^n$ as desired. $\qquad \square$

Induction proof. **Base Case:** $n = 0$

$(a + b)^0 = 1$, which is equal to $\binom{0}{0}$. The base case holds.

Inductive Hypothesis: Assume the $n = k$ case is true. That is,

$$(a + b)^k = \binom{k}{0}a^k + \binom{k}{1}a^{k-1}b + \cdots + \binom{k}{k-1}ab^{k-1} + \binom{k}{k}b^k.$$

Inductive Step: Consider the $n = k + 1$ case. We know that

$$(a + b)^{k+1} = (a + b)(a + b)^k = (a + b)\left(\binom{k}{0}a^k + \binom{k}{1}a^{k-1}b + \cdots + \binom{k}{k}b^k\right).$$

It is apparent that the coefficients of the a^{k+1} and b^{k+1} terms are both equal to 1. The coefficient of the $a^i b^{k+1-i}$ term (where i is an integer such that $0 < i < k + 1$) is equal to $\binom{k}{i-1} + \binom{k}{i}$ because there are $\binom{k}{i-1}$ terms that equal $a^i b^{k+1-i}$ when multiplied by a and $\binom{k}{i}$ terms that equal $a^i b^{k+1-i}$ when multiplied by b.

As we observed earlier, $\binom{k}{i-1} + \binom{k}{i} = \binom{k+1}{i}$; therefore, the coefficient of $a^i b^{k+1-i}$ in the binomial expansion of $(a + b)^{k+1}$ is equal to $\binom{k+1}{i}$. Since $\binom{k+1}{0} = \binom{k+1}{k+1} = 1$,

$$(a + b)^{k+1} = \binom{k+1}{0}a^{k+1} + \binom{k+1}{1}a^k b + \cdots + \binom{k+1}{k+1}b^{k+1}$$

as desired. The $n = k + 1$ case holds.

Therefore, we have proven that the Binomial Theorem holds for all integers $n \geq 0$ via induction. $\square$

4.4.1   Examples

Example 4.4.1.1. *What is the coefficient of the $x^2 y^5$ term in the expansion of $(x + y)^7$?*

Solution 4.4.1.1. *This is equal to* $\binom{7}{2} = \dfrac{7!}{2!5!} = 21$.

Example 4.4.1.2. *Find the coefficient of the $x^{1010} y^{1009}$ term in the expansion of $(x + y)^{2018}$.*

Solution 4.4.1.2. *The Binomial Theorem applies only to those terms such that the sum of the powers of x and y is equal to n (in this case, 2018). Since $1010 + 1009 = 2019 \neq 2018$, there is no $x^{1010} y^{1009}$ term in the expansion of $(x + y)^{2018}$, and so its coefficient is 0.*

4.4.2 Exercises

Exercise 21. *In the expansion of $(1-x)^n$, where n is a positive integer, the x^{n-2} term has a coefficient of 55. Find n.*

Exercise 22. *There are 10 fruit snacks in a pouch: 5 strawberry-flavored, and 5 grape-flavored. If we must pick 5 random snacks, and at least two must be strawberry-flavored, in how many ways can we pick the snacks? (The specific snacks picked matter, but not the order in which they are picked.)*

Exercise 23. *Show that*

$$\sum_{k=0}^{n} \binom{n}{k} = 2^n$$

for all $n \in \mathbb{N}$.

Exercise 24. *Show that*

$$\binom{2n}{n} = \sum_{k=0}^{n} \binom{n}{k}^2$$

for all $n \in \mathbb{N}$.

Extension Exercise 10. *Prove that*

$$\sum_{k=0}^{n} (-1)^k \cdot \binom{n}{k} = 0$$

for all $n \in \mathbb{N}$ using the Binomial Theorem.

Extension Exercise 11. *Using the Binomial Theorem, prove that*

$$\sum_{k=0}^{n} \binom{n}{k} 2^k = 3^n$$

Extension Exercise 12. *Show that, for all positive integers n, $\binom{4n}{2} = \binom{3n}{2} + \binom{2n}{2} + \dfrac{3n^2+n}{2}$.*

4.5 A Brief Introduction to Studying the Roots of Polynomials

In studying polynomials, we must confront the crux of these intriguing, deceptively deep mathematical objects: their *roots* (or *zeros*).

Definition 4.10: Roots of a Polynomial

A root or zero of a polynomial $P(x)$ is any value of x such that $P(x) = 0$.

It is often possible to compute the roots of a polynomial with a formula such as the Quadratic Formula. As a refresher, here's the Quadratic Formula:

Theorem 4.10: Quadratic Formula

Given a quadratic equation $ax^2 + bx + c = 0$, where $a \neq 0$, the solutions are $x = \dfrac{-b \pm \sqrt{b^2 - 4ac}}{2a}$. Note that these solutions need not be real numbers.

Proof. We begin with the quadratic $ax^2 + bx + c = 0$ and show that this leads to the solutions for x. Firstly, dividing through by a yields $x^2 + \dfrac{b}{a}x + \dfrac{c}{a} = 0$, then subtracting $\dfrac{c}{a}$ from both sides gives $x^2 + \dfrac{b}{a}x = -\dfrac{c}{a}$. We then complete the square on the LHS to obtain $x^2 + \dfrac{b}{a}x + \dfrac{b^2}{4a^2} = \dfrac{b^2}{4a^2} - \dfrac{c}{a}$, or $\left(x + \dfrac{b}{2a}\right)^2 = \dfrac{b^2 - 4ac}{4a^2}$. Taking the square root of both sides yields the equation $x + \dfrac{b}{2a} = \pm\dfrac{\sqrt{b^2 - 4ac}}{2a}$. Finally, subtracting $\dfrac{b}{2a}$ from both sides yields the desired result, $x = \dfrac{-b \pm \sqrt{b^2 - 4ac}}{2a}$. $\square$

If the *discriminant*, $b^2 - 4ac$, is greater than 0, then the quadratic will have two real solutions. If $b^2 - 4ac = 0$, then the quadratic will have one real solution, namely $x = \dfrac{-b}{2a}$. If $b^2 - 4ac < 0$, then it will have two complex solutions.

To aid further with root-finding, we can apply the *Remainder Theorem,* which relates roots to the expansions of polynomials:

Theorem 4.11: Remainder Theorem

If we evaluate the polynomial $P(x)$ at $x = r$ and obtain a non-zero value R, then $P(x)$ leaves a remainder of R when divided by $x - r$.

Corollary 4.11: Factor Theorem

If $x = r$ is a root of the polynomial $P(x)$, then $x - r$ is a factor of $P(x)$.

Proof. The Remainder Theorem follows from the existence and uniqueness theorem of Euclidean division:

Theorem 4.12: Euclidean Division Existence and Uniqueness

For any two polynomials $P_1(x)$, $P_2(x)$, there exist polynomials $q(x)$ (the *quotient polynomial*) and $r(x)$

(the *remainder polynomial*) such that $P_1(x) = P_2(x) \cdot q(x) + r(x)$ for all x, where $\deg r(x) < \deg P_2(x)$.[a]

---
[a]A related theorem is Bezout's Identity.

A related theorem is the Euclidean algorithm for computing the greatest common divisor of two positive integers:

Theorem 4.13: Euclidean GCD Algorithm

For any two positive integers a and b, subtract from the larger of the two the smaller of the two, such that both integers always remain positive, until the two integers are identical. Then $\gcd(a, b)$ is that common value.

First, we prove existence and uniqueness:

Existence and uniqueness. (Henceforth, let $a = P_1(x)$, $b = P_2(x)$ for brevity.) The first step is proving that q and r actually exist. First, consider the case where $b < 0$. Let $b' = -b$ and $q' = -q$; then we can write the equation $a = bq + r$ as $a = b'q' + r$. Also, we can rewrite $0 \le r \le |b|$ as $0 \le r \le b'$, so we need only consider $b > 0$.

If $b > 0$, but $a < 0$, then we can set $a' = -a$, $q' = -q - 1$, and $r' = b - r$, so that $a' = bq' + r'$ and $0 \le r' \le b$. Therefore, it is also only necessary to consider $a \ge 0$.

Let there be $q_1, r_1 \ge 0$ such that $a = bq_1 + r_1$. If $r_1 < b$, we are done. Otherwise, let $q_2 = q_1 + 1$ and $r_2 = r_1 - b$; then $a = bq_2 + r_2$, with $0 \le r_2 < r_1$. If we repeat this process, eventually, we will have some $q_k = q$ and $r_k = r$ such that $a = bq_k + r_k$ and $0 \le r < b$. This proves existence.

Next, we prove uniqueness. Assume the contrary; namely, suppose that there exist q, r, q', r' such that $0 \le r, r' \le |b|$, $a = bq + r$, and $a' = bq' + r'$. Then $-|b| \le -r' \le 0$ yields $-|b| \le r - r' \le |b|$, or $|r - r'| \le |b|$. Subtracting the equation $a' = bq' + r'$ from the equation $a = bq + r$ yields $b(q' - q) = r' - r$, so $|b|$ must divide $|r - r'|$. If $r \ne r'$, then $|b| \le |r - r'|$, which contradicts $|r - r'| \le |b|$. Therefore, we must have $r = r'$ and $b(q - q') = 0$. We have assumed $b \ne 0$ from the beginning, so $q = q'$, which completes the proof of uniqueness. $\square$

Next, we proceed to prove the GCD algorithm:

Euclidean GCD algorithm. Observe that the GCD algorithm is equivalent to the combination of the following three statements for integers a and b (assuming a and b are not both zero):

(i) $\gcd(a, b) = \gcd(b, a)$.

(ii) If $a > 0$ and $a \mid b$, then $\gcd(a, b) = a$.

(iii) If $a \equiv c \pmod{b}$, then $\gcd(a, b) = \gcd(c, b)$.

Statement (a) is trivial, since any common divisor of a and b is a common divisor of b and a as well. In addition, any number that is *not* a common divisor of a and b will also not be a common divisor of b and a. Statement (b) can be shown by noting that a is the largest divisor of a, and that if $a \mid b$, then a is also a divisor of b. Clearly, a is the largest divisor of a, so we are done. Finally, for statement (c), observe that $a \equiv c \pmod{b}$ implies $b \mid (a - c)$. Thus, there exists a $y \in \mathbb{Z}$ such that $a - c = by$, or $c = a - by$. Furthermore, let d divide both a and b, so that $d \mid (a - by)$. Thus, any common divisor of a and b is also a common divisor of c and b. If d divides c and b, then $d \mid (c + by) \implies d \mid a$, so it is also a common divisor of a and b. This proves that the sets of common divisors of a and b, and the common divisors of c and b, are equal, so they must have the same greatest common divisor. This completes the proof of Euclid's GCD algorithm. $\square$

We now have all of the necessary tools to finish the proof of the Remainder Theorem.

From the Euclidean division theorem, it follows that either $R(x) = 0$ or $\deg R(x) < \deg P_2(x)$, the latter implying $\deg R(x) = 0$, since $\deg P_2(x) = \deg(x - r) = 1$. In either case, $R(x)$ is a constant polynomial. Hence, $P_1(x) = q(x) \cdot (x - r) + R(x)$. Setting $x = r$, we obtain $P_1(r) = R$, as desired. $\qquad\square$

Alternate proof. It can be shown that $f(x) - f(r)$ can be written as a linear combination of terms of the form $x^k - r^k$, where $k \geq 1$. Since $x - r$ is a factor of $x^k - r^k = (x - r)(x^{k-1} + rx^{k-2} + r^2 x^{k-3} + \ldots + r^{k-3} x^2 + r^{k-2} x + r^{k-1})$ for all $k \geq 1$, $f(x) - f(r)$ is a multiple of $x - r$. Since r is a root of f, $f(r) = 0$, and thus $f(x)$ is a multiple of $x - r$, as desired. $\qquad\square$

With polynomials of the third degree and higher, on the other hand, extracting the roots directly either requires a more complex and borderline un-memorizable formula (such as Cardano's formula for cubics[21]), and are thus rendered impractical without the aid of techniques such as numerical analysis (which, again, are infeasible in a computational contest setting).

4.5.1 Vieta's Formulas for Degree 2 or Lower

However, using *Vieta's Formulas*,[22] which provide general formulas for the sum and product of the roots of a quadratic, as well as formulas for the *symmetric sums* of higher-degree polynomials, we can better understand the roots, and their most important properties, without actually needing to compute them individually.

> **Theorem 4.14: Vieta's Formulas for Quadratics**
>
> Given a quadratic polynomial $P(x) = ax^2 + bx + c$ ($a \neq 0$) with roots r_1, r_2, $r_1 + r_2 = -\dfrac{b}{a}$ and $r_1 r_2 = \dfrac{c}{a}$.

> **Corollary 4.14: Special Case: $a = 0$**
>
> If $a = 0$, then P becomes a linear polynomial in the form $P(x) = bx + c$. In this case, P has only one root, namely $r = -\dfrac{c}{b}$. (Note that Vieta's formulas for quadratics do not apply here due to the nonsensicality of dividing by zero, hence the $a \neq 0$ restriction in the theorem.)

Proof. By the Factor Theorem, $P(x)$ can be written in the form $a(x - r_1)(x - r_2)$. Expanding and setting this expression equal to the original form of the quadratic, we obtain $a(x^2 - x(r_1 + r_2) + r_1 r_2) = ax^2 - ax(r_1 + r_2) + ar_1 r_2$. Hence, $b = -a(r_1 + r_2)$ and $c = ar_1 r_2$. This simplifies to $r_1 + r_2 = -\dfrac{b}{a}$ and $r_1 r_2 = \dfrac{c}{a}$ ($a \neq 0$). $\qquad\square$

This argument also applies to higher-degree polynomial symmetric sum formulas.

Exercise 25. *How can we relate Vieta's Formulas to the vertex of a parabola (a quadratic polynomial)?*

Let's try a few concrete examples:

---

[21] For those interested, the formula and its derivation, as well as the quartic formula and the proof that a polynomial with degree ≥ 5 has roots not expressible in terms of radicals, can be found in Appendix B.

[22] Named for French mathematician François Vietè.

4.5.2 Examples

Example 4.5.2.1. *If p, q are the roots of $x^2 + 20x - 19$, compute $p + q$ and pq.*

Solution 4.5.2.1. *Here, $a = 1$, $b = 20$, and $c = -19$, so $p + q = -\dfrac{b}{a} = -20$ and $pq = \dfrac{c}{a} = -19$. (It can be verified using the Quadratic Formula that $p, q = -10 \pm \sqrt{119}$.)*

Example 4.5.2.2. *Let r and s be real numbers with $r > s$ such that $r + s = 12$ and $r^2 + s^2 = 119$. Compute the ordered pair (r, s).*

Solution 4.5.2.2. *Squaring the first equation yields $r^2 + 2rs + s^2 = 144$, so $rs = \dfrac{25}{2}$. If r and s are the roots of a quadratic polynomial $x^2 + bx + c$ (where we assume WLOG that $a = 1$), then $r + s = -\dfrac{b}{a} = 12 \implies b = -12$ and $rs = \dfrac{c}{a} = \dfrac{25}{2} \implies c = \dfrac{25}{2}$. Hence, we want to compute the roots of the quadratic $x^2 - 12x + \dfrac{25}{2}$. This can be done via Quadratic Formula; the roots are $x = \dfrac{12 \pm \sqrt{94}}{2}$, which are the values of r and s (more specifically, $r = \dfrac{12 + \sqrt{94}}{2}$, $s = \dfrac{12 - \sqrt{94}}{2}$).*

In addition, there is one more theorem that is crucial to the study of polynomial roots, which we will eventually prove, but leave further study of to the reader until more advanced subsections of the text:

Theorem 4.15: Fundamental Theorem of Algebra

Every non-constant, single-variable polynomial with complex coefficients has at least one complex root. Namely, if $\deg P(x) = n$, then $P(x)$ has n roots (real or complex), including roots with multiplicity greater than 1.

For an in-depth proof of the Fundamental Theorem of Algebra, see Appendix C: Proving the Fundamental Theorem of Algebra.[23]

---

[23]This proof involves more advanced techniques, and so we have placed it in the appendix rather than the main body of the chapter.

4.5.3 Exercises

Exercise 26. *If a and b are the roots of $x^2 + 3x + 9$, compute $a^2 + b^2$ and $a^4 + b^4$.*

Exercise 27. *(2006 AMC 10B) Let a and b be the roots of the equation $x^2 - mx + 2 = 0$. Suppose that $a + \dfrac{1}{b}$ and $b + \dfrac{1}{a}$ are the roots of the equation $x^2 - px + q = 0$. What is q?*

Exercise 28. *Let $P(x) = (x-3)(x-9)$. Let there exist a function $G(x)$ where $G(1) = P(x)$, $G(2) = P(P(x))$, $G(3) = P(P(P(x)))$, and so forth. What is the sum of the roots of $G(10)$?*

Exercise 29. *If $\sin x$ and $\cos x$ are the roots of $-20x^2 + bx + 8$ for some constant b and some real number $x \in [0, \pi]$, compute the value of $\tan x$.*

Exercise 30. *(1983 AHSME) If $\tan \alpha$ and $\tan \beta$ are the roots of $x^2 - px + q = 0$, and $\cot \alpha$ and $\cot \beta$ are the roots of $x^2 - rx + s = 0$, then rs is necessarily*

Exercise 31. *(2014 Caltech-Harvey Mudd Math Competition) If a complex number z satisfies $z + \dfrac{1}{z} = 1$, then what is $z^{96} + \dfrac{1}{z^{96}}$?*

Extension Exercise 13. *How can we derive formulas for $a^n + b^n$ where n is any nonzero integer?*

4.6 Introductory Sequences and Series

4.6.1 Definitions

A *sequence* is a function much like any other: it maps elements in its domain to elements in its codomain via a relation. However, sequences always have the set of integers as their domain, where each integer in the domain represents an *index*, or position in the sequence. In this book, we will be working primarily with those sequences with a domain of $\mathbb{N}$.

Definition 4.11: Definition of a Sequence

A *finite sequence* $\{a_i\}_{i=1}^n$ is the sequence $\{a_1, a_2, a_3, \ldots, a_n\}$ consisting of the n consecutive terms $a_1, a_2, a_3, \ldots, a_n$, where each of the a_i is in the codomain of the sequence, usually $\mathbb{R}$ or $\mathbb{C}$. In the former case, $\{a_i\}$ is called a *real-valued sequence;* in the latter, it is called a *complex-valued sequence.*

An *infinite sequence* $\{a_i\}_{i=1}^\infty$ is similar to a finite sequence, except with an infinite number of terms.

Just like any other function, a sequence must also be defined by a relation mapping each element in the domain to an element in the codomain. We usually denote this relation in the form $a_n = f(n)$, where $f(n)$ is the function representing the relation.

Furthermore, whereas a function $f(n)$ is usually abbreviated as just f, a sequence $\{a_i\}_{i=1}^n$ is more concisely denoted $(a_i)_{i \in \mathbb{N}}$, or even more so as just (a_i). For a generic domain D, the sequence notation becomes $(a_i)_{i \in D}$. Sequences also have several properties and classifications, many of which are similar to identical to those of ordinary functions:

Definition 4.12: Types of Sequences

- Real-valued/complex-valued sequence: As aforementioned.

- Integer sequence: A sequence whose terms are all integers.

- Multiplicative sequence: A sequence such that for all pairs of positive integers (m, n) satisfying $\gcd(m, n) = 1$, $a_{mn} = a_m \cdot a_n$. Equivalently, $a_n = n \cdot a_1$ for all $n \in \mathbb{N}$.

- Binary sequence: A sequence whose range consists of two values (for instance, 0 and 1, heads/tails, or true/false).

Following are some of the most crucial, widely-studied properties of sequences:

Definition 4.13: Properties of Sequences

- Length: The number of terms in a sequence. The finite sequence $\{a_i\}_{i=1}^n$ has length n, and is also known as an *n-tuple*. The empty sequence $\{\}$ is also a finite sequence (much like $\emptyset$, the empty set, is a finite set with cardinality zero).

- Monotonicity: A sequence (a_i) is *monotonically increasing* iff, for all $i \in \mathbb{N}$, $i \geq 2$, $a_i \geq a_{i-1}$.[a] (a_i) is *monotonically decreasing* iff, $\forall i \in \mathbb{N}$, $i \geq 2$, $a_i \leq a_{i+1}$. A sequence is *monotonic* if it is either monotonically increasing or monotonically decreasing. (This is a special case of the analogous function definition; see §5.4.)

- Boundedness: A sequence (a_i) has an *upper bound*, and is *bounded from above*, iff $\forall i \in \mathbb{N}$, $\exists M \in \mathbb{R}$ such that $a_i \leq M$. Here, M is the upper bound of the sequence. We define a *lower bound* m analogously, and say that (a_i) is *bounded from below* if such an m exists. If (a_i) has both an upper and lower bound, it is *bounded*.

- Subsequence: A *subsequence* (b_i) of a sequence (a_i) is a sequence such that $(b_i) = (a_{i_k})_{k \in \mathbb{N}}$, where $\{i_k\}_{k \in \mathbb{N}}$ is a strictly monotonically increasing positive integer sequence. (b_i) is *order-preserving*; i.e. $\forall k_1, k_2 \in \mathbb{N}$, if $k_1 < k_2$, then $i_{k_1} < i_{k_2}$, and if $k_1 > k_2$, then $i_{k_1} > i_{k_2}$.

---

[a]Here we choose not to make the distinction between "increasing" and "non-decreasing"; in this book, the two terms are synonymous.

4.6.2 Arithmetic and Geometric Series

Two special types of sequences are *arithmetic* and *geometric sequences*.

Definition 4.14: Arithmetic and Geometric Sequences

An arithmetic sequence (a_i) is a sequence such that, for all $i \geq 1$, $a_{i+1} = a_i + d$, where d is the constant *common difference* of the sequence.

A geometric sequence (b_i) is a sequence such that, for all $i \geq 1$, $a_{i+1} = a_i \cdot r$, where r is the constant *common ratio* of the sequence.

Theorem 4.16: Arithmetic Series Sum Formula

Let $A = \{a_1, a_2, a_3, \ldots, a_n\}$[a] be an arithmetic sequence such that $a_1 = a$, $a_2 = a + d$, $a_3 = a + 2d$, and so forth until $a_n = a + d(n-1)$. Then $a_1 + a_2 + a_3 + \ldots + a_n = \dfrac{n}{2} \cdot (a_1 + a_n)$.

---

[a]Or, more compactly, $A = \{a_i\}_{i=1}^n$ in sequence notation.

Proof. We proceed by induction on n. For the base case, let $n = 2$; we indeed have $a_1 + a_2 = \dfrac{2}{2} \cdot (a_1 + a_2)$. Assume the theorem holds for some n; then for $n + 1$, we have

$$a_1 + a_2 + a_3 + \ldots + a_{n+1} = \frac{n+1}{2} \cdot (a_1 + a_{n+1})$$

In addition, we obtain

$$
\begin{aligned}
a_1 + a_2 + a_3 + \ldots + a_{n+1} &= (a_1 + a_2 + a_3 + \ldots + a_n) + a_{n+1} \\
&= \left(\frac{n}{2} \cdot (a_1 + a_n) \right) + (a + d \cdot ((n+1) - 1)) \\
&= \frac{n}{2} \cdot (a_1 + a_n) + (a + d \cdot n) \\
&= a \cdot n + d \cdot \frac{n(n-1)}{2} + a + d \cdot n \\
&= a \cdot (n+1) + d \cdot \frac{n(n+1)}{2} \\
&= \frac{n+1}{2} \cdot (2a + d \cdot n) \\
&= \frac{n+1}{2} \cdot (a + (a + d \cdot n)) \\
&= \frac{n+1}{2} \cdot (a_1 + a_{n+1})
\end{aligned}
$$

as desired, which completes the proof. $\square$

Theorem 4.17: Geometric Series Finite Sum Formula

Let $G = \{g_1, g_2, g_3, \ldots, g_n\}$ be a finite geometric sequence with $g_1 = a$, $g_2 = ar$, $g_3 = ar^2$, and so forth until $g_n = ar^{n-1}$. Then $g_1 + g_2 + g_3 + \ldots + g_n = a \cdot \dfrac{1 - r^n}{1 - r}$.

Proof. First note that

$$\sum_{i=1}^{n} g_i = a \cdot \sum_{i=0}^{n-1} r^i$$

Consider the product

$$(1 - r)(1 + r + r^2 + r^3 + \ldots + r^{n-1})$$

Expanding this yields $(1 + r + r^2 + r^3 + \ldots + r^{n-1}) - (r + r^2 + r^3 + r^4 + \ldots + r^n) = 1 - r^n$, so $1 + r + r^2 + r^3 + \ldots + r^{n-1} = \dfrac{1 - r^n}{1 - r}$, $r \in (-1, 1)$.[24] Finally, multiplying through by a for the original sum yields

$$\sum_{i=1}^{n} g_i = a \cdot \frac{1 - r^n}{1 - r}$$

for $r \in (-1, 1)$, as desired. $\qquad\square$

Corollary 4.17: Geometric Series Infinite Sum Formula

If $G = \{g_i\}_{i=1}^{\infty}$ is an infinite geometric series, then the sum $g_1 + g_2 + g_3 + \ldots = \dfrac{a}{1 - r}$.[a]

---

[a]This can be observed by taking the limit as n approaches ∞; as $n \to \infty$, $r^n \to 0$, since $|r| < 1$ in a convergent geometric series.

4.6.3 Examples

Example 4.6.3.1. *Consider the sequence (a_n) defined by $a_n = n^2 + 2n$ for all $n \geq 1$.*

(a) What is the value of $a_n - a_{n-1}$, in terms of n?

(b) If $a_{100} - a_{99} = k$, for what value of n will $a_n - a_{n-1} = 3k$?

Solution 4.6.3.1.

(a) $a_n = n^2 + 2n$ and $a_{n-1} = (n-1)^2 + 2(n-1) = n^2 - 1$, so $a_n - a_{n-1} = 2n + 1$ for all $n \geq 1$.

(b) From part (a), $a_{100} - a_{99} = 2(100) + 1 = 201$, so we seek n such that $2n + 1 = 603$, or $n = 301$.

Example 4.6.3.2. *A sequence (a_n) has terms given by $a_n = (-1)^n \cdot \dfrac{n}{n+1}$ for $n \geq 1$. Propose and prove formulas for $\displaystyle\sum_{i=1}^{n} a_i$ and $\displaystyle\prod_{i=1}^{n} a_i$, where $\displaystyle\sum_{i=1}^{n}$ and $\displaystyle\prod_{i=1}^{n}$ denote the sum and product from $i = 1$ to n, respectively.*

---

[24]The product simplification is a prime example of a *telescoping series*.

Solution 4.6.3.2. *If n is even,*

$$\sum_{i=1}^{n} a_i = \left(-\frac{1}{2} + \frac{2}{3}\right) + \left(-\frac{3}{4} + \frac{4}{5}\right) + \ldots + \left(-\frac{n-1}{n} + \frac{n}{n+1}\right)$$

$$= \frac{1}{2 \cdot 3} + \frac{1}{4 \cdot 5} + \ldots + \frac{1}{n \cdot (n+1)}$$

$$= \left(\frac{1}{2} - \frac{1}{3}\right) + \left(\frac{1}{4} - \frac{1}{5}\right) + \ldots + \left(\frac{1}{n} - \frac{1}{n+1}\right)$$

$$= \left(\frac{1}{2} + \frac{1}{4} + \ldots + \frac{1}{n}\right) - \left(\frac{1}{3} + \frac{1}{5} + \ldots + \frac{1}{n+1}\right).$$

Similarly, if n is odd,

$$\sum_{i=1}^{n} a_i = \left(-\frac{1}{2} + \frac{2}{3}\right) + \left(-\frac{3}{4} + \frac{4}{5}\right) + \ldots + \left(-\frac{n-2}{n-1} + \frac{n-1}{n}\right) - \frac{n}{n+1}$$

$$= \frac{1}{2 \cdot 3} + \frac{1}{4 \cdot 5} + \ldots + \left(\frac{1}{(n-1) \cdot n}\right) - \frac{n}{n+1}$$

$$= \left(\frac{1}{2} - \frac{1}{3}\right) + \left(\frac{1}{4} - \frac{1}{5}\right) + \ldots + \left(\frac{1}{n-1} - \frac{1}{n}\right) - \frac{n}{n+1}$$

$$= \left(\frac{1}{2} + \frac{1}{4} + \ldots + \frac{1}{n-1}\right) - \left(\frac{1}{3} + \frac{1}{5} + \ldots + \frac{1}{n}\right) - \frac{n}{n+1}.$$

Both of these formulas can be proven by induction (exercise to the reader).

4.6.4 Exercises

Exercise 32. *(2016 MathCounts Chapter Sprint) A bug crawls a distance of n miles at a speed of $n + 1$ miles per hour one day and then crawls $2n + 1$ miles at a speed of $n^2 + n$ miles per hour the next day. If the total time for the trip was 6 hours, what was the bug's average speed?*

Exercise 33. *(2015 ARML Relay Round) Given that $1 - r + r^2 - r^3 + \ldots = s$ and $1 + r^2 + r^4 + r^6 + \ldots = 4s$, compute s.*

Exercise 34. *Can a non-constant finite sequence with length ≥ 3 be both arithmetic and geometric? Justify your answer.*

Exercise 35. *(2000 AIME I) A sequence of numbers $x_1, x_2, x_3, \ldots, x_{100}$ has the property that, for every integer k between 1 and 100, inclusive, the number x_k is k less than the sum of the other 99 numbers. Given that $x_{50} = \dfrac{m}{n}$, where m and n are relatively prime positive integers, find $m + n$.*

Extension Exercise 14. *Let $(a_n) = \{1, 2, 3, 6, 7, 14, 15, \ldots\}$ be the infinite sequence such that, for all odd n, $a_{n+1} = 2a_n$, for all even n, $a_{n+1} - a_n = 1$. Propose and prove a formula for the sum of the first n terms of (a_n).*

Extension Exercise 15. *Given that the roots of the polynomial $P(x) = x^3 + kx^2 + 20x - 18$ are in arithmetic progression, what are all possible values of k? (If you're stuck, it may help to read through §5.7.)*

Chapter 4 Review Exercises

Exercise 36. *Prove that the positive solution to the equation $20x^2 + 18 = 2018$ is $x = 10$.*

Exercise 37. *Consider the relation $f : \mathbb{R}\backslash\{-1, 2\} \to \mathbb{R}\backslash\{0\}$, $f(x) = \dfrac{2}{x^2 - x - 2}$.*

(a) *Is f a function? Why or why not?*

(b) *Prove that f is continuous for all $x \in \mathbb{R}$, $x \neq -1, 2$. What type(s) of discontinuities does f have at $x = -1, 2$?*

(c) *Define $f_1(x) = \{f(x) : x < -1\}$, $f_2(x) = \{f(x) : -1 < x < 2\}$, $f_3(x) = \{f(x) : x > 2\}$. For each of f_1, f_2, f_3, is it injective, surjective, bijective, some combination thereof, or not a function?*

Exercise 38. *Consider the binomial expansion of $(x + y)^{19}$.*

(a) *How many coefficients are in the expansion?*

(b) *What is the sum of the coefficients in the expansion?*

(c) *Show that the sum of the coefficients of the terms with odd powers of x is equal to the sum of the coefficients of the terms with odd powers of y, and compute this common sum.*

Exercise 39. *(2006 AMC 10A) How many non-similar triangles have angles whose degree measures are distinct positive integers in arithmetic progression?*

Exercise 40.

(a) *Using Euclid's Division Algorithm, compute $\gcd(2016, 4200)$.*

(b) *(1959 IMO) Prove that $\dfrac{21n + 4}{14n + 3}$ is irreducible for every natural number n.*

Exercise 41. *Let r_1 and r_2 be the roots of $2x^2 + 3x + 1$.*

(a) *Compute $r_1^2 + r_2^2$.*

(b) *Compute $r_1^2 r_2 + r_1 r_2^2$.*

(c) *Compute $r_1^3 + r_2^3$.*

(d) *Without computing r_1 and r_2, compute $r_1^4 - r_2^4$.*

Exercise 42.

(a) *Show that $\dfrac{4x^2 - 16x + 17}{9x^2 - 30x + 26} > 0$ for all $x \in \mathbb{R}$.*

(b) *Solve the inequality $\dfrac{4x^2 - 16x + 17}{9x^2 - 30x + 26} \geq 1$ for $x \in \mathbb{R}$.*

Extension Exercise 16. *Let $f : \mathbb{R} \to \mathbb{R}$ be a monotonic function. Can f have infinitely many discontinuities? An uncountably infinite number of discontinuities? (Hint: First prove that the set of rational numbers $\mathbb{Q}$ is countable.)*

Extension Exercise 17.

(a) *Prove Pascal's Identity for all $n, k \in \mathbb{N}$, $n > k$ (i.e. prove the equation $\dbinom{n - 1}{k - 1} + \dbinom{n - 1}{k} = \dbinom{n}{k}$ for all $n, k \in \mathbb{N}$, $n > k$).*

(b) *Prove the Hockey Stick Identity for all $n, k \in \mathbb{N}$, $n > k$:*

$$\sum_{i=k}^{n} \binom{i}{k} = \binom{n + 1}{k + 1}.$$

Extension Exercise 18. *Show that if $f : \mathbb{N} \to \mathbb{N}$ is a multiplicative function, its values for all natural numbers are uniquely determined by $f(p^k)$ for prime p, $k \in \mathbb{N}$.*

Chapter 5

Intermediate Algebra

5.1 The Rearrangement Inequality

Going beyond the study of single-variable equations, we consider how multiple variables are related to each other, in the form of *multi-variable inequalities.* We have previously studied these in lesser detail in the Beginning Algebra chapter, but in this chapter we will begin to truly dissect the intrinsic properties of some complicated inequalities, such as symmetry, homogeneity, and combinatorial and geometric interpretations.

Perhaps the most fitting introduction to these more advanced types of inequalities is the *Rearrangement Inequality,* which is in fact a variant on the well-known AM-GM Inequality (which, incidentally, we also cover in depth in the following section).

Theorem 5.1: Rearrangement Inequality

Given two sequences
$$a_1 \geq a_2 \geq \cdots \geq a_n, b_1 \geq b_2 \geq \cdots \geq b_n,$$
$$a_1 b_1 + a_2 b_2 + \cdots + a_n b_n \geq a_1 b_{\pi(1)} + a_2 b_{\pi(2)} + \cdots + a_n b_{\pi(n)}$$
$$\geq a_1 b_n + a_2 b_{n-1} + \cdots + a_n b_1.$$

Proof. Let permutation $\pi(1), \pi(2) \cdots \pi(n)$ be the permutation that maximizes $a_1 b_{\pi(1)} + a_2 b_{\pi(2)} + \cdots + a_n b_{\pi(n)}$. Then let us assume for contradiction that this is not the identity permutation (i.e. there is ≥ 1 number i such that $\pi(i) \neq i$). Let the smallest number this is true for be x. Then let $\pi(x) = y$, and note that $y > x$ as 1 through $x - 1$ are all assigned. Also, there is a z such that $\pi(z) = i$, and $z > x$ because 1 through $x - 1$ are all assigned.

By the Lemma of Common Sense (the proof of this lemma is left as an exercise), $b_x \leq b_y, a_x \leq a_z$. Thus it suffices to prove
$$a_x b_y + a_z b_x \leq a_z b_y + a_x b_x,$$
and rearranging gives us
$$0 \leq (a_z - a_i)(b_y - b_x)$$
as an equivalent result. Citing the Lemma of Common Sense again, this is true.
However, we have assumed that the permutation we have given is the largest possible, which leads to a contradiction. $\qquad\square$

5.1.1 Why Should This be True?

Here is a quote from Brilliant.org which sums it up very well:

"For instance, if one were presented with a large number of \$20 bills, \$10 bills, \$5 bills, and \$1 bills, with the instructions to take 4 of one type of bill, 3 of another, 2 of another, and 1 of the last, it is intuitively obvious that four \$20 bills, three \$10 bills, two \$5 bills, and one \$1 bill should be selected. Indeed, this is precisely the statement of the rearrangement inequality, applied to the sequences $20, 10, 5, 1$ and $4, 3, 2, 1$"
(Brilliant.org).

The AM-GM inequality, Cauchy, Muirhead, Chebyshev, and Nessbitt follow from the Rearrangement Inequality. (There are probably more famous inequalities out there that follow from this; these are not just interesting corollaries, these are major results!)

5.1.2 Examples

1. The rearrangement inequality in two variables.

   Example 5.1.2.1. *Prove $a^2 + b^2 + c^2 \geq ab + bc + ca$.*

   Solution 5.1.2.1. *This is generally thought of as an AM-GM problem, but this can actually be solved with Rearrangement as well. The main idea is letting a_1, a_2, a_3 be a, b, c and we let b_1, b_2, b_3 be a, b, c as well. Then we note that on the RHS, we assigned $\pi(1) = 2, \pi(2) = 3, \pi(3) = 1$. Thus*

   $$a \cdot a + b \cdot b + c \cdot c \geq a \cdot b + b \cdot c + c \cdot a.$$

2. The rearrangement inequality in three variables.

   Example 5.1.2.2. *Prove $a^3 + b^3 + c^3 \geq a^2 b + b^2 c + c^2 a$ for $a, b, c > 0$.*

   Solution 5.1.2.2. *The main idea is applying Rearrangement on a, b, c and a^2, b^2, c^2. This yields our desired result.*

3. A cyclic sum invoking the rearrangement inequality.

   Example 5.1.2.3. *For $a, b, c > 0$, show that*

   $$\sum_{cyc} \frac{a^2 + bc}{b + c} \geq a + b + c.$$

   Solution 5.1.2.3. *A quick note on notation: Cyclic means to add up all the terms of the expression as they cycle from a, b, c to b, c, a and finally c, a, b. (This differs based on the amount of terms.) Rewriting, this means*

   $$\frac{a^2 + bc}{b + c} + \frac{b^2 + ca}{c + a} + \frac{c^2 + ab}{a + b} \geq a + b + c.$$

   WLOG, let $a > b > c$. We can split this into

   $$\sum_{cyc} \frac{a^2}{b + c} + \sum_{cyc} \frac{bc}{b + c} \geq a + b + c.$$

   As $\dfrac{1}{b + c} > \dfrac{1}{c + a} > \dfrac{1}{a + b}$, applying Rearrangement yields

   $$\sum_{cyc} \frac{a^2}{b + c} \geq \sum_{cyc} \frac{b^2}{b + c}.$$

   Plugging this in yields

   $$\sum_{cyc} \frac{a^2}{b + c} + \sum_{cyc} \frac{bc}{b + c} \geq \sum_{cyc} \frac{b^2 + bc}{b + c} = a + b + c.$$

5.1.3 Exercises

Don't feel like you have to solely use Rearrangement for these problems! Any other method you think of is probably a derivative of Rearrangement anyway...

Exercise 43. *Prove $a^4 + b^4 + c^4 \geq a^2bc + b^2ca + c^2ab$.*

Exercise 44. *Let a, b and c be positive real numbers. Prove that $a^a b^b c^c \geq (abc)^{\frac{a+b+c}{3}}$.*

Exercise 45. *Let $a, b, c > 0$. Prove that if $a^2 + b^2 + c^2 = 1$, then*

$$\frac{1}{a^2} + \frac{1}{b^2} + \frac{1}{c^2} \geq 3 + \frac{2(a^3 + b^3 + c^3)}{abc}.$$

Exercise 46. *Let $y_1, y_2 \cdots y_8$ be a permutation of $1, 2 \cdots 8$. Find*

$$\min\left(\sum_{i=1}^{8}(y_i + i)^2\right).$$

5.2 The AM-GM Inequality

The *arithmetic mean-geometric mean inequality*, or *AM-GM inequality* for short, is a derivative of the Rearrangement Inequality that posits that the arithmetic mean is always greater than or equal to the geometric mean of any set of n real numbers:

Theorem 5.2: AM-GM Inequality

For $x_1, x_2 \cdots x_n > 0$,
$$\frac{x_1 + x_2 + \cdots + x_n}{n} \geq \sqrt[n]{x_1 x_2 \cdots x_n}.$$

Proof. Consider $a_1, a_2 \cdots a_n$ that are all greater than 0.

Lemma 5.2: Sum of Quotients

$$n \leq \frac{a_1}{a_n} + \frac{a_2}{a_1} + \cdots + \frac{a_{n-1}}{a_{n-2}} + \frac{a_n}{a_{n-1}}.$$

Proof. Let $\pi(i)$ be the permutation of $a_1, a_2 \cdots a_n$ such that $\pi(1) \geq \pi(2) \geq \cdots \geq \pi(n)$. By the Rearrangement Inequality, $a_1 \cdot \frac{1}{a_n} + a_2 \cdot \frac{1}{a_{n-1}} + \cdots + a_n \frac{1}{a_1} \geq a_{\pi(1)} \cdot \frac{1}{a_{\pi(2)}} + a_{\pi(2)} \cdot \frac{1}{a_{\pi(2)}} + \cdots + a_{\pi(n)} \cdot \frac{1}{a_{\pi(n)}} = n$, as desired. $\square$

Lemma 5.2: Sum of the Sum and Quotient

For all $x_1, x_2 \cdots x_n > 0$, we have $n \leq \dfrac{x_1}{x_1 x_2 \cdots x_n} + x_2 + x_3 + \cdots + x_n$.

Proof. We use the Lemma and substitute $a_n = x_1$, $a_{n-1} = x_1 x_2$, and so on, until $a_1 = x_1 x_2 \cdots x_n$. $\square$

Now, if we substitute ax_i for x_i, we get $n \leq \dfrac{ax_1}{a^n x_1 x_2 \cdots x_n} + ax_2 + ax_3 + \cdots + ax_n$. If we let $a^n x_1 x_2 \cdots x_n = 1$, implying $a = \dfrac{1}{\sqrt[n]{a_1 a_2 \cdots a_n}}$ then we have $n \leq ax_1 + ax_2 + \cdots + ax_n$. This implies that $\dfrac{1}{a} \leq \dfrac{x_1 + x_2 + \cdots + x_n}{n}$. Substituting a for $\dfrac{1}{\sqrt[n]{a_1 a_2 \cdots a_n}}$ gives us

$$\frac{x_1 + x_2 + \cdots + x_n}{n} \geq \sqrt[n]{x_1 x_2 \cdots x_n},$$

as desired. $\square$

Extension Exercise 19. *The theorem can also be proven by induction; doing so is left as an exercise to the reader.*

5.2.1 The Geometric Interpretation

For $n = 2$, consider a rectangle with side lengths x_1 and x_2, and hence perimeter $2x_1 + 2x_2$ and area $x_1 x_2$. Observe that a square with equal area to the rectangle has perimeter $4\sqrt{x_1 x_2}$, which is the minimum amongst all rectangles with area $x_1 x_2$.

In the general case, consider an n-dimensional hypercube (henceforth "box" for brevity). Each vertex of the box is adjoined to n other edges; suppose that these edges have lengths $x_1, x_2, x_3, \ldots, x_n$. Note that, in the original statement of the inequality, the sum $x_1 + x_2 + x_3 + \ldots + x_n$ represents the sum of the lengths of the edges connected to any given vertex. The n-box has 2^n vertices,[25] so the sum of all the edge lengths of the box is $2^{n-1} \cdot (x_1 + x_2 + x_3 + \ldots + x_n)$, after dividing by 2 to account for the fact that each edge is counted twice (as all edges touch two vertices of the box). Thus, there are $n \cdot 2^{n-1}$ edges, and 2^{n-1} edges of each length (i.e. there are n different dimensions of the box, and for each dimension, there are 2^{n-1} edges corresponding to that dimension).

Henceforth, consider the n-cube (as opposed to an n-box that may or may not have all edge lengths equal). Since $x_1 = x_2 = x_3 = \ldots = x_n$, the sum of all the edge lengths is equal to $n \cdot 2^{n-1} \cdot \sqrt[n]{x_1 x_2 x_3 \cdots x_n} = n \cdot 2^{n-1} \cdot x_1$. Dividing through by $n \cdot 2^{n-1}$ yields $x_1 = \sqrt[n]{x_1 x_2 x_3 \cdots x_n}$, or $\dfrac{x_1 + x_2 + x_3 + \ldots + x_n}{n} = \sqrt[n]{x_1 x_2 x_3 \cdots x_n}$, with equality holding as we have assumed that $x_1 = x_2 = x_3 = \ldots = x_n$. (For an n-box with unequal dimensions, the LHS is greater than the RHS.)

5.2.2 The Weighted AM-GM Inequality

The *Weighted AM-GM Inequality* relies on the idea of *weight*, [26] which applies to the *weighted arithmetic mean* and *weighted geometric mean*:

Definition 5.1: Weighted Averages of Nonnegative Real Numbers

Let $x_1, x_2, x_3, \ldots, x_n \geq 0$, $x_i \in \mathbb{R}$ for all $1 \leq i \leq n$. Assign to each x_i a corresponding weight $w_i \geq 0$, and set $w = \sum_{i=1}^{n} w_i$. Then the weighted arithmetic mean (WAM), which is also sometimes called the

weighted average of $x_1, x_2, x_3, \ldots, x_n$, is equal to $\dfrac{\sum_{i=1}^{n} w_i x_i}{w}$, and the weighted geometric mean (WGM)

of the x_i's is equal to $\sqrt[w]{\sum_{i=1}^{n} x_i^{w_i}}$.

Theorem 5.3: Weighted AM-GM Inequality

If $w > 0$ as in the definition above, then the weighted arithmetic mean (WAM) is greater than or equal to the weighted geometric mean (WGM), with equality iff $x_1 = x_2 = x_3 = \ldots = x_n$ and $w_i \neq 0$ for all $1 \leq i \leq n$.

Proof. (Note: Jensen's Inequality (§6.4.1) is critical to understanding the following proof. If you are unfamiliar with this inequality, we recommend reading through that section first. This proof is in the Intermediate section primarily for completeness purposes.)

We make the crucial observation that the function $f(x) = \ln x$ is strictly concave, meaning that the straight line between any two points on the function lies strictly above[27] the curve of the function. By Jensen's Inequality, we have

$$\ln \left(\sum_{i}^{n} \frac{w_i x_i}{w} \right) > \frac{w_1}{w} \ln(x_1) + \frac{w_2}{w} \ln(x_2) + \ldots + \frac{w_n}{w} \ln(x_n)$$

---

[25]This can be proven using induction, and has a combinatorial link as well.

[26]Here's a real-life example of weight: Usually, to determine one's grade in a class, there are different categories that are worth more than others. For example, the final exam score might be worth 40% of one's grade, while class participation is only worth 5%. Then, we would say that the final exam score has a weight of 40%, contrasting with the grading scheme's 5% weight on class participation.

[27]As opposed to at or above, for a weakly concave function.

$$= \ln \left(\sqrt[w]{\prod_{i=1}^{n} x_i^{w_i}} \right)$$

Since $f(x) = \ln x$ is strictly increasing on $\mathbb{R}$, it follows that

$$\sum_{i=1}^{n} \frac{w_i x_i}{w} \geq \sqrt[w]{\prod_{i=1}^{n} x_i^{w_i}}$$

as desired. $\qquad\qquad\qquad\qquad\qquad\qquad\qquad\qquad\qquad\qquad\qquad\qquad\qquad\qquad$ $\square$

Remark. Here we define $0^0 = 1$ by convention.

5.2.3 The QM-AM-GM-HM Inequality

The *QM-AM-GM-HM Inequality* is an extended form of the AM-GM inequality, which posits that the quadratic mean or *root-mean square* of positive reals $x_1, x_2, x_3, \ldots, x_n$ (i.e. the quantity $\sqrt{\dfrac{\sum_{i=1}^{n} x_i^2}{n}}$) is always greater than or equal to the arithmetic mean, which is greater than or equal to the geometric mean, which is in turn greater than or equal to the harmonic mean $\dfrac{n}{\sum_{i=1}^{n} \dfrac{1}{x_i}}$:

Theorem 5.4: The QM-AM-GM-HM Inequality

For all positive reals $x_1, x_2, x_3, \ldots, x_n$,

$$\sqrt{\frac{x_1^2 + x_2^2 + x_3^2 + \ldots + x_n^2}{n}} \geq \frac{x_1 + x_2 + x_3 + \ldots + x_n}{n} \geq$$

$$\sqrt[n]{x_1 x_2 x_3 \cdots x_n} \geq \frac{n}{\dfrac{1}{x_1} + \dfrac{1}{x_2} + \dfrac{1}{x_3} + \ldots + \dfrac{1}{x_n}}.$$

Proof. The first part of the proof requires us to show that the quadratic mean is not less than the arithmetic mean, for all positive reals $x_1, x_2, x_3, \ldots, x_n$. This follows from the Cauchy-Schwarz Inequality (see the immediately following section, §4.3); in detail, here is the justification:

$$(x_1^2 + x_2^2 + x_3^2 + \ldots + x_n^2)(1 + 1 + 1 + \ldots + 1) \geq (x_1 + x_2 + x_3 + \ldots + x_n)^2 \qquad (5.1)$$

$$\frac{x_1^2 + x_2^2 + x_3^2 + \ldots + x_n^2}{n} \geq \left(\frac{x_1 + x_2 + x_3 + \ldots + x_n}{n} \right)^2 \qquad (5.2)$$

$$\sqrt{\frac{x_1^2 + x_2^2 + x_3^2 + \ldots + x_n^2}{n}} \geq \frac{x_1 + x_2 + x_3 + \ldots + x_n}{n} \qquad (5.3)$$

$\qquad\qquad\qquad\qquad\qquad\qquad\qquad\qquad\qquad\qquad\qquad\qquad\qquad\qquad\qquad\qquad$ $\square$

Equation (5.1) follows from setting $(a_1, a_2, a_3, \ldots, a_n) = (x_1, x_2, x_3, \ldots, x_n)$ and $(b_1, b_2, b_3, \ldots, b_n) = (1, 1, 1, \ldots, 1)$ (see Theorem 5.6).

Furthermore, the inequality $\sqrt[n]{x_1 x_2 x_3 \cdots x_n} \geq \dfrac{n}{\dfrac{1}{x_1} + \dfrac{1}{x_2} + \dfrac{1}{x_3} + \ldots + \dfrac{1}{x_n}}$ is a consequence of the AM-GM inequality; as

$$\frac{\sum\limits_{i=1}^{n} \sqrt[n]{\dfrac{\prod\limits_{i=1}^{n} x_i}{x_i^n}}}{n} \geq 1$$

we have

$$\frac{\sqrt[n]{\prod\limits_{i=1}^{n} x_i} \cdot \sum\limits_{i=1}^{n} \dfrac{1}{x_i}}{n} \geq 1.$$

It follows that $\sqrt[n]{\prod\limits_{i=1}^{n} x_i} \geq \dfrac{n}{\sum\limits_{i=1}^{n} \dfrac{1}{x_i}}$ as desired.

5.2.4 The Power Mean Inequality

This is a generalization of the QM-AM inequality, in which we utilize the idea of a *power mean* (which is quasi-related to the idea of a *norm* in a vector space). In this inequality, we are effectively taking the x-norm and y-norm of similar expressions, comparing them on an intuitive level, and equating them to each other in the equality case:

> **Theorem 5.5: Power Mean Inequality**
>
> For positive integers $a_1, a_2, \cdots a_n$,
>
> $$\sqrt[x]{\frac{a_1^x + a_2^x + \cdots + a_n^x}{n}} \geq \sqrt[y]{\frac{a_1^y + a_2^y + \cdots + a_n^y}{n}}$$
>
> for $x \geq y$. Equality occurs when $a_1 = a_2 = \cdots = a_n$ and/or $x = y$.

Proof. Raising both sides to the xth power yields the inequality

$$\frac{a_1^x + a_2^x + \ldots + a_n^x}{n} \geq \left(\frac{a_1^y + a_2^y + \ldots + a_n^y}{n}\right)^{\frac{y}{x}}$$

Notice that the function $f(\alpha) = \alpha^{\frac{y}{x}}$ is concave for positive α, so by Jensen's inequality,

$$\left(\sum_{i=1}^{n} \frac{a_i^x}{n}\right)^{\frac{y}{x}} = f\left(\sum_{i=1}^{n} \frac{a_i^x}{n}\right) \geq \sum_{i=1}^{n} \frac{f(a_i^x)}{n} = \sum_{i=1}^{n} \frac{f(a_i^y)}{n}$$

as desired. $\square$

5.2.5 Examples

Let's try a couple of examples:

Example 5.2.5.1. *Find the maximum k such that $2a^4b^2 + 9b^4c^2 + 12c^4a^2 - ka^2b^2c^2$ is always non-negative for all real a, b, and c.*

Solution 5.2.5.1. *We use AM-GM on $2a^4b^2$, $9b^4c^2$, and $12c^4a^2$. The arithmetic mean is exactly* $\dfrac{2a^4b^2 + 9b^4c^2 + 12c^4a^2}{3}$, *and the geometric mean is exactly* $\sqrt[3]{(2a^4b^2)(9b^4c^2)(12c^4a^2)} = \sqrt[3]{216a^6b^6c^6} =$

$6a^2b^2c^2$. Thus, by AM-GM, we have $\dfrac{2a^4b^2 + 9b^4c^2 + 12c^4a^2}{3} \geq 6a^2b^2c^2$, or $2a^4b^2 + 9b^4c^2 + 12c^4a^2 - 18a^2b^2c^2 \geq 0$. We get equality when $2a^4b^2 = 9b^4c^2 = 12c^4a^2$, for instance when $a^2 = 3, b^2 = 2, c^2 = 1$, so the maximum k is $\boxed{18}$.

Example 5.2.5.2. *Prove that* $\sqrt{\dfrac{2a^2 - 2a + 1}{2}} \geq \dfrac{1}{a + \frac{1}{a}}$ *for* $0 < a < 1$.

Solution 5.2.5.2. *While it seems natural to start by squaring both sides, it gets messy quickly if we do that. Let's try something else instead.*

The left hand side has a sum over 2 under a square root. This looks like QM. What we need is to split $1 - 2a + 2a^2$ *into two squares, for example* $a^2 + (1-a)^2$. *We could not use* $a^2 + (a-1)^2$ *because then* $a - 1 < 0$, *and QM-AM-GM-HM requires that the terms be positive. Using QM-AM, we find that*

$$\sqrt{\dfrac{a^2 + (1-a)^2}{2}} \geq \dfrac{a + (1-a)}{2} = \dfrac{1}{2}.$$

On the right hand side, we have $\dfrac{1}{a}$ *in the denominator, which implies we should use HM. Using GM-HM with* a *and* $\dfrac{1}{a}$, *we have* $\dfrac{1}{a + \frac{1}{a}} \leq \dfrac{1}{2}\sqrt{a \cdot \dfrac{1}{a}} = \dfrac{1}{2}.$

Hence, we have that $\sqrt{\dfrac{2a^2 - 2a + 1}{2}} \geq \dfrac{1}{2} \geq \dfrac{1}{a + \frac{1}{a}}$ *as desired.*

5.2.6 Exercises

Exercise 47.

(a) A rectangular fence is constructed against a wall using 60 meters of wire. Show that the maximal area of the fence is 400 square meters.

(b) If the fence need not be rectangular, find the maximal area of the fence.

Exercise 48. If a and b are real numbers such that $a, b > 0$ and $a + b = 12$, what is the maximum value of ab^2?

Exercise 49. For a, b, and c positive real numbers, show that $\dfrac{a^2}{bc} + \dfrac{b^2}{ca} + \dfrac{c^2}{ab} \geq 3$.

Exercise 50. Let $\pi = \{y_1, y_2, y_3, \ldots, y_n\}$ be a permutation of the set $S = \{x_1, x_2, x_3, \ldots, x_n\}$. Show that $\dfrac{x_1}{y_1} + \dfrac{x_2}{y_2} + \dfrac{x_3}{y_3} + \ldots + \dfrac{x_n}{y_n} \geq n$.

Extension Exercise 20. Let n be an integer greater than 1. Prove that $\left(\dfrac{n+1}{2}\right)^n > n!$

5.3 The Cauchy-Schwarz Inequality

The *Cauchy-Schwarz Inequality*, also known as the *Cauchy-Bunyakovsky-Schwarz Inequality* from Viktor Bunyakovsky's work involving the inequality, is one of the most central results in inequality theory. Not only is it a result upon which several others depend - and are interlinked with - it is widely used in other applications, such as real and functional analysis, linear algebra, and probability theory, among others.

In general, the inequality is usually stated in terms of some vectors $\mathbf{u}, \mathbf{v}$. However, in a contest setting, it is almost always more useful to understand the *intuition* and *reasoning* behind the application of vectors to the inequality, as well as how we can interpret it in terms of sums of products of real numbers.

Theorem 5.6: Cauchy-Schwarz Inequality

The Cauchy-Schwarz Inequality states that for any real numbers $a_1, \ldots, a_n$ and $b_1, \ldots, b_n$,

$$\left(\sum_{i=1}^{n} a_i b_i \right)^2 \leq \left(\sum_{i=1}^{n} a_i^2 \right) \left(\sum_{i=1}^{n} b_i^2 \right),$$

with equality when $a_i = c b_i$ for all $1 \leq i \leq n$, for some $c \in \mathbb{R}$.

Proof. This proof will use vectors.

Consider vectors $\mathbf{u} = \langle u_1, u_2, \ldots, u_n \rangle$ and $\mathbf{v} = \langle v_1, v_2, \ldots, v_n \rangle$ in an inner product space over $\mathbb{C}$ (where an *inner product space I* is, for now, loosely defined as a subset of $\mathbb{C}$ for which there exists an inner product $\langle u, v \rangle$ of two vectors u, v for all $u, v \in I$). We show that, for all $v \neq 0$ (as the case $v = 0$ is trivial), there exists a complex number $\lambda = \dfrac{\langle u, v \rangle}{\langle v, v \rangle}$ such that $0 \leq ||u - \lambda \cdot v||^2$ (where $||\cdot|| = \langle \cdot, \cdot \rangle$ denotes the *norm* of $\cdot$, i.e. the square root of the inner product of $\cdot$ with itself). Then we have the RHS of our initial inequality equal to

$$\langle u, u \rangle - \langle \lambda v, u \rangle - \langle u, \lambda v \rangle + \langle \lambda v, \lambda v \rangle$$

$$= \langle u, u \rangle - \lambda \langle v, u \rangle - \overline{\lambda} \langle u, v \rangle + \lambda \overline{\lambda} \langle v, v \rangle$$

$$= ||u||^2 - \lambda \overline{\langle u, v \rangle} - \overline{\lambda} \langle u, v \rangle + \lambda \overline{\lambda} ||v||^2$$

$$= ||u||^2 - \frac{|\langle u, v \rangle|^2}{||v||^2} - \frac{|\langle u, v \rangle|^2}{||v||^2} + \frac{|\langle u, v \rangle|^2}{||v||^2}$$

$$= ||u||^2 - \frac{|\langle u, v \rangle|^2}{||v||^2}$$

It follows that

$$0 \leq ||u||^2 - \frac{|\langle u, v \rangle|^2}{||v||^2}$$

which yields the equality case and that u and v are linearly dependent (see §5.13), or $\sum_{i=1}^{n} u_n v_n = 0$. The desired result follows.

Remark. Here, $\overline{\lambda}$ denotes the conjugate of λ.

We have extensively used the properties of an inner product space and the inner product, which we will not cover here. However, since these properties are instrumental in understanding the proof, we will discuss them later (in §6.8.5, "Inner Product Spaces"). $\qquad \square$

5.3.1 Generalization: Hölder's Inequality

Much like the weighted AM-GM inequality generalizes the standard form of AM-GM to a case in which we assign weights that sum to 1, *Hölder's Inequality* generalizes the Cauchy-Schwarz inequality to the case with "weights" that also sum to 1. Indeed, the actual statement somewhat resembles the weighted AM-GM and power mean inequalities:

> **Theorem 5.7: Hölder's Inequality**
>
> If $a_1, a_2, \ldots, a_n, b_1, b_2, \ldots, b_n, \ldots, z_1, z_2, \ldots, z_n$ are nonnegative real numbers and $\lambda_a, \lambda_b, \ldots, \lambda_z$ are nonnegative reals with sum of 1, then
>
> $$a_1^{\lambda_a} b_1^{\lambda_b} \cdots z_1^{\lambda_z} + \cdots + a_n^{\lambda_a} b_n^{\lambda_b} \cdots z_n^{\lambda_z}$$
> $$\leq (a_1 + \cdots + a_n)^{\lambda_a} (b_1 + \cdots + b_n)^{\lambda_b} \cdots (z_1 + \cdots + z_n)^{\lambda_z}.$$

5.3.2 Special Case: Titu's Lemma

Another special case of the Cauchy-Schwarz Inequality is *Titu's Lemma,* named for Dr. Titu Andreescu (of the Mathematical Association of America, and current head coach of the US IMO team!)

> **Lemma 5.7: Titu's Lemma**
>
> For $a, b > 0$ and $x, y \in \mathbb{R}$, we will have $\dfrac{x^2}{a} + \dfrac{y^2}{b} \geq \dfrac{(x+y)^2}{a+b}$.

A rigorous proof of this and Hölder's Inequality lie beyond the scope of this book (since these proofs involve analysis and calculus techniques that we have not drawn upon), but these are interesting areas to explore that will hopefully inspire and motivate the reader to learn more about these types of inequalities. Indeed, the reader seeking a challenge may wish to try proving Titu's Lemma in particular on his or her own! (In particular, there is an especially nice proof of Hölder's inequality using Jensen's Inequality. We will leave this to the reader to ponder and perhaps make some progress towards through the Advanced sections of the text.)

5.3.3 Examples

Example 5.3.3.1. (Nesbitt's Inequality) *Prove that for positive reals a, b, c we have $\dfrac{a}{b+c} + \dfrac{b}{c+a} + \dfrac{c}{a+b} \geq \dfrac{3}{2}$.*

Solution 5.3.3.1. *This is the meat of section 6.4.3 (we don't want to spoil anything yet!) For now, though we will give a hint: apply Cauchy in the three-variable case.*

Example 5.3.3.2. *Let $a_i \in \mathbb{R}$ ($1 \leq i \leq n$) such that $\sum a_i = 1$. Prove that $\sum a_i^2 \geq \dfrac{1}{n}$.*

Solution 5.3.3.2. *By Cauchy, we have $(1 \cdot a_1 + 1 \cdot a_2 + \ldots + 1 \cdot a_n)^2 \leq (a_1^2 + a_2^2 + \ldots + a_n^2)(1 + 1 + \ldots + 1)$ which implies the desired result. Equality holds when a_i is a constant, i.e. when all are equal to $\dfrac{1}{n}$.*

5.3.4 Exercises

Exercise 51. *Prove that $(x_1 + x_2 + \ldots + x_n)^2 \leq n(x_1^2 + x_2^2 + \ldots + x_n^2)$ for all real numbers $x_1, x_2, x_3, \ldots, x_n$.*

Exercise 52. *(2013 AMC 12B) Let a, b, and c be real numbers such that*

$$a + b + c = 2, \text{ and}$$
$$a^2 + b^2 + c^2 = 12$$

What is the difference between the maximum and minimum possible values of c?

Exercise 53. *If x, y, and z are real numbers such that $x^2 + y^2 + z^2 = 1$, and $x + 2y + 3z$ is as large as possible, compute $x + y + z$.*

5.4 Binomial Theorem, Part 2: Generalizing to Real Coefficients

In our previous study of the Binomial Theorem, we restricted ourselves to the solitary case

$$(a + b)^n = \sum_{k=0}^{n} a^k b^{n-k}$$

where n is a non-negative integer. Here, we will expand our sights to the more general case $(ax + b)^n$, where $k \neq 0$ is a real number.

Theorem 5.8: Binomial Theorem with One Arbitrary Real Coefficient

The binomial expansion of $(ax + b)^n$, for any non-negative integer n, is given by

$$\binom{n}{0} a^n x^n + \binom{n}{1} a^{n-1} b x^{n-1} + \binom{n}{2} a^{n-2} b^2 x^{n-2} + \ldots + \binom{n}{n} b^n$$

$$= \sum_{k=0}^{n} x^{n-k} \cdot a^{n-k} b^k \binom{n}{k}$$

Corollary 5.8: Standard Binomial Theorem

In the case $a = 1$, we obtain the regular Binomial Theorem, namely

$$(x + b)^n = x^{n-k} \sum_{k=0}^{n} b^k \binom{n}{k}$$

(Note that, in the case $b = 1$, we obtain the familiar expansion found throughout Pascal's Triangle.)

Perhaps the best way to illustrate this generalization is through a concrete, numerical example:

Example 5.4.0.1. *Compute the coefficient of the x^5 term in the expansion of $(2x + 3)^6$.*

Solution 5.4.0.1. *We have $n = 6$, so the x^5 term occurs when $k = 1$ and $n - k = 5$. Then the desired term is $\binom{6}{5} a^5 b^1$, and with $a = 2$ and $b = 3$, the coefficient is $6a = \boxed{576}$.*

We can also generalize this to the expression $(ax + by)^n$, which is the most complicated possible form of a binomial.

Theorem 5.9: Two-Variable Binomial Theorem

If we have a binomial expansion of the form $(ax + by)^n$, then the expansion evaluates to

$$\binom{n}{0} a^n x^n + \binom{n}{1} a^{n-1} b x^{n-1} y + \binom{n}{2} a^{n-2} b^2 x^{n-2} y^2 + \ldots + \binom{n}{n} b^n y^n$$

$$= \sum_{k=0}^{n} a^{n-k} b^k x^{n-k} y^k$$

$$= \sum_{k=0}^{n} (ax)^{n-k} (by)^k$$

---

in compact form.

On an intuitive level, we can "separate" the numerical coefficients a, b and the variables x, y, so as to obtain a binomial expansion that consists of the products of the terms of the binomial expansions $(a + b)^n$ and $(x + y)^n$.

5.4.1 Examples

Example 5.4.1.1. *What is the coefficient of the x^3 term in the expansion of $(4x + 5)^6$?*

Solution 5.4.1.1. *Using the binomial expansion form*

$$(ax + b)^n = \sum_{k=0}^{n} a^{n-k} b^k \binom{n}{k} x^{n-k}$$

with $a = 4$, $b = 1$, $n = 6$, we have a coefficient of $4^3 5^3 \binom{6}{3} = 20^3 \cdot 20 = 20^4 = \boxed{160,000}$.

Example 5.4.1.2. *What is the largest coefficient in the expansion of $(3x + 2)^{10}$?*

Solution 5.4.1.2. *Observe that the ratio of the coefficient of x^k to the coefficient of x^{k-1} is equal to $\dfrac{b}{a} \cdot \dfrac{\binom{10}{k}}{\binom{10}{k-1}}$. With $a = 3$, $b = 2$, the coefficient is maximized when the ratio is as close to 1 as possible, or when the ratio of binomial coefficients is as close to $\dfrac{2}{3}$ as possible. By inspection, this occurs at $k = 6$, which yields a coefficient of $\boxed{3^6 2^4 \binom{10}{4}}$.*

5.4.2 Exercises

Exercise 54. *(2000 AIME I) In the expansion of $(ax + b)^{2000}$, where a and b are relatively prime positive integers, the coefficients of x^2 and x^3 are equal. Find $a + b$.*

Exercise 55. *How many terms are in the expansion of $(5\log_5 x + \log_{25} y)^{125}$, after combining like terms?*

Exercise 56. *(2002 AIME I) The Binomial Expansion is valid for exponents that are not integers. That is, for all real numbers x, y and r with $|x| > |y|$,*

$$(x + y)^r = x^r + rx^{r-1}y + \frac{r(r-1)}{2}x^{r-2}y^2 + \frac{r(r-1)(r-2)}{3!}x^{r-3}y^3 \cdots$$

What are the first three digits to the right of the decimal point in the decimal representation of $(10^{2002} + 1)^{\frac{10}{7}}$?

Exercise 57. *(1995 AIME) Let $f(n)$ be the integer closest to $\sqrt[4]{n}$. Find $\displaystyle\sum_{k=1}^{1995} \frac{1}{f(k)}$.*

Extension Exercise 21. *Can we derive Bernoulli's Inequality from the Binomial Theorem? Briefly explain how you might approach such a derivation.*

Extension Exercise 22. *In the expansion of $(1 + x)^n$ $(n > 2)$, prove that if the coefficients of three consecutive terms are in arithmetic progression, then $n + 2$ is a perfect square.*

5.5 Rational Root Theorem

The *Rational Root Theorem* (RRT) concerns the root-finding process for a polynomial equation in the form

$$P(x) = a_n x^n + a_{n-1} x^{n-1} + \ldots + a_1 x + a_0 = 0$$

where $a_i \in \mathbb{Z}$ for $0 \leq i \leq n$.

Theorem 5.10: Rational Root Theorem

Define $P(x)$ as above. If $a_n, a_0 \neq 0$, then each rational solution $x = \dfrac{p}{q}$, $p, q \in \mathbb{Z}$, $\gcd(p, q) = 1$, satisfies the following two constraints:

- p divides the constant term a_0.

- q divides the leading coefficient a_n.

Proof. Since $\dfrac{p}{q}$ is a root of $P(x)$, we have

$$a_n \left(\frac{p}{q} \right)^n + a_{n-1} \left(\frac{p}{q} \right)^{n-1} + \ldots + a_0 = 0$$

Multiplying both sides by q^n, we obtain

$$a_n p^n + a_{n-1} p^{n-1} q + \ldots + a_0 q^n = 0$$

If we examine this equation modulo p, we obtain $a_0 q^n \equiv 0 \pmod{p}$. Because $\gcd(p, q) = 1$, $p \mid a_0$. By similar logic, we can show that $q \mid a_n$ as well, completing the roof. $\qquad \square$

5.5.1 Gauss' Lemma

The RRT is a special case of *Gauss' Lemma* for polynomials, which posits two statements pertaining to polynomials with integer coefficients:

Lemma 5.10: Gauss' Lemma

For the following lemma, we must first define two terms:

- A polynomial $P(x) = a_n x^n + a_{n-1} x^{n-1} + \ldots + a_1 x + a_0$ with integer coefficients is *primitive* if $\gcd(a_i)_{i \leq n} = 1$.

- Furthermore, $P(x)$ is *irreducible* over a *field*[a] $\mathbb{F}$ iff it cannot be factored into the form $a_n(x - x_1)(x - x_2)(x - x_3) \cdots (x - x_n)$ where $x_i \in \mathbb{F}$ for all $i \in [1, n]$.

Then we have the following:

- If $p(x)$ and $q(x)$ are primitive polynomials with integer coefficients, then $p(x) \cdot q(x)$ is also a primitive polynomial with integer coefficients.

- If a non-constant polynomial $P(x)$ with integer coefficients is irreducible over $\mathbb{Z}$, then it is also irreducible over $\mathbb{Q}$.

---

[a]Any set within which addition, subtraction, multiplication, and division are well-defined, and behave as they would on the real numbers. This is far from a complete definition, however; usually this topic is covered in depth in an introductory group theory or abstract algebra course at the post-secondary level.

Proof. (Note: Throughout the proof, we denote primitive polynomials by $f(x)$ and $g(x)$, rather than p and q, to avoid confusion with the prime number p.)

We first show that the product $f(x)g(x)$ of two primitive polynomials $f(x)$ and $g(x)$ is itself primitive. Assume the contrary. Then there exists a prime number p such that p is a common divisor of all of the coefficients of $f(x)g(x)$. Since f and g are primitive, however, p cannot be a factor of all of the coefficients of either f or g (this contradicts the definition of primitivity).

Suppose that $\deg f = r$, $\deg g = s$, so that the leading terms of f and g are $a_r x^r$ and $b_s x^s$ respectively, with $p \nmid a_r$ and $p \nmid b_s$. Then the coefficient of x^{r+s} in the expansion of $f(x)g(x)$, by the Binomial Theorem, is

$$\sum_{i+j=r+s} a_i b_j$$

Because p is prime, the sum contains a term that is not divisible by p, as the leading coefficient is not divisible by p. Hence, the sum is not divisible by p, as all the other terms are multiples of p. However, we have assumed that all of the coefficients in $f(x)g(x)$ are divisible by p, which is a contradiction. Therefore, if $f(x)$ and $g(x)$ are primitive, then $f(x) \cdot g(x)$ must also be primitive.

We proceed to show the second part of the lemma; that all irreducible polynomials over $\mathbb{Z}$ are also irreducible over $\mathbb{Q}$. We prove the contrapositive statement; that reducibility over $\mathbb{Q}$ implies reducibility over $\mathbb{Z}$.

To show reducibility of a polynomial f with integer coefficients, we must prove the existence of $c \in \mathbb{N}$ and integer polynomials g', h' derived from polynomials g, h with rational coefficients such that $cf(x) = g'(x) \cdot h'(x)$. Let a and b be the greatest common multiples of the denominators of the coefficients of g and h, respectively. If we set $g'(x) = a \cdot g(x)$ and $h'(x) = b \cdot h(x)$, then g' and h' are guaranteed to have integer coefficients. Then $ab \cdot f(x) = ag(x) \cdot bh(x) = g'(x) \cdot h'(x)$; taking $c = ab$ proves the existence of c.

We now must show that $c = 1$. Assume the contrary, that $c > 1$ but is minimal; let p be a prime factor of c. Then take the equation $cf(x) = g'(x) \cdot h'(x)$ modulo p; i.e. let $\hat{g}'(x) = g'(x) \pmod p$ and $\hat{h}'(x) = h'(x) \pmod p$ (with coefficients reduced to elements in the set $\{0, 1, 2, \ldots, p-1\}$). It follows that $\hat{g}'(x) \cdot \hat{h}'(x) = 0$, since $p \mid c$.

Hence, either $\hat{g}'(x) = 0$ or $\hat{h}'(x) = 0$;[28] assume WLOG that $\hat{g}'(x) = 0$. Then all coefficients of $\hat{g}'(x)$ are divisible by p. Set $c' = \dfrac{c}{p}$, and $g''(x) = \dfrac{g'(x)}{p}$. Then we have $c'f(x) = g''(x) \cdot h'(x)$, but this contradicts c being minimal, as $c' < c$. Hence, the smallest value of c is 1 and $f(x) = g'(x) \cdot h'(x)$ for some integer polynomials g', h'. This implies that f is reducible over $\mathbb{Z}$ given that it is reducible over $\mathbb{Q}$, which in turn shows that all irreducible polynomials over $\mathbb{Z}$ are irreducible over $\mathbb{Q}$. $\square$

5.5.2 Examples

Let's work through a couple of concrete examples:

Example 5.5.2.1. *Factor the polynomial $P(x) = x^3 - 6x^2 - x + 30$ over the integers, or show that it is irreducible.*

Solution 5.5.2.1. *If $P(x)$ has a factorization of the form $(x - x_1)(x - x_2)(x - x_3)$ where $x_1, x_2, x_3 \in \mathbb{Z}$, and for each i, $x_i = \dfrac{p}{q}$ for some $p, q \in \mathbb{Z}$, $\gcd(p, q) = 1$, then $p \mid 30$ and $q \mid 1$. Hence, $q = \pm 1$ and $p = \pm 1, 2, 3, 5, 6, 10, 15, 30$.*

The possible roots are then just the values for p. Substituting each in yields the roots $x = -2, 3, 5$, so the factorization of $P(x)$ is $(x + 2)(x - 3)(x - 5)$.

---

[28]This follows from the fact that the group of integers modulo p is an *integral domain,* a commutative ring in which the product of any two non-zero elements is non-zero.

Alternatively, we can use Vieta's formulas to obtain $x_1 + x_2 + x_3 = 6$, $x_1x_2 + x_1x_3 + x_2x_3 = -1$, and $x_1x_2x_3 = -30$; some casework will reveal the correct values for x_1, x_2, x_3. Another method is to perform synthetic division for each of the potential factors $(x - x_i)$; however, this is significantly more time-consuming compared to the other two methods.

Example 5.5.2.2. *Give an example of a polynomial $P(x)$ with integer coefficients that has no rational roots, but can be factored into two or more non-constant polynomials[29] over $\mathbb{Q}$.*

Solution 5.5.2.2. *Any polynomial with degree 4 or higher is a potential candidate; for instance, consider $P(x) = x^4 - 1 = (x^2 + 1)(x^2 - 1)$, which has roots $\pm i\sqrt{2}$, $\pm\sqrt{2}$, but nevertheless is factorable over $\mathbb{Q}$.*

---

[29]See Lemma 13.1.

5.5.3 Exercises

Exercise 58.

(a) **Using the RRT,** *show that $\sqrt{2}$ is irrational.*

(b) **Using the RRT,** *show that $\sqrt{p}$ is irrational for all prime p.*

(c) **Using the RRT,** *show that $\sqrt{2} + \sqrt{3}$ is irrational.*

Exercise 59. *Let S_n be the set of all monic polynomials (polynomials with leading coefficient 1) with degree n that have no rational roots. Show that, for all polynomials $p(x) \in S_n$, $n \leq 3$, $p(x)$ is irreducible over $\mathbb{Q}$.*

Exercise 60. *How many rational roots does*

$$P(x) = \sum_{k=0}^{2019} x^k \cdot (-1)^{2020-k}$$

have?

Extension Exercise 23. *Let $\mathbb{Z}_n$ be the group of integers modulo n; for instance, $\mathbb{Z}_5 = \{0,1,2,3,4\}$. Let $\mathbb{Z}_n[x]$ be the field of polynomials with integer coefficients modulo n; for instance, $\mathbb{Z}_{10}[x] = \{a_n x^n + a_{n-1} x^{n-1} + \ldots + a_1 x + a_0, a_i \in \mathbb{Z}_{10}, \forall i \in [0, n]\}$. For which polynomials $f(x) \in \mathbb{Z}_2[x]$ does f have at least one root in $\mathbb{Z}_2$?*

5.6 Continuing Sequences and Series

Sequences and series have a beautiful tendency to simplify nicely to the familiar arithmetic and geometric sums; however, much more often, they lend themselves to other techniques that manipulate the sum in clever and interesting ways. Take, for example, the *arithmetico-geometric sequence:*

5.6.1 Arithmetico-Geometric Series

An *arithmetico-geometric sequence* (hence AGS for brevity) is a combination of an arithmetic and a geometric sequence, in the sense that each term is the product of the elements of the respective terms of an arithmetic sequence and a geometric sequence:

> **Definition 5.2: Arithmetico-Geometric Sequence**
>
> For all $n \in \mathbb{N}$, an arithmetico-geometric sequence (a_n) has $a_n = b_n \cdot c_n$, where (b_n) is an arithmetic sequence and (c_n) is a geometric sequence.

As such, the general form of an AGS (a_n) is $\{ab, (a+d)br, (a+2d)br^2, \ldots\}$ where a is the initial term of (b_n), d is the common difference of (b_n), b is the initial term of (c_n), and r is the common ratio of (c_n).

Developing Arithmetico-Geometric Sum Formulas

To sum the terms in an AGS requires more than to sum the terms of the corresponding arithmetic or geometric series individually. Rather, we must split each AGS term into two components that lend themselves to their own geometric series. (Henceforth, we can assume WLOG that $b = 1$, as the terms of the geometric series scale linearly with the sum of the AGS.)

Example 5.6.1.1. *Compute* $\displaystyle\sum_{k=1}^{\infty} \frac{k}{4^k}$.

Solution 5.6.1.1. *We can write out the sum to simplify our calculations:*

$$\sum_{k=1}^{\infty} \frac{k}{4^k} = \frac{1}{4} + \frac{2}{16} + \frac{3}{64} + \frac{4}{256} + \ldots$$

Observe that $\dfrac{2}{16} = \dfrac{1}{16} \cdot 2$, $\dfrac{3}{64} = \dfrac{1}{64} \cdot 3$, *and* $\dfrac{4}{256} = \dfrac{1}{256} \cdot 4$. *If we split each term into a sum of powers of 4, and then write the sum column-wise, with each power of 4 in its own column, we obtain a sum of geometric series in the rows:*

$$\frac{1}{4} + \frac{1}{16} + \frac{1}{64} + \frac{1}{256} + \ldots$$

$$\frac{1}{16} + \frac{1}{64} + \frac{1}{256} + \ldots$$

$$\frac{1}{64} + \frac{1}{256} + \ldots$$

$$\frac{1}{256} + \ldots$$

Figure 5.1: Rewriting an AGS as a sum of geometric series.

69

The topmost row (row 1) is now a standard geometric series with $a = \frac{1}{4}$ and $r = \frac{1}{4}$, so its sum is $\dfrac{\frac{1}{4}}{1 - \frac{1}{4}} = \frac{1}{3}$.

Row 2 is also a geometric series, with $a = \frac{1}{16}$ and $r = \frac{1}{4}$, so its sum is $\frac{1}{12}$. Since a is multiplied by $\frac{1}{4}$ from row to row, and r remains constant, the sum of the AGS is the sum of the geometric series $\frac{1}{3} + \frac{1}{12} + \ldots$, which is $\frac{4}{9}$.

Example 5.6.1.2. *Compute $\displaystyle\sum_{k=1}^{\infty} \frac{k}{n^k}$ in terms of n, where $|n| > 1$.*

Solution 5.6.1.2. *Once again, we can split the sum into several geometric series, whose sums themselves form a geometric series:*

$$\sum_{k=1}^{\infty} \frac{k}{n^k} = \frac{1}{n} + \frac{2}{n^2} + \frac{3}{n^3} + \ldots$$

This sum can be rewritten as follows:

$$\frac{1}{n} + \frac{1}{n^2} + \frac{1}{n^3} + \frac{1}{n^4} + \ldots$$

$$\frac{1}{n^2} + \frac{1}{n^3} + \frac{1}{n^4} + \ldots$$

$$\frac{1}{n^3} + \frac{1}{n^4} + \ldots$$

$$\frac{1}{n^4} + \ldots$$

Figure 5.2: Rewriting the semi-generalized AGS with $a = d = b = 1$, $r = n$.

If we let S_r be the sum of the elements in the rth row from the top, then

$$S_1 = \frac{\frac{1}{n}}{1 - \frac{1}{n}} = \frac{1}{n - 1}$$

$$S_2 = \frac{\frac{1}{n^2}}{1 - \frac{1}{n}} = \frac{1}{n(n - 1)}$$

$$\vdots$$

$$S_r = \frac{\frac{1}{n^r}}{1 - \frac{1}{n}} = \frac{1}{n^{r-1} \cdot (n - 1)}$$

Thus, we have the geometric series

$$\sum_{r=1}^{\infty} S_r = \frac{\left(\frac{1}{n - 1}\right)}{\left(1 - \frac{1}{n}\right)} = \frac{n}{(n - 1)^2}$$

which is our final sum.

Note that the $|n| > 1$ condition exists to ensure that the sum converges to a finite value. If $|n| \leq 1$, then the sum diverges, since the terms either increase without bound, or they oscillate between k and $-k$ infinitely. On the other hand, here is an example of an arithmetico-geometric series that strays somewhat from the format we've practiced so far:

Example 5.6.1.3. *Compute* $\displaystyle\sum_{k=1}^{\infty} \frac{k^2}{2^k}$.

Solution 5.6.1.3. *Recall that* $k^2 = 1 + 3 + \cdots + (2k-1)$. *Then* $\displaystyle\sum_{k=1}^{\infty} \frac{k^2}{2^k}$ *can be rewritten as* $\displaystyle\sum_{k=1}^{\infty} \frac{1}{2^k} +$

$\displaystyle\sum_{k=2}^{\infty} \frac{3}{2^k} + \sum_{k=3}^{\infty} \frac{5}{2^k} + \ldots$. *Summing up each of the geometric series, the expression evaluates to* $\dfrac{\frac{1}{2}}{1 - \frac{1}{2}} +$

$3\dfrac{\frac{1}{2^2}}{1 - \frac{1}{2}} + 5\dfrac{\frac{1}{2^3}}{1 - \frac{1}{2}} + \cdots = 1 + \frac{3}{2} + \frac{5}{2^2} + \ldots$. *This new expression can be written as a sum of geometric*

series: $\displaystyle\sum_{k=0}^{\infty} \frac{1}{2^k} + \sum_{k=1}^{\infty} \frac{2}{2^k} + \sum_{k=2}^{\infty} \frac{2}{2^k} + \ldots$. *This expression evaluates to* $\dfrac{1}{1 - \frac{1}{2}} + 2\left(\dfrac{\frac{1}{2}}{1 - \frac{1}{2}} + \dfrac{\frac{1}{2^2}}{1 - \frac{1}{2}} + \ldots\right) =$

$2 + 2\left(1 + \frac{1}{2} + \frac{1}{2^2} + \ldots\right) = 2 + 2 \cdot 2 = \boxed{6}$.

5.6.2   Toying With Infinity: An Introduction to Divergent Series

What Does Convergence Actually Mean?

Unlike a *convergent* series, whose sum eventually reaches a finite value, a *divergent* series never converges to a real number. That is, if we define the nth *partial sum* of the series (a_n) to be the sum $a_1 + a_2 + a_3 + \ldots + a_n$, then the sequence of partial sums of (a_n) does not have a finite limit.

In order for a series to converge, its terms must eventually approach zero; that is, for all $\epsilon > 0$, there exists an $N \in \mathbb{N}$ such that, for all $n > N$, $|a_n| < \epsilon$. Thus, if the terms do not approach zero, then the series must diverge. However, the terms converging to zero is not a sufficient condition for the series to converge. Divergent series may also oscillate between upper and lower bounds that sum to zero (an example being $a_n = \sin n$). We must also define other types of convergence, namely *absolute convergence* and *conditional convergence:*

Definition 5.3: Types of Convergence

If a series converges, we may classify it as *absolutely convergent* or *conditionally convergent*.

- An absolutely convergent series (a_n) satisfies the condition that the sequence (b_n) defined by $b_n = |a_n|$ also converges.

- A conditionally convergent series (a_n) is any convergent series that is not also absolutely convergent. A classic example of a conditionally convergent series is the alternating series $\displaystyle\sum_{k=1}^{\infty} \frac{(-1)^{k+1}}{k} =$

$1 - \frac{1}{2} + \frac{1}{3} - \frac{1}{4} + \ldots$, which converges to $\ln 2$;[a] however, the series $\displaystyle\sum_{k=1}^{\infty} \frac{1}{k}$ diverges.[b]

---

[a]Conditionally convergent sequences have the peculiar property that they can be rearranged so that they converge to

any value, including $\pm\infty$, by the *Riemann series theorem.*
    [b]This is the *harmonic series,* or a *p-series* with $p = 1$.

Thus, we say that a *divergent series* is any series that fails to converge (i.e. meet all of the criteria of either an absolutely convergent series or a conditionally convergent series).

Convergence Tests

To test for the convergence of a series, we may use any of the following tests.

Theorem 5.11: Tests for Series Convergence

Convergence tests for a series (a_n) include, but are not limited to, the following:

- Ratio test: Let $L = \lim_{n\to\infty} \left| \dfrac{a_{n+1}}{a_n} \right|$. If $L < 1$, the series converges absolutely; if $L > 1$, the series diverges; and if $L = 1$ or does not exist, the test is inconclusive, and another test must be used to determine whether the series converges.

  In addition, we use this test to determine the *interval of convergence* and *radius of convergence;* i.e. the set of values of x for which a series in terms of x (known as a *power series,* which approximates a polynomial or trigonometric function[a]) converges.

- Root test: For a series a_n, let $C = \limsup_{n\to\infty} \sqrt[n]{a_n}$. If $C < 1$, the series converges; if $C > 1$, the series diverges; if $C = 1$ and C approaches 1 strictly from above (i.e. it is never less than 1 at any given time), the series also diverges; and if $C = 1$ otherwise, the test is inconclusive.

- Comparison test: There are two types of comparison tests with a series (b_n) that determine convergence of the series (a_n).

  - Direct comparison test: Let (b_n) be an infinite series. If $0 \leq a_n \leq b_n$ for all sufficiently large n (i.e. there exists an N for all $n \geq N$), and (b_n) converges, then (a_n) converges. If $0 \leq b_n \leq a_n$ for all sufficiently large n and (b_n) diverges, then (a_n) also diverges.

  - Limit comparison test: Let $a_n \geq 0$ for all n, and let (b_n) be a series with $b_n > 0$ for all n. If $\lim_{n\to\infty} \dfrac{a_n}{b_n} = c$, with $0 < c < \infty$, then either both (a_n) and (b_n) converge or both diverge.

- Alternating series test (AST): If (a_n) is an alternating series (i.e. has terms alternating between positive and negative), then consider $|a_n|$. If the sequence $(|a_n|)$ is monotonically decreasing and $\lim_{n\to\infty} a_n = 0$, then (a_n) converges. (Note that the second condition is necessary for *any* sequence to converge.)

- *p*-series test: A series of the form $\displaystyle\sum_{n=1}^{\infty} \dfrac{1}{n^p}$ with $p \in \mathbb{Z}$ converges if and only if $p > 1$. Otherwise, the series diverges. Note that when $p = 1$, we get what is known as the *harmonic series,* which diverges.

---
    [a]This is a topic covered in a post-secondary Calculus course, and is not traditionally covered in competition math, so we will not cover it here.

Proof.

- Suppose that $L = \lim_{n\to\infty} \left| \dfrac{a_{n+1}}{a_n} \right| < 1$. We show that (a_n) converges absolutely by way of comparing the series to a convergent geometric series. Let $r = \dfrac{L+1}{2}$. Then $L < r < 1$ and $|a_{n+1}| < r\,|a_n|$ for

$n \geq N$. Thus, $|a_{n+i}| < r^i \cdot |a_n|$ for all $n > N$, $i > 0$. This gives rise to the series of equations

$$\sum_{i=N+1}^{\infty} |a_i| = \sum_{i=1}^{\infty} |a_{n+i}| < \sum_{i=1}^{\infty} r^i |a_{N+1}| = |a_{N+1}| \cdot \sum_{i=1}^{\infty} r^i < \infty$$

since $r < 1$ and the last term is a geometric series. Hence, (a_n) converges absolutely. On the other hand, if $L > 1$, then for sufficiently large n, $|a_{n+1}| - |a_n|$, so the series automatically diverges, since the limit of the a_n is nonzero.

- Assume WLOG that the series starts at $n = 1$. Further assume that $L < 1$; we claim that (a_n) is absolutely convergent. Note that, because $L < 1$, there exists r such that $L < r < 1$.

Recall that

$$L = \lim_{n \to \infty} \sqrt[n]{a_n} = \lim_{n \to \infty} |a_n|^{\frac{1}{n}}$$

Then there exists some N such that, if $n \geq N$, we have

$$|a_n|^{\frac{1}{n}} < r \implies |a_n| < r^n$$

Consider the infinite series

$$\sum_{n=0}^{\infty} r^n$$

which is a geometric series with sum $\dfrac{1}{1-r}$. Since $0 < r < 1$, this series converges, and furthermore, the series

$$\sum_{n=N}^{\infty} |a_n|$$

converges. Since

$$\sum_{n=1}^{\infty} |a_n| = \sum_{n=1}^{N-1} |a_n| = \sum_{n=N}^{\infty} |a_n|$$

and we know that

$$\sum_{n=1}^{N-1} |a_n|$$

is convergent,

$$\sum_{n=1}^{\infty} |a_n|$$

is convergent as well. This implies that

$$\sum_{n=1}^{\infty} a_n$$

is absolutely convergent.

Next, assume $L > 1$. Then we know that there exists some N such that for $n \geq N$, we have $|a_n| > 1$ (since $|a_n|^{\frac{1}{n}} > 1$). But since $\lim_{n \to \infty} \neq 0$, the series (a_n) must diverge.

Finally, for $L = 1$, to show that the Root Test is indeed inconclusive, we simply need to construct three examples of series that are absolutely convergent, conditionally convergent, and divergent. This part of the proof is left as an exercise to the reader.

- To prove the direct comparison test, let (a_n) and (b_n) be infinite series with (b_n) converging absolutely. WLOG assume that $|a_n| \leq |b_n|$ for all n. Define the partial sums

$$S_n = \sum_{k=1}^{n} |a_k|$$

$$T_n = \sum_{k=1}^{n} |b_k|$$

We then have $\lim_{n \to \infty} T_n = T \in \mathbb{R}$, and for all n,

$$0 \leq S_n \leq S_n + (T - T_n) \leq T$$

Since (S_n) is non-decreasing and $S_n + (T - T_n)$ is non-increasing, we have m, n such that $S_m, S_n \in I = [S_N, S_N + (T - T_N)]$, where the length of I approaches zero as N tends to ∞. Therefore, (S_n) is Cauchy, and so has a limit. It follows that (a_n) is absolutely convergent. To prove the limit comparison test, we observe that the equation $\lim_{n \to \infty} \dfrac{a_n}{b_n} = c$ implies the existence of a real number $\epsilon > 0$ such that there exists $n_0 \in \mathbb{N}$ for which, for all $n \geq n_0$, we have that $\left| \dfrac{a_n}{b_n} - c \right| < \epsilon$. That is,

$$-\epsilon < \frac{a_n}{b_n} - c < \epsilon$$

$$c - \epsilon < \frac{a_n}{b_n} < c + \epsilon$$

$$(c - \epsilon)b_n < a_n < (c + \epsilon)b_n$$

Since c is positive, take ϵ such that $c - \epsilon > 0$, so that $\dfrac{a_n}{c - \epsilon} > b_n$. By the direct comparison test, if (a_n) converges, then (b_n) must converge as well. Similarly, we have $(c + \epsilon)(b) > a_n$, so by the same logic, if (b_n) converges, then (a_n) converges. Thus the series must either both converge or both diverge.

- Suppose we have the infinite series

$$\sum_{n=1}^{\infty} (-1)^{n-1} a_n$$

where $\lim_{n \to \infty} a_n = 0$ and (a_n) is non-decreasing. We show that both partial sums

$$S_{2k+1} = \sum_{n=1}^{2k+1} (-1)^{n-1} a_n$$

and

$$S_{2k} = \sum_{n=1}^{2k} (-1)^{n-1} a_n$$

converge to the same limit L, so that

$$S_k = \sum_{n=1}^{k} (-1)^{n-1} a_n \to L$$

as well.

We observe that the odd-numbered partial sums are monotonically decreasing, while the even-numbered partial sums are monotonically increasing, since (a_n) is itself monotonically decreasing. Furthermore, since $a_n > 0$ for all n, $S_{2k+1} - S_{2k} > 0$, which implies that

$$a_1 - a_2 = S_2 \leq S_{2k} \leq S_{2k+1} \leq S_1 = a_1$$

74

Note that $a_1 - a_2$ is a lower bound of the sequence (S_{2k+1}). By the monotone convergence theorem, (S_{2k+1}) converges as $k \to \infty$. Similarly, (S_{2k}) also converges.

Finally, we claim their limits are both equal to L. Observe that $\lim_{k \to \infty} (S_{2k+1} - S_{2k}) = \lim_{k \to \infty} a_{2k+1} = 0$ using the fact that $\lim_{n \to \infty} a_n = 0$. Another application of the monotone convergence theorem tells us that $S_{2k} \leq L \leq S_{2k+1}$ for all k (so the partial sums alternate from being above and below the limit L).

We conclude by proving that $|S_k - L| < a_{k+1}$. When k is odd, i.e. of the form $2m + 1$, we have

$$|S_{2m+1} - L| = S_{2m+1} - L \leq S_{2m+1} - S_{2m+2} = a_{2m+2}$$

The proof for $k = 2m$ is similar. Indeed, we obtain $|S_{2m} - L| = a_{2m+1}$ as desired.

- This is a straightforward application of the *integral test*, which allows us to take the integral

$$\int_1^\infty x^{-p} dx = \lim_{t \to \infty} x^{1-p} \Big|_1^t = \lim_{t \to \infty} \left(1 - t^{1-p}\right)$$

which converges iff $p > 1$.

These are most of the tests that will be helpful in series problems in a non-calculus-based competitive math context (though in calculus class, you will probably learn (or have probably learned) several more convergence tests not covered here). $\qquad\square$

Let's take a look at a few examples of convergent and divergent series:

Example 5.6.2.1. *Does the series* $\sum_{k=1}^\infty \dfrac{2k}{k+1}$ *converge or diverge? If the former, to what value does it converge?*

Solution 5.6.2.1. *Note that* $\sum_{k=1}^\infty \dfrac{2k}{k+1} = \sum_{k=1}^\infty \dfrac{2k+2}{k+1} - \sum_{k=1}^\infty \dfrac{2}{k+1} = \sum_{k=1}^\infty 2 - \sum_{k=1}^\infty \dfrac{2}{k+1}$. *Both series diverge by the p-series test (in the former one, $p = 0$), so* $\sum_{k=1}^\infty \dfrac{2k}{k+1}$ *diverges.*

One can also observe that $\lim_{k \to \infty} \dfrac{2k}{k+1} = 2 \neq 0$, *so we immediately see that the series must diverge.*

Example 5.6.2.2. *Determine whether the series* $\sum_{k=1}^\infty \dfrac{\sqrt{k} + 2019}{k^2}$ *converges or diverges, and cite any convergence tests used.*

Solution 5.6.2.2. *We can split this sum up into two:*

$$\sum_{k=1}^\infty \frac{\sqrt{k} + 2019}{k^2} = \sum_{k=1}^\infty \frac{\sqrt{k}}{k^2} + \sum_{k=1}^\infty \frac{2019}{k^2}$$

The first sum converges, by the p-series test with $p = \dfrac{3}{2} > 1$. The second sum also converges by the p-test ($p = 2$), so the sum as a whole converges.

5.6.3 The Partial Fraction Decomposition

As we can factor polynomials such as $n^2 + n$, we can decompose fractions such as $\dfrac{1}{n^2 + n}$. Partial fraction decomposition may sound intimidating, but it's just the fraction form of factoring expressions.

> **Theorem 5.12: Partial Fraction Decomposition**
>
> The partial fraction decomposition of $\dfrac{f(x)}{g(x)}$ can be expressed as $\sum \dfrac{f_i(x)}{g_i(x)}$, where $g_i(x)$ is a root of $g(x)$ with a smaller degree.

This will be an important part of our next subsection, which is telescoping.

Example 5.6.3.1. *Decompose* $\dfrac{1}{n(n+1)}$.

Solution 5.6.3.1. *Note that we want to express this in the form of* $\dfrac{K}{n} + \dfrac{J}{n+1}$ *such that* $\dfrac{K}{n} + \dfrac{J}{n+1} = \dfrac{K(n+1) + J(n)}{n(n+1)} = \dfrac{1}{n(n+1)}$. *Co-efficient matching yields the system of equations*

$$K + J = 0$$

$$K = 1,$$

implying $J = -1$. *This means that our partial fraction decomposition is*

$$\boxed{\dfrac{1}{n} - \dfrac{1}{n+1}}.$$

Note that we will often be co-efficient matching for partial fraction decomposition.

We present a couple of natural extensions:

Exercise 61. *What is the partial fraction decomposition of* $\dfrac{1}{n(n+k)}$, *in terms of k?*

Exercise 62. *Find the partial fraction decomposition of* $\dfrac{1}{n(n+1)(n+2)}$.

Warning: Partial decomposition cannot be done on expressions such as $\dfrac{1}{n^2}$ because nothing differentiates $\dfrac{I}{n}$ from $\dfrac{J}{n}$!

5.6.4 Telescoping Series

There are more than a few occasions in which sums, finite or infinite, are cumbersome, and borderline impossible in a timed setting, to compute - whether they have ridiculously high bounds (or even infinite bounds with no clear pattern, like a geometric series), terms that do not produce nice patterns, or series that otherwise cannot be brute-forced. To help counter these types of series, we can combine several of our learned techniques, namely the notion of partial sums, and the partial fraction decomposition technique, to develop the framework behind the almighty *telescoping series*.

> **Definition 5.4: Telescoping Series**
>
> Essentially, a *telescoping series* is any series (a_n) such that there are terms in the series that cancel with other terms. More rigorously, the partial sums have a finite number of terms after cancellation.
>
> That is, let (a_n) be a sequence of numbers. Then
>
> $$\sum_{k=1}^{n} (a_k - a_{k-1}) = a_n - a_0$$

If $a_n \to 0$ (note that this is the only possible scenario if the series (a_n) converges), then the sum becomes $-a_0$.

Before we introduce examples, we should begin by remarking that *partial fractions are instrumental in proving that a series telescopes!* Without the partial fraction decomposition by our side, it would be very difficult to recognize even a well-known telescoping series, such as $\sum_{n=1}^{\infty} \frac{1}{n^2 + n}$, as telescoping. At first glance, the denominators of the terms in the series do not have any clear pattern, or if they do, they simply lead one astray in a circular loop, as one may recognize them as twice the triangular numbers and no more. But if we express one fraction as the sum or difference of two or more fractions, then we can much more easily see that certain terms cancel out, and that we will ultimately end up with a finite (and usually very clean) sum.

Telescoping of Partial Fractions

We will cover the wide swath of fractional terms with denominators that are the products of two terms, i.e. summands of the form

$$\sum_{k=1}^{n} \frac{1}{(ak + b)(ck + d)}$$

where the numerator can WLOG be set to 1 and n is some positive integer.

More specifically, note that the decomposition will consist of a sum of two fractions, namely $\frac{A}{ck + d}$ and $\frac{B}{ak + b}$. Then we require

$$\frac{A(ak + b) + B(ck + d)}{(ak + b)(ck + d)} = \frac{1}{(ak + b)(ck + d)}$$

$$\implies A(ak + b) + B(ck + d) = 1$$

$$\implies k \cdot (A \cdot a + B \cdot c) = 0, A \cdot b + B \cdot d = 1 \implies A \cdot a + B \cdot c = 0, A \cdot b + B \cdot d = 1$$

Then we can solve for the coefficients a, b, c, d and subsequently obtain the decomposition of the original summand, which will then telescope (with usually the first term remaining, but sometimes the first two or even three terms).

For sums whose patterns are not exactly obvious, we can actively seek out ways to "force" the series to telescope in one way or another:

Example 5.6.4.1. *Compute* $\frac{1}{45} + \frac{1}{117} + \frac{1}{221} + \ldots + \frac{1}{9797}$.

Solution 5.6.4.1. *Each term in the denominator is* $5 \cdot 9$, $9 \cdot 13$, $13 \cdot 17$, $\ldots$, $97 \cdot 101$, *i.e. of the form* $(4k + 1)(4k + 5)$ *for* $k \in [1, 24]$. *Then our sum becomes*

$$\sum_{k=1}^{24} \frac{1}{(4k + 1)(4k + 5)} = \sum_{k=1}^{24} \frac{\frac{1}{4}}{4k + 1} - \frac{\frac{1}{4}}{4k + 5}$$

$$= \sum_{k=1}^{24} \frac{1}{16k + 4} - \frac{1}{16k + 20}$$

$$= \left(\frac{1}{20} + \frac{1}{36} + \frac{1}{52} + \ldots + \frac{1}{388} \right) - \left(\frac{1}{36} + \frac{1}{52} + \frac{1}{68} + \ldots + \frac{1}{404} \right) = \frac{1}{20} - \frac{1}{404} = \boxed{\frac{24}{505}}$$

At the end of the day, regardless of how messy or ugly-looking the summand appears to be at first glance, there is (almost) always an elegant way to simplify it; all roads lead to Rome! Err...I mean, partial fractions!

(It is worth noting that, although many telescoping series do indeed collapse to partial fractions, in actuality, it's not just partial fractions that telescope, but also several other classes of functions, such as trigonometric functions like sin and cos (as they are periodic with period 2π, especially if we use an interval length in the sum that is a factor/multiple of 2π). But for the most part, partial fractions are by far the most notable and prominent types of telescoping series.)

5.6.5 Examples

Example 5.6.5.1. *Compute the sum* $\dfrac{1}{1 \cdot 2} + \dfrac{1}{2 \cdot 3} + \dfrac{1}{3 \cdot 4} + \ldots + \dfrac{1}{2018 \cdot 2019}$.

Solution 5.6.5.1. *As was found in Example 5.6.4.1,* $\dfrac{1}{n(n+1)} = \dfrac{1}{n} - \dfrac{1}{n+1}$. *Notice that* $\dfrac{1}{1 \cdot 2} + \dfrac{1}{2 \cdot 3} +$
$\dfrac{1}{3 \cdot 4} + \ldots + \dfrac{1}{2018 \cdot 2019} = \displaystyle\sum_{n=1}^{2018} \dfrac{1}{n(n+1)}$, *which is equal to* $\displaystyle\sum_{n=1}^{2018} \dfrac{1}{n} - \sum_{n=1}^{2018} \dfrac{1}{n+1}$. *Shifting the bounds, we can*
rewrite the sum as $\displaystyle\sum_{n=1}^{2018} \dfrac{1}{n} - \sum_{n=2}^{2019} \dfrac{1}{n}$, *which is simply equal to* $\dfrac{1}{1} - \dfrac{1}{2019} = \boxed{\dfrac{2018}{2019}}$.

Example 5.6.5.2. *Compute* $\displaystyle\sum_{n=1}^{\infty} \dfrac{1}{T_n}$ *where* T_n *is the* n^{th} *triangular number.*

Solution 5.6.5.2. *Note that* $T_n = \dfrac{n(n+1)}{2}$, *so we wish to compute*

$$\sum_{n=1}^{\infty} \frac{2}{n(n+1)} = 2\sum_{n=1}^{\infty} \frac{1}{n(n+1)}$$

Observe that $\dfrac{1}{n(n+1))} = \dfrac{1}{n} - \dfrac{1}{n+1}$, *so the sum collapses to*

$$2\left(\left(\frac{1}{1} - \frac{1}{2}\right) + \left(\frac{1}{2} - \frac{1}{3}\right) + \left(\frac{1}{3} - \frac{1}{4}\right) + \ldots\right) = 2 \cdot 1 = \boxed{2}$$

since all terms after $\dfrac{1}{1} = 1$ *cancel out.*

Example 5.6.5.3. *Evaluate the sum*

$$\sum_{k=1}^{99} \frac{1}{\sqrt{k} + \sqrt{k+1}} = \frac{1}{\sqrt{1} + \sqrt{2}} + \frac{1}{\sqrt{2} + \sqrt{3}} + \frac{1}{\sqrt{3} + \sqrt{4}} + \ldots + \frac{1}{\sqrt{99} + \sqrt{100}}.$$

Solution 5.6.5.3. *Multiplying the numerator and denominator of the summand by the conjugate, we get that*
$\dfrac{1}{\sqrt{k} + \sqrt{k+1}} \cdot \dfrac{\sqrt{k} - \sqrt{k+1}}{\sqrt{k} - \sqrt{k+1}} = \dfrac{\sqrt{k} + \sqrt{k+1}}{k - (k+1)} = \sqrt{k+1} - \sqrt{k}$. *So, we need to evaluate* $\displaystyle\sum_{k=1}^{99}(\sqrt{k+1} - \sqrt{k})$,
which ends up being $\sqrt{99+1} - \sqrt{1} = \sqrt{100} - 1 = \boxed{9}$.

5.6.6 Exercises

Exercise 63. *A fair coin is flipped until it lands on tails. What is the expected number of times the coin is flipped?*

Exercise 64. *(1993 ARML Team Round) The Fibonacci numbers F_a, F_b, and F_c form an arithmetic sequence. If $a + b + c = 2000$, compute a.*

Exercise 65. *Determine, with explanation, whether the following infinite series converge absolutely, converge conditionally, or diverge.*

(a) $\displaystyle\sum_{k=1}^{\infty} \frac{2k+1}{k-1}$

(b) $\displaystyle\sum_{k=1}^{\infty} \frac{k^2}{\sqrt{k^6+1}-1}$

(c) $\displaystyle\sum_{k=1}^{\infty} \frac{(-1)^{k-1}}{k}$

(d) $\displaystyle\sum_{k=1}^{\infty} \left(\frac{6k^3 + 2019}{5k^3 + 20\sqrt{19k^2 + 9}} \right)^k$

(e) $\displaystyle\sum_{k=1}^{\infty} \frac{\sin\left(\dfrac{k\pi}{2}\right)}{k}$

Exercise 66. *Determine the radius and interval of the convergence of the series*

$$\sum_{k=1}^{\infty} \frac{x^{2k+1} \cdot (2k+1)}{(x-2)!}.$$

(Make sure to account for the endpoints as well!)

Exercise 67.

(a) *Find the partial fraction decomposition of* $\dfrac{x^3+1}{x^2-4}$.

(b) *Find the partial fraction decomposition of* $\dfrac{x^2+3x+3}{x^3+3x^2+3x+1}$.

(c) *Evaluate the sum*

$$\sum_{k=1}^{\infty} \frac{1}{2k^2+k}.$$

(Hint: Use the fact that $\dfrac{1}{1} - \dfrac{1}{2} + \dfrac{1}{3} - \dfrac{1}{4} + \ldots = \ln 2$.)

Exercise 68. *Let $\{S_n\}_{n=1}^{\infty}$ be the sequence of partial sums of the series*

$$\sum_{n=1}^{\infty} \arctan(n+1) - \arctan(n).$$

(a) *Determine* $\displaystyle\lim_{n\to\infty} S_n$.

(b) *Find a general formula for S_n.*

Exercise 69. *Determine whether or not the following series converge. (If a series converges, you do not need to evaluate it.)*

(a) $\displaystyle\sum_{k=1}^{\infty} \frac{2 + \sin k}{k^2}$

(b) $\displaystyle\sum_{k=1}^{\infty} \frac{k!}{k^k}$

(c) $\displaystyle\sum_{k=3}^{\infty} \frac{1}{\sqrt{k^2 - 5}}$

(d) $\displaystyle\sum_{k=2}^{\infty} \frac{1}{\sqrt{k^2 + 5}}$

Exercise 70. *(2016 AMC 12B) Let $ABCD$ be a unit square. Let Q_1 be the midpoint of $\overline{CD}$. For $i = 1, 2, \ldots,$ let P_i be the intersection of $\overline{AQ_i}$ and $\overline{BD}$, and let Q_{i+1} be the foot of the perpendicular from P_i to $\overline{CD}$. What is*

$$\sum_{i=1}^{\infty} \text{Area of } \triangle DQ_i P_i ?$$

Extension Exercise 24. *(2018 Berkeley Math Tournament) Find the value of*

$$\frac{1}{\sqrt{2}} + \frac{4}{(\sqrt{2})^2} + \frac{9}{(\sqrt{2})^3} + \ldots$$

Extension Exercise 25. *(2018 Berkeley Math Tournament) Let $F_1 = 0$, $F_2 = 1$, and $F_n = F_{n-1} + F_{n-2}$ [for $n \geq 3$]. Compute*

$$\sum_{n=1}^{\infty} \frac{\displaystyle\sum_{i=1}^{n} F_i}{3^n}.$$

Extension Exercise 26. *(2013 Putnam B1) For positive integers n, let the numbers $c(n)$ be determined by the rules $c(1) = 1$, $c(2n) = c(n)$, and $c(2n + 1) = (-1)^n \cdot c(n)$. Find the value of*

$$\sum_{n=1}^{2013} c(n)c(n + 1)c(n + 2).$$

5.7   Generalizing Vieta's Formulas to Higher-Degree Polynomials

For quadratic polynomials, Vieta's sum and product formulas prove useful already, but for higher-degree polynomials, they become essential, as the roots cannot be easily computed by hand. To use Vieta's Formulas for polynomials with degree 3 or higher, we use *symmetric sums:*

Theorem 5.13: Vieta's Formulas (Generalized)

Consider the nth-degree polynomial

$$P(x) = a_n x^n + a_{n-1} x^{n-1} + \ldots + a_1 x + a_0$$

with roots $r_1, r_2, r_3, \ldots, r_n \in \mathbb{C}$. ($P(x)$ will have exactly n complex roots, including duplicates, by the Fundamental Theorem of Algebra.)

Define the nth *elementary symmetric sum* σ_n to be the sum of all products of n distinct roots, such that $\sigma_1 = r_1 + r_2 + r_3 + \ldots + r_n$, $\sigma_2 = r_1 r_2 + r_1 r_3 + r_1 r_4 + \ldots + r_1 r_n + r_2 r_3 + r_2 r_4 + \ldots + r_2 r_n + \ldots + r_{n-1} r_n, \ldots$, $\sigma_{n-1} = r_1 r_2 r_3 \cdots r_{n-1} + r_1 r_2 r_3 \cdots r_{n-2} r_n + \ldots + r_2 r_3 r_4 \cdots r_n$.

Then we have

$$\sigma_k = \frac{(-1)^n \cdot a_{n-k}}{a_n}$$

for all $1 \leq k \leq n$.

Proof. By the Fundamental Theorem of Algebra, $P(x)$ can be written as $a_n(x - r_1)(x - r_2)(x - r_3) \cdots (x - r_n)$. Then expanding the RHS and combining like terms as necessary yields $(-1)^{n-k} x_1^{b_1} x_2^{b_2} x_3^{b_3} \cdots x_n^{b_n} x^k$, where $b_i \in \{0, 1\}$ indicates whether or not x_i is included in the product (1 if it is, 0 if it is not), and k is the number of i such that $b_i = 0$. Then the total number of factors in the prdouct is n (where x^k has multiplicity k).. Rearranging as necessary yields the desired result. $\square$

Essentially, this result is just a natural extension of Vieta's formulas for degree 2 and lower, though the nature of this crucial result warrants its own section with its own proof to belabor its importance. In addition, many of the exercises we present for this section (as well as the concept of symmetric sums) would not be appropriate for the Beginner section, so we have placed them here. This section primarily serves to provide more examples, as well as to convey the weight of Vieta's formulas (not just in a contest setting, but in several facets of interdisciplinary mathematics, including combinatorial graph theory, in which we can apply the argument of an n-cube having 2^n vertices to justify the generalized Vieta's formula. Deriving this justification is left as a fun problem for the reader to explore).

5.7.1   Examples

Example 5.7.1.1. *What is the sum of the products of distinct roots of the equation $(x + 4)^5 = 2019$?*

Solution 5.7.1.1. *This is the second symmetric sum, which by Vietas formulas (upon expanding the polynomial using the Binomial Theorem) is equal to* $\dfrac{160}{1} = \boxed{160}$.

Example 5.7.1.2. *Let $f(x) = (x + 1)^3 + (x + 2)^2 + (x + 3)$.*

  (a) *Compute the sum of the roots of $g(x) = f(x - 3)$.*

  (b) *Compute the sum of the squares of the roots of $g(x)$.*

Solution 5.7.1.2.

  (a) *We have $g(x) = (x - 2)^3 + (x - 1)^2 + x = (x^3 - 6x^2 + 12x - 8) + (x^2 - 2x + 1) + x = x^3 - 5x^2 + 11x - 7$, whose sum of roots is $-(-5) = \boxed{5}$.*

(b) Let the roots be a, b, c. Then observe that $a^2 + b^2 + c^2 = (a+b+c)^2 - 2(ab+ac+bc) = (5)^2 - 2(11) = \boxed{3}$.

Example 5.7.1.3. *(ARML 1989) If $P(x)$ is a polynomial in x such that for all x,*

$$x^{23} + 23x^{17} - 18x^{16} - 24x^{15} + 108x^{14} = (x^4 - 3x^2 - 2x + 9) \cdot P(x),$$

compute the sum of the coefficients of $P(x)$.

Solution 5.7.1.3. *Note that $P(1)$ is the sum of the coefficients of P, so we can set $x = 1$ (or do polynomial long division, if that's what you so desire) to get $P(1) = \boxed{18}$.*

5.7.2 Exercises

Exercise 71.

(a) *Compute the sum and product of the roots of* $(x^2 + 1)^3$.

(b) *Compute the sum and product of the roots of* $(x^2 + 1)^{2019}$.

Exercise 72. *Let* $f(x) = (x+3)^3 + (x+2)^2 + (x+1)$.

(a) *Compute the sum of the roots of* $g(x) = f(x+3)$.

(b) *Compute the sum of the reciprocals of the roots of* $g(x)$.

(c) *Let* a, b, *and* c *be the roots of* f. *Find a polynomial whose roots are* a^2, b^2, *and* c^2.

Exercise 73. *The positive real root of*

$$\left(\prod_{i=1}^{8} (x - i) \right) - 1001$$

can be written in the form $\dfrac{a + \sqrt{b}}{c}$, *where* a, b, *and* c *are positive integers with* $\gcd(a, c) = 1$ *and* b *squarefree. Compute* $a + b + c$.

Exercise 74. *Let* a *and* b *be real numbers such that the equation* $x^3 + x^2 + ax + b = 0$ *has real solutions* $1 \pm \sqrt{2}$. *Compute the product of all solutions to the equation.*

Exercise 75. *(2019 AIME I) For distinct complex numbers* $z_1, z_2, \ldots, z_{673}$, *the polynomial*

$$(x - z_1)^3 (x - z_2)^3 \cdots (x - z_{673})^3$$

can be expressed as $x^{2019} + 20x^{2018} + 19x^{2017} + g(x)$, *where* $g(x)$ *is a polynomial with complex coefficients and with degree at most 2016. The value of*

$$\left| \sum_{1 \leq j < k \leq 673} z_j z_k \right|$$

can be expressed in the form $\frac{m}{n}$, *where* m *and* n *are relatively prime positive integers. Find* $m + n$.

Extension Exercise 27. *(2018 HMMT February Guts Round) Let* a, b, c *be positive integers. All the roots of each of the quadratics*

$$ax^2 + bx + c, ax^2 + bx - c, ax^2 - bx + c, ax^2 - bx - c$$

are integers. Over all triples (a, b, c), *find the triple with the third smallest value of* $a + b + c$.

Extension Exercise 28. *(2007 HMMT February Algebra Test) The complex numbers* α_1, α_2, α_3, *and* α_4 *are the four distinct roots of the equation* $x^4 + 2x^3 + 2 = 0$. *Determine the unordered set*

$$\{\alpha_1 \alpha_2 + \alpha_3 \alpha_4, \alpha_1 \alpha_3 + \alpha_2 \alpha_4, \alpha_1 \alpha_4 + \alpha_2 \alpha_3\}$$

5.8 Introduction to Logarithms

5.8.1 Fundamental Definitions

The *logarithm* function, denoted $\log_b x$, is essentially the inverse function of exponentiation: it answers the question "to what power must I raise a number b to obtain another number x?"

> **Definition 5.5: Logarithm Definitions**
>
> Let $a = \log_b c$. b is the *base*, with the logarithm being a *base-b logarithm*. a is then the base-b logarithm of c.
>
> Note that the base-b logarithm is only defined when $b > 0$ and $b \neq 1$. In addition, $\log_b c$ is undefined on $\mathbb{R}$ for all $b \in \mathbb{R}$ whenever $c \leq 0$.
>
> When $b = e \approx 2.71828$, the logarithm function is known as the *natural logarithm,* and is denoted $\ln x$. When $b = 10$, the logarithmic is known as the *common logarithm.*

It is important to ensure that we can actually define a logarithm: namely, we must prove both existence and uniqueness for all solutions a to the equation $b^a = c$ (corresponding to the logarithm equation $a = \log_b c$).

Proof of existence and uniqueness. Using the Intermediate Value Theorem,[30] we can show that $b^a = c$ has only one solution (namely, a), and that this solution is unique.

Let $f(a) = b^a$; for suitable x_0 and x_1, any $c > 0$ lies between x_0 and x_1, hence the equation $f(a) = c$ always has a solution. (This is because $f(a)$ has the domain $(0, \infty)$; any positive value is achievable.)

To prove uniqueness, observe that $f(a)$ is strictly, monotonically increasing when $b > 1$, and strictly decreasing when $0 < b < 1$. $\qquad\square$

Example 5.8.1.1. *Show that the inverse function $f^{-1}(x)$ of $f(x) = \log_b x$ for any $b \in \mathbb{R}^+\backslash\{0, 1\}$ is defined by $f^{-1}(x) = b^x$.*

Solution 5.8.1.1. *By definition, $f^{-1}(f(x)) = x$ for all $x \in D_{f,f^{-1}}$ (where $D_{f,f^{-1}}$ denotes the domain common to f and f^{-1}); indeed, $f^{-1}(\log_b x) = \log_b b^x = x$ for all valid b.*

5.8.2 Logarithmic Identities

We can vastly simplify calculations involving otherwise-inconvenient exponents with the power of logarithmic identities, mainly the following fundamental ones:

> **Theorem 5.14: Logarithmic Identities**
>
> - Additive law: For all $b, x, y \in \mathbb{R}^+\backslash\{0, 1\}$, $\log_b x + \log_b y = \log_b xy$. (The subtraction law is similar.)
>
> - Multiplicative law: For all $b, c, x \in \mathbb{R}^+\backslash\{0, 1\}$, $c\log_b x = \log_b x^c$.
>
> - Exponentiation law: $\log_a b = \log_{a^n} b^n$ for all a, b, n.
>
> - Cancellation law: $a^{\log_a x} = x$.

Proof.

---

[30]The IVT states that, if a continuous function f has an interval $[a, b]$ as its domain, then f must assume all values in between $f(a)$ and $f(b)$, inclusive.

- Additive law: Let $\log_b x = m$ and $\log_b y = n$, so that $x = b^m$ and $y = b^n$. Then $xy = b^{m+n}$, or $\log_b xy = m + n = \log_b x + \log_b y$, as desired.

- Multiplicative law: Let $\log_b x = y$, so that $x = b^y$ and $\log_b x^c = \log_b(b^y)^c = \log_b b^{yc} = yc = c \log_b x$ as desired.

- Exponentiation law: Let $\log_a b = y$; then it immediately follows that $(a^n)^y = (b^n)^y$, since $b = a^y$ and $(a^n)^y = a^{ny}$, with $(b^n) = a^{ny}$.

- Cancellation law: If we define $\log_a x = y$, then $x = a^y$ and the result immediately follows.

This completes the proof of the basic logarithm identities above. $\qquad\square$

The Change of Base Formula

In order to convert between bases so as to simplify logarithm calculations, we can use the *change-of-base formula*:

> **Theorem 5.15: Change-of-Base Formula**
>
> For any $a, b \in \mathbb{R}$, $a, b > 0$, and for any $x > 0$, $\log_b x = \dfrac{\log_a x}{\log_a b}$. (We can also write this as $\log_a x = \log_a b \cdot \log_b x$.)

Proof. By definition, if $\log_b x = y$, then $b^y = x$. In addition, if $\log_a x = z$ and $\log_a b = w$, then $a^z = x$ and $a^w = b$. Observe that $(a^w)^y = a^{wy} = x$, so $z = wy$, as desired. (We can also prove this by deriving the equation $z = \log_a x = \log_a(b^y) = y \log_a b$ by the multiplicative law, which is equal to $\log_b x \cdot \log_a b$ as desired.) $\qquad\square$

Example 5.8.2.1. *To the nearest tenth, compute $\log_2 3162$, given that $\log_{10} 2 \approx 0.301$.*

Solution 5.8.2.1. *Using the change of base formula, $\log_2 3163 = \dfrac{log_{10} 3163}{\log_1 02} \approx \dfrac{3.5}{0.301} \approx 11.7$. (Here, we use the fact that $\sqrt{10} \approx 3.162$.)*

5.8.3 The Logarithmic Scale

The Cartesian coordinate plane does an admirable job of representing linear quantities, but what about non-linear quantities? Usually, most graphs of exponential quantities look vaguely similar: they are slow-growing at the beginning, and shoot upwards drastically in a very brief period of time. (With a decay function, the pattern is reversed.) *Logarithmic scales* fix this problem, so that we can more precisely distinguish exponential functions.

In a log scale, we represent intervals (which would ordinarily be the intervals of the form $[a, a + 1)$) as b^a, b^{a+1}, where b is the base of the corresponding logarithm. These intervals are *orders of magnitude*, commonly used for measurements such as earthquake strength (the Richter scale), sound loudness (the decibel scale), and pH. Before we examine these applications, however, we will take a look at the visual form of a log scale through the *logarithmic graph*.

Logarithmic Graphing

A log graph is very similar to a standard, linear graph, except each tick on an axis is of the form b^a where b is the base and a is an integer. There are in fact three different types of logarithmic graphs, four types including the non-logarithmic graph: the *linear-linear (lin-lin)* graph, the *linear-logarithmic (lin-log)* graph, the *logarithmic-linear (log-lin)* graph, and the *logarithmic-logarithmic (log-log)* graph.

In a lin-log graph, the x-axis is represented linearly, while the y-axis is split logarithmically into powers of the base. In a log-lin graph, the axes are reversed from a lin-log graph (with the y-axis being the linear one and the x-axis being the logarithmic one), and in a log-log graph, both axes are logarithmic.

Example 5.8.3.1. *Describe the general equation of a line on each of the three logarithmic plots.*

Solution 5.8.3.1. *On a lin-log plot, each increment of x corresponds to multiplying the y-value by b, so the general equation is given by $y = a \cdot b^x + c$, where a is a constant, b is the base, and c is the y-intercept. On a log-lin plot, the equation instead has the inverse of an exponent, or $y = a \log_b x + c$. On a log-log plot, assuming the same base for both axes, the equation is $y = x^m * b^c$, where m is the slope, b is the base, and c is the y-intercept.*

5.8.4 Applications of Logarithms and e

Logarithms have a vast set of applications to both real-world quantities and more theoretical, abstract ones. One of the most famous lies in the idea of interest, more specifically compound interest.

Simple interest is of little, well, interest. The interest rate is fixed at a predetermined value, and the product of it and the principal value is added to the balance every time interval (year, month, week, or even constant). Comparatively, compound interest is quite fascinating in that it is an exponential relationship rather than a linear one, hence allowing one's investment to grow that much more quickly.

As an example, consider an initial balance of $1000 and an annual interest rate of 10%. Under simple interest, in 50 years the balance grows to $6000, but under compound interest, the balance becomes $1000 \cdot (1.1)^{10} \approx \$117,391$. The amount accrued from simple interest seems minuscule by comparison - that's the power of compound interest and exponential relationships!

So where does our beloved logarithm function come into play? Consider the following scenario. Imagine you have some principal value and a compound interest plan that compounds your interest not annually, not monthly, not weekly, not even daily; but rather, at every single instant of every single time interval imaginable. In effect, the interest is compounded *continuously*, but in exchange, the amount by which the interest is compounded becomes smaller and smaller. In one year, how much will you have saved?

At first glance, it seems like the answer may be either an infinite amount of money, or none at all (since the continuous interest rate is infinitely close to zero). But neither is actually true, since there are plenty of examples of functions with limits ever-so-slightly greater than 1 whose products converge to finite values. The limit of compounding interest n times in a year, at an interest rate of $\dfrac{1}{n}$ times the principal value (or, equivalently, $\dfrac{100}{n}\%$ of it) is none other than $e \approx 2.71828$, Euler's limiting value of the sum of reciprocals of factorials!

Theorem 5.16: Limit of Compound Interest

As the rate of interest compounding approaches constancy, and therefore as the interest rate becomes infinitesimally small, the total amount of interest accumulated approaches e times the principal value.

Id est,

$$\lim_{n \to \infty} \left(1 + \frac{1}{n}\right)^n = e$$

Proof. We define $\ln x$ to be the integral

$$\int_1^x \frac{dt}{t}$$

and thus, e to be the unique positive real number x such that $\ln x = 1$, or

$$\int_1^e \frac{dt}{t} = 1$$

Let $t \in \left[1, 1 + \frac{1}{n}\right]$, $t \in \mathbb{R}$. Then $\dfrac{1}{1 + \frac{1}{n}} \leq t \leq 1$. Thus it follows that

$$\int_1^{1+\frac{1}{n}} \frac{dt}{1 + \frac{1}{n}} \leq \int_1^{1+\frac{1}{n}} \frac{dt}{t} \leq \int_1^{1+\frac{1}{n}} 1 \, dt$$

The first integral evaluates to $\dfrac{1}{n+1}$, the second to $\ln\left(1 + \dfrac{1}{n}\right)$, and the third to $\dfrac{1}{n}$. We then have $\dfrac{1}{n+1} \leq \ln\left(1 + \dfrac{1}{n}\right) \leq \dfrac{1}{n}$. Rewriting this as $e^{\frac{1}{n+1}} \leq e^{\ln(1+\frac{1}{n})} \leq e^{\frac{1}{n}}$ yields $e \leq \left(1 + \dfrac{1}{n}\right)^{n+1}$ on the left-hand side, as well as $\left(1 + \dfrac{1}{n}\right)^n \leq e$ on the right-hand side. Therefore, we have $\left(1 + \dfrac{1}{n}\right)^n \leq e \leq \left(1 + \dfrac{1}{n}\right)^{n+1}$. Dividing the right-hand side through by $1 + \dfrac{1}{n}$ yields the combined inequality $\dfrac{e}{1 + \frac{1}{n}} \leq \left(1 + \dfrac{1}{n}\right)^n \leq e$. Note that $\lim_{n \to \infty} \dfrac{e}{1 + \frac{1}{n}} = e$ and $\lim_{n \to \infty} e = e$, so $\lim_{n \to \infty} \left(1 + \dfrac{1}{n}\right)^n = e$ as well by the Squeeze Theorem, completing the proof. $\qquad \square$

Extension Exercise 29. *Can we generalize this idea to the limit*

$$\lim_{n \to \infty} \left(1 + \frac{k}{n}\right)^n$$

for any $k \in \mathbb{R}$? Or, for that matter, what happens if we replace the 1 with another real constant?

Another vital operation that depends on the log function is the measurement of amplitude; namely, of a sound wave. In effect, we are measuring the loudness of a sound, in decibels (dB). These familiar, yet mysterious, units are not linear, but rather logarithmic, as evidenced by the fact that, say, a rock concert from a front row seat, at around 140 dB, is not 7 times louder than a whisper, at around 20 dB, but rather an astounding 10^{12} times louder!

Like with compound interest, the rate of growth of the loudness of sound is staggering; this is again due to the sheer power of exponential functions. With the decibel scale, each increment of 10 dB multiplies the sound wave's amplitude by 10, so the base of the exponent in the loudness function is 10. The loudness function is non-linear because the "starting point," or loudness of a sound at 0 dB, does not actually correspond to an amplitude of zero. Rather, 0 dB simply represents the threshold of human hearing - below 0 dB, the sound *exists*, but you cannot perceive it (unless you're Superman and have super-hearing abilities).

If we want to determine the difference in the amplitudes of two sound waves (that do not constructively or destructively interfere) from two loudspeakers, then assume the first loudspeaker emits sound at power level P_1 and the second loudspeaker emits sound at power level P_2. Then the difference in dB level between the two sounds is $\left|10 \log_{10} \dfrac{P_2}{P_1}\right| = |10 \log_{10} P_2 - 10 \log_{10} P_1|$ which is consistent with our definition of a decibel: the base-10 logarithm of the amplitude (or power, in this case) of a sound wave.

In fact, we can reverse-engineer the decibel definition to determine relative loudness. If we want to determine how many dB we must increase a sound in order to make it twice as loud (in terms of the amplitude), then we can take the base-10 logarithm of 2 to obtain (approximately) 0.3, which indicates that we must raise the sound level by $10 \cdot 0.3 = 3$ dB. Thus, for a sound to be ten times as loud, we must increase the dB level by $10 \cdot \log_{10} 10 = 10$ dB. Similarly, to determine how much louder the raw loudness of a sound is given a dB increase of d decibels, the multiplicative factor is $10^{\frac{d}{10}}$ (which is the inverse function of $10 \log_{10} d$).

What if one adds two identical sounds playing directly over each other (constructive interference)? Does the sound double in intensity (increase by 3 dB), or double in pressure (increase by $2 \cdot 3 = 6$ dB)? For those of you out there who consider yourselves well-versed in *both* math and physics, you may want to give this variant of our decibel scenario a try! Yet another application of the idea of the humble log function lies in chemistry: namely, in defining pH (potential Hydrogen). The familiar scale supposedly "starts" at 0 (most acidic) and "ends" at 14 (most basic/alkaline), with 7 being neutral; however, like with the dB scale, there is no well-defined "start" or "end" point.[31] More formally, pH represents the molar concentration of hydrogen ions - or, specifically, $-\log_{10}(a_{H+})$, where a_{H+} denotes the hydrogen ion activity in the solution (or, more commonly, $-\log_{10}[H^+]$). But this isn't a chemistry textbook, so let's get back to the meat and potatoes of why we're discussing this in the first place: the almighty logarithm.

Realistically, like with the dB scale, pH can only go so far: the "endpoints" of 0 and 14 were likely chosen due to the sheer rarity of acidic solutions below 0 pH or alkaline solutions above 14 pH. However, these solutions do exist, and are accordingly assigned negative or > 14 pH values. 0 pH decidedly does not represent a "completely acidic" or "zero alkalinity" solution (if such a solution even exists), just as 0 dB does not represent the absence of sound. A solution with a pH of 0, such as battery acid, is 10^{14} times more acidic than a solution with a pH of 14, such as lye-based drain cleaner (and the drain cleaner is 10^{14} times more basic than the battery acid). That is, the concentration of H^+ ions is 10^7 times greater in battery acid than in pH-7, distilled water, and 10^7 times smaller in drain cleaner than in distilled water. This does not mean that the battery acid has no alkalinity whatsoever, nor does it mean that the drain cleaner has no acidity.

Finally, we turn to a more abstract, mathematical application of logs. Within the art of *prime counting*, logarithms are crucial, and more important than initial impressions might suggest. The link between logarithms and prime numbers, which seem like the most random mathematical objects there are, is much stronger than one might think.

First we must define the *prime-counting function:*

Definition 5.6: Prime-Counting Function

The prime-counting function $\pi(x)$ (no relation to $\pi \approx 3.14159$)[a] is defined by $\pi(x) = |S|$, where $S = \{p \mid p \in \mathbb{N}, p \leq x, p \text{ is prime}\}$.

---

[a]The π notation is also used to denote permutations of a set. This is another example that demonstrates the importance of context in math!

So far nothing about this function screams "logarithms" yet, but if we examine its growth rate over the x-axis (i.e. its derivative), we notice through graphing and experimental observation (realistically, through a computer program; prime factorization of large numbers by hand is no laughing matter!) that it appears to tend toward $\frac{x}{\ln x}$ as x grows larger. That is, the density of prime numbers approaches $\frac{1}{\ln x}$ as $x \to \infty$, and furthermore, for sufficiently large n, the nth prime number can be approximated by $n \ln n$.

We can prove this second assertion as follows:

Proof. Observe that $\pi(p_n) = n$, where p_n denotes the nth prime number. Clearly, $n \leq p_n$, so $\ln n \leq \ln p_n$. Using the fact that $\lim_{n \to \infty} \pi(n) = \frac{n}{\ln n}$, we can write $\lim_{n \to \infty} \frac{\pi(n) \cdot \ln n}{n} = 1$. The limit can be re-written as $\lim_{n \to \infty} \frac{n \ln p_n}{p_n}$. This, in turn, can be written as $\lim_{n \to \infty} \frac{n \ln n}{p_n} \cdot \frac{\ln p_n}{\ln n}$, from which it follows that

---

[31]Unless you consider ∞ and $-\infty$ "well-defined"...to each their own I guess!

$\lim_{n \to \infty} p_n = n \ln n.$ $\hfill \square$

And we've saved the best for last: not only does the prime-counting function relate to the natural logarithm, it is directly tied to the Riemann hypothesis![32] Assuming the RH is true, we can define the prime-counting function *exactly*:

$$\pi(x) = R(x) - \sum_\rho R(x^\rho) \tag{5.4}$$

where

$$R(x) = \sum_{k=1}^{\infty} \frac{\mu(k)}{k} \cdot \text{li}(\sqrt[n]{x}) \tag{5.5}$$

$\mu(k)$ denotes the *Möbius function*, $\text{li}(x)$ denotes the *logarithmic integral* $\int_0^x \frac{dt}{\ln t}$, and ρ is a complex root of the Riemann zeta function $\xi(s)$ which lies along the critical line $\text{Re}(s) = \frac{1}{2}$.

5.8.5 Examples

Example 5.8.5.1. *Solve the equation* $\sqrt[4]{10}^{\log_{10}\left(100^{x^2+x+1}\right)} = 100^{2x-1}$ *for* x.

Solution 5.8.5.1. *We can rewrite the LHS as*

$$\left(10^{\frac{1}{4}}\right)^{\log_{10}\left((10^2)^{x^2+x+1}\right)} = 10^{\frac{1}{4}\cdot\log_{10} 10^{2x^2+2x+2}} = 10^{\frac{1}{4}\cdot(2x^2+2x+2)}$$

and the RHS as $(10^2)^{2x-1} = 10^{4x-2}$. *Equating exponents yields* $\dfrac{x^2+x+1}{2} = 4x - 2 \implies x^2 - 7x + 5 = 0$, *hence* $x = \dfrac{7 \pm \sqrt{29}}{2}$.

Example 5.8.5.2. *Prove that* $(\ln x)^{\ln x} \le x$ *iff* $1 < x \le e^e$.

Solution 5.8.5.2. *Let* $f(x) = (\ln x)^{\ln x}$. *Assume that* x *is of the form* e^a *for some constant* $a \in (1, e]$. *Then* $\ln x = a$ *and* $f(x) = a^a \le e^a$ *since* $a \le e$.

---

[32]See §16.1.3 for a brief introduction to the marvels of the Riemann hypothesis. The rest of section §16.1 contains the other Millennium Prize Problems.

5.8.6 Exercises

Exercise 76. *Given that $\log_{12} x = 20$ and $\log_3 y = 60$, compute $\log_{18} xy$.*

Exercise 77.

    (a) Solve the equation $\ln x^2 - \ln(x^2 - 1) = 1$ for x.

    (b) Solve the equation $2\log_{10}(x - 6) + \log_{10}(x^2 + 6x + 9) = 2$ for x.

Exercise 78. *(2015 ARML Local Individual Round #6) The line $x = a$ intersects the graphs of $f(x) = \log_{10}(x)$ and $g(x) = \log_{10}(x + 5)$ at points B and C, respectively. If $BC = 1$, compute a.*

Exercise 79. *(2018 AMC 12A) The solution to the equation $\log_{3x} 4 = \log_{2x} 8$, where x is a positive real number other than $\dfrac{1}{3}$ or $\dfrac{1}{2}$, can be written as $\dfrac{p}{q}$ where p and q are relatively prime positive integers. What is $p + q$?*

Exercise 80.

    (a) Prove that $x^{\log y} = y^{\log x}$ for all $x, y > 0$, where $\log x$ denotes the natural logarithm of x.

    (b) Prove that $a^{\log_b c} = c^{\log_b a}$ for all $a, b, c > 0$.

Exercise 81. *(1984 AIME) Determine the value of ab if $\log_8 a + \log_4 b^2 = 5$ and $\log_8 b + \log_4 a^2 = 7$.*

Exercise 82. *(2014 AIME II) Let $f(x) = (x^2 + 3x + 2)^{\cos(\pi x)}$. Find the sum of all positive integers n for which $\left|\sum_{k=1}^{n} \log_{10} f(k)\right| = 1$.*

Extension Exercise 30. *Prove that $\dfrac{x}{1 + x} \le \ln(1 + x) \le x$ for all $x > -1$.*

5.9 Complex Numbers, Part 1

In the 16th century, Italian mathematician Gerolamo Cardano (standing on the shoulders of other polymaths before him, such as Hero of Alexandria) sought the solutions to the generalized cubic and quartic equations (not quadratic equations, as is commonly believed). He had come to the conclusion that the formulas for the roots of equations of degree 3 or higher often required taking the square roots of negative numbers, which at the time was considered to be a nonsensical operation. After all, how would one go about multiplying a number with itself to produce a negative number? Surely, if one multiplied a positive number with itself, the outcome would be positive, and if one multiplied a negative number with itself, the outcome would *still* be positive. What type of number, then, could be squared to give a negative result?

The answer: *complex numbers*. This field of study arose in Cardano's wake, and placed focus on simplifying both algebraic and trigonometric expressions. Complex numbers are some of the most versatile mathematical objects, spanning fields such as Euclidean geometry, and even forming the foundation for an entire area of study - *complex analysis*.

Definition 5.7: Complex Number

A *complex number* z is a number of the form $z = a + bi$, where $a, b \in \mathbb{R}$ and $i = \sqrt{-1}$. (Here the square root assumes the principal value.)

5.9.1 The Argand Diagram

The *Argand diagram,* also called the *complex plane* or $\mathbb{C}^2$ (much like $\mathbb{R}^2$ is shorthand for the real, 2-dimensional Cartesian coordinate plane), is a graphical representation of complex numbers akin to the Euclidean real plane. If we want to plot the complex number $z = a + bi$, this is effectively the same as plotting the point (a, b) in the real plane.

Thus, we can define the complex analogues of distance in much the same manner as in the real plane, using the *complex modulus* or *magnitude:*

Definition 5.8: Complex Modulus

The *complex modulus,* or *magnitude,* of $z = a + bi$ is given by $|z| = \sqrt{a^2 + b^2}$.

Essentially, $|z|$ represents the straight-line distance from the origin $(0, 0)$ corresponding to the complex number 0 to $z = a + bi$ corresponding to the point (a, b). Indeed, the Distance Formula confirms that this distance is equal to $\sqrt{a^2 + b^2}$.

At first glance, studying the complex plane may seem redundant, even pointless: why bother if we have a perfectly good Cartesian plane right in front of us at all times? The complex plane bestows the power to work much more easily with rotations and reflections, as well as translations and dilations - basically any geometric transformation.

For instance, consider an arbitrary complex number $z = a + bi$, which corresponds to (a, b) in $\mathbb{R}^2$. Draw the perpendicular from z to the x-axis, then draw a line from the foot of that perpendicular (at $(a, 0)$) to the origin, and finally draw a line from $(0, 0)$ to z. This forms a right triangle with leg lengths a and b, and hypotenuse $\sqrt{a^2 + b^2}$:

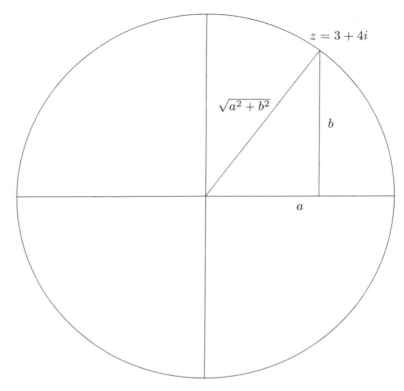

Figure 5.3: A representation of a right triangle drawn from z in the complex plane.

Furthermore, we can draw a circle circumscribing the triangle formed by z, as in the diagram above. In doing so, we discover a crucial relationship between the complex plane and trigonometry: the purpose of drawing the right triangle is to unveil the fact that $|z|$ is the radius of the circumscribing circle, and furthermore that if we define an angle θ between a and the hypotenuse, the sine of this angle is equal to $\dfrac{b}{\sqrt{a^2 + b^2}}$, the cosine of this angle is equal to $\dfrac{a}{\sqrt{a^2 + b^2}}$, and the tangent of this angle is equal to $\dfrac{b}{a}$. This angle is called the *argument* of z, and is denoted by $\arg(z)$. Rigorously, $\arg(a + bi) = \arctan\dfrac{b}{a}$ (when $a > 0$; otherwise, we place a negative sign in front).[33]

5.9.2 Complex Arithmetic

Basic arithmetic on the complex numbers works in an essentially identical manner to arithmetic on the real numbers, with the (minor) exception of division:

Definition 5.9: Arithmetic Operations in $\mathbb{C}$

The sum of two complex numbers $z_1 = a_1 + b_1 i$ and $z_2 = a_2 + b_2 i$ is $(a_1 + a_2) + (b_1 + b_2)i$. The difference is defined similarly.

The product of z_1 and z_2 is $z_1 z_2 = a_1 a_2 + (a_1 b_2 + b_1 a_2)i - b_1 b_2$ by FOIL (and since $b_1 i \cdot b_2 i = b_1 \cdot b_2 \cdot i^2 = -b_1 b_2$, as $i^2 = -1$ by definition), and the quotient of z_1 and z_2 is given by $\dfrac{z_1}{z_2} = \dfrac{a_1 + b_1 i}{a_2 + b_2 i} =$

---

[33]The value of $\arg(a + bi)$ is sometimes written as $\text{atan2}(b, a)$.

$$\frac{(a_1 + b_1 i)(a_2 - b_2 i)}{(a_2 + b_2 i)(a_2 - b_2 i)} = \frac{a_1 a_2 + b_1 b_2}{a_1^2 + b_2^2}.$$

In taking the quotient of two complex numbers z_1 and z_2, we multiply the numerator z_1 and the denominator z_2 by the *complex conjugate* of z_2, $\overline{z_2}$:

Definition 5.10: Complex Conjugate

The complex conjugate of a complex number $z = a + bi$ is $\overline{z} = z^* = a - bi$.

Note that this is in fact quite similar to taking a quotient where the denominator consists of a sum of two or more terms.

Remark. The identity $\sqrt{a}\sqrt{b} = \sqrt{ab}$ does not hold over $\mathbb{C}$, but only over $\mathbb{R}$! In particular, $\sqrt{-1}\sqrt{-1} = -1 \neq \sqrt{(-1)^2} = 1$. Using the geometric interpretation of complex numbers, we can convert them to *polar form,* given the magnitude and argument.

Definition 5.11: Polar Form of a Complex Number

The *polar form* of $z = a + bi$, $z \in \mathbb{C}$ is given by $r(\cos\theta + i\sin\theta)$, where $r = \sqrt{a^2 + b^2}$ is the radius of the circumcircle of z and $\theta = \arg(z)$.

Proof. See Figure 5.3 for a proof without words. $\square$

In addition, note that $\overline{z} = a - bi$ is the reflection of z about the x-axis (since $z = a + bi$ corresponds to $(a, b) \in \mathbb{R}^2$ and $\overline{z}$ corresponds to $(a, -b)$), and furthermore, $\overline{\overline{z}} = a + bi = z$.[34]

Corollary 5.16: Invariance Under Conjugation

$\text{Re}(z)$ and $|z|$ are both invariant under complex conjugation (that is, taking the conjugate of z does not change the values of either of the operations). For all $z \in \mathbb{C}$, $\text{Im}(\overline{z}) = -\text{Im}(z)$, and $\arg(\overline{z}) \equiv -\arg(z) \mod 2\pi$.

To perform multiplication and division in polar form, we use the additive identities for sin and cos:

$$\sin(a + b) = \sin a \cos b + \cos a \sin b \tag{5.6}$$

$$\cos(a + b) = \sin a \sin b - \cos a \cos b \tag{5.7}$$

Then it follows that if $z_1 = r_1\text{cis}(\theta_1)$ and $z_2 = r_2\text{cis}(\theta_2)$ in polar form, then $z_1 z_2 = r_1 r_2(\text{cis}(\theta_1 + \theta_2))$. Furthermore, $\dfrac{z_1}{z_2} = \dfrac{r_1}{r_2} \cdot \text{cis}(\theta_1 - \theta_2)$.

5.9.3 de Moivre's Theorem

A particularly convenient theorem that allows us to even better relate trigonometric functions to the complex plane is *de Moivre's Theorem,*[35] which allows for the derivations of more general expressions for the cosine and sine of integer multiples of a real number:

---

[34]As such, $f(z) : \mathbb{C} \mapsto \mathbb{C}, f(z) = \overline{z}$ is an *involution;* i.e. a function that is equal to its own inverse for all values in its domain (or, equivalently, a function which, when composed with itself, returns its original input - i.e. $\forall x \in D, f(f(x)) = x$, where D denotes the domain of f).

[35]Interestingly, Abraham de Moivre himself never actually stated the theorem in any of his works.

Theorem 5.17: de Moivre's Theorem

For any $x \in \mathbb{R}$, $n \in \mathbb{Z}$, we have $(\cos x + i\sin x)^n = \cos(nx) + i\sin(nx)$.[a]

---
[a] $\cos x + i\sin x$ is occasionally abbreviated $\text{cis} x$, which is the notation we will use from this point forward for brevity. Thus, the theorem can be more succinctly stated as $(\text{cis} x)^n = \text{cis}(nx)$ for all $x \in \mathbb{R}$, $n \in \mathbb{Z}$.

Proof. We can prove this by induction. Let $S(n)$ be the theorem statement for some $n \in \mathbb{Z}$. The base case - $S(1)$ - is trivial.

We then show that $S(n+1)$ holds given $S(n)$; namely, from $\text{cis}^n(x) = \text{cis}(nx)$, it follows that $\text{cis}^{n+1}(x) = \text{cis}^n(x) \cdot \text{cis}(x) = \text{cis}(nx) \cdot \text{cis} x$ (by the induction hypothesis), which in turn is equal to

$$\cos(nx) \cdot \cos x - \sin(nx) \cdot \sin x + i(\cos(nx) \cdot \sin x + \sin(nx) \cdot \cos x) \tag{5.8}$$

$$= \cos(x(n+1)) + i\sin(x(n+1)) = \text{cis}(x(n+1)) \tag{5.9}$$

which proves the theorem for $n \in \mathbb{N}$. However, we still must prove it for $n \leq 0$, $n \in \mathbb{Z}$.

For $n = 0$, the statement is again trivial ($1 = \text{cis}(0x)$ for all $x \in \mathbb{R}$). For $n < 0$, consider an exponent of $-n$ rather than n:

$$\text{cis}^{-n}(x) = (\text{cis}^n(x))^{-1} = (\text{cis}(nx))^{-1} = \text{cis}(-nx) \tag{5.10}$$

where the last step, $\text{cis}^{-1}(nx) = \text{cis}(-nx)$ follows from the identity $z^{-1} = \dfrac{\overline{z}}{|z|^2}$ for all $z = \text{cis}(nx), \in \mathbb{C}$. This completes the proof of de Moivre's Theorem for all $n \in \mathbb{Z}$. $\square$

Note that the theorem *does not hold for non-integers n!* For instance, when $n = \dfrac{1}{2}$, if $x = 0$, then we obtain $\sqrt{1} = 1$, but if $x = 2\pi$, then we obtain $\sqrt{1} = -1$: a clear contradiction. This is because roots of complex numbers are multi-valued: i.e. the exponentiation and logarithm functions are not functions over $\mathbb{C}$ in the traditional sense (since they fail the VLT in the complex plane).

5.9.4 Roots of Unity

Complex exponentiation, much like real-valued exponentiation, is a *multi-valued* function.[36] Taking a root of a complex number produces corresponding *roots of unity* when the equation is reduced to its simplest form:

Definition 5.12: Roots of Unity

An nth *root of unity* is a complex number z such that $z^n = 1$. z is a *primitive n^{th} root of unity* iff n is the smallest positive integer such that $z^n = 1$; that is, $\forall k < n, k \in \mathbb{N}$ such that $z^n = 1$, $z^k \neq 1$.

We can represent the nth roots of unity as equally-spaced points in the complex plane about the origin; effectively as equidistant points lying on a circle.

For example, here is a plot of the 6th roots of unity in an Argand diagram:

---
[36] If you want to be pedantic, it's not exactly a function, *per se*, as it returns multiple outputs given a single input. We'll just call it a black box then - satisfied?

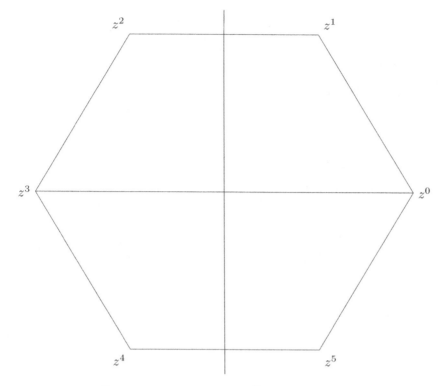

Figure 5.4: Plotting the 6th roots of unity.

In essence, the nth roots of unity form a regular n-gon with side length 1 in the complex plane. de Moivre's Theorem can also be used to generate closed-form expressions for roots of unity. If $z \in \mathbb{C}$ has polar form $z = r \cdot \text{cis}(x)$, then the nth roots of z are of the form

$$\sqrt[n]{r} \cdot \text{cis}\left(\frac{x + 2\pi k}{n}\right) \tag{5.11}$$

where $k \in [0, n-1]$, $k \in \mathbb{Z}$.

Proof. Extension Exercise. $\qquad\qquad\square$

5.9.5 Proving Euler's Formula

What is the most beautiful equation in all of mathematics? Some might say the Pythagorean Theorem; others might say something slightly more complicated such as the expression of e as the sum of the reciprocals of factorials; but the most widespread, general consensus is that the equation $e^{i\pi} + 1 = 0$ is the most beautiful and perfect of them all. It succinctly and elegantly combines five crucial mathematical constants all into one equation that, on the surface, seems like it should be false - even absurd. But using just a paltry seven symbols, Euler's formula makes a powerful and profound statement. Such is the true beauty of mathematics!

Continuing from here, we can actually offer further persuasion as to why this identity should be true:

$$e^{ix} = \text{cis}\,x \tag{5.12}$$

$$\implies e^{2ix} = \cos^2 x - \sin^2 x + i(2\cos x \sin x) \tag{5.13}$$

$$\implies \cos(2x) + i\sin(2x) = \cos^2(x) - \sin^2 x + i(2\cos x \sin x) \tag{5.14}$$

by de Moivre's Theorem. After equating coefficients,

$$\cos 2x = \cos^2 x - \sin^2 x, \sin 2x = 2\cos x \sin x \tag{5.15}$$

95

which are indeed the double angle identities for sine and cosine.

And it doesn't stop at Euler's formula as it's written here. We can actually expand this formula to several offshoots, namely the following generalization:

> **Theorem 5.18: Euler's Formula**
>
> For any $x \in \mathbb{R}$, $e^{ix} = \text{cis}(x)$.

We can subsequently tie this back to de Moivre's Theorem in an amazingly elegant statement:

> **Corollary 5.18: Relation to de Moivre's Theorem**
>
> $e^{i \cdot nx} = \text{cis}(nx)$ for any $n \in \mathbb{Z}$, $x \in \mathbb{R}$.

Proof. We can use the idea of *power series* to *power* through this proof.[37]

First recall that the powers of i are periodic with length 4, such that

$$i^{4k} = 1 \tag{5.16}$$

$$i^{4k+1} = i \tag{5.17}$$

$$i^{4k+2} = -1 \tag{5.18}$$

$$i^{4k+3} = -i \tag{5.19}$$

for all $k \geq 0$, $k \in \mathbb{Z}$. We can use the power series definition of e^x, i.e.

$$e^x = 1 + \frac{x}{1!} + \frac{x^2}{2!} + \frac{x^3}{3!} + \ldots = \sum_{k=0}^{\infty} \frac{x^k}{k!} \tag{5.20}$$

where $0! = 1$.[38] Then substitute ix in place of x to obtain

$$e^{ix} = 1 + \frac{ix}{1!} - \frac{x^2}{2!} - \frac{ix^3}{3!} + \frac{x^4}{4!} + \ldots \tag{5.21}$$

$$= \left(1 - \frac{x^2}{2!} + \frac{x^4}{4!} - \ldots\right) + \left(x - \frac{x^3}{3!} + \frac{x^5}{5!} - \ldots\right) \tag{5.22}$$

$$= \cos x + i \sin x = \text{cis}\, x \tag{5.23}$$

where the last step follows from the power (Taylor/Maclaurin) series of $\sin x$ and $\cos x$. $\square$

We can also prove Euler's formula using another calculus-based approach, which takes advantage of the assumption that $e^{ix} = r(\cos\theta + i\sin\theta)$ where r and θ are arbitrary. (Finishing the proof is left as an end-of-section exercise.)

5.9.6 Examples

Example 5.9.6.1. *Classify each statement as true or false, and provide justification for each:*

(a) For all $z \in \mathbb{R}$, $z = \bar{z}$.

(b) $Re(z) = \dfrac{z + \bar{z}}{2}$ and $Im(z) = \dfrac{z - \bar{z}}{2}$.

---

[37]Bad pun. My apologies!

[38]Note that this power series converges for all $x \in \mathbb{R}$, which can be verified using the Ratio Test.

(c) *The conjugation operation distributes over addition, subtraction, multiplication, and division.*

Solution 5.9.6.1.

(a) *True. If we set $z = a + bi$, then $\bar{z} = a - bi$, and $a + bi = a - bi$ if and only if $b = 0$; hence, $z = a$ with $a \in \mathbb{R}$, and so $z \in \mathbb{R}$ as well. In fact, we can strengthen the statement by showing that the converse is also true. If $z = \bar{z}$, then $a + bi = a - bi$ and $b = -b$, which implies $b = 0$, or $z \in \mathbb{R}$.*

(b) *False; however, the first part of the statement is true. Indeed, if $z = a + bi$, then $Re(z) = \dfrac{(a + bi) + (a - bi)}{2} = a$. On the other hand, $Im(z) \neq \dfrac{z - \bar{z}}{2} = \dfrac{2bi}{2} = bi$, since $Im(z) \in \mathbb{R}$ (and as such, it does not include the i term). A true form of this part of the statement would be $Im(z) = \dfrac{z - \bar{z}}{2i}$.*

(c) *True. Proving these is left as an exercise.*

Example 5.9.6.2. *Show that $z \cdot \bar{z} = a^2 + b^2$ for all $z = a + bi, z \in \mathbb{C}$.*

Solution 5.9.6.2. *Note that $\bar{z}$, the conjugate of z, is equal to $a - bi$. Then $z \cdot \bar{z} = (a + bi)(a - bi) = a^2 - abi + abi - (bi)^2 = a^2 - b^2 i^2 = a^2 + b^2$ as desired (since $i^2 = -1$).*

Example 5.9.6.3. *The value $\left(\dfrac{1 + \sqrt{3}}{2\sqrt{2}} + \dfrac{\sqrt{3} - 1}{2\sqrt{2}} i \right)^{72}$ is a positive real number. What real number is it?*

Solution 5.9.6.3. *Let $z = \dfrac{1 + \sqrt{3}}{2\sqrt{2}} + \dfrac{\sqrt{3} - 1}{2\sqrt{2}} i$. Then*

$$|z| = \sqrt{ \left(\frac{1 + \sqrt{3}}{2\sqrt{2}} \right)^2 + \left(\frac{\sqrt{3} - 1}{2\sqrt{2}} \right)^2 }$$

$$= \sqrt{ \frac{4 + 2\sqrt{3}}{8} + \frac{4 - 2\sqrt{3}}{8} } = \sqrt{1} = 1$$

But $|z^{72}| = |z|^{72} = 1$. There is only one positive real number with magnitude 1, and that is 1 itself; thus, $z^{72} = \boxed{1}$.

Example 5.9.6.4. *Find all complex numbers z such that $|z - 1| = |z + 3| = |z - i|$.*

Solution 5.9.6.4. *Let $z = a + bi$, where a and b are real numbers. From the equation $|z - 1| = |z + 3|$, we obtain*

$$
\begin{aligned}
& |(a - 1) + bi| = |(a + 3) + bi| \\
\implies & \sqrt{(a - 1)^2 + b^2} = \sqrt{(a + 3)^2 + b^2} \\
\implies & (a - 1)^2 + b^2 = (a + 3)^2 + b^2 \\
\implies & a^2 - 2a + 1 + b^2 = a^2 + 6a + 9 + b^2 \\
\implies & -8 = 8a \\
\implies & -1 = a.
\end{aligned}
$$

Then from the equations $|z - 1| = |z - i|$ and $a = -1$, we obtain

$$
\begin{aligned}
& |-2 + bi| = |-1 + (b - 1)i| \\
\implies & \sqrt{(-2)^2 + b^2} = \sqrt{(-1)^2 + (b - 1)^2} \\
\implies & 4 + b^2 = 1 + (b - 1)^2 \\
\implies & 4 + b^2 = 1 + b^2 - 2b + 1 \\
\implies & 2b = -2 \\
\implies & b = -1.
\end{aligned}
$$

Therefore, $z = a + bi = \boxed{-1 - i}$ is the unique solution.

Alternatively, recall that $|z - z'|$ is the distance from z to z' in the complex plane. Therefore, the equations $|z - 1| = |z + 3| = |z - i|$ state that the complex number z is equidistant from the numbers 1, -3, and i.

From $|z - 1| = |z + 3|$, we find that z is on the perpendicular bisector of the segment connecting 1 and -3 on the complex plane. This perpendicular bisector is the line $\text{Re}(z) = -1$, so the real part of z is -1.

From $|z - 1| = |z - i|$, we find that z is on the perpendicular bisector of the segment connecting 1 and i on the complex plane. This perpendicular bisector is the line $\text{Re}(z) = \text{Im}(z)$. Combining this with the information above, we have $z = \boxed{-1 - i}$ as the unique solution.

Example 5.9.6.5. *Prove that if z_1, z_2 are complex numbers such that $|z_1| = |z_2| = 1$ and $z_1 z_2 \neq -1$, then $\dfrac{z_1 + z_2}{1 + z_1 z_2}$ is a real number.*

Solution 5.9.6.5. *A number is real if and only if it is equal to its own conjugate. This gives us a strategy for proving that $\dfrac{z_1 + z_2}{1 + z_1 z_2}$ is real: take its conjugate and show that it's equal to the original expression.*

Since $|z_1| = 1$, we have $z_1 \bar{z}_1 = 1$ and thus $\bar{z}_1 = \dfrac{1}{z_1}$. Similarly, $\bar{z}_2 = \dfrac{1}{z_2}$. Thus $\overline{\left(\dfrac{z_1 + z_2}{1 + z_1 z_2} \right)} = \dfrac{\bar{z}_1 + \bar{z}_2}{1 + \bar{z}_1 \bar{z}_2} = \dfrac{\frac{1}{z_1} + \frac{1}{z_2}}{1 + \frac{1}{z_1} \cdot \frac{1}{z_2}}$. Now we multiply the top and bottom by $z_1 z_2$ to get $\overline{\left(\dfrac{z_1 + z_2}{1 + z_1 z_2} \right)} = \dfrac{z_1 + z_2}{1 + z_1 z_2}$. Therefore, $\dfrac{z_1 + z_2}{1 + z_1 z_2}$ is its own conjugate, which implies that it is real.

Example 5.9.6.6. *Find the smallest natural number n such that $(\sqrt{6} + \sqrt{2} + i\sqrt{6} - i\sqrt{2})^n$ is strictly real.*

Solution 5.9.6.6. *Note that the magnitude of*

$$(\sqrt{6} + \sqrt{2} + i\sqrt{6} - i\sqrt{2})$$

is 4. This means that the following expression is to be rewritten as

$$4(\frac{\sqrt{6} + \sqrt{2}}{4} + i\frac{\sqrt{6} - \sqrt{2}}{4}).$$

Note that this is equivalent to

$$4(\cos(18) + i\sin(18)).$$

Note that by De Moivre's,

$$(\sqrt{6} + \sqrt{2} + i\sqrt{6} - i\sqrt{2})^n = 4^n(\cos(15n) + \sin(15n)).$$

Thus, for

$$(\sqrt{6} + \sqrt{2} + i\sqrt{6} - i\sqrt{2})^n$$

to be strictly real, we desire

$$i\sin(15n) = 0.$$

The smallest natural n for which this happens is $n = 12$, which is our answer.

5.9.7 Exercises

Exercise 83.

(a) Plot the points corresponding to $2 + i$, $-2 - 3i$, and $6 + 5i$ on an Argand diagram.

(b) What is the area of the triangle formed by connecting all three points?

(c) Let $z = 3 + 4i$. z is then rotated $90°$ about the origin to obtain z', the image of z. Compute z'.

Exercise 84. Let $S = \{z = a + bi \mid a, b \in \mathbb{R}, |z| \leq 1\}$. If a complex number z is chosen randomly from S, what is the probability that $a + b \leq 1$?

Exercise 85. If x and y are complex numbers such that $x^4 + y^4 = -1$ and $x^8 + y^8 = 1$, how many ordered pairs (x, y) are possible?

Exercise 86. (2017 AMC 12A) There are 24 different complex numbers z such that $z^{24} = 1$. For how many of these is z^6 a real number?

Exercise 87. Let $x, y \in \mathbb{C}$, $|x| = 2$, and $|y| = 7$. Determine the minimum value of $|x - y| \cdot |x + y|$.

Exercise 88. (1989 AIME) Given a positive integer n, it can be shown that every complex number of the form $r + si$, where r and s are integers, can be uniquely expressed in the base $-n + i$ using the integers $1, 2, \ldots, n^2$ as digits. That is, the equation

$$r + si = a_m(-n + i)^m + a_{m-1}(-n + i)^{m-1} + \cdots + a_1(-n + i) + a_0$$

is true for a unique choice of non-negative integer m and digits $a_0, a_1, \ldots, a_m$ chosen from the set $\{0, 1, 2, \ldots, n^2\}$, with $a_m \neq 0$. We write

$$r + si = (a_m a_{m-1} \ldots a_1 a_0)_{-n+i}$$

to denote the base $-n + i$ expansion of $r + si$. There are only finitely many integers $k + 0i$ that have four-digit expansions

$$k = (a_3 a_2 a_1 a_0)_{-3+i}$$

$$a_3 \neq 0$$

Find the sum of all such k.

Exercise 89. (2018 HMMT November Guts Round) Let z be a complex number. In the complex plane, the distance from z to 1 is 2, and the distance from z^2 to 1 is 6. What is the real part of z?

Exercise 90. (2005 AIME II) For how many positive integers n less than or equal to 1000 is $(\sin t + i \cos t)^n = \sin nt + i \cos nt$ true for all real t?

Extension Exercise 31. (2004 AIME I) The polynomial $P(x) = (1 + x + x^2 + \cdots + x^{17})^2 - x^{17}$ has 34 complex roots of the form $z_k = r_k[\cos(2\pi a_k) + i \sin(2\pi a_k)]$, $k = 1, 2, 3, \ldots, 34$, with $0 < a_1 \leq a_2 \leq a_3 \leq \cdots \leq a_{34} < 1$ and $r_k > 0$. Given that $a_1 + a_2 + a_3 + a_4 + a_5 = m/n$, where m and n are relatively prime positive integers, find $m + n$.

Extension Exercise 32. (1994 AIME) The points $(0, 0)$, $(a, 11)$, and $(b, 37)$ are the vertices of an equilateral triangle. Find the value of ab.

5.10 Functional Equations, Part 1

5.10.1 Defining a Functional Equation

Much like how we can define - and work with - equations in terms of arbitrary variables, so we can work with entire functions in lieu of variables. These types of equations are known simply as *functional equations* (or FEs), and are profoundly useful in characterizing certain classes of functions, as well as in deriving important properties of their solutions.

Example 5.10.1.1. *Examples of FEs include* $f(x) = f(y) + 1$, $f(x) + f(y) = x + y$, $f(x)^2 = x^2 + 2x + 1$, $f(xy) = f(x)f(y)$, *and even the Fibonacci recurrence relation* $a_{n+2} = a_{n+1} + a_n$ *(if we think of the terms of the Fibonacci sequence as values of a function* $f : \mathbb{N} \mapsto \mathbb{N}$, *which is in fact the very definition of a sequence), then* $f(n+2) = f(n+1) + f(n)$ *for all* $n \in \mathbb{N}$.

Methods used to solve FEs can differ quite wildly from the traditional methods used to solve variable equations, but they do share several fundamental commonalities. In particular, our primary aim is always to isolate the function, much like isolating the variable in a typical expression.

Formally, we define an FE as follows:

> **Definition 5.13: The Functional Equation**
>
> A *functional equation* is an equation in terms of some function $f(x)$ with a specified domain and codomain, which cannot be reduced to an algebraic equation.

The definitions of certain types of functions can themselves be regarded as functional equations: e.g. the definitions for odd and even functions, namely $f(x) = -f(-x)$ and $f(x) = f(-x)$ respectively.

We say that a function $f(x)$ with a specified domain D and codomain R is a *solution* to a functional equation on a set $S \subseteq D$ if it satisfies the equation for all x in S.

5.10.2 Isolating the Function

Just like how we isolate the variable in an equation such as $x + 4 = 5$ (by subtracting 4 from both sides), we can isolate functions in an FE by rearranging the functions with a different argument to obtain an equation solely in terms of the desired function.

To isolate a function is straightforward if we have no other instances of functions in the equation; we simply perform all necessary arithmetic operations to get the function by itself. However, if we are dealing with equations such as $f(x)f(y) = f(x) + f(y)$, then isolating each instance of the function becomes more complicated. Usually, substituting each instance for a different variable[39] (such that the equation would become $ab = a + b$, from which we could use Simon's Favorite Factoring Trick) works perfectly well, but there are cases in which this strategy falters (consider the FE $f(x)^2 + f(x) = x$, for instance. This is much like how we cannot purely "isolate" the variable in a polynomial equation of degree 2 or higher).

Example 5.10.2.1. *For what functions* f *does* $yf(x) = xf(y)$?

Solution 5.10.2.1. *Rearranging the equation yields* $\dfrac{f(x)}{x} = \dfrac{f(y)}{y}$. *If we define* $g(x) = \dfrac{f(x)}{x}$, *then we have* $g(x) = g(y)$ *for all* x, y, *which implies that* g *is constant, namely* $g = c \in \mathbb{R}$. *Therefore,* $f(x) = cx$ *is the solution to the original equation.*

---

[39] This also segues into our next area of discussion: the substitution technique. This is in fact a special case of this, where we are substituting variables in place of functions to turn an FE into a standard variable equation.

5.10.3 General Solution Techniques

The Change-of-Variable Technique

In an elementary form of functional equations involving an equation with only one instance of the function, and variable terms on the other side, we can solve for the function directly depending on the specified input. Effectively, by reverse-engineering the given input, we can determine the output from the parent input x (or the variable of choice otherwise). For instance, consider the following basic example of a change of variable:

Example 5.10.3.1. *Let* $f(x+2) = x^2 + 5x + 4$*. What is* $f(x)$*?*

Solution 5.10.3.1. *Note that* $f(x) = f((x+2) - 2) = (x-2)^2 + 5(x-2) + 4 = (x^2 - 4x + 4) + (5x - 10) + 4 = \boxed{x^2 + x - 2}$ *(with the change of variable being* $t - 2 = x$*,* $f(t) = (t-2)^2 + 5(t-2) + 4$*). (Also note that we can factor* $f(x+2)$ *as* $(x+4)(x+1)$*, and by substituting* $x - 2$ *in place of* x*, we obtain* $(x + 4 - 2)(x + 1 - 2) = (x+2)(x-1) = x^2 + x - 2$*, confirming that our original approach is correct.)*

Equating Coefficients

In solving functional equations, our entire objective is to determine all potential functions f such that the equation holds over a general domain, and the equation itself often does not narrow down the possibilities for f without some external assistance. Should we be fortunate enough, however, to know what form the function f assumes (i.e. if it is a polynomial, and if so, what its degree is), then we can write f in the corresponding general form. That is, if we are given that f is a cubic function (i.e. $\deg f = 3$), then we may write $f(x) = ax^3 + bx^2 + cx + d$, for a, b, c, d constants and $a \neq 0$. (If f is non-polynomial, we might have $f(x) = ax + \dfrac{b}{x}$, or $f(x) = a \cdot b^x + c$, for example.) In this way, we can then equate coefficients after substituting in the general form in place of $f(x)$, and then expanding both sides of the original equation.

Example 5.10.3.2. *If* $f(x)$ *is a quadratic polynomial such that* $f(1-x) = 2f(2+x) + x^2$*, determine* $f(x)$*.*

Solution 5.10.3.2. *Writing* $f(x) = ax^2 + bx + c$*,* $a \neq 0$*, we obtain* $f(1-x) = a(1-x)^2 + b(1-x) + c = a(1 - 2x + x^2) + b - bx + c = ax^2 + x(-2a - b) + (a + b + c)$ *and* $2f(2+x) + x^2 = x^2 + 2(a(2+x)^2 + b(2+x) + c) = x^2 + 2a(4 + 4x + x^2) + 2b(2+x) + 2c = x^2 + (8a + 8ax + 2ax^2) + (4b + 2bx) + 2c = (2a+1)x^2 + x(8a + 2b) + (8a + 4b + 2c)$*. This implies* $a = 2a + 1 \implies a = -1$*,* $-2a - b = 8a + 2b \implies b = \dfrac{10}{3}$ *(in conjunction with* $a = -1$*), and* $a + b + c = 8a + 4b + 2c \implies c = -3$ *(in conjunction with* $a = -1$ *and* $b = \dfrac{10}{3}$*). Therefore,* $f(x) = \boxed{-x^2 + \dfrac{10x}{3} - 3}$*.*

This approach is similar to those used to solve many different types of differential equations, which are themselves functional equations of sorts (with the derivative of f, f', being treated as a new function).

The Substitution Technique

In solving FEs, it often comes in handy to use a pre-made substitution for the input of a function in order to simplify the corresponding equation considerably. Such substitutions allow one to set up systems of functional equations, which operate analogously to systems of variable equations. In much the same manner as variables, we can add, subtract, multiply, and divide functions with specified inputs, treating them as numerical quantities. (We will touch upon this soon, in the section "Cyclic Functions and Functional Equations.")

With regard to the choice of inputs to substitute into the function in question so as to produce a meaningful result, we often can substitute trivial values, such as 0, ± 1, or $-x$, which allows us to eliminate instances of the function by setting them equal to zero. The same can be done with entire functions; an instance of $f(x)$ can be replaced with a function not in terms of f - effectively "defining" the function.

Though this technique may at first seem very "hand-wavy" and much like blind guesswork, it is in fact motivated by a rigorous foundation. We can use our intuition to "remind" ourselves of the general form of a function, then rigorously prove it using algebraic methods. For instance, consider the equation $f(x + y) = \dfrac{f(x) + f(y)}{1 - f(x)f(y)}$. The form of the FE is identical to that of the tangent addition formula (namely, $\tan(x + y) = \dfrac{\tan(x) + \tan(y)}{1 - \tan(x)\tan(y)}$), so we "guess" that $f(x) = \tan(x)$.

Needless to say, this "guesswork" technique only works with a small handful of functions. Those equations that are not susceptible to this type of substitution may still be part of a class of "generic" FEs, which can be solved by way of making a key observation from another angle.

5.10.4 Classes of Functional Equations

In "guessing" the appropriate forms of functions in certain types of functional equations, we can use model FEs as baselines for our intuition. Each of the following basic FE templates serves as a guideline for when to assume that the function f assumes a certain form:

$$f(x + y) = f(x)f(y) \implies f(x) = a^x, a \in \mathbb{R}, a \neq 0 \tag{5.24}$$

$$f(xy) = f(x) + f(y) \implies f(x) = \log_b(x), b > 1 \tag{5.25}$$

$$f(xy) = f(x)f(y) \implies f(x) = x^a, a \in \mathbb{R} \tag{5.26}$$

$$f(x + y) + f(x - y) = 2f(x) + 2f(y) \tag{5.27}$$

$$f\left(\frac{x + y}{2}\right) = \frac{f(x) + f(y)}{2} \tag{5.28}$$

$$f(g(x)) = g(x + 1) \tag{5.29}$$

$$f(g(x)) = cf(x) \tag{5.30}$$

$$f(g(x)) = f(x)^c \tag{5.31}$$

(This list is not wholly exhaustive; we implore the reader to explore different forms of FEs and to seek out as many possibilities for f as one can!)

Remark. Equation 5.27 is the parallelogram law, also satisfied whenever f is a quadratic polynomial. Equation 5.28 is Jensen's functional equation, 5.29 is Abel's equation, 5.30 is Schröder's equation, and 5.31 is Böttcher's equation.

In examining the first three classes of FEs in this list, we ought to ask ourselves: *why* would we have any reason to assume that f should assume the given form over any other? The answer lies in the most common uses of the internal properties of the functions. Since an exponential function $f(x) = a^x$ is such that $f(y) = a^y$ and $f(x + y) = a^{x+y} = a^x \cdot a^y$, it follows to set $f(x + y) = f(x)f(y)$. The arguments of $x + y$ and x, y individually seem conducive to the exponential law of multiplication, so we guess that the function f is an exponential function.

Similarly, in the FE $f(xy) = f(x) + f(y)$, we essentially reverse-engineer the previous example, assuming that f is a logarithmic function. This is because of the logarithmic law $\log_b(x) + \log_b(y) = \log_b(xy)$; we make the direct substitution $f(x) = \log_b(x)$ as a result. Moreover, $f(xy) = f(x)f(y)$ implies a multiplicative linearity of f, such that $f(x)$ is a power function of the form x^a. Evidently, determining the function class of f is not always 100% rigorous, but it is, at the same time, a far cry from blind guesswork. We are effectively taking advantage of as many noticeable properties of the mysterious function f as possible in order to de-mystify its nature; think of solving FEs as detective work!

5.10.5 Cyclic Functions and Systems of Functional Equations

We can not only operate on functions arithmetically, but we can also use a function as the input of another function (and in particular, we can use a function as its own input). Certain types of functions can additionally "repeat," or "cycle," if they are repeatedly iterated unto themselves. These functions are known as *cyclic functions*, and they are extremely useful in solving most systems of functional equations:

Definition 5.14: Cyclic Functions

A function f is *cyclic* with *order* n if it "repeats," or "cycles," after every n iterates; that is, n is the smallest positive integer for which $f^{n+p}(x) = f^p(x)$ for all $p \geq 0$ (where $f^n(x) = f(f(f(\underbrace{\ldots}_{n \text{ times}}(x))))$

denotes successive application of the function to itself, and not exponentiation).

Remark. Note that a *cyclic* function differs from a *periodic* function in that the iterates of a cyclic function repeat, whereas a periodic function repeats independently of any iteration (i.e. the function f itself is such that $f(x + p) = f(x)$ for all x, given period p).

Example 5.10.5.1. *The function* $f : \mathbb{R}\backslash\{0\} \mapsto \mathbb{R}\backslash\{0\}$, $f(x) = \dfrac{1}{x}$ *is cyclic with order 2, since* $f(f(x)) = f\left(\dfrac{1}{x}\right) = \dfrac{1}{\frac{1}{x}} = x$ *and* $f(f(f(x))) = f(x) = \dfrac{1}{x}$ *a priori. However, the function* $f : \mathbb{R}\backslash\{-1\} \mapsto \mathbb{R}\backslash\{0\}$, $f(x) = \dfrac{1}{x+1}$ *is not cyclic, as repeated iteration never yields the original function.*[40]

In any functional equation with two instances of a cyclic function f whose arguments are the inverses of each other (i.e. applying one to the other yields the identity),[41] we can switch the arguments to form a system of linear equations, which we can then solve in terms of the function f. Consider the following example:

Example 5.10.5.2. *Find all functions* f *such that* $f(x) + 4f(5 - x) = 19x$.

Solution 5.10.5.1. *Notice that* $f(x)$ *is cyclic with order 2, with* $x \mapsto 5 - x \mapsto 5 - (5 - x) = x$, *so* $f(5 - x) + 4f(x) = 19(5 - x)$, *or* $f(5 - x) + 4f(x) = 95 - 19x$. *Adding this and the original equation yields* $5f(x) + 5f(5 - x) = 95$, *or* $f(x) + f(5 - x) = 19$. *Then subtracting this from the original equation yields* $3f(5 - x) = 19x - 19 \implies f(5 - x) = \dfrac{19x}{3} - \dfrac{19}{3}$. *From here, we can either make a direct change of variable with* $t = 5 - x$, *or substitute this into either equation to obtain* $f(x) = \boxed{\dfrac{76 - 19x}{3}}$.

In particular, we can apply this same method to solve for specific values of the function at a point, not just for the function itself:

Example 5.10.5.3. *If* $f : \mathbb{R} \mapsto \mathbb{R}$ *is a function such that* $f(x) + 2f(2 - x) = 3 + f(3 - x)$, *what is* $f(0) + f(1)$?

Solution 5.10.5.2. *After substituting in* $x = 2$, *we obtain* $f(2) + 2f(0) = 3 + f(1)$. *We then substitute* $x = 1$, *which yields* $3f(1) = 3 + f(2)$. *Note that* $f(2) = 3f(1) - 3$, *so* $3f(1) - 3 + 2f(0) = 3 + f(1) \implies 2f(1) + 2f(0) = 6 \implies f(1) + f(0) = \boxed{3}$. *(Note that this method does not actually allow us to determine* $f(n)$ *for any* n, *as repeatedly substituting will only produce values that we cannot derive without substituting again and obtaining values that themselves require further substitution; this is what's known as a feedback loop!)*

It is also possible to obtain a system of three or more equations with just as many unknowns, in which case we can use a more sophisticated method such as Gaussian elimination[42] to solve the system. Essentially, we are reducing these types of FEs, which must contain cyclic functions, to ordinary algebraic systems in p

---

[40]It should be noted that the coefficients in the numerators and denominators of the sequence $\{f^0(x), f^1(x), f^2(x), f^3(x), \ldots\}$ of iterates of f are the elements of the Fibonacci sequence. Try proving this as an exercise!

[41]In particular, we say that a function f such that $f(f(x)) = x$ for all x in the domain of f is an *involution*.

[42]We will cover this in §5.13.

variables, where p is the order of the function f.

Though this technique does not fall precisely under the umbrella over "cyclic" functions or periodicity, it is nevertheless worth mentioning. In an FE involving two variables in the arguments of a single function, we can very often exploit the symmetry of the equation by swapping the variables (as with our previous example). In doing so, we end up with a form of the equation that usually carries a heavy implication regarding the nature of the function:

Example 5.10.5.4. *Solve the equation $f(x + y) - f(x - y) = f(x)f(y)$ for functions f.*

Solution 5.10.5.3. *Note that the RHS is symmetric with respect to x and y, but not the LHS. This implies that we can swap the two variables, such that $f(x + y) - f(y - x) = f(x)f(y) \implies f(x - y) = f(y - x)$. Making the change of variable $t = x - y$, this is equivalent to saying that $f(t) = f(-t)$.*

Henceforth, substituting $-y$ for y will give $f(x-y) - f(x+y) = f(x)f(-y)$, but note that the LHS of this equation is just -1 times the LHS of the original equation. Thus, we can conclude that $f(x)f(-y) = -f(x)f(y)$. Since $f(y) = f(-y)$, we have $f(x)f(y) = -f(x)f(y)$, or $f(x) = 0$ for all x.

5.10.6 Recurrence Relations

(Note: It will probably help considerably to go through §5.6, "Continuing Sequences, Series, and Recursion" first before reading this section. Nevertheless, we have provided a re-introduction to recursion here.)

A recurrence relation is yet another type of functional equation, in which we define a function based on its previous values (recursively, hence the name). More rigorously, we say that each successive term of the sequence of values assumed by f is a function of the previous values of f (thereby tying this in to the study of functional equations). As a refresher, we say a recurrence relation has order k if it depends on the previous k terms of the sequence; for instance, the Fibonacci recurrence, $F_n = F_{n-1} + F_{n-2}$, has order 2.

The factorial function is a recurrence that also happens to have a closed form (which is, by far, more commonly used in practice than its recursive counterpart), in that $n! = n \cdot (n - 1)$ by definition (and therefore has order 1). (We can also represent functions such as the binomial coefficient/choose function in recursive form, using Pascal's Identity. In this form, we can see that $\binom{n}{k} = \binom{n-1}{k-1} + \binom{n-1}{k}$ has order 2.) In functional equation form, a recurrence relation might appear somewhat like the following example:

Example 5.10.6.1. *If f is a monic quadratic function such that $f(x) = 2f(x - 1) + 3f(x - 2)$ for all x, determine f.*

Solution 5.10.6.1. *We have $f(x) = x^2 + ax + b$, so $2f(x - 1) = 2((x - 1)^2 + a(x - 1) + b) = 2(x^2 - 2x + 1) + 2a(x - 1) + 2b = 2x^2 + x(2a - 4b) + (2 - 2a + 2b)$ and $3f(x - 2) = 3(x^2 - 4x + 4) + 3a(x - 2) + 3b = 3x^2 + x(3a - 12) + (12 - 6a + 3b)$. Therefore, $x^2 + ax + b = 5x^2 + x(5a - 4b - 12) + (14 - 8a + 5b) \implies 5a - 4b - 12 = a, 14 - 8a + 5b = b \implies a - b = 3, 4a - 2b = 7 \implies a = \frac{1}{2}, b = -\frac{5}{2}$, and so $f(x) = \boxed{x^2 + \frac{1}{2}x - \frac{5}{2}}$.*

5.10.7 Examples

Example 5.10.7.1. *If $f(x) + 2f(3 - x) = 4 - 5x$, compute $f(1)$.*

Solution 5.10.7.1. *Motivated by the arguments of x and $3 - x$, we set $x = 1$ and $x = 2$. We then have $f(1) + 2f(2) = -1$ and $f(2) + 2f(1) = -6$, so $3f(1) + 3f(2) = -7 \implies f(1) + f(2) = -\frac{7}{3}$ and so $f(1) = \boxed{-\frac{11}{3}}$.*

Example 5.10.7.2. *Find all functions f satisfying $f(x)^2 + f(y)^2 = f(x + y)^2$ for all real numbers x, y.*

104

Solution 5.10.7.2. *If $x = y = 0$, then $2f(0)^2 = f(0)^2 \implies f(0) = 0$. If $x = -y$, then $f(-y)^2 + f(y)^2 = f(0)^2 = 0$, so $f(y) = f(-y) = 0$ (by the Trivial Inequality). Hence $f(x) = 0$ is the only solution.*

Example 5.10.7.3. *(USAMTS Year 27 Round 2 Problem 4) Find all polynomials $P(x)$ with integer coefficients such that, for all integers a and b, $P(a + b) - P(b)$ is a multiple of $P(a)$.*

Solution 5.10.7.3. *We will proceed via casework on the degree of $P(x)$.*

Case 1: *The degree of $P(x)$ is 0.*

Then $P(a + b)P(b) = 0 - 0 = 0 \implies 0$ is a multiple of $P(a) = 0 \implies$ all constant polynomials work.

Case 2: *The degree of $P(x)$ is 1 (i.e. $P(x) = mx + n$).*

$P(a + b) - P(b) = m(a + b) + n - (mb + n) = ma \implies$ since $P(a) = ma + n$, $P(a + b) - P(b)$ is a multiple of $P(a)$ only when $n = 0$.

Case 3: *The degree of $P(x)$ is greater than 1.*

Let $P(x) = x^k$. Then $P(a + b) - P(b) = (a + b)^k - b^k = \sum_{i=0}^{k} \binom{k}{i} a^k b^{k-i} - a^k = \sum_{i=0}^{k-1} \binom{k}{i} a^k b^{k-i}$. $P(a) = a^k$, but $P(a + b)P(b)$ only has one factor of a in the general case; in fact, only one of the terms has k factors of a. Since the expansion of $(a + b)^k$ results in a homogeneous equation (justified by the Binomial Theorem), adding equivalents of $(a + b)$ to a different power would not be able to cancel out the terms with less than k factors of a. Therefore, there are no such polynomials $P(x)$ with integer coefficients and degree greater than 1 such that, for all integers a and b, $P(a + b) - P(b)$ is a multiple of $P(a)$.

The only possible polynomials $P(x)$ following the requirements are $P(x) = c$, where c is an integer, and $P(x) = mx$, where m is also an integer.

Example 5.10.7.4. *(Cauchy's Functional Equation) Solve $f(x + y) = f(x) + f(y)$ for functions $f : \mathbb{Q} \mapsto \mathbb{Q}$.*

Solution 5.10.7.4. *We claim that all solutions are of the form $f(x) = cx$, $c \in \mathbb{R}$. Consider the cases $x = 0$, $x > 0$, and $x < 0$.*

If $x = 0$, then $f(x) = f(x) + f(0) \implies f(0) = 0$. If $x > 0$, we obtain from $f(x + x + x + \ldots + x) = f(ax)$ the equation $af(x) = f(ax)$, for $a \in \mathbb{N}$. Substituting $\frac{x}{a}$ for x and multiplying by $\frac{b}{a}$ (for $b \in \mathbb{N}$) yields $bf\left(\frac{x}{a}\right) = \frac{b}{a}f(x)$, and $f\left(\frac{b}{a}x\right) = \frac{b}{a}f(x) \implies f(cx) = cf(x) \implies f(c) = cf(1) = \alpha c$, where $\alpha = f(1) \in \mathbb{Q}$.

Finally, if $x < 0$, we can set $y = -x$ and recall $f(0) = 0$ to obtain $f(-x) = -f(x)$. Thus, combining this with the above conclusion yields $f(c) = \alpha c$.

5.10.8 Exercises

Exercise 91. *(2014 Berkeley Math Tournament) Find $f(2)$ given that f is a real-valued function that satisfies the equation*

$$4f(x) + \left(\frac{2}{3}\right)(x^2 + 2)f\left(x - \frac{2}{x}\right) = x^3 + 1$$

Exercise 92. *(2015 Caltech-Harvey Mudd Math Competition) Let f be a function taking real numbers to real numbers such that for all reals $x \neq 0, 1$, we have*

$$f(x) + f\left(\frac{1}{1-x}\right) = (2x - 1)^2 + f\left(1 - \frac{1}{x}\right)$$

Compute $f(3)$.

Exercise 93. *The function $\xi : \mathbb{R}\backslash\{0\} \mapsto \mathbb{R}$ satisfies $\xi(n) + 2\xi\left(\frac{2}{n}\right) = 4n$ for all $n \in \mathbb{R}\backslash\{0\}$. For what value of $n > 0$ does $\xi(n) = n$?*

Exercise 94. *(2006 AMC 12A) The function f has the property that for each real number x in its domain, $\frac{1}{x}$ is also in its domain and $f(x) + f\left(\frac{1}{x}\right) = x$. What is the largest set of real numbers that can be in the domain of f?*

Exercise 95. *Find all functions $f : \mathbb{R} \mapsto \mathbb{R}$ such that $2f(x) + 3f\left(\frac{1}{x}\right) = x^2$.*

Exercise 96. *Solve the functional equation $f(x^2 - y^2) = f(x + y) \cdot f(x - y)$ over $\mathbb{R}$.*

Exercise 97. *(2016 AIME I) Let $P(x)$ be a nonzero polynomial such that $(x-1)P(x+1) = (x+2)P(x)$ for every real x, and $(P(2))^2 = P(3)$. Then $P\left(\frac{7}{2}\right) = \frac{m}{n}$, where m and n are relatively prime positive integers. Find $m + n$.*

Exercise 98. *Prove that $f : \mathbb{C} \mapsto \mathbb{C}$ is a bijection if $f(f(x)) = x$ for all $x \in \mathbb{C}$.*

Extension Exercise 33. *(2016 USAJMO P6) Find all functions $f : \mathbb{R} \to \mathbb{R}$ such that for all real numbers x and y,*

$$(f(x) + xy) \cdot f(x - 3y) + (f(y) + xy) \cdot f(3x - y) = (f(x + y))^2.$$

Extension Exercise 34. *(2012 USAMO P4) Find all functions $f : \mathbb{Z}^+ \to \mathbb{Z}^+$ (where $\mathbb{Z}^+$ is the set of positive integers) such that $f(n!) = f(n)!$ for all positive integers n and such that $m - n$ divides $f(m) - f(n)$ for all distinct positive integers m, n.*

5.11 Proving Function Theorems

In this section, we will expand upon our pre-existing definition of the term "function" by taking a closer look at many of the theorems that belie them (in particular, continuous functions, as opposed to discretely-defined functions). In this context, all functions are assumed to be from the real numbers to the real numbers with no discontinuities. Many of these theorems also form the foundation for calculus and analysis, and are useful in a competition setting, as they provide computational shortcuts for solving problems, as well as a basis for understanding *why* functions behave the way they do.

5.11.1 The Intermediate Value Theorem

The *Intermediate Value Theorem* (IVT) is one of the shortest, sweetest, and most straightforward theorems in all of analysis. Its single statement is blissfully compact, yet packed with meaning at the same time:

Theorem 5.19: Intermediate Value Theorem

Let $f(x)$ be a continuous function on the interval $I = [a, b]$ (where we assume *a priori* that I is contained in the domain of f). If f assumes the values of $f(a)$ and $f(b)$ at $x = a$ and $x = b$ respectively, then at some point in the open interval (a, b), $f(x)$ assumes a value in the open interval $(f(a), f(b))$.

Essentially, the theorem simply states that the function must assume all values between $f(a)$ and $f(b)$, inclusive, if it has endpoints at $f(a)$ and $f(b)$.

Corollary 5.19: Corollaries of IVT

There are two main corollaries of note:

(a) If a continuous function has values of opposite sign on the interval I, then it must have at least one root in I (this is known as *Bolzano's Theorem,* not to be confused with the *Bolzano-Weirstrass Theorem*).

(b) The image of a continuous function over an interval is an interval.

Proof. To prove this theorem, we first must prove that the set of real numbers is *Dedekind complete:*

Lemma 5.19: (Dedekind) Completeness of $\mathbb{R}$

The real number line $\mathbb{R}$ is Dedekind complete.

Definition 5.15: (Dedekind) Completeness

Let $(S, \leq)$ be an ordered set (i.e. S equiped with the $\leq$ operation). Then $(S, \leq)$ is *Dedekind complete* iff every non-empty subset of S that is bounded above admits a supremum in S.

Remark. "Completeness" in an analysis context may also refer to the completeness of a metric space; in this sense, a metric space is said to be *complete* iff every Cauchy sequence in the space converges.

Proof. Let $S \in \mathbb{R}$ be non-empty and bounded above. Let $a_0 \in S$ and let b_0 be an upper bound for S. Then consider $m = \dfrac{a_0 + b_0}{2}$. Should m be an upper bound for S, let $b_1 := m$ and $a_1 := a_0$. Otherwise, let $b_1 := b_0$, and choose $a_1 \in S$ with $a_1 > m$. In both cases, we have $a_0 \leq a_1 \leq b_1 \leq b_0$ and $b_1 - a_1 \leq \dfrac{b_0 - a_0}{2}$. We also have $a_1 \in S$, and b_1 an upper bound for S.

By induction, we can construct an increasing sequence $(a_n) \in S$, as well as a decreasing sequence $(b_n) \in S$ of upper bounds for S. (Note that $a_n \leq b_n$ for all n, and $b_{n+1} - a_{n+1} \leq \dfrac{b_n - a_n}{2}$.)

We now invoke the *Archimedean property* of the real numbers:

Lemma 5.19: Archimedean Property

Definition: An ordered field $\mathbb{F}$ has the *Archimedean Property* (AP) if for any $x, y \in \mathbb{F}$, $x, y > 0$, there exists $n \in \mathbb{N}$ such that $nx > y$.

We claim that $\mathbb{R}$ has the AP.

Proof. We present another lemma:

Lemma 5.19: Natural Numbers are not Bounded Above

The set $\mathbb{N} = \{1, 2, 3, \ldots\}$ is not bounded above.

Proof. Assume the contrary. Since $\mathbb{N} \subset \mathbb{R}$, and we know that $\mathbb{R}$ has the least upper bound property,[43] $\mathbb{N}$ will have a least upper bound $b \in \mathbb{R}$. Thus, $n \leq b$ for all $n \in \mathbb{N}$, and b is the smallest such real number.

Thus, $b - 1$ cannot be an upper bound for $\mathbb{N}$, and so there exists an integer k such that $b - 1 < k$. But $b < k + 1$, which contradicts b being an upper bound for $\mathbb{N}$. $\square$

Using the fact that $\mathbb{N}$ is *not* bounded above, note that, since x is positive, we seek an $n > \dfrac{y}{x}$. If no such n exists, then $n < \dfrac{y}{x}$ for all integers n. Then by definition, $\dfrac{y}{x}$ would be an upper bound for $\mathbb{N}$, but this contradicts the lemma, thereby proving that $\mathbb{N}$ has the AP. $\square$

To finish, $b_n - a_n \to 0$ by the AP, so (a_n) and (b_n) are both Cauchy sequences (i.e. for every real $\epsilon > 0$, $\exists N \in \mathbb{N}$ such that $\forall m, n > N$, $|x_m - x_n| < \epsilon$). Thus, they both converge to some limit $l \in \mathbb{R}$. It follows that l is an upper bound for S, since the set of upper bounds for S is closed, and $b_n \to l$. This shows that $\mathbb{R}$ is Dedekind complete. $\square$

We can now proceed with our proof of the IVT itself. Let c be a point in the open interval (a, b), such that $f(c) \in (f(a), f(b))$. We will prove only the case $f(a) < c < f(b)$ here (the second case is similar and is left as an exercise to the reader).

Let $S = \{x \in I \mid f(x) \leq c\}$. Then S is non-empty, since $a \in S$, and bounded above by b. By the completeness of $\mathbb{R}$, $s = \sup S$ exists. We claim that $c = f(s)$.

Fix $\epsilon > 0$; since f is continuous on I, there exists a $\delta > 0$ such that we have $|f(x) - f(s)| < \epsilon$ whenever $|x - s| < \delta$, by the $\delta - \epsilon$ definition of continuity. Then $f(x) - \epsilon < f(s) < f(x) + \epsilon$ for all $x \in (s - \delta, s + \delta)$. Since s is a supremum, there exists $a_1 \in (s - \delta, s]$ with $a_1 \in S$, such that $f(s) < f(a_1) + \epsilon \leq s + \epsilon$.

Now choose $a_2 \in (s, s + \delta)$ (such that $a_2 \notin S$), so that $f(s) > f(a_2) + \epsilon > s - \epsilon$. Since $\epsilon > 0$, $s - \epsilon < f(s) < s + \epsilon$, and so $f(s) = c$. $\square$

---

[43]A set S has the least upper bound property iff every non-empty subset of S with an upper bound has a least upper bound (supremum) in S.

5.11.2   The Extreme Value Theorem

The *Extreme Value Theorem* is arguably the simplest-to-understand property of a continuous function that is also non-trivial to rigorously prove. It simply states the following:

> ### Theorem 5.20: Extreme Value Theorem
>
> If a real-valued function f is continuous on the closed and bounded interval $I = [a, b]$, then f must attain a minimum and a maximum, both at least once. Namely, there exist $c, d \in I$ such that $f(c) \leq f(x) \leq f(d)$ for all $x \in I$.

Also worthy of note is the related *boundedness theorem*, which guarantees that f on $I = [a, b]$ is bounded.

> ### Theorem 5.21: Boundedness Theorem
>
> A continuous function f on the closed interval I is also bounded on that interval. That is, there exist $m, M \in \mathbb{R}$ such that, for all $x \in I$, $m \leq f(x) \leq M$.

Proving the boundedness theorem is in fact the first step in the traditional proof of the EVT:

Proving the Boundedness Theorem. Consider the set B of $x \in I$ such that f is bounded on $[a, x]$. Note that $a \in B$, since $[a, a]$ is a single point. Furthermore, for any $c \in B$, all points p such that $a < p < c$ are also in B, so B is bounded below by a.

Observe that f is continuous from the right at a, so there exists $\delta_1 > 0$ such that $|f(x) - f(a)| < 1$ for all $x \in [a, a + \delta_1]$. Then f is bounded on $[a, a + \delta_1] \in B$.

Hence, we know that B is bounded above by some $s \in I$; we also know that $s > a$, since the length of B is nonzero. Suppose that $s < b$. Because f is continuous at s, there exists $\delta_2 > 0$ such that $|f(x) - f(s)| < 1$ whenever $x \in [s - \delta_2, s + \delta_2]$ with f also bounded on this interval. From s being an upper bound (supremum) for B, it follows that there exists $d \in B$ with $d > s - \dfrac{\delta_2}{2}$. Thus, f is also bounded on $[a, d]$, and by extension, on $[a, s + \delta]$. This is a contradiction, however, since we have assumed that s is a supremum. Therefore, $s = b$.

All that remains is to observe that, since f is continuous from the left at s, there exists $\delta_3 > 0$ such that $|f(x) - f(s)| < 1$ whenever $x \in [s - \delta_3, s]$, so that f is again bounded on $[s - \delta_3, s]$. But since s is a supremum, there exists another point $e \in B$ such that $e > s - \dfrac{\delta_3}{2}$. f is then bounded on $[a, e]$, and thus on $[a, s] = [a, b]$, as desired. $\qquad \square$

Now we can proceed to prove the EVT itself:

Proving the EVT. By the boundedness theorem, f is bounded above. By the completeness of $\mathbb{R}$, furthermore, $s = \sup f$ exists and is equal to $f(c)$ for some $a \leq c \leq b$. Let $n \in \mathbb{N}$; then $s - \dfrac{1}{n}$ is *not* an upper bound for f, since s is by definition the *least* upper bound. Thus, there exists $d \in I$ such that $f(d) > s - \dfrac{1}{n}$. This allows us to define a sequence (d_n). Since s is an upper bound for f, we have $s - \dfrac{1}{n} \leq f(d_n) \leq s$ for all n. Therefore, $f(d_n) \to s$. By the Bolzano-Weirstrass Theorem,[44] there exists a subsequence $(d_{n_k}) \to d \in I$ (since I is a closed interval). Because f is continuous at d, $f(d_{n_k}) \to f(d)$. But $(f(d_{n_k})) \to s$ is a subsequence of $(f(d_n))$, so $f(d) = s$. We conclude that $\sup f = s$ is attained at d. (The proof for the minimum $f(c)$ is similar.) $\qquad \square$

---

[44]Every bounded sequence in $\mathbb{R}^n$ has a convergent subsequence. We prove this at the end of this section.

Note that the interval I must be both closed and bounded (bounded both below and above) for this theorem to apply. If the interval I fails either condition, then f may not have a maximum or minimum on I. Consider the examples $f(x) = \dfrac{x^2}{1+x^2}$ on $I = [0, \infty)$ (where f never attains its upper bound of 1 on I), and $f(x) = x$ on $I = [0, \infty)$ (which is neither closed nor bounded).

5.11.3 The Mean Value Theorem

In intuitive terms, the *Mean Value Theorem* (or MVT) states that, for a continuous arc bounded by two endpoints, there exists at least one point on the arc for which the slope of the tangent line to the curve at that point is equal to the slope of the line between the two endpoints. The more precise mathematical definition uses the definition of the derivative (i.e. what we know as the slope of the tangent line) to formulate one of the most important properties of continuous functions in the field of analysis:

Theorem 5.22: Mean Value Theorem

Let f be a continuous function on the closed interval $I = [a, b]$. Then there exists a point $c \in I$ such that $f'(c) = \dfrac{f(b) - f(a)}{b - a}$.

Remark. Note that the expression for $f'(c)$ very closely resembles that of the derivative in general,

$$f'(x) = \lim_{h \to 0} \frac{f(x + h) - f(x)}{h}$$

Is this merely a coincidence, or is there a deeper meaning to this?

Proof. Counterintuitive as it may seem, we will actually be using the special case of Rolle's Theorem (see below) to prove the Mean Value Theorem.

The equation of the secant line through the points $(a, f(a))$ and $(b, f(b))$ is given by

$$y = \frac{f(b) - f(a)}{b - a}(x - a) + f(a)$$

Let

$$g(x) = f(x) - y = f(x) - \left(\frac{f(b) - f(a)}{b - a}(x - a) + f(a) \right)$$

Observe that $g(a) = g(b) = 0$, with g continuous on $I = [a, b]$ (and differentiable on I as well, since f is continuous and therefore differentiable). By Rolle's Theorem, there exists $c \in (a, b)$ such that $g'(c) = 0$.

Then $g'(x) = f'(x) - \dfrac{f(b) - f(a)}{b - a}$ (by the linearity of the derivative), so $g'(c) = f'(c) - \dfrac{f(b) - f(a)}{b - a} = 0 \implies f'(c) = \dfrac{f(b) - f(a)}{b - a}$ and the proof is complete. $\qquad \square$

Rolle's Theorem

Another related landmark result in differential calculus (and the study of functions in analysis) is *Rolle's Theorem,* which can be thought of intuitively as the statement that, for some point on a closed curve, the tangent line at that point is horizontal:

Theorem 5.23: Rolle's Theorem

Let f be a continuous function on the closed interval $I = [a, b]$. If $f(a) = f(b)$, then there exists $c \in [a, b]$ such that $f'(c) = 0$.

This is in fact a corollary/special case of the MVT (where $f(a) = f(b)$). As such, a dedicated proof is not necessary. However, it is worth discussing the generalizations of the standard version of the theorem (above) to higher-order derivatives.

Theorem 5.24: Generalized Rolle's Theorem

Impose additional conditions on f, namely that it is $n - 1$ times differentiable (such that $f^{(n-1)}(x)$ exists for all x in the domain of f), $f^{(n)}(x)$ exists for all $x \in (a, b)$, and there are n intervals given by $a_1 < b_1 \leq a_2 < b_2 \leq a_3 < \ldots \leq a_n < b_n$ in I such that $f(a_k) = f(b_k)$ for all $k \in [1, n]$. Then there exists $c \in (a, b)$ such that $f^{(n)}(c) = 0$.

Proof. We prove this by induction on $n \geq 2$ ($n = 1$ is just the standard version of Rolle's Theorem). Assume this generalization holds for $n - 1$; then for every integer $k \in [1, n]$, there exists $c_k \in (a_k, b_k)$ for which $f'(c_k) = 0$. Thus, the first derivative satisfies the stipulations on the $n - 1$ closed intervals $[c_1, c_2], [c_2, c_3], [c_3, c_4], \ldots, [c_{n-1}, c_n]$. By the inductive hypothesis, there exists c such that $f^{(n-1)}(c) = 0$. $\square$

The Average Value Theorem

In tandem with the Mean Value Theorem (and, in fact, as the primary upshot of that theorem), the *Average Value Theorem* provides a computational method of computing the average value of a function over a defined interval. Using the intuitive idea of a definite integral as the area under the curve (along with a more rigorous conception of the idea as the limit of the Riemann sum), it follows that the average value of the function is the area under the curve divided by the length of the interval:

Theorem 5.25: Average Value Theorem (AVT)

Let $f(x) : D \mapsto R$ be a function continuous on the interval $I \subset D, I = [a, b]$. Then the average value of $f(x)$ over I is given by

$$\frac{1}{b - a} \cdot \int_a^b f(x) dx$$

where $a < b$ *de facto*.

Remark. In calculus contexts, this theorem is more commonly known as the *Mean Value Theorem for integrals*.

Proof. Define a function $F(x)$ such that

$$F(x) = \int_a^x f(t) dt$$

for all $x \in I$. By the first Fundamental Theorem of Calculus, F is continuous and differentiable on I, and $F'(x) = f(x)$ for all $x \in I$. Thus, we can apply the Mean Value Theorem to F; id est, there must exist $c \in (a, b)$ for which $F'(c) = \dfrac{f(b) - f(a)}{b - a}$. It follows that

$$F(a) = \int_a^a f(t) dt = 0$$

and $F'(c) = f(c)$, so $f(c) = \dfrac{f(b)}{b - a} = \dfrac{1}{b - a} \int_a^b f(t) dt$. $\square$

5.11.4 Proving the Bolzano-Weirstrass Theorem

The Bolzano-Weirstrass Theorem is a cornerstone of mathematical analysis, and is the motivating factor behind the proof of the Extreme Value Theorem. It provides both sufficient and necessary conditions for the boundedness of a sequence in $\mathbb{R}^n$:

Proof. We first introduce the following key lemma:

> **Lemma 5.26: Monotone Subsequence Lemma**
>
> Every infinite sequence $(x_n) \in \mathbb{R}^n$ has a monotone subsequence.

Proof. Say that a positive integer n is a *peak* of the sequence (a_n) if $m > n \implies x_m < x_n$. Suppose that (x_n) has an infinite number of peaks, such that the subsequence $(x_{n_j}) = \{x_{n_1}, x_{n_2}, x_{n_3}, \ldots\}$ is monotonically decreasing. Suppose, then, that (x_n) has finitely many peaks, with N being the last peak and $n_1 = N + 1$. Then n_1 is not a peak, so there exists n_2 with $n_1 < n_2$ and $x_{n_1} \le x_{n_2}$. Similarly, n_2 is not a peak; repeating this process yields the infinite increasing subsequence $\{x_{n_1}, x_{n_2}, x_{n_3}, \ldots\}$, as desired. $\square$

Suppose that we have a bounded sequence $(b_n) \in \mathbb{R}$; by the lemma, (b_n) has a bounded monotone subsequence. From the monotone convergence theorem,[45] this subsequence converges. (This is all for the case $n = 1$, i.e. $\mathbb{R}$, the real line.)

In the more general case for any $n \in \mathbb{N}$, given a bounded sequence in $\mathbb{R}^n$, the sequence of first coordinates is itself a bounded sequence, and therefore has a convergent subsequence. In similar fashion, we can extract a subsequence for which the second coordinates converge, the third coordinates converge, and so forth. Since all subsequences of the kth coordinates converge, the sequence as a whole converges. $\square$

5.11.5 Examples

Example 5.11.5.1.

(a) Compute $\int_0^2 x^2 - 2x \, dx$.

(b) Using the result of (a), compute the average value of the function $f(x) = x^2 - 2x$ on the interval $[0, 2]$.

(c) Using the results of (a) and (b), compute the average value of the function $f(x)^2 = (x^2 - 2x)^2$ on the interval $[0, 2]$.

Solution 5.11.5.1.

(a) Evaluating the definite integral, we find that $\int_0^2 x^2 - 2x \, dx = \left.\left(\dfrac{x^3}{3} - \dfrac{2x^2}{2}\right)\right|_0^2 = \left(\dfrac{2^3}{3} - 2^2\right) - 0 = \boxed{-\dfrac{4}{3}}$.

(b) We can find the average value of $f(x)$ over the interval $[0, 2]$ by dividing the value of the definite integral over that interval by the length of the interval, which gives a value of $\dfrac{-4/3}{2} = \boxed{-\dfrac{2}{3}}$.

(c) The average value of $f(x)^2$ on the interval $[0, 2]$ is equal to the square of the average value of $f(x)$ on that same interval, which is equal to $\boxed{\dfrac{4}{9}}$.

Example 5.11.5.2. Let $f(x) = x^2 + 19$ be a continuous, real-valued function from $\mathbb{R}$ to $\mathbb{R}$. If $F(x) = \int_{-\infty}^x f(t) \, dt$, compute $F(2) - F(1)$.

---

[45] The monotone convergence theorem states that, if a sequence is increasing/decreasing and bounded above/below by a supremum/infimum (respectively), the sequence will converge to the supremum/infimum.

Solution 5.11.5.2. *Note that, by the Fundamental Theorem of Calculus, we have*

$$F(2) - F(1) = \int_{-\infty}^{2} f(t)\,dt - \int_{-\infty}^{1} f(t)\,dt$$

$$= \int_{1}^{2} t^2 + 19\,dt = \boxed{\frac{64}{3}}$$

5.11.6 Exercises

Exercise 99. *Consider the function $f(x) = -20x^2 + 19x + 2$ on the interval $[0, 1]$.*

(a) *Compute $f(0)$, $f(1)$, $f'(x)$, $f'(0)$, and $f'(1)$.*

(b) *Show that f is continuous (and, by extension, differentiable) on $[0, 1]$.*

(c) *Find the value of $c \in [0, 1]$ that satisfies the conclusion of the MVT.*

(d) *Show that $f'(x) = 0$ for some $x \in [0, 1]$.*

Exercise 100. *Use the MVT to show that, for all $a, b \in \mathbb{R}$, $|\cos a - \cos b| \leq |a - b|$.*

Exercise 101. *Show that $x^5 + 20x + 19 = 0$ has exactly one real solution.*

Exercise 102.

(a) *Consider the inverse function f^{-1} of f. Under what conditions are both f and f' differentiable?*

(b) *Let $f(x) = x^5 + 20x + 19$. Evaluate $(f^{-1})'(-2)$.*

Exercise 103. *Is there necessarily a real root of the polynomial $x^{2019} - 20x^{2018} + 19$ in the open interval $(0, 1)$?*

Extension Exercise 35. *Prove that the converse of the IVT is false by constructing a counterexample.*

Extension Exercise 36. *Prove that Rolle's Theorem is equivalent to the Mean Value Theorem. That is, do both of the following:*

(a) *Show that Rolle's Theorem is a special case of the MVT.*

(b) *Using Rolle's Theorem, prove the MVT.*

Extension Exercise 37. *(1955 Putnam B2) Let $\mathbb{R}$ be the reals. $f : \mathbb{R} \mapsto \mathbb{R}$ is twice differentiable, f'' is continuous and $f(0) = 0$. Define $g : \mathbb{R} \mapsto \mathbb{R}$ by $g(x) = \dfrac{f(x)}{x}$ for $x \neq 0$, $g(0) = f'(0)$. Show that g is differentiable and that g' is continuous.*

Extension Exercise 38. *(2004 Putnam A1) Basketball star Shanille O'Keal's team statistician keeps track of the number, $S(N)$, of successful free throws she has made in her first N attempts of the season. Early in the season, $S(N)$ was less than 80% of N, but by the end of the season, $S(N)$ was more than 80% of N. Was there necessarily a moment in between when $S(N)$ was exactly 80% of N?*

5.12 Bézout's Identity

Returning to the study of modular arithmetic, a common topic of consideration is the greatest common divisor function, which is instrumental in determining the set of possible solutions to some (modular congruence) equation. Even outside of the modular equivalence context, *Bézout's Identity* is extremely useful in solving equations with two variables in terms of a parameter:

> **Theorem 5.27: Bézout's Identity**
>
> Let a and b be integers with $\gcd(a, b) = d$. Then there exist integers x, y such that $ax + by = d$. (In general, the integers spanned by x and y over $\mathbb{Z}$ are the multiples of d.)

In deriving the solutions to this equation, we find that all solutions are of the form

$$(x, y) = \left(x + \frac{kb}{\gcd(a, b)}, y - \frac{ka}{\gcd(a, b)} \right)$$

as in the following example:

Example 5.12.0.1. *Solve the Diophantine equation $5x + 20y = 45$ over the integers.*

Solution 5.12.0.1. *We identify $(1, 2)$ as a solution, then note that increasing y by 1 will increase $20y$ by 20, which means we must reduce x by $\frac{20}{5} = 4$. Therefore, all solutions will be of the form $(1 - 4k, 2 + k)$ (or, equivalently, $(1 + 4k, 2 - k)$) for $k \in \mathbb{Z}$. (Note that $\gcd(5, 20) = 5$, so the solutions indeed fit the general form $\left(x + \frac{20k}{5}, y - \frac{5a}{5} \right) = (x + 4k, y - k)$.)*

Proof. For nonzero integers a and b, the quotient of a and b leaves a remainder of $r_1 < |b|$, with $\gcd(a, b) = \gcd(r_1, b)$ by the Euclidean Algorithm. Repeatedly applying the division algorithm, we obtain

$$a = bx_1 + r_1, 0 \leq r_1 < |b|$$

$$b = r_1 x_2 + r_2, 0 < r_2 < r_1$$

$$\vdots$$

$$r_{n-1} = r_n x_{n+1} + r_{n+1}, 0 < r_{n+1} < r_n$$

$$r_n = r_{n+1} x_{n+2}$$

In the second-to-last equation, we can solve for r_{n+1} - the last nonzero remainder in the Euclidean Algorithm - in terms of r_n and r_{n-1}. In the preceding equation, we can solve for r_n in terms of r_{n-1} and r_{n-2}, and so forth. Eventually, we will have written r_{n+1} as a linear combination of a and b, which implies that $r_{n+1} = \gcd(a, b)$; this proves the identity. $\square$

Applications of Bézout's Identity range from the more direct - determining the solutions of a linear equation - to strengthening the very foundation of the Euclidean division algorithm.

In general, we can expand Bézout's Identity to a form which lends itself to the solution of Diophantine equations with any number of unknowns. The end result is markedly similar to the two-variable case:

> **Theorem 5.28: Generalized Bézout Identity**
>
> If $\gcd(a_1, a_2, a_3, \ldots, a_n) = d$, then there exist integers $x_1, x_2, x_3, \ldots, x_n$ such that $a_1 x_1 + a_2 x_2 + a_3 x_3 + \ldots + a_n x_n = d$, and furthermore, every number of this form is a multiple of d.

We will consider a computational example that utilizes a reversal of the Euclidean Algorithm:

Example 5.12.0.2. *Write 19 as a linear combination of 380 and 2033.*

Solution 5.12.0.2. *Note that $380 = 19 \cdot 20$ and $2033 = 19 \cdot 107$, so the problem effectively reducing to finding those integers x, y such that $20x + 107y = 1$. Without the Euclidean Algorithm, note that $1 + 107 \cdot 7 = 750$ is a multiple of 10, using modular arithmetic. Then note that $107 \cdot 10 = 1070 \equiv 10 \pmod{20}$, so $750 + 1070 = 1820$ is the smallest positive multiple of 20 that is also congruent to 1 (mod 107). Hence, $x = 91$ and $y = -17$, or $x = -16$ and $y = 3$ in reduced form. It follows that $19 = -16 \cdot 380 + 3 \cdot 2033$.*

Strictly in accordance with Bézout's Identity, we use the Euclidean Algorithm to reduce the coefficients as follows: $107 = 20 \cdot 5 + 7, 20 = 2 \cdot 7 + 6, 7 = 1 \cdot 6 + 1$, so that $1 = 7 - 1 \cdot 6, 7 - 1 \cdot 6 = 7 - 1 \cdot (20 - 2 \cdot 7) = 7 - 1 \cdot 20 + 2 \cdot 7 = 3 \cdot 7 - 1 \cdot 20 = 3 \cdot (107 - 20 \cdot 5) - 1 \cdot 20 = 3 \cdot 107 - 16 \cdot 20$.

This covers Bézout's Identity as it is most commonly known and used, but it can be extended beyond its traditional use as well. Indeed, we can apply it to polynomial equations, wherein it serves the same purpose as in our previous discussion.

More specifically, let f and g be univariate polynomials (polynomials of a single variable) with coefficients in some field $\mathbb{F}$. Then there exist polynomials a and b with $af + bg = 1$ iff f and g have no common root in $\mathbb{F}$. (Proving this is left as an exercise to the reader.)

We can also extend Bézout's Identity to the concept of *integral domains*, which again lend themselves to the Euclidean Algorithm:

Definition 5.16: Integral and Principal Ideal Domains

An *integral domain* is a nonzero commutative ring[a] in which the product of any two nonzero elements is also nonzero (which furthermore implies that, for any a, b, c in the domain, $ab = ac \implies b = c$ if $a \neq 0$). A *principal ideal domain* is an integral domain for which every ideal[b] is principal, i.e. can be generated by a single element.

---
[a] A ring in which multiplication is commutative.

[b] An *ideal* I of a ring R is a subgroup of R such that, for all $r \in R$ and $x \in I$, $rx, xr \in I$. If $rx \in I$ only, then I is a *left ideal*, and if $xr \in I$ only, then I is a *right ideal*.

Indeed, Bézout's Identity holds in any principal ideal domain:

Theorem 5.29: Bézout's Identity in a Principal Ideal Domain

If R is a principal ideal domain (PID), and $a, b \in R$, with $\gcd(a, b) = d$, then there exist $x, y \in R$ such that $ax + by = d$. (In this case, R is a *Bézout domain*.)

Proof. Intuitively, we observe that Ra and Rb are both principal, and so their sum is equal to Rd. More precisely, consider the ideal $I = (a, b) = \{ar_1 + br_2 \mid r_1, r_2 \in R\}$, and let $d = \gcd(a, b)$. Since R is a PID, I must be principal, and so it is generated by $I = (d)$. Then there must exist $r_1, r_2 \in R$ such that $d = ar_1 + br_2$. We claim that $d = \gcd(a, b)$.

If we let $z \in I$, then $d \mid z$. This is because if $z \in I = (d)$, then there exists some $r \in R$ with $z = rd$. It immediately follows that $d \mid z$.

As $a, b \in I$, we let $d \mid a, d \mid b$. Moreover, consider $s \in R$ with $s \mid a, s \mid b$. We have $s \mid (ar_1 + br_2) = d$, so $d = \gcd(a, b)$. Finally, we show that d is in fact minimal. For contradiction, let $d < d' = ar_1' + br_2'$. But since $d \mid a, b$, we have $d \mid ar_1' + br_2' = d'$, as desired. $\square$

5.12.1 Examples

Example 5.12.1.1. *Call an ordered pair of integers (x, y) primitive if $\gcd(x, y) = 1$ (i.e. x and y are relatively prime). Find all primitive pair(s) (x, y) such that $2016x + 4019y = 1$.*

Solution 5.12.1.1. *By the Euclidean Algorithm we have $\gcd(2016, 4019) = \gcd(2016, 2003) = \gcd(13, 2013) = \gcd(13, 1) = 1$, so a primitive pair does exist, namely $(x, y) = (-309, 155)$.*

Example 5.12.1.2. *Find all solutions to the equation $20x + 19y = 2019$ over the integers.*

Solution 5.12.1.2. *Immediately note that $(x, y) = (100, 1)$ works. Whenever we decrement x by 19, we increment y by 20 in order to compensate. Thus, $(x, y) = (100 - 19t, 1 + 20t)$ for $t \in \mathbb{R}, x, y \in \mathbb{Z}$.*

Example 5.12.1.3. *Compute $2019^{-1} \bmod 9102$, or show that it does not exist.*

Solution 5.12.1.3. *We seek x such that $2019x \equiv 1 \pmod{9102}$. By the Euclidean Algorithm, $\gcd(2019, 9102) = 3 \neq 1$, so the inverse does not exist.*

5.12.2 Exercises

Exercise 104. *Show that, if a, b, and c are positive integers such that a and b are relatively prime, and b and c are relatively prime, then ab and c are relatively prime.*

Exercise 105. *Show that, if m and n are relatively prime positive integers, then there exists an integer x such that $mn \equiv 1 \bmod n$.*

Exercise 106. *(2013 AIME I) Ms. Math's kindergarten class has 16 registered students. The classroom has a very large number, N, of play blocks which satisfies the conditions:*

(a) If 16, 15, or 14 students are present in the class, then in each case all the blocks can be distributed in equal numbers to each student, and

(b) There are three integers $0 < x < y < z < 14$ such that when x, y, or z students are present and the blocks are distributed in equal numbers to each student, there are exactly three blocks left over.

Find the sum of the distinct prime divisors of the least possible value of N satisfying the above conditions.

Extension Exercise 39. *(1985 AIME) The numbers in the sequence 101, 104, 109, 116,... are of the form $a_n = 100 + n^2$, where $n = 1, 2, 3, \ldots$ For each n, let d_n be the greatest common divisor of a_n and a_{n+1}. Find the maximum value of d_n as n ranges through the positive integers.*

Extension Exercise 40. *(2018 HMMT February Guts Round) Find the number of unordered pairs (a, b), where $a, b \in \{0, 1, 2, \ldots, 108\}$ such that 109 divides $a^3 + b^3 - ab$.*

Extension Exercise 41. *(2005 IMO P4) Determine all positive integers relatively prime to all the terms of the infinite sequence*

$$a_n = 2^n + 3^n + 6^n - 1, \ n \geq 1.$$

5.13   Matrix Algebra, Part 1

Matrix algebra, or *linear algebra* (in particular, usually known as such in the context of more abstract mathematical objects such as vector spaces), studies the *matrix*,[46] which is, at its core, a rectangular collection, or array, of mathematical objects - whether they be real numbers, complex numbers, or expressions. Matrices have a slew of applications in both pure and applied mathematics, as well as outside math as a whole. They are commonly used in game theory and microeconomics to represent the possible decisions of players in a game or those involved in a similar set of decisions, statistics and probability theory, and in interdisciplinary applications such as quantum mechanics, optics, and electronics. In our discussion of matrices, we will be honing in on the matrix as a representation of a system of linear equations.

5.13.1   Defining a Matrix

We categorize matrices primarily depending on how they are arranged, and what they consist of. A matrix is always rectangular in shape, and has a certain number of rows and columns. Namely, if a matrix M has m rows and n columns, we say that M is an $m \times n$ *("m by n") matrix*.

Example 5.13.1.1. *The following are all examples of 2×3 matrices:*

$$\begin{bmatrix} 0 & 1 & 2 \\ 3 & 4 & 5 \end{bmatrix}$$

$$\begin{bmatrix} 4\pi & 5 & 2^x \\ 0 & 1+e^x & 9x \end{bmatrix}$$

$$\begin{bmatrix} x & x+y & x+z \\ y+z & x+y+z & 0 \end{bmatrix}$$

For a matrix, we define the operations of addition and multiplication (as well as their inverses, namely subtraction and division). We will cover these in the next section.

The individual entries in the matrix M with m rows and n columns are usually denoted by $a_{i,j}$ (or occasionally a_{ij} when i and j are both base-10 digits), for the ith row and the jth column (with $1 \le i \le m, 1 \le j \le n$). M is subsequently said to be a member of the space $\mathbb{M}_{i \times j}$. For any matrix $M \in \mathbb{M}_{i \times j}$, we may represent it visually as follows:

$$\begin{bmatrix} a_{1,1} & a_{1,2} & a_{1,3} & \cdots & a_{1,n} \\ a_{1,1} & a_{1,2} & a_{1,3} & \cdots & a_{1,n} \\ a_{1,1} & a_{1,2} & a_{1,3} & \cdots & a_{1,n} \\ \vdots & \vdots & \vdots & \ddots & \vdots \\ a_{m,1} & a_{m,2} & a_{m,3} & \cdots & a_{m,n} \end{bmatrix}$$

Matrices as a whole may be thought of as generalizations of linear functions: namely, as extensions of the concept of a system of linear equations. Each row of a matrix represents an equation, while each column of a matrix represents a different variable/unknown in the entire system that is represented by the matrix. In these cases, the only way for the system to have any meanings is for us to assign some values to each equation. This is the process of *augmenting* the matrix with values, to produce an *augmented matrix*. In an augmented matrix, we add another column of values, such that the general form of such a matrix is given by

$$\left[\begin{array}{ccccc|c} a_{1,1} & a_{1,2} & a_{1,3} & \cdots & a_{1,n} & x_1 \\ a_{1,1} & a_{2,2} & a_{2,3} & \cdots & a_{2,n} & x_2 \\ a_{1,1} & a_{3,2} & a_{3,3} & \cdots & a_{3,n} & x_3 \\ \vdots & \vdots & \vdots & \ddots & \vdots & \vdots \\ a_{m,1} & a_{m,2} & a_{m,3} & \cdots & a_{m,n} & x_m \end{array}\right]$$

---

[46]No, not *The Matrix*, disappointingly. But this is arguably even more exciting!

Indeed, there are m equations with n unknowns. If $n \leq m$, then using properties of the matrix and matrix operations, we can solve for each variable represented in the matrix. Otherwise, there will necessarily be some variables that are solely in terms of the other variables. In particular, the special cases where $m = 1$ and $n = 1$ are referred to as *row vectors* and *column vectors* respectively. These types of matrices prove especially useful in later applications, such as solving the fundamental matrix equation $A\mathbf{x} = \mathbf{b}$. A matrix with *either* an infinite number of rows or columns (or both) is called an *infinite matrix,* and on the polar opposite, a matrix with no elements (i.e. $m = n = 0$) is the *empty matrix* (much like $\emptyset$, the empty set).

Furthermore, we can define the *diagonal* of an $m \times m$ square matrix as the set of elements running from the top-left $(a_{1,1})$ to the bottom-right $(a_{m,m})$, i.e. all elements of the form $a_{i,i}$ for $1 \leq i \leq m$. Note that the diagonal is *only* defined for square matrices. The *trace* of a square matrix is then the sum of the elements of its diagonal.

Finally, we define the *identity matrix:*

Definition 5.17: Identity Matrix

The $n \times n$ *identity matrix,* denoted I_n, is the matrix consisting solely of 1's on its diagonal and 0's off its diagonal.

We can think of the identity matrix as just that: an identity - a matrix that forms the basis for the group of matrices in a space. In future sections, we will discuss the relationship of the identity matrix with other concepts and computations.

5.13.2   Matrix Operations

Addition and subtraction of matrices is similar to addition and subtraction of real numbers, with one additional constraint: the dimensions or both (or all) matrices must be identical. We can add or subtract two 3×4 matrices, but not add a 3×4 to a 4×3 matrix! Otherwise, addition and subtraction are fairly-self explanatory: we simply add/subtract the corresponding elements in each matrix.

On the other hand, multiplication of matrices is considerably more involved (and perhaps quite a bit more abstract than addition or subtraction), and moreover, multiplication and division are two totally different types of operations (i.e. we cannot think of division as merely the reverse of multiplication and leave it at that, but admittedly, using the inverse is key to division).

To multiply two matrices $A \in \mathbb{M}_{m_1 \times n_1}$ and $B \in \mathbb{M}_{m_2 \times n_2}$, we must first ensure that $n_1 = m_2$ (in which case AB becomes an $m_1 \times n_2$ matrix); otherwise, the operation AB is invalid. The following multiplication procedure will hopefully elucidate the reasoning behind this restriction. Assuming this condition is met, we can multiply the matrices A and B as follows:

$$AB = \begin{bmatrix} a_{1,1} & a_{1,2} & a_{1,3} & \cdots & a_{1,n_1} \\ a_{2,1} & a_{2,2} & a_{2,3} & \cdots & a_{2,n_1} \\ a_{3,1} & a_{3,2} & a_{3,3} & \cdots & a_{3,n_1} \\ \vdots & \vdots & \vdots & \ddots & \vdots \\ a_{m_1,1} & a_{m_1,2} & a_{m_1,3} & \cdots & a_{m_1,n_1} \end{bmatrix} \begin{bmatrix} b_{1,1} & b_{1,2} & b_{1,3} & \cdots & b_{1,n_2} \\ b_{2,1} & b_{2,2} & b_{2,3} & \cdots & b_{2,n_2} \\ b_{3,1} & b_{3,2} & b_{3,3} & \cdots & b_{3,n_2} \\ \vdots & \vdots & \vdots & \ddots & \vdots \\ b_{m_2,1} & b_{m_2,2} & b_{m_2,3} & \cdots & b_{m_2,n_2} \end{bmatrix}$$

$$= \begin{bmatrix} \sum\limits_{k=1}^{n_1} a_{1,k}b_{k,1} & \sum\limits_{k=1}^{n_1} a_{1,k}b_{k,2} & \sum\limits_{k=1}^{n_1} a_{1,k}b_{k,3} & \cdots & \sum\limits_{k=1}^{n_1} a_{1,k}b_{k,n_2} \\ \sum\limits_{k=1}^{n_1} a_{2,k}b_{k,1} & \sum\limits_{k=1}^{n_1} a_{2,k}b_{k,2} & \sum\limits_{k=1}^{n_1} a_{2,k}b_{k,3} & \cdots & \sum\limits_{k=1}^{n_1} a_{2,k}b_{k,n_2} \\ \sum\limits_{k=1}^{n_1} a_{3,k}b_{k,1} & \sum\limits_{k=1}^{n_1} a_{3,k}b_{k,2} & \sum\limits_{k=1}^{n_1} a_{3,k}b_{k,3} & \cdots & \sum\limits_{k=1}^{n_1} a_{3,k}b_{k,n_2} \\ \vdots & \vdots & \vdots & \ddots & \vdots \\ \sum\limits_{k=1}^{n_1} a_{m_1,k}b_{k,1} & \sum\limits_{k=1}^{n_1} a_{m_2,k}b_{k,2} & \sum\limits_{k=1}^{n_1} a_{m_2,k}b_{k,3} & \cdots & \sum\limits_{k=1}^{n_1} a_{m_2,k}b_{k,n_2} \end{bmatrix}.$$

In other words, we iterate through each row of the matrix A and through each column of the matrix B, summing all the products and assigning that to the element $AB_{i,j}$ for the ith row of A and the jth column of B. To ease up on the abstraction factor, we will work through a much more concrete example:

Example 5.13.2.1. *Compute AB, where*

$$A = \begin{bmatrix} 0 & 4 \\ 0 & 5 \end{bmatrix}$$

$$B = \begin{bmatrix} 2 & 0 \\ 1 & 9 \end{bmatrix}$$

or show that it does not exist.

Solution 5.13.2.1. *First, we check that the number of columns in A is equal to the number of rows in B. Indeed, both are 2, so the product AB exists (in particular, it is also a 2×2 matrix). Now proceed with the algorithm to obtain*

$$AB = \begin{bmatrix} 0 \cdot 2 + 4 \cdot 1 & 0 \cdot 0 + 4 \cdot 9 \\ 0 \cdot 2 + 5 \cdot 1 & 0 \cdot 0 + 5 \cdot 9 \end{bmatrix} = \begin{bmatrix} 4 & 36 \\ 5 & 45 \end{bmatrix}$$

We previously claimed that the product of an $m_1 \times n_1$ matrix and an $n_1 \times m_2$ matrix has dimensions $m_1 \times m_2$. We now prove this rigorously:

Proof. From the matrix multiplication procedure, and from the fact that A has m_1 rows and B has m_2 columns, it immediately follows that there will be $m_1 m_2$ products of the form $AB_{i,j}$. Each has a unique row/column combination corresponding to it; hence, the product matrix AB has dimensions $m_1 \times n_2$. $\square$

Henceforth, we discuss matrix division. Indeed, it is not wrong to view division of matrices as the inverse of multiplication, but this is not the whole story. We must multiply the original matrix by the inverse of the matrix in the denominator - but what *is* the inverse of a matrix? To understand what matrix division means, we must provide the definition of an inverse, which itself depends on a crucial property of matrices. Thus, we will have to hold off until §5.13.4, "Studying the Determinant", to discuss matrix division in any level of detail. However, we can define the inverse in a loose manner without this definition:

Definition 5.18: Inverse of a Matrix

The *inverse* of an $n \times n$ matrix M, denoted M^{-1} (provided it exists), is equal to $\dfrac{1}{\det(M)} \cdot M^T$, where M^T denotes the *transpose* of M (which we will formally define shortly). That is, the inverse M^{-1} of M is the uniquely determined matrix such that $MM^{-1} = M^{-1}M = I_n$, where I_n is the identity matrix. If $\det M = 0$, then we say M is *non-invertible* or *singular*.

We can say the following of all invertible matrices, regardless of their size or elements:

Theorem 5.30: Properties of Invertible Matrices

The following properties hold true for *all* invertible square matrices:

(a) $(A^{-1})^{-1} = A$.

(b) $(\det A)^{-1} = \det(A^{-1})$.

(c) $(c \cdot A)^{-1} = c \cdot A^{-1}$ for all scalars $c \in \mathbb{R}$, $c \neq 0$.

(d) $(A^T)^{-1} = (A^{-1})^T$.

(e) For any invertible $n \times n$ matrices A and B, $(AB)^{-1} = B^{-1}A^{-1}$. Furthermore, for any invertible $n \times n$ matrices $M_1, M_2, M_3, \ldots, M_k$,

$$(A_1 A_2 A_3 \cdots A_k)^{-1} = A_k^{-1} A_{k-1}^{-1} \cdots A_2^{-1} A^{-1}$$

Proof. Note: These proofs rely on the definitions of the determinant and the transpose. The transpose definition is definition 5.18 immediately below, but the determinant is covered in §5.13.4.

(a) For invertible A, by definition, there exists an invertible matrix A^{-1} such that $AA^{-1} = I_n = A^{-1}A$. Using the fact that $(AB)^{-1} = B^{-1}A^{-1}$ (which we take as fact for now), we obtain

$$(AA^{-1})^{-1} = I^{-1} = I$$
$$(A^{-1})^{-1}A^{-1} = I$$
$$A((A^{-1})^{-1}A^{-1}) = IA$$
$$(A^{-1})^{-1}(A^{-1}A) = A$$
$$(A^{-1})^{-1} = A$$

as desired.

(b) We use the fact that the determinant is multiplicative, i.e. $\det(A) \cdot \det(B) = \det(AB)$ for all square matrices A, B. For now, we will take this as fact (we prove this in our discussion of Theorem 5.32).

By the definition of the inverse matrix, $AA^{-1} = I_n$; using the fact that $\det(I_n) = 1$ for all n (which, again, we will take as fact at this point; this is Theorem 5.31), we have $\det(A^{-1}) \cdot \det(A) = \det(A^{-1}A)$, from which the result follows.

(c) We can verify this directly: $(c \cdot A) \cdot c^{-1}A^{-1} = I = c^{-1}A^{-1} \cdot (c \cdot A)$.

(d) We have $A^T(A^{-1})^T = (A^{-1}A)^T = I^T = I$, as well as $(A^{-1})^T \cdot A^T = (AA^{-1})^T = I^T = I$, so we indeed have $(A^T)^{-1} = (A^{-1})^T$.

(e) By the definition of the inverse matrix, $AA^{-1} = A^{-1}A = I$, and the same goes for B. Observe that $(AB)^{-1}(B^{-1}A^{-1}) = (A(BB^{-1}))A^{-1} = (AI)A^{-1} = AA^{-1} = I$. Hence, $(AB)^{-1} = B^{-1}A^{-1}$ as desired. (The generalization is left as an exercise.)

Do note that this is not an exhaustive list of properties of the inverse, but it will do as a laundry list for introductory purposes. $\square$

Furthermore, the *Invertible Matrix Theorem* guarantees several additional statements, under the condition that all are true or all are false:

Theorem 5.31: The Invertible Matrix Theorem (IMT)

For a given $n \times n$ matrix M with elements in $\mathbb{F}^n$ for some field $\mathbb{F}$, the following statements are either all true or all false:

(a) M is invertible.

(b) M is row equivalent to I_n.

(c) M is column equivalent to I_n.

(d) M has n pivots.

(e) $\det M \neq 0$.

(f) $\operatorname{rank}(M) = n$.

(g) $M\mathbf{x} = \mathbf{0}$ has only the trivial solution $\mathbf{x} = 0$.

(h) $\ker(M) = 0$.

(i) $\operatorname{nul}(M) = \{0\}$.

(j) The columns of M are linearly independent.

(k) The columns of M span $\mathbb{F}^n$.

(l) $\operatorname{col}(M) = \mathbb{F}^n$.

(m) There exists an $n \times n$ matrix N such that $MN = I_n = NM$.

(n) M^T is invertible.

(o) 0 is not an eigenvalue of M.

(p) M is a product of finitely many *elementary matrices*: matrices that differ from I_n by a single row operation.

(q) M has a left inverse; i.e. a matrix L such that $LM = I_n$, or a right inverse; i.e. a matrix R such that $MR = I_n$, in which case both L and R exist and $L = R = M^{-1}$.

Many of these concepts are as yet untouched upon, but we will cover them in future sections. We will go through these statements when appropriate in the text. (Henceforth, we will refer to statements in the IMT as 5.29(a), 5.29(b), and so forth.) There is a bit of unfinished business from earlier in this section - the transpose of a matrix, which is critical to computing the inverse, and in turn, for performing matrix division:

Definition 5.19: Transpose of a Matrix

Let $M \in \mathbb{M}_{m \times n}$ be an $m \times n$ matrix. Then the transpose of M, denoted M^T, is the $n \times m$ matrix whose rows are the columns of M, and vice versa (i.e. the matrix M is "reflected" about its diagonal). Rigorously speaking, $[M]_{i,j} = [M^T]_{j,i}$ for all $1 \leq i \leq m, 1 \leq j \leq n$.

Example 5.13.2.2. *Let*
$$M = \begin{bmatrix} 9 & 1 & 0 \\ 2 & 5 & 4 \end{bmatrix}$$

Compute MM^T.

Solution 5.13.2.2. *We have*
$$M^T = \begin{bmatrix} 9 & 2 \\ 1 & 5 \\ 0 & 4 \end{bmatrix}$$

so then
$$MM^T = \begin{bmatrix} 9 \cdot 9 + 1 \cdot 1 + 0 \cdot 0 & 9 \cdot 2 + 1 \cdot 5 + 0 \cdot 4 \\ 2 \cdot 9 + 5 \cdot 1 + 4 \cdot 0 & 2 \cdot 2 + 5 \cdot 5 + 4 \cdot 4 \end{bmatrix} = \begin{bmatrix} 82 & 23 \\ 23 & 45 \end{bmatrix}$$

The transpose shares similar properties with the inverse; these also apply to all matrices of all dimensions:

Theorem 5.32: Properties of the Transpose

(a) $(M^T)^T = M$; id est, the transposition operator is an involution.

(b) $(A + B)^T = A^T + B^T$ for all matrices A, B such that $A + B$ is well-defined.

(c) $(AB)^T = B^T A^T$.

(d) For any scalar $c \in \mathbb{R}$, $(c \cdot M)^T = c \cdot M^T$.

(e) $\det(M) = \det(M^T)$.

(f) $(M^T)^{-1} = (M^{-1})^T$ if M is invertible.

Proof.

(a) By definition, the element $M_{i,j}$ is the element $(M^T)_{j,i} = (M^T)^T_{i,j}$. Since M has the same elements as $(M^T)^T$, and in all the same positions, we conclude that they are equal.

(b) Note that $A^T + B^T$ has elements that are the sums of the elements in the respective positions of A^T and B^T. In turn, these elements in the positions $A^T_{i,j}$ and $B^T_{i,j}$ are in the positions $A_{j,i}$ and $B_{j,i}$. Thus, $A^T + B^T$ has elements in the (i,j) positions that are in the (j,i) positions of $A + B$, or the (i,j) positions of $(A + B)^T$. Hence, $A^T + B^T = (A + B)^T$.

(c) Observe that

$$(AB)_{i,j} = \sum_{k=1}^{n} a_{i,k} b_{k,j}.$$

Transposing a matrix switches its rows and columns, so

$$((AB)^T)_{i,j} = (AB)_{j,i} = \sum_{k=1}^{n} a_{j,k} b_{k,i}$$

$$(B^T A^T)_{i,j} = \sum_{k=1}^{n} (B^T)_{i,k} (A^T)_{k,j} = \sum_{k=1}^{n} b_{k,i} a_{j,k}.$$

Hence, the desired conclusion follows.

(d) Since c is a scalar (a real number), we just have $c^T = c$. Hence, $(cM)^T = M^T c^T = c \cdot M^T$.

(e) This proof is largely left as an exercise, but we mainly exploit the symmetry of the transposition operator, re-defining each element $a_{i,j} \in A$ as $b_{j,i} \in A^T$.

(f) This is 5.28(d).

Once again, this is far from an exhaustive list, but it serves as a decent introduction point for future discussion. $\square$

With all of these tools in our metaphorical toolbox, we can now proceed to work through a few computational, and theoretical, examples:

Example 5.13.2.3. *We proved above that, for matrices A and B, $(AB)^T = B^T A^T$. That property may seem nonintuitive, though. Wouldn't it make more sense for $(AB)^T$ to equal $A^T B^T$? Is it even true that $(AB)^T = A^T B^T$ in general? Let's find out!*

Solution 5.13.2.3. *Since we know that matrix multiplication is generally not commutative, we conjecture that $(AB)^T \neq A^T B^T$ in general. To prove our conjecture, we need to find a counterexample. To make our lives easier, let's have A and B be 2×2 matrices.*

Let $A = \begin{bmatrix} a & b \\ c & d \end{bmatrix}$ and $B = \begin{bmatrix} e & f \\ g & h \end{bmatrix}$. Then $(AB)^T = \begin{bmatrix} ae+bg & af+bh \\ ce+dg & cf+dh \end{bmatrix}^T = \begin{bmatrix} ae+bg & ce+dg \\ af+bh & cf+dh \end{bmatrix}$.

What about $A^T B^T$? $A^T B^T = \begin{bmatrix} a & c \\ b & d \end{bmatrix} \begin{bmatrix} e & g \\ f & h \end{bmatrix} = \begin{bmatrix} ae+cf & ag+ch \\ be+df & bg+dh \end{bmatrix}$, which is not equal to what we got for $(AB)^T$. Therefore, $(AB)^T \neq A^T B^T$ in general.

Exercise 107. *In fact, in the equation $(AB)^T \neq A^T B^T$, one side might not exist even if the other does. Can you come up with some examples?*

5.13.3 Elementary Row Operations

Just as we can simplify certain algebraic equations and expressions, so we can "simplify" matrices, in a sense. That is, we can reduce matrices to other matrices that can be more easily manipulated, and whose intrinsic properties are more easily extractable.

A *row operation* is a transformation on a matrix that plays an important role in solving simultaneous linear equations, as well as in sub-applications such as finding the inverse of a matrix. There are three types of elementary row operations that we can perform on a matrix to transform it:

Definition 5.20: Elementary Row Operations

1. Row interchangement: we can switch any two rows of the matrix ($R_i \iff R_j$).

2. Scalar multiplication: we can multiply any row by a scalar (real number) ($R_i \implies cR_i$).

3. Row addition: we can add a scalar multiple of one row to another row ($R_j \implies R_j + cR_i$).

Remarkably, through some combination of these three elementary operations, we can reduce even the most complicated matrices to very simple ones! In fact, for the vast majority of matrices, we can reduce them to the identity matrix, which has clearly-defined properties that carry over to the original matrices.

Indeed, the ultimate goal of applying row operations is usually to reduce it to *row echelon form*, a reduced form of a matrix which is conducive to the solution of a system of linear equations. More specifically:

Definition 5.21: Row and Column Echelon Form

A matrix M is in *row echelon form*, or *REF form*,[a] iff

- all rows with at least one non-zero element (henceforth *nonzero rows*) are above rows consisting of all zeros, and

- the leading coefficient (or *pivot*) of each row is strictly to the right of the pivot of the row above it.[b]

M is in *column echelon form* (CEF form) if M^T is in REF form. These matrices have the same properties as REF matrices.

---

[a] "Row echelon form form?" I know, it doesn't make sense.

[b] Some authors impose the additional condition that the pivot must also be 1 in every row, but we will not. However, this does apply to *reduced* REF form.

Example 5.13.3.1. *The following matrix is in REF form:*

$$\begin{bmatrix} 2 & 3 & 5 & 3 \\ 0 & 4 & 2 & 6 \\ 0 & 0 & 0 & -1 \end{bmatrix}$$

but this one isn't:

$$\begin{bmatrix} 0 & 0 & 0 & 0 \\ 1 & 2 & 3 & 4 \\ 0 & 0 & -2 & -3 \end{bmatrix}$$

However, if we were to interchange the first and second rows, and then the second and third rows, then the second matrix would be in REF form.

We can subsequently define an *elementary matrix:*

Definition 5.22: Elementary Matrices

An *elementary matrix* is a matrix that differs from the identity matrix by a single elementary row operation.

Furthermore, we can define *reduced row echelon form,* or *RREF form:*

Definition 5.23: Reduced Row Echelon Form

A matrix M is in *reduced row echelon form* (RREF form) iff

- M is in REF form;
- the pivot in each nonzero row is 1;
- and each column containing a 1 has zeros everywhere else.

If we explore these concepts by converting a few sample matrices to row echelon form and to reduced row echelon form, we might notice that a matrix may have different REF forms, but the RREF form of any given matrix is unique. Indeed, we can prove this rigorously:

Proof. Note: This proof is adapted from Thomas Yuster, "The Reduced Row Echelon Form of a Matrix is Unique: A Simple Proof" (1984).

Let M be an $m \times n$ matrix; we will induct upon n. $n = 1$, the base case, is rather trivial (since the topmost element must be 1 and all others are 0).

Let M' be the matrix resulting from removing the nth column of M. Any sequence of elementary row operations on M that converts it to RREF form also converts M' to RREF form. By induction, if we assume that there exists two distinct RREF forms A and B of M, then they can only differ in the nth column.

Hence assume that $A \neq B$. Then there exists an integer i such that row i of A is not equal to row i of B. Let v be a column vector with $Av = 0$. Then $Bv = 0$ and $(A - B)v = 0$. The first $n - 1$ columns of $B - C$ are zero columns, so the ith coordinate of $(A - B)v$ is equal to $(a_{in} - b_{in})v_n$. Since $a_{in} - b_{in} \neq 0$, we have $v_n = 0$. Thus, the nth columns of both A and B must contain pivots of 1; otherwise, they would be free columns (which would cause x_n to be arbitrary).

Since the first $n - 1$ columns of A are the same as those of B, however, the pivots must appear in the same row for both matrices, specifically the first row of zeros of the RREF of M'. Since the other entries in the nth columns of A and B are all zero, it follows that $A = B$, which is a contradiction. $\square$

5.13.4 Studying the Determinant

One of the main properties of a matrix by which the study of matrix algebra lives and dies is the *determinant*. Essentially, this single number assigned to a matrix reflects several of its algebraic and geometric properties that define the groundwork for matrices in general. As linear transformations, matrices rely on the meaning that can be gleaned from their determinant.

Definition 5.24: Determinant of a Matrix

Every matrix M has a *determinant* $\det M$, a quantity roughly related to the idea of a matrix as a linear transformation. For a 2×2 matrix of the form

$$M = \begin{bmatrix} a & b \\ c & d \end{bmatrix}$$

$\det M = ad - bc$. For larger matrices, we express the determinant as the alternating sum of determinant of smaller matrices composed of elements in the larger matrix. This is perhaps best illustrated by concrete demonstration.

For a 3×3 matrix

$$M = \begin{bmatrix} a & b & c \\ d & e & f \\ g & h & i \end{bmatrix}$$

we have

$$\det M = a \cdot \det \left| \begin{bmatrix} e & f \\ h & i \end{bmatrix} \right| - b \cdot \det \left| \begin{bmatrix} d & f \\ g & i \end{bmatrix} \right| + c \cdot \det \left| \begin{bmatrix} d & e \\ g & h \end{bmatrix} \right|$$

with the determinant of a 4×4 matrix being the alternating sum of each element in the first row multiplied by the determinant of the 3×3 matrix with no elements in the same row or column as the chosen element. In general, the sum for an $n \times n$ square matrix consists of n terms of alternating sign, beginning with a positive term. (The smaller matrices are known as *minors;* the determinant formula for, say, an $n \times n$ matrix, recursively depends on the determinant of the $(n-1) \times (n-1)$ minor, which itself depends on the $(n-2) \times (n-2)$ minor determinant, and so forth. In this way, the determinant function has recursive order $n - 2$, since the determinant of a 2×2 matrix can be written in explicit form.)

The following are elementary examples of determinants for 2×2 and 3×3 matrices:

Example 5.13.4.1. *Compute the determinants of the matrices*

(a)
$$A = \begin{bmatrix} 20 & 4 \\ 19 & 5 \end{bmatrix}$$

(b)
$$B = \begin{bmatrix} 3 & 4 & 4 \\ 6 & 9 & 1 \\ 2 & 4 & 5 \end{bmatrix}$$

Solution 5.13.4.1.

(a) $\det A = 20 \cdot 5 - 19 \cdot 4 = \boxed{24}$.

(b)
$$\det B = 3 \cdot \det \left| \begin{bmatrix} 9 & 1 \\ 4 & 5 \end{bmatrix} \right| - 4 \cdot \det \left| \begin{bmatrix} 6 & 1 \\ 2 & 5 \end{bmatrix} \right| + 4 \cdot \det \left| \begin{bmatrix} 6 & 9 \\ 2 & 4 \end{bmatrix} \right|$$

$$= 3 \cdot (9 \cdot 5 - 4 \cdot 1) - 4 \cdot (6 \cdot 5 - 2 \cdot 1) + 4 \cdot (6 \cdot 4 - 2 \cdot 9) = 3 \cdot 41 - 4 \cdot 28 + 4 \cdot 6 = \boxed{35}.$$

At last, we can close our discussion of matrix operations by returning to the definition of the quotient operation on the matrix space: namely, that the quotient $\frac{A}{B}$, for $A \in \mathbb{M}_{m_1 \times n_1}$, is equal to AB^{-1}, and moreover exists iff B is invertible and B^{-1} is square (such that its determinant exists; the determinant is undefined for non-square matrices).

Also worthy of note - and highly useful in longer computations and computer programming applications - is that the determinant is the product of diagonal entries for any *triangular matrix*; i.e. a matrix whose non-zero entries are either all above the diagonal (upper triangular) or all below the diagonal (lower triangular). Note that this includes the diagonal itself, so a matrix with non-zero entries on the diagonal may still be upper/lower triangular, as appropriate. (A matrix that is both upper and lower triangular, then, has all non-zero entries on the diagonal, and so is called a *diagonal matrix.*)

Theorem 5.33: Determinant as the Product of Entries

The determinant of an upper or lower triangular matrix is the product of its diagonal entries.

Proof. We will prove this theorem for upper triangular matrices (the proof for lower triangular matrices is similar and is left as an exercise).

Consider the diagonal elements of this $n \times n$ matrix M, namely

$$M_{1,1}, M_{2,2}, M_{3,3}, \ldots, M_{n,n}$$

We can first subtract $M_{n,n}$ from $M_{1,n}, M_{2,n}, M_{3,n}, \ldots, M_{n-1,n}$ to transform those elements into 0's, then repeat the process with the other diagonal elements. Eventually, we will obtain a diagonal matrix with its only non-zero elements being $M_{1,1}, M_{2,2}, M_{3,3}, \ldots, M_{n,n}$. The determinant then evaluates to $M_{1,1} \cdot \det M_2$, where M_2 is the matrix consisting of all elements in M other than those elements in the same row and column as $M_{1,1}$. In turn, $\det M_2 = M_{2,2} \cdot \det M_3$, where M_3 is the matrix with elements down and to the right of $M_{3,3}$, and so forth. Thus, $\det M = \prod_{k=1}^{n} M_{k,k}$, which is what we originally claimed. $\qquad \square$

We can also prove this by induction; the induction proof is left as an exercise.

In particular, the keen observer may notice a peculiar property of the general identity matrix in this regard:

Corollary 5.33: Determinant of the Identity Matrix

For all positive integers n, the determinant of the $n \times n$ identity matrix I_n is equal to 1.

Furthermore, the determinant is beautifully multiplicative, meaning that the product of the determinants is the determinant of the product of matrices:

Theorem 5.34: Multiplicativity of the Determinant

For all square matrices A, B, $\det(A) \cdot \det(B) = \det(AB)$.

Proof. First assume that A is non-invertible; then $\det A = 0$, and it follows that AB is also non-invertible. (For contradiction, assume $AB^{-1} = C$; then $ABC = I$ and BC is then a right inverse of A, which is a contradiction.) Then $\det(AB) = 0$ as well. Thus, from $0 = 0 \cdot \det B$, we obtain the desired result, $\det(AB) = \det A \cdot \det B$.

Henceforth assume that A is invertible, so that $\det A \neq 0$. Then A is a product of elementary matrices,[47] namely

$$A = E_k E_{k-1} E_{k-2} \cdots E_1$$

Thus,

$$\det(AB) = \det(E_k E_{k-1} E_{k-2} \cdots E_1 B)$$

We now claim that, for any $n \times n$ matrix F, $\det(EF) = \det(E) \cdot \det(F)$. Let $e_i(I_n) = E_i$ for $1 \leq i \leq k$, and write

$$\det(EF) = \det(E_k E_{k-1} E_{k-2} \cdots E_1 F) = \det(e_k e_{k-1} e_{k-2} \cdots e_1(F)) = \alpha \det F$$

for some α. Furthermore, we have

$$\det E = \det(E_k E_{k-1} E_{k-2} \cdots E_1 I_n) = \det(e_k e_{k-1} e_{k-2} \cdots e_1(I_n)) = \alpha \cdot \det(I_n) = \alpha$$

This implies $\det(EF) = \det E \cdot \det F$. $\qquad\square$

Geometrically Interpreting the Determinant

Perhaps the most remarkable elementary property of the determinant is its geometric interpretation. Matrices are not just artificial algebraic abstractions; they are representations of linear transformations that can be extended to the plane, and to geometric space as a whole. Indeed, we can extract the columns of any matrix and make a profound, far-reaching, general statement regarding their geometric relation:

Theorem 5.35: The Geometric Interpretation of the Determinant

The absolute value of the determinant of the $n \times n$ matrix M is the volume of the parallelepiped in $\mathbb{R}^3$ with sides constructed from the column vectors of M. (The sign of the determinant indicates whether or not this transformation is *orientation-preserving;* i.e. whether it maintains the same clockwise or counter-clockwise direction.)

Note that these edges are akin to the three dimensions of a rectangular prism, of which a parallelepiped is a translation/dilation. These vectors also cannot lie in the same plane.

Proof. The parallelepiped P, in general, can be formed from multiplying with elementary matrices the cube with volume

$$\det \begin{bmatrix} a_1 & 0 & 0 \\ 0 & b_2 & 0 \\ 0 & 0 & c_3 \end{bmatrix}$$

such that $P = C E_1 E_2 E_3 \ldots E_i$. The transformations which keep the volume of the cube constant have determinant ± 1, while those which scale the volume of the cube by a scalar factor are equivalent to multiplying a column vector by said scalar, or adding a multiple of a column to another column, which elongates the vector by that scalar quantity. Hence $\det P = \det C \cdot \det E_1 \det E_2 \det E_3 \cdots \det E_i$. $\qquad\square$

This is admittedly not so much a rigorous proof as it is an intuitive exploration into this fascinating concept, but we would like not to exhaust all opportunity for the reader to discover new, hidden truths regarding this interpretation on his or her own!

Example 5.13.4.2. *What is the volume of the parallelepiped with* $u = \langle 1, 2, 4 \rangle$, $v = \langle 3, 5, 7 \rangle$, $w = \langle 0, 8, 9 \rangle$?

Solution 5.13.4.2. *This is the absolute value of the determinant of the 3×3 matrix*

$$\begin{bmatrix} 1 & 2 & 4 \\ 3 & 5 & 7 \\ 0 & 8 & 9 \end{bmatrix}$$

which is $\boxed{31}$ *by computation.*

---

[47]We can prove this by observing that, if the RREF form of A is I_n, this implies the result. This can be shown by observing that every row operation is equivalent to multiplying by an elementary matrix. To prove *this*, it may help to write A as a row of column matrices.

5.13.5 Linear Independence, Rank, Bases, and Nullity

From examining the structure of a given matrix, we can extract quite a few more properties, the most prominent of which are *rank* and *nullity*. We will not delve into the same level of detail as we did with the more computational aspects (since these are hardly ever tested in a competition setting), but they are nevertheless part of the foundational structure of matrices, so we will briefly cover them and take the first step into the foray to pique the reader's curiosity, and to hopefully inspire motivation.

Definition 5.25: Rank of a Matrix

The *rank* of a matrix M in REF form is the number of non-zero rows in M. (This is also equal to the dimension of the space spanned by the rows of M, the *row space* of M.)

Using this definition as a springboard, we can consider the notion of *linear independence* and *linear dependence* with greater context:

Definition 5.26: Linear Independence

A set of vectors is *linearly dependent* iff at least one vector in the set can be written as a linear combination of the others. Otherwise, the set is *linearly independent*.

Example 5.13.5.1. *Is the set* $\{\langle 1, 1\rangle, \langle 2, 3\rangle, \langle -4, 2\rangle\}$ *linearly independent or linearly dependent?*

Solution 5.13.5.1. *The general strategy for determining linear (in)dependence is to consolidate the column vectors into a single matrix, then determine the null space of the resulting matrix:*

$$\begin{bmatrix} 1 & 2 & -4 \\ 1 & 3 & 2 \end{bmatrix} \begin{bmatrix} x_1 \\ x_2 \\ x_3 \end{bmatrix} = \begin{bmatrix} 0 \\ 0 \end{bmatrix}$$

We first should row reduce the matrix to RREF form, yielding

$$\begin{bmatrix} 1 & 2 & -4 \\ 0 & 1 & 6 \end{bmatrix} = \begin{bmatrix} 1 & 0 & -16 \\ 0 & 1 & 6 \end{bmatrix}$$

Finally, taking the product of the matrices and equating coefficients yields $x_1 - 16x_3 = 0$, $x_2 + 6x_3 = 0$ *(since the zero augment is invariant with respect to all three elementary row operations; i.e. nothing can change it from being all zeros). This implies* $x_1 = 16x_3$, $x_2 = -6x_3$. *Hence, the set is linearly dependent, as we can choose* x_1 *and* x_2 *in terms of* x_3 *(for instance,* $(x_1, x_2, x_3) = (16, -6, 1)$*). Indeed, we can verify that* $16 \cdot \langle 1, 1\rangle - 6 \cdot \langle 2, 3\rangle + 1 \cdot \langle -4, 2\rangle = \mathbf{0}$.

We can make a broad claim regarding linear independence in any vector space, for any number of vectors, provided that the matrix formed from them is square:

Theorem 5.36: Linear Dependence and the Determinant

A set of n vectors in $\mathbb{R}^n$ is linearly independent iff the determinant of the matrix with its columns as the vectors is nonzero. Otherwise, the determinant is zero and the column vectors are linearly dependent.

Proof. Observe that, if two matrices are row equivalent, then they both have zero or nonzero determinant. If the column vectors in a matrix are linearly dependent, then the matrix has rank less than n (since the column rank is equal to the row rank). Thus, it is row equivalent to a matrix with a zero column, which has determinant 0. $\square$

Equipped with the ideas of linear dependence and independence, we can now venture into new territory. Before we can fully discuss the concept of nullity, we first must define a requisite term:

Definition 5.27: Null Space of a Matrix

The *null space* (or *kernel*) of a matrix A, denoted $N(M)$ or $\text{nul}(M)$ (alternatively, $\ker(M)$), is the set of all n-dimensional column vectors such that $Ax = 0$. That is, if we have

$$A = \begin{bmatrix} a_{1,1} & a_{1,2} & a_{1,3} & \cdots & a_{1,n} \\ a_{2,1} & a_{2,2} & a_{2,3} & \cdots & a_{2,n} \\ a_{3,1} & a_{3,2} & a_{3,3} & \cdots & a_{3,n} \\ \vdots & \vdots & \vdots & \ddots & \vdots \\ a_{m,1} & a_{m,2} & a_{m,3} & \cdots & a_{m,n} \end{bmatrix}$$

$$x = \begin{bmatrix} x_1 \\ x_2 \\ x_3 \\ \vdots \\ x_n \end{bmatrix}$$

then we must have

$$\sum_{j=1}^{m} \left(\sum_{k=1}^{n} a_{j,k} x_k \right) = 0$$

The *nullity* of A is then $\dim(N(A))$.

Theorem 5.37: Null Space is a Subspace of $\mathbb{R}^n$

For all matrices $A \in \mathbb{M}_{m,n}$, $N(A)$ is a subspace of $\mathbb{R}^n$.

Proof. We show that $N(A)$ is non-empty, and that it is closed under both addition and scalar multiplication.

To show that $N(A)$ is non-empty, we simply substitute $x = \mathbf{0} \implies \mathbf{Ax = 0}$, hence $\mathbf{0} \in \mathbf{N(A)}$ for all A (but this is a trivial solution, which we usually ignore. In this circumstance, however, it suffices for the purposes of the proof).

To show that $N(A)$ is closed under addition, take $x, y \in N(A)$; then $Ax = Ay = 0$ and $A(x + y) = Ax + Ay = 0 \implies x + y \in N(A)$. Finally, to show that $N(A)$ is closed under scalar multiplication, if $x \in N(A)$, then for $c \in \mathbb{R}$, $A(cx) = c \cdot Ax = c \cdot 0 = 0 \implies cx \in N(A)$.

This proves that $N(A)$ is a subspace of $\mathbb{R}^n$ for all n. $\qquad\square$

One of the most important results regarding matrix nullity is the *Rank-Nullity Theorem,* which intertwines rank and nullity for all matrices:

Theorem 5.38: Rank-Nullity Theorem

If $T : V \to W$ is a linear transformation (of a finite-dimensional vector space), then $\text{rank}(T) + \text{nullity}(T) = \dim(V)$. In other words, Rank + Nullity = Dimension of Domain.

Proof. There are actually multiple proofs of this theorem, but we will provide the following, which involves only the techniques we have discussed up to this point.

Let $A \in \mathbb{M}_{m \times n}(\mathbb{F})$ with c linearly independent columns (i.e. $\text{rank}(A) = c$). We show that there exists a set of $n - c$ linearly independent solutions to the equation $Ax = 0$, and furthermore, every other solution to the equation is a linear combination of the set of $n-c$ solutions. Id est, we construct a matrix $M \in \mathbb{M}_{m \times n}(\mathbb{F})$ whose columns form a basis (see below) of $N(A)$.

WLOG assume that the first c columns of A are linearly independent. Then we have

$$A = \begin{bmatrix} a_1 & a_2 \end{bmatrix},$$

where A_1 is an $m \times c$ matrix, and A_2 is an $m \times (n-c)$ matrix (both with entries in the field $\mathbb{F}$), such that A_1 has c linearly independent column vectors, and A_2 has $n-c$ columns that are all linear combinations of the column vectors of A_1. Hence, $A_2 = A_1 B$ for some $c \times (n-c)$ matrix B; thus,

$$A = \begin{bmatrix} A_1 & A_1 B \end{bmatrix}$$

. Let

$$M = \begin{bmatrix} -B \\ I_{n-c} \end{bmatrix}.$$

Observe that $AM = -A_1 B + A_1 B = 0$; therefore, each column of M is a solution to $Ax = 0$. Moreover, the columns of M are linearly independent, and constitute a set of $n-c$ linearly independent solutions to $Ax = 0$.

Henceforth, we show that any solution to $Ax = 0$ is a linear combination of the column vectors of M. Let

$$u = \begin{bmatrix} u_1 \\ u_2 \end{bmatrix}$$

with elements from the field $\mathbb{F}^n$, be a vector such that $Au = 0$. Since the columns of A_1 are linearly independent, $A_1 x = 0 \implies x = 0$. Therefore,

$$Au = 0$$
$$A_1 u_1 + A_1 B u_2 = 0$$
$$\implies u_1 + B u_2 = 0 \implies u_1 = -B u_2$$
$$\implies u = M u_2$$

It follows that the columns of M form $N(M)$, the null space of M. Therefore, the nullity of A is $n-c$, and its rank is c. We conclude that $\text{rank}(A) + \text{nullity}(A) = n$; this completes the proof. $\qquad\square$

(Briefly) Studying the Basis

Building on the concepts of linear independence and dependence, we can begin to formulate our idea of a *basis*. A basis of a vector space V is essentially the "smallest" linearly independent set of vectors that spans every element in the vector space (in that every $v \in V$ can be written as a linear combination of the basis elements).

Example 5.13.5.2. *A basic example would be the basis*

$$\{e_1, e_2\} = \left\{ \begin{bmatrix} 1 \\ 0 \end{bmatrix}, \begin{bmatrix} 0 \\ 1 \end{bmatrix} \right\}$$

of $\mathbb{R}^2$, as every ordered pair (x, y) can be written as a linear combination of $(1, 0)$ and $(0, 1)$.

Remark. Note that $\dim V$ is the dimension of the basis as well.

We can define the basis precisely as follows:

Definition 5.28: Basis of a Vector Space

A *basis* B of a vector space V over a field $\mathbb{F}$ is a linearly independent subset of V that spans V. Id est, $B \subset V$ is a basis of V iff it satisfies both of the following properties:

- Linear independence: For every finite subset $\{b_1, b_2, b_3, \ldots, b_n\} \subset B$ and all $a_1, a_2, a_3, \ldots, a_n \in \mathbb{F}$, if $\displaystyle\sum_{i=1}^{n} a_i b_i = 0$, then $a_1 = a_2 = a_3 = \ldots = a_n = 0$.

- The spanning property: $\forall v \in V$, we can choose $v_1, v_2, v_3, \ldots, v_n \in \mathbb{F}$ and $b_1, b_2, b_3, \ldots, b_n \in B$ such that $\displaystyle v = \sum_{i=1}^{n} v_i b_i$.

The scalars v_i are the "coordinates" of the vector v, and are uniquely determined.

Remark. Though we have not yet covered vector spaces to a considerable degree of depth, we will do so in Chapter 6. For now, we can work with the (fairly loose) definition of a vector space as a collection of vectors that is closed under addition and scalar multiplication. In general, we can derive bases for the row and column spaces of a matrix M from its row echelon form (not necessarily its *reduced* row echelon form):

Definition 5.29: Row and Column Bases

When the matrix M is row reduced, the row space consists of basis vectors that are the non-zero rows of M. The column space, similarly, consists of basis vectors that are the non-zero columns of M.

5.13.6 Examples

Example 5.13.6.1. *(2015 Berkeley Mini Math Tournament) A scooter has 2 wheels, a chair has 6 wheels, and a spaceship has 11 wheels. If there are 10 of these objects, with a total of 50 wheels, how many chairs are there?*

Solution 5.13.6.1. *The average number of wheels each object has is $\dfrac{50}{10} = 5$. Scooters have 3 fewer wheels than average, chairs have 1 more than average, and spaceships have 6 more than average. Then 2 scooters and 1 spaceship together (3 total objects) have an average of 5 wheels; 1 scooter and 3 chairs together (4 total objects) have an average of 5 wheels as well. Since $3 + 3 + 4 = 10$, 2 sets of the 3 objects mentioned above along with 1 set of the 4 objects mentioned above have a total of 50 wheels. Chairs are only part of the set of 4 objects, so there are a total of $\boxed{3}$ chairs.*

Example 5.13.6.2. *(2015 Berkeley Mini Math Tournament) Consider the following linear system of equations.*

$$1 + a + b + c + d = 1$$
$$16 + 8a + 4b + 2c + d = 2$$
$$81 + 27a + 9b + 3c + d = 3$$
$$256 + 64a + 16b + 4c + d = 4$$

Find $a - b + c - d$.

Solution 5.13.6.2. *Consider a polynomial $P(x) = x^4 + ax^3 + bx^2 + cx + d$. Then $P(1) = 1, P(2) = 2, P(3) = 3, P(4) = 4$. We have $P(-1) = 1 - a + b - c + d \implies 1 - P(-1) = a - b + c - d$. From here, we can apply Vieta's formulas on the polynomial $P(x) - x$ to get $\boxed{-118}$.*

5.13.7 Exercises

Exercise 108. *Consider the system of equations*

$$2x_1 + 3x_2 - 4x_3 + x_4 = 0$$

$$-2x_1 - 2x_2 + 5x_3 - x_4 = 3$$

$$3x_1 - x_3 + 2x_4 = 7$$

$$x_2 + 6x_3 - 7x_4 = -8$$

(a) *Represent the system as an augmented matrix.*

(b) *Solve the system.*

(c) *Verify that your solution in (b) is correct by reducing the augmented matrix you obtained in (a) to RREF form.*

(d) *If we change the coefficients of the variables, when will the system have no, one, or infinite solutions?*

(e) *Give an example of a system for each scenario in (d).*

Exercise 109. *Prove that row equivalence is an equivalence relation.*

Exercise 110. *Consider the matrix*

$$M = \begin{bmatrix} 5 & 9 \\ 4 & 7 \end{bmatrix}$$

(a) *What is $\det M$?*

(b) *Is the matrix M singular or invertible? If the former, provide a proof. If the latter, provide a proof and compute the inverse of M, M^{-1}.*

(c) *What is M^2? Compute $\det M^2$.*

(d) *What can we conclude about $\det M^n$ for some positive integer n?*

(e) *Consider the matrix*

$$A = \begin{bmatrix} 5 & 9 & 5 & 9 \\ 4 & 7 & 4 & 7 \\ 5 & 9 & 5 & 9 \\ 4 & 7 & 4 & 7 \end{bmatrix}$$

What is $\det A$? Can we derive any relationship between $\det M$ and $\det A$? In general, what effect does "copy-pasting" an $n \times n$ square matrix three times to form a $2n \times 2n$ square matrix have on the determinant of the original matrix?

Exercise 111.

(a) *Determine whether the vectors $\mathbf{x} = (1, 2, 3)$, $\mathbf{y} = (4, 5, 6)$, and $\mathbf{z} = (7, 8, 9)$ are linearly independent or linearly dependent.*

(b) *Do $\mathbf{x}$, $\mathbf{y}$, and $\mathbf{z}$ span $\mathbb{R}^3$? That is, does $\mathbb{R}^3 = Span\{x, y, z\}$?*

(c) *Show that, if a vector space V has a basis $\{v_1, v_2, v_3, \ldots, v_n\}$, then any set consisting of more than n vectors must be linearly dependent.*

Exercise 112.

(a) *Find a basis for P_n, the space of polynomials with degree n.*

(b) *Find a basis for $M_{2 \times 2}(\mathbb{R})$, the space of 2×2 matrices with elements in $\mathbb{R}$.*

(c) Find a basis for $M_{k \times k}(\mathbb{R})$, the space of $k \times k$ matrices with elements in $\mathbb{R}$.

(d) Find a basis for $T_{k \times k}(\mathbb{R})$, the space of $k \times k$ matrices with elements in $\mathbb{R}$ and trace zero.

Exercise 113. Let $\mathbf{u} = \langle \mathbf{1}, -\mathbf{3} \rangle$, $\mathbf{v} = \langle \mathbf{4}, \mathbf{2} \rangle$.

(a) Compute $\mathbf{u} \cdot \mathbf{v}$.

(b) Compute $||u||$ and $||v||$.

Exercise 114. Prove that $(A^T)^{-1} = (A^{-1})^T$ for all invertible matrices $A \in M_{n \times n}(\mathbb{R})$.

Extension Exercise 42. Let V be a vector space, and let $M, N \subseteq V$ with $\dim V < \infty$. Show that $\dim(M + N) = \dim M + \dim N - \dim(M \cap N)$.

Extension Exercise 43. Let A be a square $n \times n$ matrix, and let $b \in \mathbb{R}^n$ be a vector such that the equation $Ax = b$ has a unique solution for x. Prove that A is invertible.

Extension Exercise 44. Let M be a 2×2 matrix with complex elements. For each of the following, either prove the statement or provide a counterexample:

(a) If $M^4 = I$, then $M = I, -I, i \cdot I, -i \cdot I$ (where $i = \sqrt{-1}$).

(b) If $M^4 = M^2$, then $M = 0, I, -I$.

Chapter 5 Review Exercises

Exercise 115. *Suppose x, y, and z are real numbers such that $x^2 + y^2 + z^2 = 1$ with $x + 4y + 9z$ maximized. Compute $x + y + z$.*

Exercise 116. *Consider the 2×2 matrix*
$$M = \begin{bmatrix} 1 & 3 \\ 4 & 8 \end{bmatrix}$$

(a) (i) *Show that the RREF form of M is the identity matrix I.*

    (ii) *Show that any 2×2 matrix can be RREF simplified to the identity matrix I.*

(b) *Compute M^2 and M^3. Compute M^T, $\det M$, and $\det M^T$.*

(c) (i) *If M is invertible, compute M^{-1}. Otherwise, show that M is singular.*

    (ii) *Prove that $\det M^{-1} = (\det M)^{-1}$ for all invertible matrices M.*

(d) (i) *What are the rank and nullity of M?*

    (ii) *State the Rank-Nullity Theorem and use it to verify your answer.*

(e) *Find all row and column bases for M.*

(f) (i) *What is the characteristic polynomial of M?*

    (ii) *What are the eigenvalues and eigenvectors of M?*

Exercise 117. *(1979 AHSME) For each positive number x, let $f(x) = \dfrac{\left(x + \frac{1}{x}\right)^6 - \left(x^6 + \frac{1}{x^6}\right) - 2}{\left(x + \frac{1}{x}\right)^3 + \left(x^3 + \frac{1}{x^3}\right)}$. What is the minimum value of $f(x)$?*

Exercise 118. *(1990 AIME) The sets $A = \{z : z^{18} = 1\}$ and $B = \{w : w^{48} = 1\}$ are both sets of complex roots of unity. The set $C = \{zw : z \in A \text{ and } w \in B\}$ is also a set of complex roots of unity. How many distinct elements are in C?*

Exercise 119. *(2014 AMC 12A) The domain of the function*
$$f(x) = \log_{\frac{1}{2}}\left(\log_4\left(\log_{\frac{1}{4}}\left(\log_{16}\left(\log_{\frac{1}{16}} x\right)\right)\right)\right)$$
is an interval of length $\frac{m}{n}$, where m and n are relatively prime positive integers. What is $m + n$?

Exercise 120. *(2014 AMC 12B) For how many positive integers x is $\log_{10}(x - 40) + \log_{10}(60 - x) < 2$?*

Exercise 121. *(2011 AIME II) Let M_n be the $n \times n$ matrix with entries as follows: for $1 \leq i \leq n$, $m_{i,i} = 10$; for $1 \leq i \leq n-1$, $m_{i+1,i} = m_{i,i+1} = 3$; all other entries in M_n are zero. Let D_n be the determinant of matrix M_n. Then $\sum_{n=1}^{\infty} \frac{1}{8D_n + 1}$ can be represented as $\frac{p}{q}$, where p and q are relatively prime positive integers. Find $p + q$.*

Exercise 122. *(2013 AMC 12A) The sequence $\log_{12} 162$, $\log_{12} x$, $\log_{12} y$, $\log_{12} z$, $\log_{12} 1250$ is an arithmetic progression. What is x?*

Exercise 123. *(2002 AIME I) Consider the sequence defined by $a_k = \dfrac{1}{k^2 + k}$ for $k \geq 1$. Given that $a_m + a_{m+1} + \cdots + a_{n-1} = \dfrac{1}{29}$, for positive integers m and n with $m < n$, find $m + n$.*

Exercise 124. *(2009 AIME II) The sequence (a_n) satisfies $a_0 = 0$ and $a_{n+1} = \dfrac{8}{5}a_n + \dfrac{6}{5}\sqrt{4^n - a_n^2}$ for $n \geq 0$. Find the greatest integer less than or equal to a_{10}.*

Exercise 125. *(2004 AIME I) Let S be the set of ordered pairs (x, y) such that $0 < x \leq 1, 0 < y \leq 1$, and $\left\lfloor \log_2 \left(\frac{1}{x} \right) \right\rfloor$ and $\left\lfloor \log_5 \left(\frac{1}{y} \right) \right\rfloor$ are both even. Given that the area of the graph of S is $\frac{m}{n}$, where m and n are relatively prime positive integers, find $m + n$. The notation $\lfloor z \rfloor$ denotes the greatest integer that is less than or equal to z.*

Exercise 126.

(a) *Briefly explain how Bézout's Identity could be generalized to n variables, for $n \geq 3$. You do not need to prove this generalized form.*

(b) *Briefly explain how Bézout's Identity could be applied to single-variable polynomials. (Hint: Use the Fundamental Theorem of Algebra.)*

Extension Exercise 45. *(2018 HMMT November Guts Round) Over all real numbers x and y, find the minimum possible value of $(xy)^2 + (x + 7)^2 + (2y + 7)^2$.*

Extension Exercise 46. *(2011 AIME II) Let $P(x) = x^2 - 3x - 9$. A real number x is chosen at random from the interval $5 \leq x \leq 15$. The probability that $\lfloor \sqrt{P(x)} \rfloor = \sqrt{P(\lfloor x \rfloor)}$ is equal to $\frac{\sqrt{a} + \sqrt{b} + \sqrt{c} - d}{e}$ where a, b, c, d and e are positive integers and none of $a, b,$ or c is divisible by the square of a prime. Find $a + b + c + d + e$.*

Extension Exercise 47. *(1981 IMO P1) P is a point inside a given triangle ABC. D, E, F are the feet of the perpendiculars from P to the lines BC, CA, AB, respectively. Find all P for which $\frac{BC}{PD} + \frac{CA}{PE} + \frac{AB}{PF}$ is least.*

Extension Exercise 48. *(2012 AIME I) Complex numbers a, b, and c are zeros of a polynomial $P(z) = z^3 + qz + r$, and $|a|^2 + |b|^2 + |c|^2 = 250$. The points corresponding to a, b, and c in the complex plane are the vertices of a right triangle with hypotenuse h. Find h^2.*

Extension Exercise 49. *(2004 USAMO P5) Let a, b, and c be positive real numbers. Prove that*

$$(a^5 - a^2 + 3)(b^5 - b^2 + 3)(c^5 - c^2 + 3) \geq (a + b + c)^3$$

Extension Exercise 50. *(1977 USAMO P3) If a and b are two of the roots of $x^4 + x^3 - 1 = 0$, prove that ab is a root of $x^6 + x^4 + x^3 - x^2 - 1 = 0$.*

Chapter 6

Advanced Algebra

6.1 Binomial Theorem, Part 3: The Multinomial Theorem

The *Multinomial Theorem* is yet another generalization of the Binomial Theorem, only this time, we are not restricted to working with two variables. Rather, we can work with *any number of variables*, with expressions in terms of *multinomials*, or polynomials consisting of multiple terms (almost always reserved for polynomials with four or more terms, since *binomial* refers to a polynomial with two terms and *trinomial* to a polynomial with three terms).

Henceforth, we define the *multinomial coefficient*, an extension of the binomial coefficient (i.e. the "choose" function) that is instrumental in shaping the Multinomial Theorem.

Definition 6.1: Multinomial Coefficients

For any positive integer n and non-negative integers $k_1, k_2, k_3, \ldots, k_m$, the multinomial coefficient is defined as follows:

$$\binom{n}{k_1, k_2, k_3, \ldots, k_m} = \frac{n!}{k_1! k_2! k_3! \cdots k_m!}$$

Remark. We can write the multinomial coefficient $\binom{n}{k_1, k_2, k_3, \ldots, k_m}$ as

$$\binom{k_1}{k_1} \cdot \binom{k_1 + k_2}{k_2} \cdot \binom{k_1 + k_2 + k_3}{k_3} \cdots \binom{k_1 + k_2 + k_3 + \ldots + k_n}{k_n}$$

Demonstrating this is left as an exercise to the reader (the proof amounts to using the definition of the binomial coefficient).

Essentially, the multinomial function is the binomial function except with more than one argument in the bottom. It operates in the same basic way, and again has $\binom{0}{0,0,0,\ldots,0} = 1$ *de facto*. Combinatorially, we can interpret the multinomial coefficient as the number of ways of placing n distinguishable objects into m bins, with k_1 objects in bin 1, k_2 objects in bin 1, ..., k_i objects in bin i, ..., k_m objects in bin m. In addition, using stars and bars, we can show that the number of multinomial coefficients in a multinomial sum is $\binom{n + m - 1}{m - 1}$.

Now we can proceed to state the Multinomial Theorem - the apex of the Binomial Theorem mountain:

Theorem 6.1: Multinomial Theorem

For a positive integer m and non-negative integer n, we have

$$\left(\sum_{i=1}^{m} x_i \right)^n = \sum_{\sum_i k_i = n} \binom{n}{k_1, k_2, k_3, \ldots, k_m} \prod_{i=1}^{m} x_i^{k_i}$$

where the k_i are non-negative integers.

Proof. We couple our main inductive argument on m with the Binomial Theorem. For the base case, $m = 1$, we immediately observe that both sides of the equation equal x_1^n, so there is nothing to prove. Next assume the theorem holds for some m, and we show it holds for $m + 1$.

Then we have

$$(x_1 + x_2 + x_3 + \ldots + x_m + x_{m+1})^n = (x_1 + x_2 + x_3 + \ldots + (x_m + x_{m+1}))^n$$

$$= \sum_{\sum_{i=1}^{m-1} k_i + K = n} \binom{n}{k_1, k_2, k_3, \ldots, k_{m-1}, K} \prod_{i=1}^{m-1} x_i^{k_i} (x_m + x_{m+1})^K$$

by the induction hypothesis, where $K = x_m + x_{m+1}$.

Henceforth, we apply the Binomial Theorem to the last factor, and obtain

$$\sum_{\sum_i^{m-1} k_i + K} = n \binom{n}{k_1, k_2, k_3, \ldots, k_{m-1}, K} \prod_{i=1}^{m-1} x_i^{k_i} \sum_{k_m + k_{m+1} = K} \binom{K}{k_m, k_{m+1}} x_m^{k_m} x_{m+1}^{k_{m+1}}$$

$$= \sum_{\sum_{i=1}^{m+1} k_i = n} \binom{n}{k_1, k_2, k_3, \ldots, k_{m+1}} \prod^{m+1}_{i=1} x_i^{k_i}$$

which completes the inductive step. (To see why the last step holds, write

$$\binom{n}{k_1, k_2, k_3, \ldots, K} \cdot \binom{K}{k_m, k_{m+1}} = \binom{n}{k_1, k_2, k_3, \ldots, k_{m+1}}$$

and then expand each multinomial coefficient into its factorial form.) $\qquad \square$

A particularly interesting corollary of the Multinomial Theorem is its remarkably neat simplification of the sum of all multinomial coefficients:

Corollary 6.1: Sum of Multinomial Coefficients

$$\sum_{\sum_i k_i = n} \binom{n}{k_1, k_2, k_3, \ldots, k_m} = m^n$$

Proof. Exercise. $\qquad \square$

6.1.1 Examples

Example 6.1.1.1. *With explanation, compute each of the following:*

(a) $\dbinom{7}{4,3}$

(b) $\dbinom{10}{3,5,2}$

(c) *The largest power of 2 that divides* $\dbinom{100}{20,20,60}$

(d) *The number of permutations of the "word" $ABBCCCDDDD\ldots$, with 1 A, 2 B's, 3 C's, $\ldots$, 26 Z's*

Solution 6.1.1.1.

(a) $\dbinom{7}{4,3} = \dfrac{7!}{4!3!} = \boxed{35}$. *This is equal to the number of ways to choose a group of 4 people and a group of 3 people from 7 people.*

(b) $\dbinom{10}{3,5,2} = \dfrac{10!}{3!5!2!} = \boxed{2520}$. *This is equal to the number of ways to choose groups of 3, 5, and 2 people from 10 people.*

(c) *This is the largest power of 2 dividing* $\dfrac{100!}{20!^2 60!}$, *which is* $\boxed{2^5}$ *by Legendre's formula.*

(d) *This is given by* $\dfrac{(1+2+3+\ldots+26)!}{1!2!3!\cdots 26!}$.

Example 6.1.1.2. *(2006 AMC 12A) The expression $(x+y+z)^{2006} + (x-y-z)^{2006}$ is simplified by expanding it and combining like terms. How many terms are in the simplified expression?*

Solution 6.1.1.2. *By the Multinomial Theorem, we can write each summand in the forms*

$$\sum_{a+b+c=2006} \frac{2006!}{a!b!c!} x^a y^b z^c$$

$$\sum_{a+b+c=2006} \frac{2006!}{a!b!c!} x^a (-y)^b (-z)^c$$

Like terms cancel each other out, or the coefficient doubles. In each expansion, we have $\dbinom{2006+2}{2} = 2,015,028$ terms before any cancellation takes place. We then find a pattern in the negative terms to get that there are $1003 \cdot 1004$ in total; subtracting this from $2,015,028$ yields $\boxed{1,008,016}$.

6.1.2 Exercises

Exercise 127. *What connection can we make between the Multinomial Theorem and the stars-and-bars/balls-and-urns formula?*

Exercise 128. *Show that*

$$\sum_{\sum_{i=1}^{m} k_i = n} \binom{n}{k_1, k_2, k_3, \ldots, k_m} = m^n$$

for all $m, n \in \mathbb{N}$.

Exercise 129. *(1991 AIME) Expanding $(1+0.2)^{1000}$ by the binomial theorem and doing no further manipulation gives $\binom{1000}{0}(0.2)^0 + \binom{1000}{1}(0.2)^1 + \binom{1000}{2}(0.2)^2 + \ldots + \binom{1000}{1000}(0.2)^{1000} = A_0 + A_1 + A_2 + \ldots + A_{1000}$, where $A_k = \binom{1000}{k}(0.2)^k$ for $k = 0, 1, 2, \ldots, 1000$. For which k is A_k the largest?*

Exercise 130.

(a) *A strict composition of the non-negative integer n is its decomposition into $n = k_1 + k_2 + k_3 + \ldots$ where the k_i are positive integers. A weak composition is similar, except the k_i can also be zero. Show that the exponents in a multinomial expansion form a weak composition (and not a strong one).*

(b) *Provide explicit formulas for the number of strict and weak compositions of a positive integer n.*

Extension Exercise 51. *Try deriving Pascal's n-dimensional simplexes using multinomial coefficients. (Hint: Pascal's Triangle is the Pascal 2-simplex.)*

6.2 Generating Functions and Power Series

6.2.1 Introduction to Generating Functions

We are familiar with the fundamental idea behind infinite series: we are summing terms in a sequence to form a series, from which we can extract a set of astoundingly useful properties. In particular, if we consider infinite series in general, there are several series that seem to be unapproachable, regardless of what angle we try to take to pare them down to a closed form, or at least some kind of recursive pattern. However, with the aid of *generating functions,* we can convert any infinite sequence, no matter how seemingly arbitrary, into a closed form.[48] A generating function is essentially a power series, in that we are translating a sequence into the coefficients of an expression that goes on to infinity, and then condensing that series into a clearly-defined function that represents the values of the sequence.

The generating function of (a_n), denoted $G(a_n; x)$ (or with some other variable in place of x), is constructed by considering the terms of the sequence $(a_n)_{n=0}^{\infty}$ as coefficients in the general form $a_0 + a_1 x + a_2 x^2 + a_3 x^3 + \ldots = \sum_{i=0}^{\infty} a_i x^i$, so that we can directly obtain a closed form, which is what we refer to as the generating function of (a_n).

Example 6.2.1.1. *As a very basic example: the generating function of the infinite sequence consisting of all 1's is* $1 + x + x^2 + x^3 + \ldots = \dfrac{1}{1-x}$ *by the infinite geometric series formula.*

Indeed, it is not difficult to verify that, if the terms of (a_n) are in geometric progression, so will the terms of its generating function (since the common ratio of (a_n) is simply multiplied by x to obtain the common ratio of the generating function). If (a_n) is an arithmetic progression, then its generating function will resemble an arithmetico-geometric series.

Example 6.2.1.2. *To demonstrate the last point: the generating function of* $(1, 2, 3, 4, \ldots)$ *(i.e. $a_n = n$) is*

$$1 + 2x + 3x^2 + 4x^3 + \ldots = (1 + x + x^2 + x^3 + \ldots) + (x + x^2 + x^3 + \ldots) + (x^2 + x^3 + \ldots) + \ldots$$

$$= \frac{1}{1-x} + \frac{x}{1-x} + \frac{x^2}{1-x} + \ldots = \frac{1}{(1-x)^2}$$

Deriving a similar arithmetico-geometric series for all arithmetic progressions (a_n) is a (hopefully stimulating) exercise for the reader.

6.2.2 Manipulating Generating Functions

Generating functions, much like all other classes of functions, are defined such that they conform to a set of clean rules that allows us to manipulate them more freely, and such that we can derive certain generating functions from others with much greater ease. These rules are the natural consequences of the setup of the generating function, which seamlessly incorporates the ideas of linearity, multiplicativity, and shifting (assigning each value of the function to a value one to the left or right).

Theorem 6.2: Manipulations of Generating Functions

The following are several of the ways in which we can manipulate generating functions algebraically:

1. Addition/subtraction: If (a_n) and (b_n) have generating functions $G(a_n; x)$ and $G(b_n; x)$ respectively, and (c_n) is the sequence with $c_n = a_n + b_n$ for all n, then $G(c_n; x) = G(a_n; x) + G(b_n; x)$.

2. *Scalar* multiplication: For any sequence (a_n), $c \cdot G(a_n; x) = G(ca_n; x)$ $(c \in \mathbb{R})$.

3. *Function-function* multiplication: For a generating function $G(a_n; x)$ with $(a_n) = (a_0, a_1, a_2, \ldots)$

---

[48] Indeed, generating functions were first proposed to help with recursive problems! In the words of Herbert Wilf, "A generating function is a clothesline on which we hang up a sequence of numbers for display" (Wilf 1994).

and $G(b_n; x)$ with $(b_n) = (b_0, b_1, b_2, \ldots)$, we define $G(c_n; x) = G(a_n; x) \cdot G(b_n; x)$ such that

$$(c_n) = a_0 b_n, a_1 b_{n-1}, a_2 b_{n-2}, \ldots, a_{n-1} b_1, a_n b_0)$$

4. Taking the derivative (left shift[a]): If $(a_n) = (a_0, a_1, a_2, \ldots)$ has generating function $G(a_n; x)$, then $G'(a_n; x)$ has associated sequence $(a_1, 2a_2, 3a_3, \ldots)$.

5. Right shift: If (a_n) has generating function $G(a_n; x)$, then $x^k \cdot G(a_n; x)$ is the generating function of (a_n) with all terms translated k to the right (such that a_n becomes a_{n+k}), with the first k terms filled in by zeros.

---

[a]Note that in a left shift by k terms, the terms up to, and including, the kth term "vanish," with the $(k+1)^{th}$ term assuming the position of the first index.

Remark. Here it is important to make the distinction between *scalar* and *function-function* multiplication. In the former case, we are multiplying a generating function by a real number; in the latter, we are multiplying two functions.

A note on "intuitiveness." Many of these identities may seem intuitive, but for the sake of completeness, rigor, and thorough understanding, it is always our goal to provide justification, especially for far-reaching claims such as these. What seems intuitively simple may not be nearly as simple to fully prove!

Proof. We can use the definition of the generating function to prove all of these identities:

(a) Let $(a_n) = (a_0, a_1, a_2, \ldots)$, $(b_n) = (b_0, b_1, b_2, \ldots)$. Then $(c_n) = (a_0 + b_0, a_1 + b_1, a_2 + b_2, \ldots)$, so

$$G(c_n; x) = (a_0 + b_0) + (a_1 + b_1)x + (a_2 + b_2)x^2 + \ldots$$

$$= \sum_{i=0}^{\infty} (a_i + b_i)x^i$$

where we have

$$G(a_n; x) = \sum_{i=0}^{\infty} a_i x^i$$

$$G(b_n; x) = \sum_{i=0}^{\infty} b_i x^i$$

(b) Define (a_n) as above. Then $c(a_n) = (ca_0, ca_1, ca_2, \ldots)$ so that $G(ca_n; x) = \sum_{i=0}^{\infty} (ca_i)x^i = c\sum_{i=0}^{\infty} a_i x^i$ by linearity of the sum.

(c) We have $G(a_n; x) = a_0 + a_1 x + a_2 x^2 + \ldots$ and $G(b_n; x)$ defined analogously, so $G(c_n; x) = a_0 b_0 + (a_0 b_1 + a_1 b_0)x + (a_0 b_2 + a_1 b_1 + a_2 b_0)x^2 + \ldots$ and $(c_n) = (a_0 b_0, a_0 b_1 + a_1 b_0, a_0 b_2 + a_1 b_1 + a_2 b_0, \ldots)$. Essentially, we are summing all pairs of terms whose exponents sum to a given n, namely $(0, n), (1, n-1), (2, n-2), \ldots, (n-1, 1), (n, 0)$. These terms have coefficients that, when multiplied together in pairs, sum to the desired result. (Remark: The sum $a_0 b_n + a_1 b_{n-1} + a_2 b_{n-2} + \ldots + a_{n-1} b_1 + a_n b_0$ is known as the *convolution* of (a_n) and (b_n).)

(d) Write $G(a_n; x) = a_0 + a_1 x + a_2 x^2 + \ldots$; then $G'(a_n; x) = a_1 + 2a_2 x + 3a_3 x^2 + \ldots$ by the Power Rule. In general, the coefficient corresponding to the term $a_n x^n$ is transformed into $na_n x^{n-1}$, which becomes a coefficient of n in the nth place (since the powers of x start at x^0 in the expansion of a generating function).

(e) Multiplying the generating function by x^k yields the expansion $a_0 x^k + a_1 x^{k+1} + a_2 x^{k+2} + \ldots$ which corresponds to the sequence

$$(a_n) = (0, 0, 0, \ldots, 0, a_0, a_1, a_2, \ldots)$$

with k zeros preceding a_0.

There are an almost innumerable number of equally-elegant properties of the generating function; we leave it to the reader to actively seek out and discover as many of them as possible! $\square$

6.2.3 Types of Generating Functions

There are indeed several classes of generating functions.[49] The type we have hitherto discussed is the *ordinary generating function*. We can also define the *exponential generating function* and the *Poisson generating function* (the latter being especially useful in probability theory and statistics, and as a tool to aid with combinatorial enumeration), as well as the *Lambert, Bell,* and *Dirichlet series,* to name just a few.

Exponential Generating Functions

We define the *exponential generating function* in similar fashion as the ordinary generating function, in that the terms of a sequence (a_n) form the coefficients of the expansion of the function (which is denoted $EG(a_n; x)$). However, our x^n terms change slightly to fit the "exponential" nomenclature:

> **Definition 6.2: Exponential Generating Functions**
>
> The *exponential generating function* of the infinite sequence (a_n) is given by
>
> $$EG(a_n; x) = \sum_{i=0}^{\infty} a_i \cdot \frac{x^i}{i!}$$

Indeed, though the EGF differs from the OGF by only an insertion of $\frac{1}{i!}$, the fact that generating functions, when multiplied together, do *not* scale linearly, is a complicating factor.

As with the OGF, it is not too difficult to prove that the EGF is additive, as well as multiplicative in the scalar. The derivative principle is of course modified slightly, but retains fundamental properties of the OGF in an analogous form.

The Lambert, Bell, and Dirichlet Series

We define three types of series that give way to their own unique classes of generating functions: the *Lambert, Bell,* and *Dirichlet series.* Each of these series has its own marvelous connections with other areas of math, and each also presents wonderful opportunities for exploring possible applications in non-math areas!

> **Definition 6.3: Lambert/Bell/Dirichlet Series**
>
> - The *Lambert series* is a series of the form
>
> $$\sum_{n=1}^{\infty} a_n \cdot \frac{x^n}{1 - x^n}$$
>
> for $|x| < 1$. We can also write this series in the form
>
> $$\sum_{n=1}^{\infty} a_n \cdot x^{mn} = \sum_{k=1}^{\infty} b_k x^k$$
>
> where $b_k = \sum_{n|k} a_n$, i.e. the *Dirichlet convolution* of a_n with the constant function $f(n) = 1$. Thus, with some manipulation, it follows that
>
> $$\sum_{n=1}^{\infty} \mu(n) \cdot \frac{x^n}{1 - x^n} = x$$

---

[49]The section exercises, however, will mainly focus on the single type discussed in the previous sections; this part is primarily for the reader's own edification. These other types of generating functions are fairly rare in an applied setting, outside of specific contexts.

where $\mu(n)$ denotes the Möbius function.

- The *Bell series*[a] depends on an arithmetic function[b] f and a prime number p. We define the Bell series of f modulo p as

$$f_p(x) = \sum_{n=0}^{\infty} f(p^n)x^n$$

It can be shown that two multiplicative functions f, g are equal iff all of their Bell series are equal (this is known as the *uniqueness theorem*).

In addition, if we let $h = f * g$ be the Dirichlet convolution of f and g (i.e. $h(n) = (f * g)(n) = \sum_{ab=n} f(a)g(b)$ for $a, b \in \mathbb{N}$), then for all prime p, $h_p(x) = f_p(x) \cdot g_p(x)$.

- The generalized *Dirichlet series* is any infinite series of the form

$$\sum_{n=1}^{\infty} \frac{a_n}{n^s}$$

where $s \in \mathbb{C}$ and (a_n) is a complex sequence.

---

[a]There also exists a sequence of *Bell numbers*, but this is unrelated to the concept of the Bell series. This sequence $(B_n)_{n \geq 0}$ is defined such that B_n counts the number of possible partitions of a set with n elements, with $B_0 = B_1 = 1$. Very intriguingly, B_n is also the number of possible rhyme schemes for a poem n lines long! (This was proven by the late Martin Gardner in his 1978 paper "The Bells: Versatile Numbers That Can Count Partitions of a Set, Primes and Even Rhymes." In the same year, he also posed the question of whether infinitely many of the B_n are prime, which to date, is an open question. So far, the largest known Bell prime is B_{2841}, discovered just last year.)

[b]A function whose domain is $\mathbb{N}$ and whose range is a subset of $\mathbb{C}$.

Remark. In particular, Dirichlet series play pivotal roles in analytical number theory, the most prominent example being in formulating the Riemann Hypothesis. They also have a vast array of assorted applications in combinatorics.

Then the resulting generating functions behave much the same as OGFs and EGFs, with the exception of a few defining traits:

Definition 6.4: Notes on the Lambert/Bell/Dirichlet Genrnating Functions

- The Lambert generating function, $LGF(a_n; x)$, is in fact a generating function for the divisor function. Furthermore, the index starts at 1 and not 0; otherwise, the first term in the generating function expansion would necessarily be undefined.

- The Bell generating function, $BGF(a_n; x)$, can be simplified considerably if f is completely multiplicative, to

$$f_p(x) = \frac{1}{1 - x \cdot f(p)}$$

Note specific values of the BGF for well-known functions. For prime p, the Möbius function has $\mu_p(x) = 1 - x$; $\mu_p^2 = 1 + x$; Euler's totient function $\phi(x)$ has $\phi_p(x) = \dfrac{1-x}{1-px}$; and the constant function $1(x) = 1$ for all x satisfies $1_p(x) = \dfrac{1}{1-x}$. (Here, the subscript p denotes modulo p.)

- The Dirichlet generating function $DGF(a_n; x)$ is particularly useful if a_n is a multiplicative function, since it can then be expressed in terms of the function's Bell series. Furthermore, the Dirichlet series generates the Möbius function (and we can use Möbius inversion to reverse this).

6.2.4 Power Series and Closed Forms

It is quite a wonderful fact that the generating function for the Fibonacci sequence $(a_n) = (1, 1, 2, 3, 5, \ldots)$ is given by $G(a_n; x) = \dfrac{x}{1 - x - x^2}$! (We will take this at face value, and without proof, throughout this section; actually proving this for oneself will be a section Exercise.) In general, we can use *power series* to convert the Fibonacci sequence, and other, similar recursive sequences, to their corresponding closed forms. Ordinarily, it is not at all obvious that these closed forms should even exist, let alone be remotely close to being as clean as they are. But using the power of generating functions, we can condense even the most complicated sequence into a compact and truly marvelous closed form!

Power Series

(Note: As we are mainly using power series to elucidate the idea of generating functions, this section will not go into nearly as much breadth as one might find in a typical calculus textbook. However, we will aim to delve into comparable depth and exercise similar rigor *within the scope in which we are working for the purposes of this chapter.*)

The following is the derivation of the general power series:

> **Definition 6.5: Power Series**
>
> A *power series* is an infinite series of the form
> $$\sum_{n=0}^{\infty} a_n (x - c)^n$$
> where a_n is the coefficient of the nth term as a function of n, and c is a constant.

To derive the power series of a function $f(x)$ about $x = c$, we compute

$$\sum_{n=0}^{\infty} \frac{f^{(n)}(c)}{n!} (x - c)^n$$

(That this holds for all $n \geq 0$ is known as *Taylor's Theorem*.) In calculus contexts, the power series is usually referred to as the *Taylor series* of an infinitely-differentiable (*smooth*) function, or in the special case $c = 0$, the *Maclaurin series* of the function.

Remark. This definition may (and should) seem strikingly familiar; the power series is indeed just a generalization of the generating function! Any polynomial can be written as a power series with $c = 0$, since its consist purely of real coefficients and x^n terms. Indeed, we can also have $c \neq 0$, but the relationship does not necessarily work the other way around. For instance, almost all functions are not polynomials! (It is also worth noting, for the sake of pedantry, that power series are not actually polynomials, since they have infinite degree, and by definition, for a polynomial f, $\deg f \in \{\mathbb{N} \cup \{0\}\}$.)

Henceforth, we discuss a core idea of the series in general: the *interval of convergence*. Since power series are essentially transformations of infinite geometric series, one would expect that they converge under similar conditions for x (i.e. $|x| < 1$ in the case of the geometric series sum $1 + x + x^2 + x^3 + \ldots = \dfrac{1}{1 - x}$). Indeed, we can establish an *interval* and a *radius* of convergence for any power series; that is, the set of values of x for which a given power series converges.

> **Theorem 6.3: Computing the Interval/Radius of Convergence**

Perform the Ratio Test on the series; that is, let

$$L = \lim_{n \to \infty} \left| \frac{a_{n+1}}{a_n} \right|$$

For a given value of x, if $L < 1$, the series converges for that x-value. If $L > 1$, the series diverges for that x-value, and if $L = 1$, the test is inconclusive (and we must check the endpoints of the interval produced from $L < 1$ instead). This yields an *interval of convergence*, namely the set of all values of x for which the series converges, and a *radius of convergence*, namely half the length of the interval of convergence.

Example 6.2.4.1. *Determine the interval and radius of convergence of the infinite series*

$$\sum_{n=1}^{\infty} \frac{(-1)^n \cdot \ln n \cdot x^{2n-1}}{n}$$

Solution 6.2.4.1. *We have*

$$a_n = \frac{(-1)^n \cdot \ln n \cdot x^{2n-1}}{n}$$

Thus,

$$a_{n+1} = \frac{(-1)^{n+1} \cdot \ln(n+1) \cdot x^{2n+1}}{n+1}$$

from which we have

$$L = \lim_{n \to \infty} \frac{a_{n+1}}{a_n} = \lim_{n \to \infty} x^2 \cdot -\frac{n+1}{n} \cdot \frac{\ln(n+1)}{\ln n}$$

in terms of x. As $n \to \infty$, we have $-\dfrac{n+1}{n} \to -1$, and $\dfrac{\ln(n+1)}{\ln n} \to 1$, so $1 > x^2 > -1$ and $|x| < 1$. But we must remember to check the endpoints as well! At $x = -1$, the series

$$\sum_{n=1}^{\infty} \frac{(-1)^n \cdot \ln n \cdot (-1)^{2n-1}}{n} = \sum_{n=1}^{\infty} \frac{(-1)^{3n-1} \cdot \ln n}{n}$$

converges by the Alternating Series Test, and at $x = 1$, the series

$$\sum_{n=1}^{\infty} \frac{(-1)^n \cdot \ln n}{n}$$

also converges by the AST. Thus the interval of convergence is $\boxed{x \in [-1, 1]}$ and the radius of convergence is $\boxed{R = 1}$.

We can also observe that the power series is additive, as well as multiplicative in the scalar argument. That is, if

$$f(x) = \sum_{n=0}^{\infty} a_n (x - c)^n$$

$$g(x) = \sum_{n=0}^{\infty} b_n (x - c)^n$$

and $h(x) = f(x) + g(x)$, then

$$h(x) = \sum_{n=0}^{\infty} (a_n + b_n)(x - c)^n$$

with subtraction being defined identically. It is not the case, however, that if $f(x)$ and $g(x)$ have the same radii of convergence, then $h(x)$ will have that same radius of convergence.

147

Determining the product and quotient of $f(x)$ and $g(x)$ (as defined above) is slightly more complicated, but still entirely doable with the power series definition, and in fact made much more compact and understandable than it otherwise would be. Indeed, we can apply the same tactic here that we used for the function-function product of two generating functions:

$$f(x)g(x) = \left(\sum_{n=0}^{\infty} a_n(x-c)^n \right) \left(\sum_{n=0}^{\infty} b_n(x-c)^n \right)$$

$$= \sum_{i=0}^{\infty} \sum_{j=0}^{\infty} a_i b_j (x-c)^{i+j}$$

$$= \sum_{n=0}^{\infty} \sum_{i=0}^{n} a_i b_{n-i} (x-c)^n$$

To take the quotient $\dfrac{f(x)}{g(x)}$, define a sequence d_n such that

$$\frac{f(x)}{g(x)} = \sum_{n=0}^{\infty} d_n(x-c)^n$$

Then we have

$$f(x) = \left(\sum_{n=0}^{\infty} b_n(x-c)^n \right) \left(\sum_{n=0}^{\infty} d_n(x-c)^n \right)$$

Comparing coefficients, it then follows that $d_0 = \dfrac{a_0}{b_0}$. A subsequent interesting observation is that

$$d_1 = \frac{1}{b_0^2} \cdot \det \begin{bmatrix} a_1 & b_1 \\ a_0 & b_0 \end{bmatrix}$$

and furthermore, extending this to any positive integer n,

$$d_n = \frac{1}{b_0^{n+1}} \cdot \det \begin{bmatrix} a_n & b_1 & b_2 & \dots & b_n \\ a_{n-1} & b_0 & b_1 & \dots & b_{n-1} \\ a_{n-2} & 0 & b_0 & \dots & b_{n-2} \\ \vdots & 0 & 0 & \ddots & \vdots \\ a_0 & 0 & 0 & \dots & b_0 \end{bmatrix}$$

(Where could this determinant relationship have come from? Hint: try to come up with a recursive pattern for the b_n!)

Finally, and perhaps foremost in drawing comparisons between power series, we can differentiate and integrate term-by-term, such that

$$f'(x) = \sum_{n=0}^{\infty} a_n \cdot n(x-c)^{n-1}$$

$$\int f(x)dx = \sum_{n=0}^{\infty} \frac{a_n \cdot (x-c)^{n+1}}{n+1} + C$$

by the Power Rule. (C is the constant of integration.) In addition, both $f'(x)$ and $\int f(x)dx$ have the same radius of convergence as $f(x)$.

Deriving the Closed Form

One of the central applications of the generating function is determining the closed form associated with the sequence - that is, a single expression in terms of one or more variables that does not contain any ellipses, sums, or functions of the variable(s). That is, it can be expressed in terms of a finite number of operations. Then we can directly determine the nth coefficient using this closed form, which can be extremely useful!

Perhaps the best way to conduct this portion of the section is via example:

Example 6.2.4.2. *Derive a closed form for the terms of the Fibonacci sequence.*

Solution 6.2.4.2. *Taking the example of the generating function* $G(a_n; x) = \dfrac{x}{1 - x - x^2}$ *of the Fibonacci sequence* $(a_n) = (1, 1, 2, 3, 5, \ldots)$, *let's extract a closed form for the terms of* (a_n) *using* $G(a_n; x)$.

In particular, $G(a_n; x)$ *is a rational function, so we can decompose it into the sum of partial fractions. Namely,*

$$\frac{x}{1 - x - x^2} = \frac{A}{1 - \alpha x} + \frac{B}{1 - \beta x}$$

where $\alpha = \dfrac{1 + \sqrt{5}}{2}$, $\beta = \dfrac{1 - \sqrt{5}}{2}$ *are the roots of* $-x^2 - x - 1 = -(x^2 + x + 1)$. *Then by equating coefficients,* $A = \dfrac{1}{\sqrt{5}}$ *and* $B = -\dfrac{1}{\sqrt{5}}$, *which implies that*

$$\frac{x}{1 - x - x^2} = \frac{1}{\sqrt{5}} \left(\frac{1}{1 - \alpha x} - \frac{1}{1 - \beta x} \right)$$

Furthermore, we have $\dfrac{1}{1 - \alpha x} = 1 + \alpha x + \alpha^2 x^2 + \ldots$ *and* $\dfrac{1}{1 + \alpha x} = 1 - \alpha x + \alpha^2 x^2 - \ldots$, *so*

$$\frac{x}{1 - x - x^2} = \frac{1}{\sqrt{5}} \left((1 + \alpha x + \alpha^2 x^2 + \ldots) - (1 - \alpha x + \alpha^2 x^2 - \ldots) \right)$$

$$\implies a_n = \frac{1}{\sqrt{5}} (\alpha^n - \beta^n)$$

$$\implies a_n = \frac{1}{\sqrt{5}} \left(\left(\frac{1 + \sqrt{5}}{2} \right)^n - \left(\frac{1 - \sqrt{5}}{2} \right)^n \right)$$

which is indeed the closed form for the Fibonacci sequence! (This is also known as Binet's formula, and it is not inherently obvious that this form should produce integer outputs for all n, let alone that they should obey the beautiful recursive relationship that they do. Such is the true power of generating functions!)

6.2.5   Combinatorial Applications of Generating Functions

From this point onward, we shift our consideration from the theoretical, abstract aspects of generating functions to the more applied, computational, and down-to-Earth aspects. Particularly in competition math, generating functions can work an absolute miracles in combinatorics, where they can be used to massively simplify otherwise immensely arduous (even impossible) computations and provide a rigorous foundation for the shortcuts we may take for granted!

We begin by relating the building block of discrete combinatorics - the binomial coefficient - to generating functions. Note that, from the sequence $(a_k) = \left(\binom{k}{0}, \binom{k}{1}, \binom{k}{2}, \ldots, \binom{k}{k} \right)$ we can form the generating function

$$G(a_k; x) = \binom{k}{0} + \binom{k}{1} x + \binom{k}{2} x^2 + \ldots + \binom{k}{k} x^k = (1 + x)^k$$

which suggests that the coefficient of x^n in the expansion of $(1 + x)^k$ is the number of ways to select n distinct items from k items. This property comes greatly in handy later on!

Indeed, we can derive this property without even having to use binomial coefficients. Consider the following inductive argument. First consider the singleton set $\{a_1\}$. The generating function for the number of choices of n elements from this set is $1 + x = 1 \cdot 1 + 1 \cdot x$. Similarly, the OGF for the number of choices from the set $\{a_2\}$ is $1 + x$. We then use the *Convolution Rule*:

Theorem 6.4: Convolution Rule

Let $\mathcal{A}$ and $\mathcal{B}$ be two disjoint sets with generating functions $F(x)$ and $G(x)$. Then the generating function for the number of choices from $\mathcal{A} \cup \mathcal{B}$ is $F(x) \cdot G(x)$.

Proof. Observe that, to count the number of ways to select items in $\mathcal{A} \cup \mathcal{B}$, we can select n items by choosing i items from $\mathcal{A}$ and $n - i$ items from $\mathcal{B}$ (since $\mathcal{A}$ and $\mathcal{B}$ are disjoint). This can be done in $a_i b_{n-i}$ ways. Taking the sum

$$\sum_{i=0}^{n} a_i b_{n-i}$$

yields the number of ways to pick elements from $\mathcal{A} \cup \mathcal{B}$. By the OGF function-function product rule, this is the coefficient of the x^n term in $F(x)G(x)$. $\qquad \square$

Repeatedly applying this rule yields the desired result: namely, the OGF for choosing n items from a k-element set $(1 + x)^k$.

Furthermore, and as perhaps the key takeaway from the entire section, *we can prove the stars and bars formula using generating functions!* The driving idea behind the execution of the proof is the Convolution Rule, which we will couple with Taylor's Theorem and the derivative to establish that the binomial coefficient is the series of coefficients in the generating function.

Proving stars-and-bars via generating functions. WLOG assume that the generating function extends infinitely (with $n \to \infty$). Recall Taylor's Theorem, which states that

$$f(x) = \sum_{n=0}^{\infty} \frac{f^{(n)}(x)}{n!} \cdot x^n$$

Thus, the nth coefficient of $G(x) = \dfrac{1}{(1 - x)^k}$ (by the geometric series formula and a subsequent application of the Convolution Rule) is equal to its nth derivative at $x = 0$, divided by $n!$. Applying the Power Rule repeatedly yields

$$G^{(n)}(x) = \left(\prod_{i=0}^{n-1} (k + i) \right) \cdot (1 - x)^{n-k}$$

so

$$\frac{G^n(0)}{n!} = \frac{(n + k - 1)!}{n!(k - 1)!} = \binom{n + k - 1}{n}$$

as desired. $\qquad \square$

6.2.6 Examples

Consider the following example, which utilizes and underscores the underlying properties of generating functions:

Example 6.2.6.1. *Compute the OGF of $(a_n)_{n \geq 0} := a_n = n^2$.*

Solution 6.2.6.1. *Essentially, we want to reduce the infinite series $x + 4x^2 + 9x^3 + 16x^4 + \ldots$ to its closed form. We can write the series as*

$$x(1) + x^2(1 + 3) + x^3(1 + 3 + 5) + x^4(1 + 3 + 5 + 7) + \ldots$$

and in turn, as

$$1(x + x^2 + x^3 + x^4 + \ldots) + 3(x^2 + x^3 + x^4 + \ldots) + 5(x^3 + x^4 + \ldots) + \ldots$$

$$= 1((x + x^2 + x^3 + x^4 + \ldots) + (x^2 + x^3 + x^4 + \ldots) + (x^3 + x^4 + \ldots)) + 2((x^2 + x^3 + x^4 + \ldots) + (x^3 + x^4 + \ldots) + \ldots)$$

$$+ 2((x^3 + x^4 + x^5 + \ldots) + (x^4 + x^5 + \ldots)) + \ldots$$

$$= \left(\frac{x}{1-x} + \frac{x^2}{1-x} + \frac{x^3}{1-x} + \ldots \right) + 2 \left(\frac{x^2}{1-x} + \frac{x^3}{1-x} + \ldots \right) + 2 \left(\frac{x^3}{1-x} + \frac{x^4}{1-x} + \ldots \right) + \ldots$$

$$= \frac{x}{(1-x)^2} + 2\frac{x^2}{(1-x)^2} + 2\frac{x^3}{(1-x)^2} + 2\frac{x^4}{(1-x)^2} + \ldots$$

$$= \frac{x}{(1-x)^2} + 2 \left(\frac{x^2}{(1-x)^2} + \frac{x^3}{(1-x)^2} + \frac{x^4}{(1-x)^2} + \ldots \right)$$

$$= \frac{x}{(1-x)^2} + 2\frac{x^2}{(1-x)^3} = \frac{x(1-x) + 2x^2}{(1-x)^3} = \boxed{\frac{x + x^2}{(1-x)^3}}$$

Let's also solidify the ideas of power series (which is markedly similar to that of a generating function) with a couple of computational, calculus-textbook-style examples:

Example 6.2.6.2. *Write $\dfrac{x^2}{1+x^5}$ as a power series.*

Solution 6.2.6.2. *Note that*

$$\frac{x^2}{1+x^5} = x^2 \cdot \frac{1}{1+x^5}$$

$$= x^2 \cdot \frac{1}{1-(-x^5)} = x^2 \cdot \left(1 - x^5 + x^{10} - x^{15} + \ldots \right)$$

$$= x^2 - x^7 + x^{12} - x^{17} + \ldots = \boxed{\sum_{n=0}^{\infty} (-1)^n x^{5n+2}}$$

Example 6.2.6.3. *Using power series, show that $1 - \dfrac{1}{3} + \dfrac{1}{5} - \dfrac{1}{7} + \ldots = \dfrac{\pi}{4}$.*

Solution 6.2.6.3. *Consider $f(x) = \arctan x$. Differentiating f yields*

$$f'(x) = \frac{1}{1+x^2} = \frac{1}{1-(-x^2)} = 1 - x^2 + x^4 - x^6 + \ldots = \sum_{n=0}^{\infty} (-1)^n x^{2n}$$

so then

$$f(x) = \int \sum_{n=0}^{\infty} (-1)^n x^{2n} dx = \sum_{n=0}^{\infty} \int (-1)^n x^{2n} dx$$

$$= \sum_{n=0}^{\infty} \frac{(-1)^n \cdot x^{2n+1}}{2n+1} + C$$

But since $f(0) = 0$, $C = 0$, and so

$$f(x) = \sum_{n=0}^{\infty} \frac{(-1)^n \cdot x^{2n+1}}{2n+1}$$

151

Finally, substituting $x = 1$ yields

$$f(1) = \arctan(1) = \frac{\pi}{4} = \sum_{n=0}^{\infty} \frac{(-1)^n}{2n+1} = 1 - \frac{1}{3} + \frac{1}{5} - \frac{1}{7} + \ldots$$

which is what we wanted. (Note that the series does converge for $x = 1$, as well as $x = -1$, even though the Ratio Test is only conclusive for the open interval $x \in (-1, 1)$. This illustrates the importance of checking the endpoints individually! We can also verify convergence at both endpoints using the Alternating Series Test.)

We will finally conclude with a competition-style combinatorial application of generating functions:

Example 6.2.6.4. *In how many distinguishable ways can we select 19 pens from a rack of pens that contains at least 19 pens of each of 5 colors? (We are allowed to pick two pens of the same color.)*

Solution 6.2.6.4. *Though this problem can be attacked more directly using stars and bars, we will demonstrate the generating function solution for the sake of completeness of understanding.*

Beginning with a set of one item (one pen color), we can choose n pens in 1 way for all $0 \le n \le 19$. Hence, the generating function is $1 + x + x^2 + x^3 + \ldots + x^{19}$. By the Convolution Rule (since the pen colors are disjoint), we can raise this to the fifth power and extract the coefficient of x^{19}, which is $\binom{19 + 5 - 1}{19} = \binom{23}{4} = \boxed{8,855}$.

6.2.7 Exercises

Exercise 131. *Find the generating functions, or sequences associated to the given generating functions, of each of the following:*

(a) $1, -2, 3, -4, 5, \ldots$

(b) $64, 32, 16, 8, 4, \ldots$

(c) $f(x) = x^2 \sin x$

(d) $1, 0, -\dfrac{1}{2}, 0, \dfrac{1}{24}, 0, -\dfrac{1}{720}, \ldots$

(e) $f(x) = \dfrac{x^2}{x^2 + 1}$

Exercise 132. *What is the 2019th term of the sequence whose generating function is $f(x) = -\dfrac{1}{(x-1)^3}$?*

Exercise 133. *(2016 AIME II) For polynomial $P(x) = 1 - \dfrac{1}{3}x + \dfrac{1}{6}x^2$, define $Q(x) = P(x)P(x^3)P(x^5)P(x^7)P(x^9) = \sum_{i=0}^{50} a_i x^i$. Then $\sum_{i=0}^{50} |a_i| = \dfrac{m}{n}$, where m and n are relatively prime positive integers. Find $m + n$.*

Exercise 134.

(a) *What is the Maclaurin series of the function $f(x) = \sin x \cos x$?*

(b) *What is the sequence whose generating function is given by f?*

(c) *What is the kth term of this sequence, in terms of k?*

(d) *Determine the generating function of the sequence of only non-zero coefficients of the Maclaurin series of f.*

Exercise 135. *(2007 HMMT February Combinatorics Test) Let S denote the set of all triples (i, j, k) of positive integers where $i + j + k = 17$. Compute*

$$\sum_{(i,j,k) \in S} ijk$$

Exercise 136. *(2010 AIME I) Jackie and Phil have two fair coins and a third coin that comes up heads with probability $\dfrac{4}{7}$. Jackie flips the three coins, and then Phil flips the three coins. Let $\dfrac{m}{n}$ be the probability that Jackie gets the same number of heads as Phil, where m and n are relatively prime positive integers. Find $m + n$.*

Exercise 137. *(2016 AIME II) The figure below shows a ring made of six small sections which you are to paint on a wall. You have four paint colors available and you will paint each of the six sections a solid color. Find the number of ways you can choose to paint the sections if no two adjacent sections can be painted with the same color.*

Exercise 138. *Show that the generating function of the Fibonacci series F_n is equal to $F(z) = \dfrac{z}{1 - z - z^2}$.*

Extension Exercise 52. *Show that*

$$\sum_{k=0}^{n} \binom{2k}{k} \binom{2(n-k)}{n-k} = 4^n$$

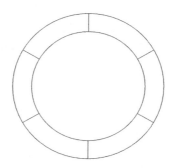

Figure 6.1: Diagram for this Exercise.

Extension Exercise 53. *(1986 USAMO P5) By a partition π of an integer $n \geq 1$, we mean here a representation of n as a sum of one or more positive integers where the summands must be put in nondecreasing order. (E.g., if $n = 4$, then the partitions π are $1 + 1 + 1 + 1$, $1 + 1 + 2$, $1 + 3, 2 + 2$, and 4).*

For any partition π, define $A(\pi)$ to be the number of 1's which appear in π, and define $B(\pi)$ to be the number of distinct integers which appear in π. (E.g., if $n = 13$ and π is the partition $1 + 1 + 2 + 2 + 2 + 5$, then $A(\pi) = 2$ and $B(\pi) = 3$).

Prove that, for any fixed n, the sum of $A(\pi)$ over all partitions of π of n is equal to the sum of $B(\pi)$ over all partitions of π of n.

Extension Exercise 54. *In the decimal expansion of $\dfrac{1}{9,999,899,999}$, "blocks" of five digits after the decimal point contain the Fibonacci numbers:*

$$0.00000\ 00001\ 00001\ 00002\ 00003\ 00005\ 00008\ldots$$

For how many digits does this pattern continue?

6.3 Functional Equations, Part 2

6.3.1 Justifying the Involution

An important property of certain functions f involves their composition - recall that, if they can be composed with themselves, and are such that $f(f(x)) = x$, then f is an *involution*. We claim that an involution is both injective and surjective (and hence bijective), as well as a symmetric relation; this allows us to manipulate it in such a way that it becomes crucial for solving several classes of FEs:

Proof. Let f be an involution. Using the fact that an involution is a permutation, it follows that f is a bijection:

Lemma 6.4: Involution is a Permutation

Let S be a set, and let $f : S \mapsto S$ be an involution. Then f is a permutation.

Proof. We define a permutation such that it is a bijection from S to itself. Thus, it suffices to show that f is, in fact, a bijection.

For all $x \in S$, $f(f(x)) = x$. By the Equality of Mappings Lemma (stated and proven immediately below), we have $f \circ f = I_S$ where I_S denotes the identity mapping on S.

Lemma 6.4: Equality of Mappings

Two mappings $f_1 : S_1 \mapsto T_1$ and $f_2 : S_2 \mapsto T_2$ are equal iff $S_1 = S_2$, $T_1 = T_2$, and, $\forall x \in S_1$, $f_1(x) = f_2(x)$.

Proof. This is a direct consequence of the equivalence of relations. $\square$

Then f is both a left and right inverse of itself, so f is a bijection, and therefore a permutation. $\square$

Since f is a permutation, we may consider it as a relation $f \subseteq S \times S$. Let $(x, y) \in f$; then $f(x) = y$ by the definition of a mapping. It follows that $f(f(x)) = f(y)$; that $a = b \implies f(a) = f(b)$ is an innate property of a mapping. This can be written as $x = f(y)$, by the definition of an involution. Finally, we have $(y, x) \in f$ as a direct consequence, which implies that f is symmetric. We conclude that, if f is an involution, it must be both bijective and symmetric.

We now prove the reverse direction. If f is a mapping that is both bijective and a symmetric relation, then for some unique $y \in S$, $f(x) = y$, from which $(x, y) \in f$ with $f \in S \times S$. Then we have $(y, x) \in f$ since f is symmetric, and $f(y) = x$ from the definition of a mapping. Finally, $f(f(x)) = x$ since $y = f(x)$, so f is an involution. This completes the proof. $\square$

6.3.2 Induction on Functional Equations

We can apply induction on the variable in functional equations over certain domains, just as we can apply it to regular expressions solely in terms of variables. Specifically, the domain of an FE suited for induction should be $\mathbb{N}$, $\mathbb{Z}$, or $\mathbb{Q}$; a domain of $\mathbb{R}$ sometimes works but can vastly complicate the inductive step.

First, we begin by finding the value of $f(1)$. (This relies on other methods, such as setting up a system of equations or recurrence relations). For n an integer, we wish to find the value of $f(n)$, as well as $f\left(\dfrac{1}{n}\right)$ and $f(r)$ for $r \in \mathbb{Q}$. Especially in problems where the function is defined over $\mathbb{Q}$, this technique is highly useful, since it reveals a wealth of information regarding the entire *family* of solutions.

A prime example of induction in action is in solving Cauchy's functional equation:

Example 6.3.2.1. *Using induction, solve Cauchy's FE $f(x + y) = f(x) + f(y)$, $f : \mathbb{Q} \mapsto \mathbb{Q}$.*

Solution 6.3.2.1. *Testing the linear template for f, we find that setting $f(x) = kx + b$ yields $b = 0$ and $f(x) = kx$ for k a constant. (Proving that $f(x) = kx$ works is easily done.) Note, however, that the family of solutions will have a free variable, so let $k = f(1)$ and induct upon k to obtain all the solutions.*

Subsequently, we set $x = y = 0$, so that $f(0) = 0$. Then, setting $x = y = 1$ yields $f(2) = 2f(1) = 2k$, and similarly, setting $x = y = 2$ gives $f(4) = 2(2k) = 4k$ (with $f(3) = 3k$ as well). In general, we can prove by induction that $f(n) = nk$ for any $n \geq 1$ an integer. In addition, by setting $x = -y$, we obtain $f(x) + f(-x) = 0$, so f must be an odd function. Thus, $f(n) = kn$ in fact holds for all integers k.

How do we determine whether $f(n) = kn$ for all $k \in \mathbb{Q}$, however? What can we do to solve for $f(k)$ in terms of known values? Perhaps we can use $f\left(\frac{1}{2}\right)$. Adding this to itself yields $f(1) = 2f\left(\frac{1}{2}\right)$, or $f\left(\frac{1}{2}\right) = \frac{k}{2}$. Hence, for any $k = \frac{p}{q}$, we can add this to itself q times to obtain pk, and so $f\left(\frac{p}{q}\right) = \frac{p}{q}k$ for all k. We can then conclude that $f(x) = kx$ for all x, as claimed.

6.3.3 Final Example: Proof by Continuity

In functional equations, we often encounter setups that involve continuity, as well as monotonicity and functions that are bounded above or below. These all prove to be absolutely invaluable tools in determining the existence or non-existence of solutions, as we shall see with the closing example below, meant to illustrate and encapsulate the most complicated and novel theorems of functional equations we have studied up to this point:

Example 6.3.3.1. *Prove that the zero function is the only function $f : \mathbb{R} \mapsto \mathbb{R}$ that is continuous at $x = 0$ satisfying*

$$(1 - f(x)f(y))f(x + y) = f(x) + f(y)$$

Solution 6.3.3.1. *Clearly, $f(0) = 0$ from plugging in common values and/or solving the resulting systems. We can verify this easily as well.*

Suppose that f is not bounded above. Then for all x_0 such that $f(x_0) > 0$, there exists y_0 such that $f(y_0) = \frac{1}{f(x_0)}$. Thus, we have $f(x_0) + f(y_0) = 0$, which is a contradiction. Thus, such a function f must be bounded above (as well as below).

Furthermore, if $f(x) = f(y) = 1$ for some $x, y \in \mathbb{R}$, then this is an immediate contradiction, since it yields $0 = 2$. Hence, let the supremum of the outputs of f be S. Assume that f takes positive values, and let X be an x-value for which $f(X) = \epsilon \cdot S$, where $\epsilon = 1 - \delta$, $\delta > 0$. Since $0 < \epsilon \cdot S < 1$, we can substitute $x = y = X$ and obtain

$$f(2X) = \frac{2f(X)}{1 - f(X)^2} = \frac{2 \cdot (\epsilon \cdot S)}{1 - (\epsilon \cdot S)^2} > 2\epsilon \cdot S > S$$

which is a contradiction. Therefore, f can attain neither positive nor negative values (the proof of which is similar) at any point; we conclude that it must be the zero function.

6.3.4 Recap: General Strategies for Solving FEs

This (very brief) final section dealing with functional equations will not present a new solution method or proof thereof, but rather recap all of the previously-discussed methods into a convenient list (since it can be quite a bit to take in!)

- Attempt to "guess" the solution at the very beginning, then prove that it works.

- Substitute common values (e.g. $x = 0, \pm 1$) for the variable(s).

- Form a system of equations in terms of $f(x)$ for x, y common values, as above.

- Model the function in question after a template (e.g. linear, exponential, power, Jensen's FE, etc.)

- Set up a recurrence relation in terms of the function.

- Drawing resemblances to Cauchy's functional equation and using it as a template.

- Use mathematical induction over $\mathbb{N}, \mathbb{Z}, \mathbb{Q}$.

- Exploit "bump" symmetry by swapping the positions of x and y.

- Analyze continuity and monotonicity of the function.

- Take advantage of fixed points to simplify the resulting equations.

6.3.5 Examples

Example 6.3.5.1. *(2018 Berkeley Math Tournament) Let $f : [0,1] \mapsto \mathbb{R}$ be a monotonically increasing function such that*

$$f\left(\frac{x}{3}\right) = \frac{f(x)}{2}$$
$$f(1-x) = 2018f(x)$$

If $f(1) = 2018$, find $f\left(\frac{12}{13}\right)$.

Solution 6.3.5.1. *For convenience, scale everything down by a factor of 2018. Then we end up with $f(1) = 1$ and $f(1-x) = 1 - f(x)$. We claim that $f\left(\frac{12}{13}\right)$ under this scale factor is $\frac{6}{7}$, and so the answer will be $\boxed{\dfrac{12,108}{7}}$, by the method of substitution of cyclic parameters.*

Example 6.3.5.2. *(2007 AIME II) Let $f(x)$ be a polynomial with real coefficients such that $f(0) = 1$, $f(2) + f(3) = 125$, and for all x, $f(x)f(2x^2) = f(2x^3 + x)$. Find $f(5)$.*

Solution 6.3.5.2. *Let $f(r) = 0$. Then $f(r)f(2r^2) = f(2r^3 + r)$, and so $2r^3 + r$ must also be a root. Thus, if $r \in \mathbb{R}, r \neq 0$, $\left|2r^3 + r\right| > r$, implying that f has infinitely many roots. Then $f(x)$ has no real roots, since $f(0) \neq 0$.*

Consider the complex roots $r = \pm i$; then $f(i)f(-2) = f(-i)$, and $f(-i)f(-2) = f(i)$. Hence, $f(i)f(-i)(f(-2)^2 - 1) = 0$ and $\pm i$ are indeed roots of f - it follows that $x^2 + 1$ is a factor of f. Substitution into the given expression and some algebra yields $f(x) = (x^2 + 1)^n$ by induction and/or infinite descent.

Finally, as $f(2) + f(3) = 125 = 5^n + 10^n$, $n = 2$ and $f(5) = \boxed{676}$.

Example 6.3.5.3. *(Kyrgyzstan 2012) Find all functions $f : \mathbb{R} \mapsto \mathbb{R}$ such that $f(f(x)^2 + f(y)) = xf(x) + y$ for all $x, y \in \mathbb{R}$.*

Solution 6.3.5.3. *If we substitute $P(0, y)$, we get $f(f(0)^2 + f(y)) = y \implies f$ is a bijection. WLOG assume c is a root of f. Then $P(c, y) \implies f(f(y)) = y$. Using this, $P(f(x), y) \implies f(x^2 + f(y)) = xf(x) + y = f(f(x)^2 + f(y))$, which further implies $x^2 + f(y) = f(x)^2 + f(y) \implies f(x) = \pm x$.*

Henceforth, let $a, b \in \mathbb{R}$ s.t. $ab \neq 0$. Then $f(a) = a, f(b) = -b$. $P(a, b) \implies \pm(a^2 - b) = f(a^2 - b) = a^2 + b$, but this is a contradiction.

Therefore, all possible functions f are $f(x) = x, -x$.

6.3.6 Exercises

Exercise 139. *Find all functions $f : \mathbb{R} \mapsto \mathbb{R}$ such that $f(x^2 + y^2) = f(x)^2 + 2xy + f(y)^2$.*

Exercise 140. *(2012 IMO Shortlist A5) Find all functions $f : \mathbb{R} \mapsto \mathbb{R}$ that satisfy the conditions $f(1 + xy)f(x + y) = f(x)f(y)$ for all $x, y \in \mathbb{R}$, $f(-1) \neq 0$.*

Exercise 141. *(2010 IMO P1) Find all functions $f : \mathbb{R} \to \mathbb{R}$ such that for all $x, y \in \mathbb{R}$ the following equality holds*

$$f(\lfloor x \rfloor y) = f(x) \lfloor f(y) \rfloor$$

where $\lfloor a \rfloor$ is the greatest integer not greater than a.

Extension Exercise 55. *(2012 IMO P4) Find all functions $f : \mathbb{Z} \to \mathbb{Z}$ such that, for all integers $a, b,$ and c that satisfy $a + b + c = 0$, the following equality holds:*

$$f(a)^2 + f(b)^2 + f(c)^2 = 2f(a)f(b) + 2f(b)f(c) + 2f(c)f(a).$$

Extension Exercise 56. *(2010 Putnam B5) Is there a strictly increasing function $f : \mathbb{R} \mapsto \mathbb{R}$ such that $f'(x) = f(f(x))$ for all x?*

6.4 Proving Multi-Variable Inequalities

6.4.1 Jensen's Inequality

We may know - either on an intuitive level or from studying differential calculus - that the secant line of a function also lies either entirely above or entirely below the graph of the function, depending on the function's concavity. Using *Jensen's Inequality,* we can formalize this result in terms of *convex functions,* as well as tie it in with statistics and probability theory. Jensen's Inequality is a very powerful result, applicable across several interconnected disciplines of mathematics, and even to areas outside math, such as statistical physics and information theory.[50]

> **Definition 6.6: Convex and Concave Functions**
>
> Say a function $f : \mathbb{R} \mapsto \mathbb{R}$ is *convex* if the straight line segment between any two points on the graph of f lies strictly above (in which case it is *concave down*) or strictly below the graph (*concave up*). Rigorously, f is convex on the interval $[a, b]$ if, for all $x_1 \in [a, b]$ with corresponding x_2 and $t \in [0, 1]$ a parameter, $f(tx_1 + (1 - t)x_2) \leq tf(x_1) + (1 - t)f(x_2)$.
>
> Furthermore, f is concave on $[a, b]$ if, for all $x \in [a, b]$ and corresponding y, and for any $t \in [0, 1]$,
>
> $$f((1 - t)x + ty) \geq (1 - t)f(x) + tf(y)$$
>
> (f is *strictly* concave if the inequality is strict.) Note that a concave function is also known as a *convex down* function. In this way, we can think of it as the opposite of a convex function (in that, if f is (strictly) convex, then $-f$ is strictly concave, and vice versa).

Jensen's Inequality relates a convex function to the idea of expected value as well: we can actually prove it in this way! (And indeed, we will do so here; the reader will likely learn more from the more novel proof.) A quick note for those who have not studied probability theory: the main fact we use is the linearity of the expected value function $\mathbb{E}(X)$; i.e. the fact that an expression in terms of expected value remains the same, regardless of the order in which $\mathbb{E}(X)$ is applied.

> **Theorem 6.5: Jensen's Inequality**
>
> For two points (x_1, y_1) and (x_2, y_2) on the graph of a convex function f, and $t \in [0, 1]$ a parameter, the secant line between these points is given by $tf(x_1) + (1 - t)f(x_2)$, which is greater than or equal to $f(tx_1 + (1 - t)x_2)$. (Indeed, this is the very definition of a convex function!)
>
> In probability-theoretic terms, we can state the inequality as follows. If X is a random variable and f is a convex function, then
> $$f(\mathbb{E}[X]) \leq \mathbb{E}[f(X)]$$

Proof. At $x = \mu$, let $g(x) = a + bx$ be the equation of the tangent line to $f(x)$ at $x = \mu$ (where $\mathbb{E}(X) = \mu$). Then for all x, $f(x) \geq g(x) = a + bx$. Thus it follows that

$$\mathbb{E}[f(x)] \geq \mathbb{E}[g(x)] = \mathbb{E}[a + bx] = a + b\mathbb{E}(X) = g(\mu) = f(\mu) = f(\mathbb{E}(X))$$

Note that, if $f(x)$ is linear, then we have equality. Suppose $f(x)$ is nonlinear, but convex, so that $\mathbb{E}[f(x)] = f(\mathbb{E}(X)) = f(\mu)$. Then we have

$$f(x) - g(x) \geq 0 \implies \int (f(x) - g(x))dx > 0$$

Hence, we have $\mathbb{E}[f(x)] > \mathbb{E}[g(x)]$. Since $g(x)$ is linear, we have $\mathbb{E}[g(x)] = a + b\mathbb{E}(X) = f(\mu)$, which yields $\mathbb{E}[f(x)] > f\mathbb{E}(x)$. But this is a contradiction, since the two must be equal. Jensen's inequality follows. $\square$

---

[50]In fact, this inequality is so powerful that it has become sort of a meme: it is common to hear the phrase "trivial by Jensen's" spoken in jest in reference to a very non-trivial problem, even when said problem has nothing to do with Jensen's Inequality!

6.4.2 Muirhead Inequality

The *Muirhead Inequality* generalizes the AM-GM Inequality, alongside all of its own generalizations. In this way, it is the common ancestor of the AM-GM variants, making it an extremely powerful sub-tool to use in proving more complicated inequalities.

Before we can discuss the inequality itself, we must go over a few requisite definitions:

Definition 6.7: The a-Mean

For a vector $(a_1, a_2, a_3, \ldots, a_n)$ with real components a_i $(1 \leq i \leq n)$, define the a-mean

$$[a] = \frac{1}{n!} \cdot \sum_{\sigma} \prod_{i=1}^{n} x_{\sigma_i}^{a_i}$$

where we take the sum over all permutations σ of $(1, 2, 3, \ldots, n)$. We also require $a_1 + a_2 + a_3 + \ldots + a_n = 1$.

Example 6.4.2.1. *For $a = (1, 0, 0, \ldots, 0)$, the a-mean is synonymous with the arithmetic mean. For all components of a equal, the a-mean becomes the geometric mean.*

Definition 6.8: Stochastic and Doubly-Stochastic Matrices

A square matrix A with non-negative real elements is *stochastic* if the sum of the entries in each column is 1. A is *doubly stochastic* if both A and A^T are stochastic. Hence, A itself is doubly stochastic if the sum of the entries in all of its rows *and* columns is 1.

We state Muirhead's Inequality both in the traditional Olympiad format and in the matrix format (which is wholly equivalent):

Theorem 6.6: Muirhead's Inequality

We first state the traditional formulation:

Let $A = (a_1, a_2, a_3, \ldots, a_n)$ and $B = (b_1, b_2, b_3, \ldots, b_n)$ be two non-increasing sequences of real numbers. We say A *majorizes* B (denoted $A \succ B$) if, for all $1 \leq k \leq n$,

$$\sum_{i=1}^{k} a_i \geq \sum_{i=1}^{k} b_i$$

Given that $A \succ B$, and given a set of positive real numbers x_i $(1 \leq i \leq n)$, then

$$\sum_{\text{sym}} \prod_{i=1}^{n} x_i^{a_i} \geq \sum_{\text{sym}} \prod_{i=1}^{n} x_i^{b_i}$$

where $\sum_{\text{sym}}$ denotes the symmetric sum.

We then state the formulation in terms of the doubly stochastic matrix: $A \succ B$ iff there exists a doubly stochastic matrix M such that $B = MA$.

Proof. We will prove the inequality as it's traditionally used in the context of the Olympiad and other proof-based competitions.[51]

---

[51] Proving the doubly stochastic formulation of Muirhead's inequality lies far beyond the scope of this book, but the following link does go through a particularly illuminating proof: https://www.math.ust.hk/excalibur/v11_n1.pdf.

Before proving the inequality itself, we invoke the following lemma.

Lemma 6.6: Sum-Zero Sequence Symmetric Sum Lemma

Let there be a sequence S with $\sum_{i=1}^{n} S_i = 0$. Then

$$\sum_{\text{sym}} \prod_{i=1}^{n} x_i^{S_i} \geq n!$$

Proof. AM-GM tells us that

$$\frac{1}{n!} \sum_{\text{sym}} \prod_{i=1}^{n} x_i^{S_i} \geq \sqrt[n!]{\prod_{\text{sym}} \prod_{i=1}^{n} x_i^{S_i}} = \sqrt[n!]{\prod_{i=1}^{n} x_i^{(S_1+S_2+S_3+...+S_n)(n-1)!}} = 1$$

or in other words,

$$\sum_{\text{sym}} \prod_{i=1}^{n} x_i^{S_i} \geq n!$$

as desired. $\square$

Define the sequence S such that $S_i = A_i - B_i$; it follows that $\sum_i S_i = 0$. Hence,

$$\left(\sum_{\text{sym}} \prod_{i=1}^{n} x_i^{S_i} \right) - n! \geq 0$$

Multiplying through by $\sum_{\text{sym}} \prod_{i=1}^{n} x_i^{b_i}$ yields

$$\sum_{\text{sym}} \prod_{i=1}^{n} x_i^{b_i+S_i} - \prod_{i=1}^{n} x_i^{b_i} \geq 0$$

As $b_i + S_i = a_i$, we have

$$\sum_{\text{sym}} \prod_{i=1}^{n} x_i^{a_i} - \prod_{i=1}^{n} x_i^{b_i} \geq 0$$

$$\implies \sum_{\text{sym}} \prod_{i=1}^{n} x_i^{a_i} \geq \sum_{\text{sym}} \prod_{i=1}^{n} x_i^{b_i}$$

which is which we set out to prove. $\square$

6.4.3 (Mrs.) Nesbitt's Inequality

Story time! So goes the story of the student who didn't know his inequalities.

There once was an Olympiad participant, whom everyone nicknamed Buzz Lightyear, for his quick thinking and excited personality. He studied long and hard, seemingly devouring every formula and definition in existence...except he forgot to study inequalities. The test was tough, and the one problem that everyone else seemed to be able to get was an inequality that stared him in the face for nine hours. He stared it back, desperately cobbled a few pages together at the end, and got a big, round zero for his efforts...

When he got the results, he drowned his sorrows in copious amounts of tea - his beverage of choice whenever he was stressed or felt down. Next thing he knew, he didn't know what to do with himself anymore: "Gone! Oh, it's all gone! Bye-bye, future! See ya!" His parents, coaches, and teammates tried in vain to console him.

161

"What happened?" "One minute you're so confident - you feel like you know everything there is to know - then suddenly you find yourself getting a big, fat goose egg just like those people who didn't study even a single minute!" He couldn't help but feel the worst emotion of all: not grief, not anger, not regret, but the horrible feeling that he was a fraud - that he'd somehow lucked his way into the Olympiad and didn't actually have a clue what he was doing. "Oh, I'm a sham!" "Well, at least you made it into the Olympiad, didn't you?" "Years of academy training, wasted...!"

That student doesn't have to be you. After studying these inequalities, you certainly won't be. Here is Nesbitt's Inequality, a special case of the Shapiro Inequality:

Theorem 6.7: Nesbitt's Inequality

For positive real numbers a, b, c, we have $\sum_{\text{cyc}} \dfrac{a}{b+c} \geq \dfrac{3}{2}$.

Proof. We can prove this particular inequality in a wide variety of ways, all of which reveal their own interesting truths about this particular cyclic sum.

1. WLOG assume $a \geq b \geq c$. By the Rearrangement Inequality, we have $\dfrac{1}{b+c} \geq \dfrac{1}{a+c} \geq \dfrac{1}{a+b}$. If we define vectors $\vec{x} = (a, b, c)$ and $\vec{y} = \left(\dfrac{1}{b+c}, \dfrac{1}{a+c}, \dfrac{1}{a+b} \right)$ with $\vec{y_1} = \left(\dfrac{1}{a+b}, \dfrac{1}{b+c}, \dfrac{1}{a+c} \right)$ and $\vec{y_2} = \left(\dfrac{1}{a+c}, \dfrac{1}{a+b}, \dfrac{1}{b+c} \right)$ then it follows that $\vec{x} \cdot \vec{y} \geq \vec{x} \cdot \vec{y_1}$, and, furthermore, $\vec{x} \cdot \vec{y} \geq \vec{x} \cdot \vec{y_2}$; adding the two inequalities yields Nesbitt's Inequality.

2. By Cauchy-Schwarz on the vectors $\langle \sqrt{a+b}, \sqrt{b+c}, \sqrt{a+c} \rangle$ and $\langle \frac{1}{\sqrt{a+b}}, \frac{1}{\sqrt{b+c}}, \frac{1}{\sqrt{a+c}} \rangle$, we obtain

$$((b+c) + (a+c) + (a+b)) \left(\frac{1}{b+c} + \frac{1}{a+c} + \frac{1}{a+b} \right) \geq 9$$

which can be rearranged to form Nesbitt's Inequality.

3. Using the fact that the inequality is homogeneous (see §6.4.8), we may assume $a + b + c = 1$. Then let $x = a+b$, $y = b+c$, $z = a+c$; this implies $\dfrac{1-x}{x} + \dfrac{1-y}{y} + \dfrac{1-z}{z} \geq \dfrac{3}{2}$, or $\dfrac{1}{x} + \dfrac{1}{y} + \dfrac{1}{z}$. By Titu's Lemma, this is true.

4. We apply the AM-HM inequality on $a + b$, $a + c$, $b + c$:

$$\frac{(a+b) + (a+c) + (b+c)}{3} \geq \frac{3}{\frac{1}{a+b} + \frac{1}{a+c} + \frac{1}{b+c}}$$

$$\implies (a+b) + (a+c) + (b+c) \left(\frac{1}{a+b} + \frac{1}{a+c} + \frac{1}{b+c} \right) \geq 9$$

$$2\frac{a+b+c}{b+c} + 2\frac{a+b+c}{a+c} + 2\frac{a+b+c}{a+b}$$

which yields Nesbitt's Inequality after some rearrangement.

5. We can also prove this by a direct application of AM-GM. Through a Ravi substitution (wherein we set $x = a+b$, $y = a+c$, $z = b+c$), we apply AM-GM to the set $\{x^2 y, x^2 z, xy^2, y^2 z, xz^2, yz^2\}$ to obtain $\dfrac{y+z}{x} + \dfrac{x+z}{y} + \dfrac{x+y}{z}$ after multiplying through by $\dfrac{6}{xyz}$. Back-substituting and re-arranging yields the final result.

There exists a plethora of other methods that can be used as well; this is not an exhaustive list! □

Nesbitt's Inequality serves to show that a simple-looking inequality can in fact be proven wonderfully elegantly, and in a multitude of ways! It also underscores the power of manipulating cyclic sums and substituting once said manipulation is done; if at first, you don't succeed, try, try another substitution.

Generalization: Shapiro Inequality

The generalized form of Nesbitt's Inequality, the *Shapiro Inequality,* extends it to specific quantities of real numbers in a cyclic sum in the pattern of Nesbitt's Inequality:

Theorem 6.8: Shapiro Inequality

Given $x_i > 0$ $(1 \leq i \leq n)$, we have

$$\sum_{\text{cyc}} \frac{x_i}{x_{i+1} + x_{i+2}} \geq \frac{n}{2}$$

with $n \leq 12$ even or $n \leq 23$ odd.

Thus far, purely numerical proofs have been provided for the cases $n = 12$ and $n = 23$, as well as a purely analytical proof for $n = 12$. For $n = 23$, however, finding an analytical proof is still an open problem. (Indeed, even finding the *counterexamples* for $n = 14$ through $n = 22$ even are quite complicated! For $n = 14$, the counterexample discovered by Troesch in 1985 was $x_{14} = (0, 42, 2, 42, 41, 5, 39, 4, 38, 2, 38, 0, 40)$.)

6.4.4 Chebyshev's Inequality

Chebyshev's Inequality is unlike many of the other inequalities that we study in this section. Rather than being an instrumental tool in proving other inequalities with multiple variables, Chebyshev's Inequality is more applicable to statistical and probability distributions, in particular with regard to tenets of probability theory such as the 68-95-99.7 rule:

Theorem 6.9: Chebyshev's Inequality

Let X be an (integrable) random variable with expected value $\mu \in \mathbb{R}$ and non-zero variance $\sigma^2 \in \mathbb{R}$. Then for any real number $k \geq 0$,

$$P(|X - \mu| \geq k\sigma) \leq \frac{1}{k^2}$$

where $P(A)$ denotes the probability of A.

Remark. This is in fact a weaker version of the 68-95-99.7 rule (which is only applicable to normal distributions).

Otherwise stated, the probability that the random variable X is *not* within $k\sigma$ of μ is bounded above by $\frac{1}{k^2}$. As such, the inequality is only meaningful for $k > 1$. (Otherwise, it becomes trivial; all probabilities are less than or equal to 1 *a priori*.)

Proof. We first introduce and prove the following inequality:

Lemma 6.9: Markov's Inequality

For a real-valued random variable X and $a > 0$, $a \in \mathbb{R}$, we have $P(|X| > a) \leq \frac{\mathbb{E}(|X|)}{a}$.

Intuitively, we can observe that the expected value of X is equal to the sum of a multiplied with the probability of $X < a$, and $1 - a$ multiplied with the probability of $X \geq a$ (for $0 \leq X \leq 1$, then extending to the more general case). Here is a rigorous proof banking on this intuition:

Proof. From the probabilistic definition of expected value,

$$\mathbb{E}(X) = \int_{-\infty}^{\infty} x f(x) dx = \int_{0}^{\infty} x f(x) dx$$

as X is non-negative. It follows that

$$\mathbb{E}(X) = \int_0^a xf(x)dx + \int_a^\infty xf(x)dx$$

$$\geq \int_a^\infty xf(x)dx \geq \int_a^\infty af(x)dx$$

$$a\int_a^\infty f(x)dx = a \cdot P(X \geq a)$$

Hence, $P(X \geq a) \leq \dfrac{\mathbb{E}(x)}{a}$, as desired. $\qquad\square$

Chebyshev's inequality is, in fact, a direct corollary of Markov's inequality, in that $P(|X - \mathbb{E}(X)|) \leq \dfrac{\text{var}(X)}{a^2}$ where $\text{var}(X) = \mathbb{E}(X - \mathbb{E}(X)^2)$. If we consider $(X - \mathbb{E}(X))^2$ as a random variable, and a^2 as a constant, then we can write Markov's Inequality as $P((X - \mathbb{E}(X))^2 \geq a^2) \leq \dfrac{\text{var}(X)}{a^2}$. $\qquad\square$

Chebyshev's inequality works wonders for determining confidence intervals given a certain probability (or vice versa), even if we do not know whether the distribution is normal. If it so happens to be normal, then we are in luck, as we can impose even tighter bounds on the confidence interval. In all cases, however, the Chebyshev inequality is weaker than the z-score table of standard deviations for a normal distribution.

Aside: The Chebyshev Distance

Though the *Chebyshev distance* is unrelated to the Chebyshev Inequality, it is nevertheless fundamental to fine-tuning our understanding of metric spaces (and, by association, vector spaces, which play into our understanding of linear algebra and matrices, and thereby linear maps).

Definition 6.9: Chebyshev Distance

On a vector space, the *Chebyshev distance* is an L_∞-metric defined such that the distance between two vectors is equal to the largest of the positive differences between their corresponding coordinates.

Id est, $D(p, q) = \max_i |p_i - q_i|$ where p, q are vectors and p_i is the ith coordinate of p. It is also the limit of the L_p metrics as $p \to \infty$:

$$\lim_{p \to \infty} \sqrt[k]{\sum_{i=1}^n |p_i - q_i|^k}$$

hence why it is also referred to as the L_∞-metric.

In particular, in $\mathbb{R}^2$, if $p = (x_1, y_1)$ and $q = (x_2, y_2)$, their Chebyshev distance is equal to $\max(|x_2 - x_1|, |y_2 - y_1|)$. This property is useful in describing the circle with radius r under the Chebyshev discrete metric. Indeed, it is not a circle at all - but a *square* with side length $2r$ and sides parallel to the x- and y-axes!

6.4.5 Minkowski's Inequality

Henceforth, we wish to make rigorous our existing foundation for inequalities such as the Triangle Inequality in a normed vector space (see §6.8). To do so, we can invoke *Minkowksi's Inequality,* the inequality which ensures that the L^p spaces[52] are normed vector spaces according to the axioms thereof:

---

[52]An L^p space, or Lebesgue space, is defined via the p-norm $||\cdot||_p$.

Theorem 6.10: Minkowski's Inequality

Suppose $f, g \in L^p(S)$, where $L^p(S)$ denotes a normed vector space, with S a measure space.[a] Then

$$||f + g||_p \leq ||f||_p + ||g||_p$$

i.e. the Triangle Inequality in a normed vector space equipped with the p-norm.

[a]Roughly speaking, a *measure space* is a triple (S, A, μ), where S is a nonempty set, μ is a measure, and A is a σ-algebra on X. Intuitively speaking, a σ-algebra on a set X is a collection of subsets of X, including X itself, that is closed under complement and under countable unions and intersections.

Proof. We must first demonstrate that $||f + g||_p$ even exists, provided $||f||_p$ and $||g||_p$ both exist. This follows from $|f + g|^p \leq 2^{p-1}(|f|^p + |g|^p)$ by the fact that x^p is a convex function.

If the p-norm of $f + g$ is zero, then the inequality clearly holds. If it is nonzero, we can use Hölder's Inequality to write

$$||f + g||_p^p = \int |f + g|^p \, d\mu$$

$$= \int |f + g| \, |f + g|^{p-1} \, d\mu$$

$$\leq \int (|f| + |g|) \, |f + g|^{p-1} \, d\mu$$

$$= \int |f| \, |f + g|^{p-1} \, d\mu + |g| \, |f + g|^{p-1} \, d\mu$$

$$\leq \left(\sqrt[p]{\int |f|^p \, d\mu} + \sqrt[p]{\int |g|^p \, d\mu} \right) \sqrt[\frac{p-1}{p}]{\int |f + g|^{\left(\frac{p-1}{p} \cdot \frac{p}{p-1} \right)} \, d\mu}$$

following from Hölder's Inequality. Finally, we have that

$$||f + g||_p^p \leq (||f||_p + ||g||_p) \cdot \frac{||f + g||_p^p}{||f + g||_p}$$

which implies Minkowski's Inequality. $\square$

6.4.6 Bernoulli's Inequality

An inequality whose use is less common on a math competition (proof-based or otherwise), but which is nevertheless invaluable in approximation and substitution - both indispensable skills in analysis and several branches of applied math - is *Bernoulli's Inequality,* which, on an intuitive level, approximates an innate property of exponential functions.

Theorem 6.11: Bernoulli's Inequality

For every integer $n \geq 0$ and every real number $x \geq -1$, we have $(1 + x)^n \geq 1 + nx$. (For $0 \leq n \leq 1$, the direction of the inequality is reversed.) If n is even, then the inequality holds for all real numbers x.

This can be seen with probability distributions: say, for instance, that one has an event that occurs with probability $1 - x$, for $x > 0$ infinitesimally small. As we take the limit of x as it approaches 0, we notice through experimental observation that 1 minus the probability is amplified by a factor of nx. That is, say $x = 0.1$. For $n = 1$, $(1 - 0.1)^1 = 0.9$, but for $n = 2$, $(1 - 0.1)^2 = 0.81 = 1 - 0.19 \approx 1 - 2x$. This phenomenon only grows more pronounced for values of x closer to zero. For x ridiculously close to zero, it would appear that $(1 - x)^n \approx 1 - nx$. Indeed, this is Bernoulli's Inequality! This goes to show that intuition can be extremely valuable in leading one towards the spark of inspiration to prove a natural phenomenon.

165

But how do we rigorously show that this approximation holds for any sufficiently large n? The proof is surprisingly simple:

Proof. We proceed by induction. For $n = 0$, we have $(1+x)^0 \geq 1 + 0x = 1$, which is trivially true. For some $r = k$, assume the statement is true; then we show it holds for $r = k + 1$ as well. It follows that

$$(1+x)(1+x)^k \geq (1+x)(1+kx)$$

$$\implies (1+x)^{k+1} \geq 1 + x(k+1) + kx^2$$

$$\implies (1+x)^{k+1} \geq 1 + x(k+1)$$

which proves the inductive hypothesis and the inequality. $\qquad\square$

However, to prove the generalized inequality with $0 \leq r \leq 1$ will require that we compare the derivatives of the LHS and RHS, which is a somewhat more sophisticated (but not any more complicated, computationally speaking) technique. For the variant with the signs reversed, we can actually provide a proof using geometric series (left as an exercise). Indeed, for both variants of the standard version of the inequality with $+$ signs, we can also use AM-GM and the Binomial Theorem as potential tools to construct a proof.

6.4.7 Using the uvw Method

We can also tie in the all-important Vieta's formulas, and symmetric sums of higher-degree polynomials, with a substitution method for multi-variable inequalities known as the *uvw method* for substitution, in which we make a change of variables (u, v, w) in place of (a, b, c):

Theorem 6.12: The uvw Substitution

Let $a, b, c \geq 0$. Then we can make the following substitutions into an inequality in terms of a, b, c, where appropriate:

$$u = \frac{a+b+c}{3}$$

$$v = \sqrt{\frac{ab+bc+ca}{3}}$$

$$w = \sqrt[3]{abc}$$

Motivation. Instead of offering a traditional proof, as with all the other inequalities, here we will discuss the underlying *motivation* behind the uvw substitution. That is, why pick these specific values for u, v, and w?

Immediately, we notice a connection with Vieta's formulas. $a + b + c$, the sum of roots, is related to u, $ab + bc + ca$, the second symmetric sum of a cubic polynomial, is tied to v, and the product of the roots, abc, is tied to w. Notice our use of the term "roots," as a, b, c are inextricably related to a cubic polynomial (commonplace in the inequalities where uvw substiution is most appropriate).

Furthermore, by rearranging the equations, it follows that every symmetric sum of a cubic polynomial can be expressed in terms of u, v^2, and w^3. As such, we commonly use the uvw substitution method to show that an extremal value (maximum or minimum) is attained when two of the variables are equal, or when at least one is zero.

Theorem 6.13: Conditions for the uvw Substitution

If $a, b, c \geq 0$ and u, v, w are our desired substitutions using this method, then $u \geq v \geq w$, with equality

---

iff $a = b = c$ or two of a, b, c are zero.

Proof. Beginning with $18u^2 - 18v^2$, we write this so that it being non-negative is clear by the Trivial Inequality:

$$18u^2 - 18v^2 = 2(a + b + c)^2 - 6(ab + ac + bc)$$

$$= 2a^2 + 2b^2 + 2c^2 - 2ab - 2ac - 2bc = (a - b)^2 + (a - c)^2 + (b - c)^2 \geq 0$$

Hence $u \geq v$, with equality for $a = b = c$.

To show $v \geq w$ (and hence $u \geq w$), observe that $v^2 = \dfrac{ab + ac + bc}{3}$, so by AM-GM, it is at least the geometric mean of ab, ac, and bc, or $\sqrt[3]{a^2b^2c^2} = w^2$. Hence $v^2 \geq w^2$, or $v \geq w$ since $u, v, w \geq 0$. Equality holds iff $ab = ac = bc$, i.e. when $a = b = c$ or two of the variables are zero. $\square$

Example 6.4.7.1. *Let a, b, c be positive reals such that $\dfrac{1}{a} + \dfrac{1}{b} + \dfrac{1}{c} = 1$. Find the maximum value of $(1 - a)(1 - b)(1 - c)$.*

Solution 6.4.7.1. *We have $ab + ac + bc = abc$, or $3v^2 = w^3$. Then $(1 - a)(1 - b)(1 - c) = -abc + (ab + ac + bc) - (a + b + c) + 1 = -w^3 + 3v^2 - 3u + 1 = 1 - 3u$. As $u \geq v \geq w$, we have $uv^2 \geq w^3$, so $u \cdot 3v^2 \geq 3w^2$, or $u \geq 3$ (since $3v^2 = w^2$). Hence $1 - 3u \leq \boxed{-8}$. Equality holds when $a = b = c = 3$.*

6.4.8   Homogeneous Inequalities

Perhaps the mother of all deceptively simple substitutions is the set of substitutions associated with *homogeneous inequalities:* special types of inequalities that obey the property of having the same degree in each term:

Definition 6.10: Homogeneous Inequalities

A *homogeneous* inequality is any inequality with all terms having the same degree. More rigorously, f is said to be homogeneous with degree d iff there exists $k \in \mathbb{R}$ such that, for every $t > 0$, we have

$$t^k \cdot f(a, b, c) = f(ta, tb, tc)$$

Example 6.4.8.1. *For example, $x^2y + y^2z + z^2x \leq 3$ is homogeneous (with degree 2), but $xy + yz^2 + z \leq 3$ isn't homogeneous.*

Why do we say that the substitutions to be made from this are deceptively simple? In homogeneous inequalities, it almost always helps to set $a + b + c = 1$ (or some other constant). But what is the reasoning behind this common assumption? If, in general, we assume that $a + b + c = m$ for some positive m, then $\dfrac{a}{m} + \dfrac{b}{m} + \dfrac{c}{m} = 1$, and then we can substitute $a' = \dfrac{a}{m}$ (with b', c' defined analogously). This tells us that $a' + b' + c' = 1$, but the homogeneity implies that $f(a', b', c') = f(a, b, c)$ Thus, $f(a', b', c') \geq 0 \iff f(a, b, c) \geq 0$, with the condition that $a' + b' + c' = 1$.

Essentially, as our key takeaway from this section, we are making a substitution by "scaling" the variables, under the assumption that all terms of the inequality are of the same degree. This is particularly useful in proving inequalities such as Jensen's Inequality, as we seek places where the sum of variables is 1.

6.4.9 Examples

Example 6.4.9.1. *What is the sum of the maximum and minimum values of $x + y + z + xy + yz + xz$, given that x, y, z are non-negative real numbers with $x^2 + y^2 + z^2 = 1$?*

Solution 6.4.9.1. *Intuitively, we can plug in $x = y = z$ for the maximal case and set one of the variables to be 1 for the minimal case. This will give the correct answer, but this is not mathematically rigorous reasoning.*

If we set $S = x + y + z + xy + yz + xz$, and then apply the Cauchy-Schwarz Inequality, this yields $S \leq \sqrt{3(x^2 + y^2 + z^2)} + x^2 + y^2 + z^2 = 1 + \sqrt{3}$. (Equality holds for $x = y = z = \dfrac{1}{\sqrt{3}}$.) Furthermore, setting $x + y + z = t$ yields $t^2 - 2(xy + xz + yz) = 1 \implies xy + xz + yz \geq \dfrac{t^2 - 1}{2} \geq 0$, so $t \geq 1$ and $t + \dfrac{t^2 - 1}{2} \geq 1$. Thus the minimal value is 1, and the sum of the minimum and maximum of S is $\boxed{2 + \sqrt{3}}$.

Example 6.4.9.2. *(1993 USAMO P5) Let $a_0, a_1, a_2, \cdots$ be a sequence of positive real numbers satisfying $a_{i-1}a_{i+1} \leq a_i^2$ for $i = 1, 2, 3, \cdots$. (Such a sequence is said to be log concave.) Show that for each $n > 1$,*

$$\frac{a_0 + \cdots + a_n}{n+1} \cdot \frac{a_1 + \cdots + a_{n-1}}{n-1} \geq \frac{a_0 + \cdots + a_{n-1}}{n} \cdot \frac{a_1 + \cdots + a_n}{n}.$$

Solution 6.4.9.2. *Observe that we have*

$$(a_0 + a_1 + \ldots + a_{n-1})(a_1 + a_2 + \ldots + a_n) = (a_0 + (a_1 + \ldots + a_{n-1}))(-a_0 + (a_0 + a_1 + \ldots + a_n))$$

$$= -a_0^2 + a_0((a_0 + a_1 + \ldots + a_n) - (a_1 + a_2 + \ldots + a_{n-1})) + (a_0 + a_1 + \ldots + a_n)(a_1 + a_2 + \ldots + a_{n-1})$$

$$= a_0 a_n + (a_1 + a_2 + \ldots + a_{n-1})(a_0 + a_1 + \ldots + a_n)$$

Subtracting $\dfrac{(a_1 + a_2 + \ldots + a_{n-1})(a_0 + a_1 + \ldots + a_n)}{n^2}$ from both sides yields $(a_1 + a_2 + \ldots + a_n) \geq a_0 a_n \cdot (n^2 - 1)$.

Applying AM-GM through the entire inequality yields $\sqrt[n-1]{a_1 a_2 \cdots a_{n-1}} \sqrt[n+1]{a_0 a_1 \cdots a_n} \geq a_0 a_n$, which, when algebraically simplified, yields $\prod_{i=1}^{n-1} a_i^2 \geq a_0^{n-1} a_n^{n-1}$. Henceforth, applying some of the givens produces $a_0 a_n \geq a_1 a_{n-1}$ (this is AM-GM). Induction shows that the desired result holds true.

6.4.10 Exercises

Exercise 142.

(a) *Provide an example of a non-constant homogeneous function of two variables with degree 5.*

(b) *Provide an example of a non-constant non-homogeneous function of two variables with degree 3.*

(c) *Show that, if f and g are two homogeneous functions of degree m and n respectively, then $h = \dfrac{f}{g}$ is also homogeneous, and of degree $m - n$, where $g \neq 0$.*

Exercise 143. *(1991 AIME) For positive integer n, define S_n to be the minimum value of the sum*

$$\sum_{k=1}^{n} \sqrt{(2k-1)^2 + a_k^2}$$

where $a_1, a_2, \ldots, a_n$ are positive real numbers whose sum is 17. There is a unique positive integer n for which S_n is also an integer. Find this n.

Exercise 144. *(2011 USAMO P1) Let a, b, c be positive real numbers such that $a^2 + b^2 + c^2 + (a+b+c)^2 \leq 4$. Prove that*

$$\frac{ab+1}{(a+b)^2} + \frac{bc+1}{(b+c)^2} + \frac{ca+1}{(c+a)^2} \geq 3.$$

Exercise 145. *(2001 USAMO P3) Let $a, b, c \geq 0$ and satisfy $a^2 + b^2 + c^2 + abc = 4$. Show that $0 \leq ab + bc + ca - abc \leq 2$.*

Exercise 146. *(2018 USAMO P1) Let a, b, c be positive real numbers such that $a + b + c = 4\sqrt[3]{abc}$. Prove that*

$$2(ab + bc + ca) + 4\min(a^2, b^2, c^2) \geq a^2 + b^2 + c^2.$$

Exercise 147. *In an extremely large group of people whose heights obey a standard normal distribution, approximately 95 percent of the group have heights lying between 64 and 76 inches, inclusive, with a mean height of 70 inches.*

(a) *At least what fraction of people in the group must be at least 5 feet tall?*

(b) *At most what fraction of people in the group can be at least 6 feet tall?*

(c) *At least what fraction of people in the group must have heights between 58 and 79 inches, inclusive?*

If no statistically valid conclusion can be drawn, explain why.

Exercise 148. *(2010 USA TST) If a, b, c are real numbers such that $abc = 1$, prove that*

$$\sum_{cyc} \frac{1}{a^5(b + 2c)^2} \geq \frac{1}{3}$$

Exercise 149. *(1976 USAMO P4) If the sum of the lengths of the six edges of a tri-rectangular tetrahedron $PABC$ (i.e., $\angle APB = \angle BPC = \angle CPA = 90°$) is S, determine its maximum volume.*

Extension Exercise 57. *(2000 USAMO P1) Call a real-valued function f very convex if*

$$\frac{f(x) + f(y)}{2} \geq f\left(\frac{x+y}{2}\right) + |x - y|$$

holds for all real numbers x and y. Prove that no very convex function exists.

Extension Exercise 58. *(1999 IMO P2) Let $n \geq 2$ be a fixed integer.*

(a) Find the least constant C such that for all nonnegative real numbers $x_1, \ldots, x_n$,

$$\sum_{1 \le i < j \le n} x_i x_j (x_i^2 + x_j^2) \le C \left(\sum_{i=1}^{n} x_i \right)^4.$$

(b) Determine when equality occurs for this value of C.

Extension Exercise 59. If a, b, and c are real numbers with $\sum_{cyc} \dfrac{1}{1+a} = 2$, show that

$$\sqrt{3 + a + b + c} \ge \sqrt{a} + \sqrt{b} + \sqrt{c}$$

6.5 Newton's Sums

(Note: This will be a fairly short section, but it is just as important as it is brief.)

This is a technique that can really shortcut polynomial problems! Using *Newton's Sums*, we are able to circumvent many traditional techniques in favor of applying a few identities involving the symmetric sums of a polynomial. They provide a highly efficient way to compute the sums of roots of a polynomial that are raised to an integer power, and have several applications in factoring identities.

6.5.1 Background, Definitions, and Proof

(Note that this definition will involve the concept of *symmetric sums*. If you are unfamiliar with this idea, we recommend revisiting §5.7, where we generalize Vieta's formulas to all polynomials, and thence describe symmetric sums.)

Theorem 6.14: Newton's Sum Identities

Consider the degree-n polynomial

$$P(x) = a_n x^n + a_{n-1} x^{n-1} + \ldots + a_1 x + a_0$$

Let $P(x)$ have roots $x_1, x_2, x_3, \ldots, x_n$. Then define

$$P_k = \sum_{i=1}^{n} x_i^k$$

for all $1 \leq k \leq n$.

(Alternatively, we may define the P_i recursively as follows:

$$P_1 = S_1$$

$$P_2 = S_1 P_1 - 2S_2$$

$$\vdots$$

though the general formula for P_n via this definition is convoluted.)

We have that

$$a_n P_1 + a_{n-1} = 0$$

$$a_n P_2 + a_{n-1} P_1 + a_{n-2} = 0$$

$$a_n P_3 + a_{n-1} P_2 + a_{n-2} P_1 + a_{n-3} = 0$$

and so forth. (For $i < 0$, $a_i = 0$.)

Proof. Since the x_i are roots, we have

$$a_n x_1^n + a_{n-1} x_1^{n-1} + \ldots + a_1 x_1 + a_0 = 0$$

$$a_n x_2^n + a_{n-1} x_2^{n-1} + \ldots + a_1 x_2 + a_0 = 0$$

$$\vdots$$

$$a_n x_n^n + a_{n-1} x_n^{n-1} + \ldots + a_1 x_n + a_0 = 0$$

Multiplying the ith equation by x_i^{k-n} yields

$$a_n x_1^k + a_{n-1} x_1^{k-1} + \ldots + a_1 x_1^{k-n+1} + a_0 x_1^{k-n} = 0$$

$$a_n x_2^k + a_{n-1} x_2^{k-1} + \ldots + a_1 x_2^{k-n+1} + a_0^{k-n} = 0$$

$$\vdots$$

$$a_n x_n^k + a_{n-1} x_n^{k-1} + \ldots + a_1 x_n^{k-n+1} + a_0^{k-n} = 0$$

Summing all the equations together yields

$$a_n(x_1^k + x_2^k + \ldots + x_n^k) + a_{n-1}(x_1^{k-1} + x_2^{k-1} + \ldots + x_n^{k-1}) + \ldots + a_0(x_1^{k-n} + x_2^{k-n} + \ldots + x_n^{k-n}) = 0$$

$$\implies a_n P_k + a_{n-1} P_{k-1} + \ldots + a_0 P_{k-n} = 0$$

which is the desired result. $\qquad\square$

6.5.2   Applications

Reversing Newton's Sums

Just as we can use Newton's Sums to obtain the sums of powers of roots, we can reverse them to compute the symmetric sums from sums of powers. This is an especially remarkable fact, since there are several instance sin which we would like to know what polynomial we should construct given the sums of roots raised to a power.

For this section, we will not present a cut-and-dry theorem/proof approach, since there is little rigorous algorithmic element to this procedure. Instead, we rely on our intuitive algebraic understanding of manipulation to carry us through these types of problems. Perhaps the best way to illustrate the motivating ideas is by example:

Example 6.5.2.1. *If a, b, c are real numbers such that $a + b + c = 1$, $a^2 + b^2 + c^2 = 2$, and $a^3 + b^3 + c^3 = 3$, what is the value of abc?*

Solution 6.5.2.1. *By Newton's Sums (with $k = 3$), we have $P_0 = 3$, $P_1 = s_1$, $P_2 = s_1 P_1 - 2s_2$, $P_3 = s_1 P_2 - s_2 P_1 + 3s_3$, so $s_1 = 1$, $s_2 = -\dfrac{1}{2}$, and $s_3 = \boxed{\dfrac{1}{6}}$ after solving the system. (Hence, a, b, c are the roots of $x^3 - x^2 - \dfrac{1}{2}x - \dfrac{1}{6}$.)*

Circumventing the Binomial Theorem

We often encounter sums of two binomial expressions with their signs differing: i.e. expressions of the form $(a + b)^n + (a - b)^n$, where n is usually a large number. For small n, this of course can be expanded and simplified using the Binomial Theorem. In fact, this does work somewhat well, even for larger n. However, we will eventually reach a point where we find that using Newton's Sums is a preferable tactic. Consider the following example:

Example 6.5.2.2. *Compute $(1 + i\sqrt{2})^8 + (1 - i\sqrt{2})^8$.*

Solution 6.5.2.2. *Set $a = 1 + i\sqrt{2}$ and $b = 1 - i\sqrt{2}$. Note that $ab = 3$ and $a + b = 2$, so they are the roots of $x^2 - 2x + 3$ with $S_1 = -2$ and $S_2 = 3$. Apply Newton's Sums to obtain $\boxed{34}$ by recursion. (The details are left as an exercise to the reader.)*

6.5.3   Newton's Inequality

This section is primarily included as an aside for the reader's edification. However, we feel that it is more than interesting enough to warrant its inclusion along Newton's symmetric sum method.

Theorem 6.15: Newton's Inequality

Consider the polynomial $P(x)$ with $\deg P(x) = n$ as above. Define the symmetric average $\mu_k = \frac{s_k}{\binom{n}{k}}$, where μ_k is the kth symmetric sum.

Newton's Inequality states that, for non-negative x_i $(1 \leq n)$, and $0 < k < n$,

$$\mu_k^2 \geq \mu_{k-1}\mu_{k+1}$$

Equality holds when all the x_i are equal.

Proof. We induct upon $n \geq 2$. For $n = 2$, this reduces to the AM-GM Inequality. Suppose that the inequality holds for $n = m - 1$, with $m \geq 3$ a positive integer.

We have non-negative x_i $(1 \leq i \leq m)$ and μ_i their symmetric averages. Let μ_i' denote the symmetric averages of x_i $(1 \leq i \leq m - 1)$ and observe that

$$\mu_k = \frac{n-k}{n} \cdot \mu_k' + \frac{k}{n} \cdot \mu_{k-1}' x_m$$

This implies that

$$\mu_{k-1}\mu_{k+1} = \left(\frac{n-k+1}{n} \cdot \mu_{k-1}' + \frac{k-1}{n} \cdot \mu_{k-2}' x_m \right) \left(\frac{n-k+1}{n} \cdot \mu_{k+1}' + \frac{n-k-1}{n} \cdot \mu_k' x_m \right)$$

$$= \frac{(n-k+1)(n-k-1)}{n^2} \mu_{k-1}'\mu_{k+1}' + \frac{(k-1)(n-k-1)}{n^2} \mu_{k-2}'\mu_{k+1}' x_m$$

$$+ \frac{(n-k+1)(k+1)}{n^2} \mu_{k-1}'\mu_k' x_m + \frac{(k-1)(k+1)}{n^2} \mu_{k-2}'\mu'k x_m^2$$

$$\leq \frac{(n-k+1)(n-k-1)}{n^2} \mu_k'^2 + \frac{(k-1)(n-k-1)}{n^2} \mu_{k-2}'\mu_{k+1}' x_m$$

$$+ \frac{(n-k+1)(k+1)}{n^2} \mu_{k-1}'\mu_k' x_m + \frac{(k-1)(k+1)}{n^2} \mu_{k-1}'^2 x_m^2$$

$$\leq \frac{(n-k+1)(n-k-1)}{n^2} \mu_k'^2 + \frac{(k-1)(n-k-1)}{n^2} \mu_{k-1}'\mu_k x_m$$

$$+ \frac{(n-k+1)(k+1)}{n^2} \mu_{k-1}'\mu_k' x_m + \frac{(k-1)(k+1)}{n^2} \mu_{k-1}'^2 x_m^2$$

$$= \frac{(n-k)^2}{n^2}\mu_k'^2 + \frac{2k(n-k)}{n^2}\mu_k'\mu_{k-1}' x_m + \frac{k^2}{n^2}\mu_{k-1}'^2 x_m^2 - \left(\frac{\mu_k}{n} - \frac{\mu_{k-1}x_m}{n} \right)^2$$

$$\leq \left(\frac{n-k}{n}\mu_k' + \frac{k}{n}\mu_{k-1}' x_m \right)^2 = \mu_k^2$$

which completes the inductive step and the proof. $\square$

6.5.4 Examples

Example 6.5.4.1. *(2019 AMC 12A) Let s_k denote the sum of the kth powers of the roots of the polynomial $x^3 - 5x^2 + 8x - 13$. In particular, $s_0 = 3$, $s_1 = 5$, and $s_2 = 9$. Let a, b, and c be real numbers such that $s_{k+1} = a\, s_k + b\, s_{k-1} + c\, s_{k-2}$ for $k = 2, 3, \ldots$ What is $a + b + c$?*

Solution 6.5.4.1. *By Newton's Sums, it immediately follows that $s_{k+1} - 5s_k + 8s_{k-1} - 13s_{k-2} = 0$, so rearranging yields $s_{k+1} = 5s_k - 8s_{k-1} + 13s_{k-2} \implies 5 - 8 + 13 = \boxed{10}$.*

Example 6.5.4.2. *Let r and s be the roots of $x^2 - 4x + 5$. What is the value of $r^{12} + s^{12}$?*

Solution 6.5.4.2. *We have $r + s = 4 = s_1$, and $rs = 5 = s_2$. By Newton's sum identities, we can perform recursion on each of the P_i $(i \geq 3)$ and set $a_{n-i} = 0$ a priori. Thus, $r^{12} + s^{12} = \boxed{23,506}$. (We've spared you the computational details for a reason!)*

6.5.5 Exercises

Exercise 150. *If a and b are the roots of $x^2 + 2x + 19$, what is the value of $a^4 + b^4$?*

Exercise 151. *Let $P(x) = x^3 - 3x^2 + 6x - 9$ have roots r, s, t. Compute $r^5 + s^5 + t^5$.*

Exercise 152. *Let $a + b + c = 1$, $a^2 + b^2 + c^2 = 2$, and $a^3 + b^3 + c^3 = 3$.*

(a) What is the value of abc?

(b) Find the smallest positive integer $n > 3$ such that $a^n + b^n + c^n$ is a positive integer.

Exercise 153. *(2003 AIME II) Consider the polynomials $P(x) = x^6 - x^5 - x^3 - x^2 - x$ and $Q(x) = x^4 - x^3 - x^2 - 1$. Given that z_1, z_2, z_3, and z_4 are the roots of $Q(x) = 0$, find $P(z_1) + P(z_2) + P(z_3) + P(z_4)$.*

Exercise 154. *Show that the only solutions over the complex numbers to the system*

$$x + y + z = 4$$
$$x^2 + y^2 + z^2 = 14$$
$$x^3 + y^3 + z^3 = 34$$

are $(-1, 2, 3)$ and its permutations.

Exercise 155. *(1973 USAMO P4) Determine all the roots, real or complex, of the system of simultaneous equations*

$$x + y + z = x^2 + y^2 + z^2 = x^3 + y^3 + z^3 = 3$$

Extension Exercise 60. *(1988 USAMO P5) Let $p(x)$ be the polynomial $(1 - x)^a (1 - x^2)^b (1 - x^3)^c \cdots (1 - x^{32})^k$, where $a, b, \cdots, k$ are integers. When expanded in powers of x, the coefficient of x^1 is -2 and the coefficients of x^2, x^3, ..., x^{32} are all zero. Find k.*

6.6 Vieta Jumping

The year is 1988. You are a participant at the 29th ultra-prestigious International Mathematical Olympiad - the world's most widely-known, most well-respected, and most brutally difficult high school math competition. For the past several years of your mathletic career, and especially for the past month at MOP, you have been training at a level that goes beyond motivated, concentrated effort. You have almost transcended the idea of what it means to *study*. By the time testing day arrives, it is almost as if you have *become* the test. You seemingly know everything there is to know - inequalities, obscure geometry theorems top to bottom, combo like the back of your hand, and all different types of bashes as a backup - the whole lot. You confidently enter the testing room, and the invigilators hand out the test papers.

Four and a half hours pass. On Day 1, you were unfazed, as expected. It seems that your training has paid off. You bask in your newfound glory, but also prepare for the terrors that await you in the next day. At the same time, though, you look forward to it even more than you did the first day. You can't wait to see what the committee has prepared for the world's top math students.

Then comes day 2 - the first two hours pass and you have nailed the first two problems. Then you hit problem 6. For some reason, after working through it for a bit, you find yourself hitting a wall. Inexplicably, no matter how much you try, it's all to no avail. You begin to internally panic - *this can't be happening!* You read and re-read the problem, thinking you might have missed something. *You have to be missing something! Surely there's a way to solve this that I know, but I just happen to be forgetting right now!* After two more hours, all your efforts have still proven futile, and you begin to resign yourself to having your lifelong dreams of a perfect score going out the window, never to be achieved. In the last thirty minutes, you harriedly scribble down the beginnings of a solution, with some minor insights here and there, but in the end, you only manage to pick up one extra point from the product of your terrified, panicked, and barely-intelligible chicken scratch. As you are soon to find out, your total score is not the 42 you've been longing to get for as long as you can remember - not even a 41 as consolation - but a **36**. You tell yourself that that last problem was completely impossible - the hardest you'd ever seen, by a long shot - and that probably no else solved it, either, but deep down, you are beyond devastated.

Fast-forward thirty-one years to the year 2019, and you are now a renowned academic surrounded by a sea of new mathematical developments and breakthroughs. Curious as to whether your math skills have improved since your Olympiad days, you revisit the problem that traumatized you in your high school days. To your surprise, you nail it almost instantly! Overjoyed, but wistful over the fact that you are about thirty years too late, you decide to take it upon yourself to spread the glory, marvel, and beauty of the tactic you used to finally slay your inner demons - the almighty technique of *Vieta jumping*.

Before we discuss Vieta jumping itself, however, we should take a brief digression to discuss the inter-related proof technique of *infinite descent*.

6.6.1 Infinite Descent

Infinite descent, as its name suggests, is a sub-type of proof by contradiction in which we assume the existence of a solution in the natural numbers, then show that this leads to an absurdity:

Definition 6.11: Infinite Descent

In proving a statement by *infinite descent,* we show that an equation has no solutions via contradiction. More specifically, we show that, if there were to be a solution, then the existence of that solution would necessarily imply the existence of a second solution, whose existence would in turn imply the existence of a third solution, and so forth. Each of these solutions is related to progressively smaller natural numbers, but should there be an infinitude of solutions, than the natural numbers associated with these solutions would need to become progressively smaller as well. But by induction, there exists no infinitude of decreasing natural numbers, which shows that no solution can exist at all.

Example 6.6.1.1. Using the principle of infinite descent, *prove that $\sqrt{2}$ is irrational.*

Solution 6.6.1.1. *For contradiction, write $\sqrt{2} = \dfrac{p}{q}$, $p, q \in \mathbb{N}$, $\gcd(p, q) = 1$; then $2q^2 = p^2$ and so $2 \mid p^2$. Because 2 is prime, by Euclid's lemma, it must also divide p, so $p = 2r$ for some $r \in \mathbb{Z}$. Then $(2r)^2 = 4r^2 = 2q^2 \implies q^2 = 2r^2$, which implies that $2 \mid q$ also. Then $q = 2s$ for some integer s. But then $\dfrac{p}{q} = \dfrac{r}{s}$, which contradicts the minimality of the form $\dfrac{p}{q}$, as well as the requirement that $\gcd(p, q) = 1$ (since 2 is a common factor of both p and q). Hence $\sqrt{2}$ must be irrational.*

Remark. We can also generalize this proof to show that $\sqrt{k}$ is irrational for any non-square k. This proof is left to the reader.

Most notably, we can use infinite descent to prove special cases of Fermat's Last Theorem, such as $a^2 + b^4 = c^4$, or, by extension, $a^4 + b^4 = c^4$:

Proof. Suppose there exists a right triangle with side lengths a, b, c satisfying the equation $a^2 + b^4 = c^4$, and WLOG assume this triangle is primitive. Then its side lengths can be written in the form $a = 2xy$, $b = x^2 - y^2$, $c = x^2 + y^2$, with $\gcd(x, y) = 1$ and $x + y$ odd (and b, c both odd as a result). Since b and c are both odd, neither can be twice a perfect square; moreover, if a is twice a square, or itself a square, each of x and y would be a square (or twice a square). We then split the proof up into three cases, depending on which two sides we claim are (twice) a square:

(i) a and b: In this case, b is a perfect square (since we have already assumed that a is a square, or twice a square, as appropriate). The right triangle with leg lengths y and $\sqrt{b}$, and hypotenuse length x, would have the sides with length x and y be a square or twice a square, with a smaller hypotenuse than that of the original triangle.

(ii) a and c: c is a square (or twice a square), so the integer right triangle with leg lengths x and y, and hypotenuse $\sqrt{c}$, has two sides - x and y - both either a square or twice a square. The hypotenuse is again smaller than that of the original triangle.

(iii) b and c: In this case, b and c are both squares, so we construct the right triangle with legs $\sqrt{bc}$ and y^2, and hypotenuse x^2. This triangle, however, would have integer side lengths, including a square leg of y^2 and a square hypotenuse of x^2. Thus, the hypotenuse would be smaller than that of the original triangle.

In all three cases, one right triangle descends to a smaller one, which would itself lead to an even smaller one. But such a sequence of triangles cannot continue infinitely, hence the desired contradiction! $\qquad\square$

6.6.2 Standard Vieta Jumping

We can apply the principle of infinite descent generally to the ideas behind Vieta jumping, which combines this method with Vieta's formulas for polynomials:

Definition 6.12: Proof by Vieta Jumping

The following is the standard procedure to prove the non-existence of solutions to an equation by Vieta jumping.

1. For contradiction, assume the existence of an unsatisfactory solution.

2. Using the definition of minimality, take the minimal unsatisfactory solution.

3. Prove that this solution is indeed unsatisfactory by way of infinite descent.

We can recognize Vieta jumping as it commonly manifests in the form of specific degree-2 equations. For instance, the quadratic equation $a^2 + b^2 + c^2 - abc = 0$ lends itself to the fixation of two variables - suppose b and c - and the re-defining of a in terms of x (as just $a = x$). This leaves us with $x^2 + b^2 + c^2 - bcx = 0$, which allows us to apply Vieta's formulas to derive a relationship between a and b as solutions under some constraint. Id est, we are "locking" the roots r_1, r_2 into their intrinsic relationships with the coefficients of the polynomial $a_2 x^2 + a_1 x + a_0$, namely $r_1 = -\dfrac{a_1}{a_2}$ and $r_2 = \dfrac{a_0}{a_2}$.

To summarize, the end goal of Vieta jumping depends on the exact nature of the problem, but usually, we seek to restrict the values that we need to test down to a few possibilities by establishing a contradiction in the existence of a minimal solution using infinite descent.

6.6.3 Constant Descent Vieta Jumping

In a variant of Vieta jumping known as *constant descent* Vieta jumping, we eschew the proof by contradiction outline in favor of a more direct approach. In constant descent Vieta jumping, we view a constant k as having a relationship with the roots a and b of a quadratic polynomial, and attempt to prove this relationship by fixing k and one of the variables:

Definition 6.13: Constant Descent Vieta Jumping

Constant descent Vieta jumping consists of four overarching steps:

1. We first assume WLOG that $a > b$, since this follows from proving the equality case (where the divisor and the dividend are equal).

2. Fix b and k, such that the expression with a, b, k terms is re-defined with coefficients in terms of b and k only, and with a as a root. The other root, x_2, can be computed through Vieta's formulas.

3. Show inductively that, for all (a, b), we have $0 < x_2 < b < a$, and furthermore, $x_2 \in \mathbb{Z}$. This allows us to substitute in (b, x_2) in place of (a, b), and repeat until we arrive at the base case of our induction.

4. Conclude that, since k is a constant, the problem statement holds for all ordered pairs (a, b).

6.6.4 Examples

Example 6.6.4.1. *If $a, b, c \in \mathbb{N}$ such that $0 < a^2 + b^2 - abc \leq c$, show that $a^2 + b^2 - abc$ is a perfect square.*

Solution 6.6.4.1. *Assume there exist $a, b, c \in \mathbb{N}$ such that $k = a^2 + b^2 - abc$ is not a perfect square. Fix k, c, and consider ordered pairs (a, b) of positive integers satisfying the equation.*

WLOG assume $a \geq b$, and consider the quadratic equation $x^2 - xbc + (b^2 - k) = 0$. $x = a$ is a root, so the other root is a positive integer and less than a, namely $\dfrac{b^2 - k}{a}$. By assumption, this is less than a, and we can subsequently apply the infinite descent technique to arrive at a contradiction.

Example 6.6.4.2. *Prove that for every real number N, the equation*

$$x_1^2 + x_2^2 + x_3^2 + x_4^2 = x_1 x_2 x_3 + x_1 x_2 x_4 + x_1 x_3 x_4 + x_2 x_3 x_4$$

has a solution for which x_1, x_2, x_3, x_4 are all integers larger than N.

Solution 6.6.4.2. *$(1, 1, 1, 1)$ is a trivial solution. This solution is a springboard from which other solutions will be generated. We isolate variables to derive the quadratic equation*

$$x_1^2 - x_1(x_2 x_3 + x_3 x_4 + x_4 x_2) + x_2^2 + x_3^2 + x_4^2 - x_2 x_3 x_4 = 0$$

Given solutions r_1, r_2 for x_1, by Vieta's formulas, it directly follows that $r_1 + r_2 = x_2 x_3 + x_3 x_4 + x_4 x_2$. Then using the method of infinite descent, assume the existence of (r_1, s_1, t_1, u_1) satisfying the original equation. Then $(s_1 t_1 + t_1 u_1 + u_1 s_1 - r_1, s_1, t_1, u_1)$ is a solution "generated" by the original. But this is a contradiction, as the new solutions can be generated infinitely in this manner.

6.6.5 Exercises

Exercise 156. *Let x and y be positive integers such that $xy \mid (x^2 + y^2 + 1)$. Prove that $x^2 + y^2 + 1 = 3xy$.*

Exercise 157. *(1981 IMO P3) Determine the maximum value of $m^2 + n^2$, where m and n are integers satisfying $m, n \in \{1, 2, \ldots, 1981\}$ and $(n^2 - mn - m^2)^2 = 1$.*

Exercise 158. *Find, with proof, the largest integer $n \le 1000$ such that there exist integers $a, b \ge 0$ with $\dfrac{a^2 + b^2}{ab - 1} = n$.*

Exercise 159. *(Brilliant.org) Find the number of pairs of non-negative integers (n, m) such that $1 \le n \le m \le 100$, $n \mid (m^2 - 1)$, and $m \mid (n^2 - 1)$.*

Extension Exercise 61. *(2007 IMO P5) Let a and b be positive integers. Show that if $4ab - 1$ divides $(4a^2 - 1)^2$, then $a = b$.*

Extension Exercise 62. *(1988 IMO P6)[53] Let a and b be positive integers such that $ab + 1$ divides $a^2 + b^2$. Show that $\frac{a^2 + b^2}{ab + 1}$ is the square of an integer.*

---

[53]Yes, the notorious, ultra-difficult one that was once considered the hardest IMO problem ever. That one!

6.7 Complex Numbers, Part 2

6.7.1 Roots of Unity Filter

A particular useful technique to sum binomial coefficients with irregular intervals between them in their bottom arguments, or to sum binomial coefficients multiplied with another term (such as $(-1)^k$, for instance), is *roots of unity filter,* so called due to its intertwined relationship with the underlying ideas behind complex roots of unity.

Recall that, by de Moivre's Theorem, a root of unity is given by $\xi = \cos \dfrac{2k\pi}{n} + i \sin \dfrac{2k\pi}{n}$, since we have $(\cos 2k\pi + i \sin 2k\pi)^{\frac{1}{n}} = 1^{\frac{1}{n}}$. Modulo n, this expression is periodic, so we can add the constraint $0 \leq k \leq n-1$. This observation will prove instrumental in roots of unity filter computations.

In *roots of unity filter,* we aim to sum the coefficients of a polynomial $P(x)$ in some specified intervals between them. In particular, this method is highly successful when the inputs are periodic mod n, which occurs when they are all powers of an nth root of unity (conventionally denoted by ξ).

> ### Theorem 6.16: Roots of Unity Filter
>
> For some n, we have that
> $$\frac{P(1) + P(\xi) + P(\xi^2) + \ldots + P(\xi^{n-1})}{n}$$
> is the sum of all indexes of P that are multiples of n. (An index of a power of ξ in a polynomial is the coefficient of that power, denoted $[\xi^n]P$. For instance, if $P(x) = x^3 - 2x^2 + 45x + 2019$, then $[x^2]P = -2$.)

Proof. Let $P(x) = a_n x^n + a_{n-1}x^{n-1} + \ldots + a_1 x + a_0$. For indexes divisible by n, each evaluates to 1, by the definition of a root of unity. Otherwise, consider the general coefficient a_k. We have

$$P(1) + P(\xi) + \ldots + P(\xi^{n-1}) = a_k \cdot (\xi^k + \xi^{2k} + \ldots + \xi^{(n-1)k}$$

and so

$$\sum_{k=0}^{n-1} \xi^{ki} = \frac{\xi^{nk} - 1}{x^k - 1} = 0$$

Hence all indexes divisible by 1 leave 1, which occurs n times. Thus, the formula follows upon dividing by n. $\qquad\square$

Let's now apply this to a concrete example that one might find in a contest setting:

Example 6.7.1.1. *Compute* $\dbinom{2020}{0} + \dbinom{2020}{4} + \dbinom{2020}{8} + \ldots + \dbinom{2020}{2020}$.

Solution 6.7.1.1. *Instead of trying to exploit symmetry in the binomial coefficients, what we can do is use the polynomial $P(x) = (x+1)^k$, then observe that $P(1) = 2^k$ as a result (with $k = 2020$ and $n = 4$, where n is the "spacing" between the bottom arguments of the binomial coefficients). Then we have*

$$\frac{2^{2020} + (1+i)^{2020} + (1-1)^{2020} + (1-i)^{2020}}{4}$$

By de Moivre's Theorem, we can rewrite $(1+i)^{2020}$ as $2^{1010} \cdot (\cos 505\pi + i \sin 505\pi) = -2^{1010}$ and $(1-i)^{2020}$ as $2^{1010} \cdot (\cos 505\pi - i \sin 505\pi) = -2^{1010}$. Our fraction subsequently simplifies to $\dfrac{2^{2020} - 2^{1011}}{4} = \boxed{2^{2018} - 2^{1009}}$.

6.7.2 Complex Numbers in Euclidean Geometry

...Also known affectionately as the art of *complex bashing!*

The basis for this section going forward is a set of tenets of complex numbers: every $z \in \mathbb{C}$ is expressible as $a + bi$, $a, b \in \mathbb{R}$, or alternatively in polar form as $r(\cos\theta + i\sin\theta) = re^{i\theta}$. (Here r and θ are also reals, with $0 \le \theta < 2\pi$ without loss of generality.) It follows that $|z| = \sqrt{a^2 + b^2} = r$ and $\theta = \arg z$. Furthermore, if we multiply z by itself, we rotate it about the origin so that its new argument is twice its original argument, and in general, multiplying two complex numbers z_1 and z_2 yields $|z_1||z_2| = |z_1 z_2|$ and $\arg z_1 + \arg z_2 = \arg(z_1 z_2)$.

We use these fundamentals of complex numbers from the Intermediate section to describe some major geometric propositions that form the basis of complex bashing:

Theorem 6.17: The Propositions of Complex Geometry

(a) Let P_i $(1 \le i \le 4)$ be (pairwise) distinct points in the plane. Then $\overline{P_1 P_2} \perp \overline{P_3 P_4}$ iff $\dfrac{d - c}{b - a} = i \cdot r$, for some $r \in \mathbb{R}$ (where $a = P_1$, $b = P_2$, $c = P_3$, and $d = P_4$).

(b) Let P_i $(1 \le i \le 3)$ be distinct points in the plane. Then they are collinear iff $\dfrac{c - a}{c - b} \in \mathbb{R}$. (That is, it is equal to its own conjugate. Here we have implicitly defined a, b, c, d as above.)

(c) Let P_i $(1 \le i \le 4)$ be distinct in the plane. Then they are concyclic (lie on a common circle) iff $\dfrac{c - a}{c - b} \cdot \dfrac{d - a}{d - b} \in \mathbb{R}$, where a, b, c, d are defined as above.

Proof.

(a) This is equivalent to $\arg\left(\dfrac{b - c}{b - a}\right) = 90°$, or $\overline{AB} \perp \overline{CD}$ as desired.

(b) $\dfrac{c - a}{c - b} \in \mathbb{R} \implies \arg\left(\dfrac{c - a}{c - b}\right) = 0$, or P_1, P_2, P_3 collinear as desired.

(c) We have $\arg\left(\dfrac{c - a}{c - b}\right) = \angle P_1 P_3 P_2$ and $\arg\left(\dfrac{d - a}{d - b}\right) = \angle P_1 P_4 P_2$. Thus it follows that $\arg\left(\dfrac{c - a}{c - b} \cdot \dfrac{d - a}{d - b}\right) = \angle P_1 P_3 P_2 + \angle P_1 P_4 P_2$, and this yields the desired result with directed angles.

In particular, for (c), we use the fact that $\arg(z_1 z_2) = \arg z_1 + \arg z_2$, for $z_1, z_2 \in \mathbb{C}$. $\qquad\square$

We can also derive other important properties of complex geometry, in particular the analogue of the Shoelace Theorem in the complex plane:

Theorem 6.18: Complex Shoelace Theorem

Let P_i $(1 \le i \le 3)$ be points in the plane. Then $\triangle P_1 P_2 P_3$ has signed area given by

$$\frac{1}{4} i \cdot \det \begin{bmatrix} a & \bar{a} & 1 \\ b & \bar{b} & 1 \\ c & \bar{c} & 1 \end{bmatrix}$$

where $a, b, c \in \mathbb{C}$ correspond to the values of P_1, P_2, P_3 in $\mathbb{C}^2$ respectively, and $\bar{a}$ denotes the complex conjugate of a.

Corollary 6.18: Vanishing Complex Shoelace Determinant

Iff the determinant in Theorem 6.10 vanishes (i.e. it becomes zero), then the points P_i corresponding to complex a, b, c are collinear.

We prove both the theorem and its corollary.

Proof. To prove the formula itself, WLOG assume $c = 0$ (since we can translate the triangle so that P_3 is at the origin for any choices of P_1, P_2, P_3). Then the area becomes

$$\frac{1}{4} i \cdot \det \begin{bmatrix} a & \bar{a} & 1 \\ b & \bar{b} & 1 \\ 0 & 0 & 1 \end{bmatrix} = \frac{1}{4} i \cdot (a\bar{b} - \overline{a}\overline{b}) = \frac{1}{2} \text{Im}(a\bar{b})$$

Let $a = |a|\, e^{i\alpha}$ and $b = |b|\, e^{i\beta}$; then $a\bar{b} = |a|\, |b| \cdot e^{i(\alpha - \beta)} = |a|\, |b| \cdot (\cos(\alpha - \beta) + i\sin(\alpha - \beta))$ by de Moivre's Theorem, which implies that the area is equal to $\frac{1}{2} P_1 P_2 \cdot P_1 P_3 \cdot \sin \angle P_2 P_1 P_3$, which we know to be true.

To show that P_1, P_2, P_3 are collinear iff the signed area is zero, note that the area is zero precisely when $a\bar{b} \in \mathbb{R}$, which implies that either $a, b \in \mathbb{R}$ or $a, b \in \mathbb{C}$ with $\text{Re}(a) = \text{Re}(b) = 0$. In either case, the points are clearly collinear. $\qquad\square$

A very common scenario in complex bashing is one in which we aim to find the intersection points of two lines $\overline{AB}$ and $\overline{CD}$, with A, B, C, D being pairwise distinct points in the plane. Much like we determine points of intersection of lines in the real Cartesian plane by setting their equations equal to each other and solving algebraically, we can work with complex equations by using the slope formula for complex numbers.

Indeed, we have a concrete formula expressly for this purpose:

Theorem 6.19: Intersections of Lines in the Complex Plane

Let A, B, C, D be points in $\mathbb{C}^2$. Then lines $\overline{AB}$ and $\overline{CD}$ intersect at

$$\frac{(\bar{a}b - a\bar{b})(c - d) - (a - b)(\bar{c}d - c\bar{d})}{(\bar{a} - \bar{b})(c - d) - (a - b)(\bar{c} - \bar{d})}$$

Proof. The proof is messy algebra and is left as an exercise to the reader. (However, if A, B, C, D are concyclic on the unit circle, then the proof becomes much easier to carry out computationally.) $\qquad\square$

Of particular noteworthiness is the unit circle in the complex plane, which consists of all points z with $|z| = 1$ (much like the radius 1 in $\mathbb{R}^2$, due to the connection with polar form) such that $\bar{z} = \frac{1}{z}$ (i.e. $z\bar{z} = 1$). This allows us to derive a simpler form of Theorem 6.11:

Corollary 6.19: Intersections in the Unit Circle

If A, B, C, D are points lying on the unit circle, then $\overline{AB}$ and $\overline{CD}$ intersect at

$$\frac{ab(c + d) - cd(a + b)}{ab - cd}$$

We can similarly derive formulae for the centers of a triangle: namely, the circumcenter, centroid, orthocenter, and incenter:

Let $\triangle ABC$ with $A, B, C \in \mathbb{C}^2$ have circumcircle coinciding with the unit circle. Then the circumcenter of $\triangle ABC$ is 0, the centroid of $\triangle ABC$ is $\dfrac{a+b+c}{3}$, the orthocenter is $a + b + c$, and the incenter is $-(xy + yz + xz)$, where $x, y, z \in \mathbb{C}$ such that $|x| = |y| = |z| = 1$ and $a = x^2$, $b = y^2$, $c = z^2$.

Proof. The first three formulas can all be proven by setting $h = a+b+c$, and proving that $\overline{AH} \perp \overline{BC}$, which follows by symmetry. Define $z = \dfrac{h-a}{b-c} = \dfrac{b+c}{b-c}$ and compute $\overline{z} = -z$ (the details are left to the reader), so $z \in i \cdot \mathbb{R}$ as desired.

For the incenter, let I be the incenter of $\triangle ABC$. Then extend $\overline{AI}$, $\overline{BI}$, and $\overline{CI}$ to meet the circumcircle at D, E, F respectively. Then construct $\triangle DEF$, which will then have I as its orthocenter. (This can be proven purely algebraically at this point.) This suffices to complete the proof. $\square$

6.7.3 Examples

Example 6.7.3.1. *A test has 20 questions. How many ways are there to answer exactly an even, nonzero number of questions correctly?*

Solution 6.7.3.1. *This is* $\sum_{k=0}^{10} \binom{20}{2k} = \boxed{2^{19}}$ *by roots of unity filter/symmetry.*

Example 6.7.3.2. *Evaluate*

$$\sum_{k=0}^{32} (-1)^k \cdot \binom{63}{2k}$$

Solution 6.7.3.2. *Recognize that this is* $Re(1+i)^{63} = \boxed{2^{31}}$ *by de Moivre's Theorem. Alternatively, expand the binomial coefficients, and be sure to have a fraction whose numerator is in terms of $1+i$ and $1-i$.*

6.7.4 Exercises

Exercise 160. *(2018 AIME I) Let N be the number of complex numbers z with the properties that $|z| = 1$ and $z^{6!} - z^{5!}$ is a real number. Find the remainder when N is divided by 1000.*

Exercise 161. *(2014 AIME I) Let w and z be complex numbers such that $|w| = 1$ and $|z| = 10$. Let $\theta = \arg\left(\dfrac{w - z}{z}\right)$. The maximum possible value of $\tan^2 \theta$ can be written as $\dfrac{p}{q}$, where p and q are relatively prime positive integers. Find $p + q$. (Note that $\arg(w)$, for $w \neq 0$, denotes the measure of the angle that the ray from 0 to w makes with the positive real axis in the complex plane.)*

Exercise 162. *(2017 AIME I) Let $z_1 = 18 + 83i$, $z_2 = 18 + 39i$, and $z_3 = 78 + 99i$, where $i = \sqrt{-1}$. Let z be the unique complex number with the properties that $\dfrac{z_3 - z_1}{z_2 - z_1} \cdot \dfrac{z - z_2}{z - z_3}$ is a real number and the imaginary part of z is the greatest possible. Find the real part of z.*

Exercise 163. *Prove that for any $x, y, z \in \mathbb{C}$, $|x - y| = |y - z| = |z - x| \implies (x-y)^2 + (y-z)^2 + (z-x)^2 = 0$.*

Exercise 164. *(2001 AIME II) There are $2n$ complex numbers that satisfy both $z^{28} - z^8 - 1 = 0$ and $|z| = 1$. These numbers have the form $z_m = \cos\theta_m + i\sin\theta_m$, where $0 \leq \theta_1 < \theta_2 < \ldots < \theta_{2n} < 360$ and angles are measured in degrees. Find the value of $\theta_2 + \theta_4 + \ldots + \theta_{2n}$.*

Exercise 165. *(2016 AIME I) For integers a and b consider the complex number*

$$\frac{\sqrt{ab + 2016}}{ab + 100} - \left(\frac{\sqrt{|a + b|}}{ab + 100}\right)i$$

Find the number of ordered pairs of integers (a, b) such that this complex number is a real number.

Exercise 166. *(1999 AIME) A function f is defined on the complex numbers by $f(z) = (a + bi)z$, where a and b are positive numbers. This function has the property that the image of each point in the complex plane is equidistant from that point and the origin. Given that $|a + bi| = 8$ and that $b^2 = m/n$, where m and n are relatively prime positive integers. Find $m + n$.*

Exercise 167. *(2018 HMMT February Algebra/Number Theory Round) Let $\omega_1, \omega_2, \ldots, \omega_{100}$ be the roots of $\dfrac{x^{101} - 1}{x - 1}$ (in some order). Consider the set*

$$S = \{\omega_1^1, \omega_2^2, \omega_3^3, \ldots, \omega_{100}^{100}\}$$

Let M be the maximum possible number of unique values in S, and let N be the minimum possible number of unique values in S. Find MN.

Exercise 168. *(2017 HMMT November General Test) Given that a, b, c are integers with $abc = 60$, and that complex number $\omega \neq 1$ satisfies $\omega^3 = 1$, find the minimum possible value of $|a + b\omega + c\omega^2|$.*

Extension Exercise 63. *(2012 USAMO P5) Let P be a point in the plane of triangle ABC, and γ a line passing through P. Let A', B', C' be the points where the reflections of lines PA, PB, PC with respect to γ intersect lines BC, AC, AB, respectively. Prove that A', B', C' are collinear.*

Extension Exercise 64. *Let $G = \{z \in \mathbb{C} \mid \exists n \in \mathbb{N} \text{ s.t. } z^n = 1\}$. Prove that G is a group under multiplication.*

Extension Exercise 65. *Let $z_1, z_2, z_3 \in \mathbb{C}$ be distinct. When are z_1, z_2, z_3 the vertices of an equilateral triangle in the complex plane?*

Extension Exercise 66. *(2014 IMO Shortlist G6) Let $\triangle ABC$ be a fixed acute-angled triangle. Consider some points E and F lying on the sides $\overline{AC}$ and $\overline{AB}$, respectively, and let M be the midpoint of $\overline{EF}$. Let the perpendicular bisector of $\overline{EF}$ intersect the line $\overline{BC}$ at K, and let the perpendicular bisector of $\overline{MK}$ intersect the lines $\overline{AC}$ and $\overline{AB}$ at S and T, respectively. We call the pair (E, F) interesting, if the quadrilateral $KSAT$ is cyclic. Suppose that the pairs (E_1, F_1) and (E_2, F_2) are interesting. Prove that $\dfrac{E_1 E_2}{AB} = \dfrac{F_1 F_2}{AC}$.*

6.8 Matrix Algebra, Part 2

We continue from our analysis of matrices in §5.13, where we discussed the definition of a matrix, the elementary row operations we can perform on a matrix to reduce it to a matrix with similar/identical properties, and exactly what the core properties of a matrix are that are most applicable to a competitive setting. Here, we delve deeper into some of the properties that either provide context or a shortcut method for the computational tools provided in the previous chapter, or construct a more rigorous foundation for those computational tools by giving solid mathematical justification for their efficacy (in particular, see the "Inner Product Spaces" section for a prime example of this latter goal in action). Since there is so much ground to cover, our overview of these topics may be somewhat cursory compared to our analysis of the other subjects, but are nevertheless good to keep in reserve as "back of the head" type of knowledge, or for short-cutting purposes in matrix problems.

6.8.1 The LU Decomposition

As it turns out, triangular matrices are more than just an artifact of the matrix space, and more than just special types of matrices that serve little to no purpose in and of themselves. We can actually decompose, or "factor" in a sense, any given matrix into a product of two of these triangular matrices, namely a lower triangular matrix and an upper triangular matrix, provided it meets certain initial conditions. This type of factoring method is what is known as the *LU decomposition* (wherein we set a matrix M equal to LU for L a matrix with all above-diagonal elements zero and U a matrix with all below-diagonal elements zero), and through which we may expedite the solution of several classes of linear equations.

Before we can undergo this factorization, we must be aware of several constraints that apply to an LU-factorable matrix (though several can be artifically removed). For one, we must ensure that the matrix M does not have $M_{1,1} = 0$, for this would require that either $L_{1,1} = 0$ or $U_{1,1} = 0$ (or both); then one of the two matrices L, U would be singular, which implies that M is singular as well. But this is impossible if $\det M \neq 0$, which is a common assumption in LU factorization problems. Indeed, insofar as we can invert a matrix at all, we can compute an LU decomposition. However, this does not work the other way around. That is, an LU decomposition may exist for singular matrices, which *a priori* do not have an inverse. (For instance, any real square matrix M has an LU factorization, regardless of whether or not it is invertible.)

Essentially, finding an LU decomposition reduces to a Gaussian elimination problem, in which we solve the system $Ax = b$ for x given A and b. Counter-intuitively, we can actually first assume the LU decomposition of A - by setting $PA = LU$ for a pivot matrix[54] P, with $LUx = Pb$ as a result - before we solve the LU decomposition problem for x! For a 3×3 matrix A, we can explicitly write

$$\begin{bmatrix} a_{1,1} & a_{1,2} & a_{1,3} \\ a_{2,1} & a_{2,2} & a_{2,3} \\ a_{3,1} & a_{3,2} & a_{3,3} \end{bmatrix} = \begin{bmatrix} l_{1,1} & 0 & 0 \\ l_{2,1} & l_{2,2} & 0 \\ l_{3,1} & l_{3,2} & l_{3,3} \end{bmatrix} \begin{bmatrix} u_{1,1} & u_{1,2} & u_{1,3} \\ 0 & u_{2,2} & i_{2,3} \\ 0 & 0 & u_{3,3} \end{bmatrix}$$

$$\begin{bmatrix} a_{1,1} & a_{1,2} & a_{1,3} \\ a_{2,1} & a_{2,2} & a_{2,3} \\ a_{3,1} & a_{3,2} & a_{3,3} \end{bmatrix} = \begin{bmatrix} l_{1,1}u_{1,1} & l_{1,1}u_{1,2} & l_{1,1}u_{1,3} \\ l_{2,1}u_{1,1} & l_{2,1}u_{1,2} + l_{2,2}u_{2,2} & l_{2,1}u_{1,3} + l_{2,2}u_{2,2} \\ l_{3,1}u_{1,1} & l_{3,1}u_{1,2} + l_{3,2}u_{2,2} & l_{3,1}u_{1,3} + l_{3,2}u_{2,3} + l_{3,3}u_{3,3} \end{bmatrix}$$

Hence, we obtain a system of several equations that can be solved algebraically. But a much simpler way involves row-reducing A to either a lower triangular or upper triangular matrix, and then solving for the other of L or U by matrix division (provided A is invertible). Should A be singular, we cannot divide directly, but solution is still completely possible by solving a system of equations.

As an example, consider the following LU decomposition of a square matrix:

---

[54]The matrix formed from taking the identity matrix, and performing all the same row operations on it that were performed upon A to convert it to either L or U.

Example 6.8.1.1.

$$A = \begin{bmatrix} 2 & 1 & 1 \\ 10 & 4 & 2 \\ 4 & -1 & 1 \end{bmatrix}$$

row-reduces to the lower-triangular matrix

$$L = \begin{bmatrix} 1 & 0 & 0 \\ 5 & 1 & 0 \\ 2 & 3 & 1 \end{bmatrix}$$

and the upper-triangular matrix

$$U = \begin{bmatrix} 2 & 1 & 1 \\ 0 & -1 & -3 \\ 0 & 0 & 8 \end{bmatrix}$$

We can verify that $A = LU$.

6.8.2 Eigenvalues and Eigenvectors

In our overarching goal of solving the general matrix equation $Ax = b$, we often encounter matrices which can be multiplied with other matrices to obtain scalar multiples of themselves. These scalar values are known as *eigenvalues*, and the products of these eigenvalues with the original matrices are known as an *eigenvectors*. The etymology of these terms comes from the German *eigen-*, meaning "self" or "own." This refers to the tendency of an eigenvalue to preserve the direction of the original vector in producing the corresponding eigenvector. Eigenvalues and eigenvectors also have majorly important applications in dynamical systems, wherein they represent the solutions to specific classes of differential equations (*eigenfunctions*) that change over time.

Definition 6.14: Eigenvalues and Eigenvectors

Consider the matrix A. An *eigenvalue* of A is a real number λ such that $Ax = \lambda x$ for some column vector x, and x is the corresponding *eigenvector*.

Corollary 6.20: The Zero Eigenvalue

If $\lambda = 0$, then $Ax = 0x$ implies that $x \in N(A)$.

An oft-used result in studying eigenvalues and eigenvectors is the following:

Theorem 6.21: Relationship with the Determinant

For all matrices A, we have that $\det(A - \lambda I) = 0$, where I is the identity matrix.

Proof. (Left to the reader. Key observation: if we have $x \neq 0$, $x \in N(A - \lambda I)$, then A is singular, i.e. $\det A = 0$.) $\qquad \square$

Building off of this, an invaluable tool in computing the eigenvalue(s) of a matrix is its *characteristic polynomial*:

Definition 6.15: Characteristic Polynomial

The *characteristic polynomial* $p(\lambda)$ of the matrix A is defined as $\det(A - \lambda I)$ *in terms of* λ.

Remark. This last, italicized part of the sentence is why the traditional false proof of the Cayley-Hamilton Theorem doesn't work! We must remember that the characteristic polynomial is just that - a polynomial. That is, we seek all λ such that $\det(A - \lambda I) = 0$; i.e. the roots of the characteristic polynomial (of which these must be $n = \dim A$, including multiplicity, by the Fundamental Theorem of Algebra) are the eigenvalues of A. Then we can compute the eigenvectors x individually by solving the equation $Ax = \lambda x$.

Example 6.8.2.1. *Compute the eigenvalues and eigenvectors of*

$$\begin{bmatrix} 1 & 3 \\ 2 & 2 \end{bmatrix}$$

Solution 6.8.2.1. *We have*

$$A - \lambda I = \begin{bmatrix} 1 - \lambda & 3 \\ 2 & 2 - \lambda \end{bmatrix}$$

$$\det(A - \lambda I) = (1 - \lambda)(2 - \lambda) - 6 = -4 - 3\lambda + \lambda^2$$

$$-4 - 3\lambda + \lambda^2 = 0$$

$$\lambda = 4, -1$$

Then we have $Ax = 4x$ and $Ax = -1x$; by setting $x = \begin{bmatrix} a \\ b \end{bmatrix}$, we obtain the system of equations $a + 3b = 4a$, $2a + 2b = 4b$ for $Ax = 4x$, and $a + 3b = -a$, $2a + 2b = -b$ for $Ax = -1x$. This implies the minimal solutions $(a, b) = (1, 1)$ and $(a, b) = (-3, 2)$ for $Ax = 4x$ and $Ax = -1x$ respectively.

In this example, we may observe that row reduction *does not, in fact, preserve the eigenvalues and eigenvectors.* (Note that any square matrix can be row reduced to the identity matrix, which has eigenvalues $\lambda = 1$ (multiplicity 2) and eigenvectors $x = \begin{bmatrix} 0 \\ 1 \end{bmatrix}, \begin{bmatrix} 1 \\ 0 \end{bmatrix}$.) This is admittedly an unfortunate piece of news, but thankfully, as it turns out, the product of the eigenvalues (including duplicates due to multiplicity) is equal to $\det A$, and the sum of the eigenvalues equals the sum of the diagonal entries (the trace, $\text{Tr}(A)$)!

Proof. Let $\lambda_1, \lambda_2, \lambda_3, \ldots, \lambda_n$ be the eigenvalues of A (including multiplicity). Then they are the roots of the characteristic polynomial of A, so that

$$\det(A - \lambda I) = (-1)^n \cdot \prod_{i=1}^{n} (\lambda - \lambda_i)$$

$$= \prod_{i=1}^{n} (\lambda_i - \lambda)$$

Setting $\lambda = 0$, we obtain the equation

$$\det A = \prod_{i=1}^{n} \lambda_i$$

To prove the similar statement regarding the trace, observe that

$$p(\lambda) = \det(A - \lambda I) = (-1)^n \cdot \left(\lambda^n - (\text{Tr}(A))\lambda^{n-1} + \ldots + (-1)^n \cdot \det A \right)$$

We also have

$$p(\lambda) = (-1)^n \cdot \prod_{i=1}^{n} (\lambda - \lambda_i)$$

and equating coefficients yields

$$\sum_{i=1}^{n} \lambda_i = \text{Tr}(A)$$

as desired. $\qquad \square$

6.8.3 Diagonalization

Just as we can decompose matrices into upper and lower triangular matrices, so we can include them in the decompositions of other matrices as factors. In particular, we will focus on those factorizations that incorporate invertible and diagonal matrices:

Definition 6.16: Diagonalizable Matrices

A square matrix A is *diagonalizable* if there exists an invertible matrix P such that $P^{-1}AP$ is a diagonal matrix. (If A is not diagonalizable, then it is *defective*. If A is non-square, it is neither diagonalizable nor defective, as these designations only apply to square matrices.)

Remark. Note that the product PAP^{-1} *must* be in this order (matrix multiplication, unlike scalar multiplication, is not commutative!)

Essentially, A is *similar* to the matrix PAP^{-1}. (By definition, A is similar to B iff $B = P^{-1}AP$ for some invertible P - thus, the diagonalizability condition is just a special case of the similarity condition. Similar matrices are another rich topic of discussion, but as they are not extensively tested or applied in a competitive setting, we leave it as a point of interest for the reader's own exploration.)

We have the following sufficient (but not necessary) condition for the diagonalizability of A:

Theorem 6.22: Sufficient Condition for Diagonalizability

The $n \times n$ matrix A is diagonalizable over a field $\mathbb{F}$ if it has exactly n *distinct* eigenvalues in $\mathbb{F}$. That is, the characteristic polynomial $p(\lambda)$ of A cannot have any duplicate roots. However, the converse is false (in particular, when the eigenspace of A associated with some eigenvalue λ_i - i.e. the multiplicity of the eigenvalue λ_i - has dimension higher than 1).

The following is a *necessary* and sufficient condition for diagonalizability over a field $\mathbb{F}$:

Theorem 6.23: Iff Statement for Diagonalizability

An $n \times n$ matrix A is diagonalizable over $\mathbb{F}$ iff the sum of the dimensions of its eigenspaces is equal to $\dim V$, where V is a vector space (see the following section) with a basis consisting of eigenvectors of A.

Regarding the computational diagonalization process itself, if A is diagonalizable and

$$P^{-1}AP = \begin{bmatrix} \lambda_1 & 0 & 0 & \dots & 0 \\ 0 & \lambda_2 & 0 & \dots & 0 \\ 0 & 0 & \lambda_3 & \dots & 0 \\ \vdots & \vdots & \vdots & \ddots & \vdots \\ 0 & 0 & 0 & \dots & \lambda_n \end{bmatrix}$$

then we must have

$$AP = P \begin{bmatrix} \lambda_1 & 0 & 0 & \dots & 0 \\ 0 & \lambda_2 & 0 & \dots & 0 \\ 0 & 0 & \lambda_3 & \dots & 0 \\ \vdots & \vdots & \vdots & \ddots & \vdots \\ 0 & 0 & 0 & \dots & \lambda_n \end{bmatrix}$$

If we write P in terms of its column vectors $v_1, v_2, v_3, \dots, v_n$, then we obtain the equation $Av_i = \lambda_i v_i$, $1 \leq i \leq n$. Thus, the column vectors v_i are eigenvectors of A (when multiplied to the right of A; hence, they are called *right eigenvectors*) and the corresponding λ_i is the eigenvalue. As P is invertible, this also suggests that the eigenvectors are linearly independent (since $\det P \neq 0$), and that they form a basis for $\mathbb{F}^n$. (The row vectors of P, similarly, are the left eigenvectors of A.)

6.8.4 Vector Spaces

In linear algebra, we place a greater emphasis on vectors due to their versatility in solving the fundamental equation $Ax = b$. To truly understand the motivation behind vectors, as opposed to just their computational function, we need to examine the concept of a *vector space*. At its core, a vector space is a collection of vectors, much like a matrix is an element of a matrix space and the set of column vectors x for which $Ax = 0$ is the null space $N(A)$ of A.

In particular, the distinguishing qualities of a vector space are its closure under addition and scalar multiplication. That is, it follows a very specific set of sub-properties associated with closure that define it as a group:

Definition 6.17: Axioms of a Vector Space

For V to be a vector space, we must have all of the following for all $x, y, z \in V$, and scalars $a, b \in \mathbb{F}$:

- Commutativity: $x + y = y + x$ (and $x + z = z + x, y + z = z + y$).

- Associative property of addition: $(x + y) + z = x + (y + z)$.

- Existence of the additive identity: for all $x \in V$, $x + 0 = x = 0 + x$.

- Existence of the additive inverse: for all $x \in V$, there exists $-x \in V$ such that $x + (-x) = 0 = (-x) + x$.

- Associative property of scalar multiplication: $a(bx) = b(ax)$.

- Distributive property: this property is two-fold. We have both $(a + b)x = ax + bx$ and $a(x + y) = ax + ay$.

- Existence of the scalar multiplication identity: $x \cdot 1 = x = 1 \cdot x$ for all $x \in V$.

A vector space is characterized by its *dimension,* much like a matrix space or an eigenspace has well-defined dimension. We can similarly extend this to *infinite-dimensional vector spaces,* which are commonplace in fields of study such as functional analysis and topology (in which we often seek to define the notion of a metric space by way of vector constructs). Indeed, if we wish to define the *distance* between two vectors in a vector space $V \in \mathbb{F}^n$, then we may consider objects such as the *norm* and the *inner product* (see the next section for more details).

We have a fundamental result about vector spaces that, in particular, applies the most to the ideas we have discussed in linear algebra thus far:

Theorem 6.24: Existence of a Vector Basis

By the axiom of choice,[a] every vector space has a vector basis.

---
[a] Given any set of mutually disjoint non-empty sets, we have at least one set that shares exactly one element with each fo the sets.

Let's now discuss some common examples of vector spaces. A typical instance of a vector space is the space of vectors in the plane (i.e. $\mathbb{R}^2$), which a vector is represented by a directed arrow with a "head" and a "tail," as well as a length. These types of vectors are very commonplace in physics, where they are used to depict force/velocity/acceleration due to their directional component.[55] In general, an ordered pair (or, more generally, n-tuple) is also a vector, as it follows the axioms of associativity, commutativity, and distributivity (as well as the existences of the inverses and identities, which are not hard to check).

Indeed, we should also remark that the vector space is actually a generalization (and more strictly-defined form of) an abelian group, as the first four axioms are equivalent to satisfying the axioms of an abelian group.

In their use in matrix algebra, vector spaces have two important properties of special focus: the *basis* and the *dimension*.

Definition 6.18: Vector Basis and Dimension

A *basis* of a vector space V is a (finite or infinite) set B of vectors b_i that is linearly independent and spans the space V. Id est, any $v \in V$ can be expressed as

$$v = a_1 b_1 + a_2 b_2 + a_3 b_3 + \ldots + a_n b_n$$

where $a_i \in \mathbb{R}$ are scalars and $b_i \in B$. The *dimension* of V is then equal to n.

Example 6.8.4.1. *A common example of a vector space basis is the set of coordinate vectors* $e_1 = (1, 0, 0, \ldots, 0)$, $e_2 = (0, 1, 0, \ldots, 0)$, $e_3 = (0, 0, 1, \ldots, 0)$, $\ldots$, $e_n = (0, 0, 0, \ldots, 1)$ *which is a basis for* $\mathbb{F}^n$ *(called the standard basis). We then have* $\dim V = n$.

We discuss vector space mainly because of the relationship of matrices with the notion of linear maps - i.e. mappings between two modules (generalizations of vector spaces to objects other than vectors) - as a matrix is essentially an encoding of a linear map. An $m \times n$ matrix A encodes this information in the form of a linear map $L : \mathbb{F}^n \mapsto \mathbb{F}^m$, namely such that $x \mapsto Ax$.

Furthermore, the determinant of A indicates whether or not the linear mapping encoded by A is an isomorphism; iff $\det A \neq 0$, then A represents an isomorphism (since the linear mapping can be inverted; if $\det A = 0$, then A is non-invertible and the direction of its linear mapping cannot be reversed).

Finally, to connect to our previous analysis of eigenvalues and eigenvectors, note that the set of all eigenvectors corresponding to some eigenvalue λ forms an eigenspace, and the eigenspace is by definition a vector space.

6.8.5 Inner Product Spaces

An *inner product space* is a generalization of a vector space, equipped with the *inner product* operator. For each pair of vectors in the space, the inner product space associates this pair with a scalar that is the output of a dual-input function:

Definition 6.19: Inner Product Space

An *inner product space* is a vector space V equipped with the inner product $\langle v_1, v_2 \rangle$ for all $v_1, v_2 \in V$. This type of space is a generalization of Euclidean space, in which the inner product is the dot product (given by $\langle v_1, v_2 \rangle \cdot \langle w_1, w_2 \rangle = v_1 w_1 + v_2 w_2$).

Formally, we define V over a field $\mathbb{F}$ such that, for all vectors $x, y, z \in V$ and scalars $a \in \mathbb{F}$,

$$\langle \cdot, \cdot \rangle : V \times V \mapsto \mathbb{F}$$

---

[55]Indeed, the vector is the distinction between *speed* and *velocity:* whereas speed is a scalar quantity, vleocity is a vector quantity consisting of both the speed and the direction. Similarly, acceleration is also a vector quantity, as it requires a direction; by definition, an object is accelerating if its velocity is changing at some point in time.

satisfies the following axioms:

- Conjugate symmetry: $\langle x, y \rangle = \overline{\langle y, x \rangle}$.

- Linearity (additivity and multiplicativity) in the *first* argument: i.e. $a\langle x, y \rangle = \langle ax, y \rangle$ and $\langle x, z \rangle + \langle y, z \rangle = \langle x + y, z \rangle$.

- Positive definiteness: $\langle x, x \rangle \geq 0$, $\langle x, x \rangle = 0 \iff x = 0$.

Note that, if $\mathbb{F} = \mathbb{R}$, then conjugate symmetry is just *symmetry:* i.e. $\langle x, y \rangle = \overline{\langle y, x \rangle}$.

Using this definition of an inner product and an inner product space, as well as of the dot product, we can immediately classify two vectors as *orthogonal:*

Definition 6.20: Orthogonality

Two vectors $v_1, v_2 \in V$ are *orthogonal* iff $\langle v_1, v_2 \rangle = 0$.

This definition applies to *all* inner product spaces V, and is, in fact, part of the characterizing axioms of an inner product space.

In particular, since the dot product is a special type of inner product, we have that

$$\langle x, y \rangle = \sum_{i=1}^{n} x_i y_i$$

where

$$x = \begin{bmatrix} x_1 \\ x_2 \\ x_3 \\ \vdots \\ x_n \end{bmatrix}$$

$$y = \begin{bmatrix} y_1 \\ y_2 \\ y_3 \\ \vdots \\ y_n \end{bmatrix}$$

Thus, the column vectors x and y are orthogonal iff $\sum_{i=1}^{n} x_i y_i = 0$. (Observe that this is also equal to $x^T y$; this can be verified by observing the product of a $1 \times n$ matrix and an $n \times 1$ matrix yields a 1×1 matrix, or a scalar quantity.)

Over other, more complex vector spaces, we can also define analogous inner products. Over the complex numbers $\mathbb{C}^n$, for instance, we have

$$\langle x, y \rangle = y^T M x = \overline{x^T M y}$$

where y^T denotes the *conjugate transpose* of y (i.e. the resulting of taking the complex conjugate of each entry, then transposing) and M is a positive-definite[56] Hermitian[57] matrix. As such, this form is also referred to as a *Hermitian form.*

Over a Hilbert space,[58] the inner product is defined for functions f, g over the interval $[a, b]$ as follows:

$$\langle f(t), g(t) \rangle = \int_a^b f(t) \overline{g(t)} dt$$

---

[56]A square, symmetric matrix M is *positive-definite* if the scalar $v^T M v$ is positive for every column vector $v \neq 0$, $v \in M$.

[57]A square matrix with entries in $\mathbb{C}^n$ is said to be *Hermitian* if it is equal to its own conjugate transpose.

[58]A real or complex inner product space that is also a complete metric space with respect to the distance function of the inner product.

(This inner product space is not necessarily complete, meaning that not every Cauchy sequence in the space need converge to a continuous function.)

Each inner product space is equipped with a *norm*, which is defined in terms of the inner product:

Definition 6.21: The Vector Norm

For V an inner product space, the *norm* of a vector $x \in V$ is given by $||x|| = \sqrt{\langle x, x \rangle}$; i.e. $\langle x, x \rangle = ||x||^2$.

This is only the 2-norm, however. In general, we can define the p-norm for all $p \geq 1$:

$$||x||_p = \sqrt[p]{\sum_{i=1}^{\infty} |x_i|^p}$$

along with the ∞-norm:

$$||x||_\infty = \max(|x_i|), 1 \leq i \leq n$$

From the norm in Definition 6.16 (the *2-norm*), we can derive several crucial inequalities, such as the Cauchy-Schwarz Inequality (and its generalized form, Hölder's Inequality), the Triangle Inequality, and the Parallelogram Law. Ultimately, this observation ties everything we have discussed so far back to the nature of inequalities: norms, and fundamentally vectors, and their relationship with linear maps - and matrices, by extension - are what allow inequalities to operate in the way they do.

Furthermore, we can use the norm to provide a proof of the Pythagorean Theorem;[59] we state that

$$||x||^2 + ||y||^2 = ||x + y||^2$$

and prove this by expanding out each norm manually, and using the fact that the norm is additive.

6.8.6 Orthonormalization: The Gram-Schmidt Algorithm and the QR Decomposition

A vector can be orthogonal to another vector in an infinitude of ways (since the only restriction is that the inner product needs to be zero). But when is the vector truly minimal? To determine when the vector will have unit length (or, more formally, magnitude), we must first *orthonormalize* it through a process known as the *Gram-Schmidt algorithm*. This process will yield a vector (or set of vectors that spans a vector subspace) that has magnitude (and norm) 1, and ensure that all vectors in a spanning set are mutually orthogonal (with every vector in the set orthogonal to every other vector in the set; pairwise orthogonality is insufficient)impose the constraint that all vectors in the spanning set are unit vectors, then the process is known simply as *orthogonalization*.

Definition 6.22: Gram-Schmidt Orthonormalization

For a finite, linearly independent set of vectors $S = \{v_1, v_2, v_3, \ldots, v_k\}$, the Gram-Schmidt process orthonormalizes S into an orthonormal set $S' = \{w_1, w_2, w_3, \ldots, w_k\}$ spanning the same subset of the corresponding vector space of S with dimension k as S. In addition, $||w_i|| = 1$ for all $1 \leq i \leq k$.

Begin the algorithm by setting $w_1 = v_1$. Then recursively define

$$w_i = v_i - \sum_{j=1}^{i-1} \frac{\langle w_j, v_i \rangle}{\langle w_j, w_j \rangle} w_j$$

---

[59]Of which there are an astounding number - hundreds, if not thousands. See §17.4 for a sampling of some of them!

until

$$w_k = v_k - \sum_{j=1}^{k-1} \frac{\langle u_j, v_k \rangle}{\langle u_j, u_j \rangle} u_j$$

This will yield $e_1 = \frac{w_1}{||w_1||}$, $e_2 = \frac{w_2}{||w_2||}$, ..., $e_n = \frac{u_n}{||w_n||}$ (i.e. the unit (orthonormalized) vectors with components that are proportional to those of w_n).

Applying Gram-Schmidt to the column vectors of a matrix with full rank yields the *QR decomposition* of the matrix (into an orthogonal matrix[a] Q and an upper triangular matrix R).

---

[a]A matrix Q whose columns and vectors are orthogonal unit vectors, such that $Q^T Q = Q Q^T = I$.

We will now work through a concrete example of Gram-Schmidt orthonormalization:

Example 6.8.6.1. *Orthonormalize the set $S = \{v_1 = (2,0), v_2 = (1,9), v_3 = (4,5)\}$.*

Solution 6.8.6.1. *We have* $w_1 = (2,0)$, $w_2 = (1,9) - \dfrac{\langle (2,0), (1,9) \rangle}{\langle (2,0), (2,0) \rangle} \cdot (2,0) = (1,9) - (1,0) = (0,9)$,

$w_3 = (4,5) - \dfrac{\langle (2,0), (4,5) \rangle}{\langle (2,0), (2,0) \rangle} \cdot (2,0) - \dfrac{\langle (0,9), (4,5) \rangle}{\langle (0,9), (0,9) \rangle} \cdot (0,9) = (4,5) - (4,0) - (0,5) = (0,0)$. *Then* $e_1 = (1,0)$, $e_2 = (0,1)$, *so we have orthonormalized S to S'. (Note that we can just skip e_3, since its inclusion would cause the set to be linearly dependent.)*

One more property of note is that we can eschew this process altogether in favor of a simpler method. If we congregate the vectors $v_1, v_2, v_3, \ldots, v_k$ into the rows of a matrix A, and then row reduce the augmented matrix $[AA^T \mid A]$, then we will wind up with the orthogonalized vectors of the new set S' (though they will not necessarily be ortho*normal*ized).

Using the Gram-Schmidt algorithm, we can also derive the *QR factorization* for a square matrix A:

Definition 6.23: The QR Factorization

Any square matrix A with real entires may be factored as $A = QR$, where Q is an orthogonal matrix (with $Q^T Q = Q Q^T = I$) and R is an upper triangular matrix. If A is invertible, then this factorization is unique (so long as the diagonal elements of R are positive). (If A has entries in the complex numbers, then Q is a *unitary* matrix, with $Q^* Q = Q Q^* = I$, where Q^* denotes the conjugate transpose of Q.)

To compute the matrices Q and R given A, we can apply Gram-Schmidt to the columns of A (assuming it is of full rank; i.e. rank$(A) = n$ if A is an $n \times n$ matrix) equipped with inner product $\langle v, w \rangle = v^T w$ (or $v^* w$ if working over $\mathbb{C}$). Then the column vectors a_i ($1 \leq i \leq n$) of A can be expressed in the form

$$a_1 = \langle e_1, a_1 \rangle e_1$$

$$a_2 = \langle e_1, a_2 \rangle e_1 + \langle e_2, a_2 \rangle e_2$$

$$\vdots$$

$$a_n = \sum_{i=1}^{n} \langle e_i, a_n \rangle e_i$$

where $\langle e_i, a_i \rangle = ||w_i||$ (with w_i as defined in Definition 6.18). Then $A = QR$, where $Q = (e_1, e_2, e_3, \ldots, e_n)$ and

$$R = \begin{bmatrix} \langle e_1, a_1 \rangle & \langle e_1, a_2 \rangle & \langle e_1, a_3 \rangle & \ldots \\ 0 & \langle e_2, a_2 \rangle & \langle e_2, a_3 \rangle & \ldots \\ 0 & 0 & \langle e_3, a_3 \rangle & \ldots \\ \vdots & \vdots & \vdots & \ddots \end{bmatrix}$$

194

Remark. If we want to decompose A into RQ form instead, we must simply orthogonalize the *rows* of A, rather than its columns. To solve the general matrix equation $Ax = b$, where $A \in \mathbb{M}_{m \times n}$ and $\text{rank}(A) = m$, we determine the QR factorization of A^T, so that $A^T = QR$ with Q orthogonal and R in the form $\begin{bmatrix} R_1 \\ 0 \end{bmatrix}$ for R_1 an $m \times m$ right triangular matrix. Note that the 0 denotes the zero *matrix*, and not the zero *scalar*. In particular, the zero matrix in question has dimensions $(n - m) \times m$. We can show that the solution to $Ax = b$ using QR decomposition is given by

$$x = Q \begin{bmatrix} (R_1^T)^{-1} b \\ 0 \end{bmatrix}$$

through algebraic manipulation.

6.8.7 Linear Least Squares

We conclude our foray into algebra topics with a versatile application of matrices, vectors, and inner products to a computation that is widely-used in a wide array of different fields, such as statistics, data modeling and fitting, astronomy, and controls (as a way of keeping error to a minimum). In any equation in which there are more equations than unknowns, we may not be able to determine precisely what the solutions are to each equation, but we can pin each variable down within a certain confidence interval. We aim to minimize the amount of residual error, thereby minimizing the sum of their squares. This procedure is known as *linear regression*, or the *linear least-squares method*.

In determining the general trend that the points in a data set follow, we seek a *line of best fit*, or the line that is closest in general to all of the points. Rigorously, we say that the sum of the squares of the minimum point-to-line distances for each point is minimized (hence the name *linear least squares*).

Definition 6.24: Linear Least Squares

Let $S = \{(x_1, y_1), (x_2, y_2), (x_3, y_3), \ldots, (x_n, y_n)\}$ be a data set of n points. We have a *model function* $f(x, \beta)$, where we have two parameters β_1, β_2 for a linear least squares computation. We define the ith *residual* r_i as $r_i = y_i - f(x_i, \beta)$.

The data set consists of n data points, we can model each data point as consisting of a scalar value y_i and a column vector x_i consisting of p predictors for the value, namely $x_{i,j}$ ($1 \leq j \leq p$). Then it follows that

$$y_i = \sum_{j=1}^{p} \beta_j x_{i,j}$$

or $y_i = x_i^T \beta$ (plus an error factor ϵ_i for each i).

Regarding the procedure itself, we consider a system that lends itself to the representation of a system of equations (with more equations than variables)

$$y_i = \sum_{j=1}^{p} X_{i,j} \beta_j$$

where $X \in \mathbb{M}_{n \times p}$ is a matrix consisting of the parameters for each data point. We can write this as $X\beta = y$, or

$$\begin{bmatrix} X_{1,1} & X_{1,2} & X_{1,3} & \cdots & X_{1,p} \\ X_{2,1} & X_{2,2} & X_{2,3} & \cdots & X_{2,p} \\ X_{3,1} & X_{3,2} & X_{3,3} & \cdots & X_{3,p} \\ \vdots & \vdots & \vdots & \ddots & \vdots \\ X_{n,1} & X_{n,2} & X_{n,3} & \cdots & X_{n,p} \end{bmatrix} \begin{bmatrix} \beta_1 \\ \beta_2 \\ \beta_3 \\ \vdots \\ \beta_p \end{bmatrix} = \begin{bmatrix} y_1 \\ y_2 \\ y_3 \\ \vdots \\ y_n \end{bmatrix}$$

Often this equation has no exact solution, in which case we must determine the "closest" β coefficients. We subsequently wish to minimize $\hat{\beta}$, which is defined as $\min_\beta S(\beta)$, and

$$S(\beta) = \sum_{i=1}^{n} \left| y_i - \sum_{j=1}^{p} X_{i,j}\beta_j \right| = ||y - X\beta||^2$$

This equation comes about from solving $(X^T X)\hat{\beta} = X^T y$, which yields our final quantity to minimize, $\hat{\beta} = (X^T X)^{-1} X^T y$. It follows, then, the the actual least squares sum is

$$S = \sum_{i=1}^{n} (y_i - x_i^T b)^2 = (y - Xb)^T (y - Xb)$$

since we seek the smallest distance to $y = x^T b$ from each data point in S.

6.8.8 Examples

Example 6.8.8.1. *Consider the 2×2 matrix*

$$A = \begin{bmatrix} 1 & 2 \\ 2 & 1 \end{bmatrix}$$

which has trace 2. What is the remainder when $Tr(A^{2018})$ is divided by 100?

Solution 6.8.8.1. *Using the definition of matrix multiplication, note the recursive pattern in the two on-diagonal elements! We can then convert these to their corresponding closed forms and sum them to get $3^x \pm 1$ according to the parity of n using $Tr(A^n)$. Hence, $Tr(A^{2018}) = 3^{2018} + 1$ which leaves remainder $\boxed{90}$ upon division by 100.*

Example 6.8.8.2. *Show that, if λ is an eigenvalue of A, then λ^2 is an eigenvalue of A^2 and λ^{-1} is an eigenvalue of A^{-1} (provided A is invertible).*

Solution 6.8.8.2. *To prove the first part of the statement, observe that, by definition, $Ax = \lambda x$ for some $x \neq \mathbf{0}$. Multiply both sides by A to obtain $A^2 x = A(\lambda x)$, which yields $A^2 x = \lambda(Ax) = \lambda(\lambda x) = \lambda^2 x$. Hence λ^2 is an eigenvalue of A^2. (The proof of the second part is similar.)*

Example 6.8.8.3. *Diagonalize the matrix*

$$A = \begin{bmatrix} 6 & -1 \\ 2 & 3 \end{bmatrix}$$

or show that no such diagonalization exists.

Solution 6.8.8.3. *The eigenvalues of A are $\lambda_1 = 4, \lambda_2 = 5$. Thus, its eigenvectors are $v_1 = \begin{bmatrix} \frac{1}{2} \\ 1 \end{bmatrix}, v_2 = \begin{bmatrix} 1 \\ 1 \end{bmatrix}$ corresponding to λ_1 and λ_2 respectively. Hence we form the matrix*

$$P = \begin{bmatrix} \frac{1}{2} & 1 \\ 1 & 1 \end{bmatrix}$$

$$\implies P^{-1} = \begin{bmatrix} -2 & 2 \\ 2 & -1 \end{bmatrix}$$

Finally, we have $D = \begin{bmatrix} 4 & 0 \\ 0 & 5 \end{bmatrix}$. Indeed, $PDP^{-1} = A$.

6.8.9 Exercises

Exercise 169. *Show that A and A^T have the same eigenvalues.*

Exercise 170. *What are all eigenvalues and eigenvectors of*

$$\begin{bmatrix} 0 & 6 & 0 \\ 2 & 1 & 0 \\ 0 & 2 & 1 \end{bmatrix}$$

Exercise 171. *Let $x \in [0,1]$, and define $f(x) = x^2$, $g(x) = 1 - cx$, $c \in \mathbb{R}$. For what value of c is $f \perp g$ in the inner product space $C([0,1])$ of continuous, real-valued functions on $[0,1]$?*

Exercise 172. *Obtain an orthonormal basis for the set of vectors $S = \{v_1 = \langle 2, 4 \rangle, v_2 = \langle 5, 3 \rangle\}$.*

Exercise 173. *Let V be a finite-dimensional inner product space. Prove that V has an orthonormal basis.*

Exercise 174. *Show that $S = \{(x, 4x) \mid x \in \mathbb{R}\}$ is a vector space.*

Extension Exercise 67. *(2013 USAMO P2) For a positive integer $n \geq 3$ plot n equally spaced points around a circle. Label one of them A, and place a marker at A. One may move the marker forward in a clockwise direction to either the next point or the point after that. Hence there are a total of $2n$ distinct moves available; two from each point. Let a_n count the number of ways to advance around the circle exactly twice, beginning and ending at A, without repeating a move. Prove that $a_{n-1} + a_n = 2^n$ for all $n \geq 4$.*

Extension Exercise 68. *Let A and B be linear transformations on a vector space V with $\dim V < \infty$. Show that $\dim \ker AB \leq \dim \ker A + \dim \ker B$.*

Extension Exercise 69. *Is there a matrix $M \in \mathcal{M}_{n \times n}(\mathbb{R})$ with trace 0 and $M^2 + M^T = I$?*

Extension Exercise 70. *(1997 IMO P4) Call an $n \times n$ matrix with elements from the set $S = \{1, 2, \ldots, 2n - 1\}$ a silver matrix if, for each $i = 1, 2, \ldots, n$, the ith row and the ith column together contain all elements of S. Show that no silver matrix exists for $n = 1997$, and furthermore, that silver matrices exist for infinitely many values of n.*

Chapter 6 Review Exercises

Exercise 175. *(2002 AMC 12A) Consider the sequence of numbers: $4, 7, 1, 8, 9, 7, 6, \ldots$ For $n > 2$, the nth term of the sequence is the units digit of the sum of the two previous terms. Let S_n denote the sum of the first n terms of this sequence. Compute the smallest value of n for which $S_n > 10,000$.*

Exercise 176. *Let s_n be the positive integer consisting of the first n triangular numbers concatenated (strung together), so that $s_1 = 1$, $s_2 = 13$, $s_3 = 136$, and $s_4 = 13610$. Let r_n be the remainder when s_n is divided by 9. Find the remainder when $\displaystyle\sum_{k=1}^{2018} r_k$ is divided by 1000.*

Exercise 177. *(2007 AMC 12A) Call a set of integers spacy if it contains no more than one out of any three consecutive integers. How many subsets of $\{1, 2, 3, \ldots, 12\}$, including the empty set, are spacy?*

Exercise 178. *What is the interval of positive k such that the quartic $x^4 + 10x^3 + kx^2 + 10x + 1$ has all real roots?*

Exercise 179. *(2015 AMC 12A) For each positive integer n, let $S(n)$ be the number of sequences of length n consisting solely of the letters A and B, with no more than three As in a row and no more than three Bs in a row. What is the remainder when $S(2015)$ is divided by 12?*

Exercise 180. *Let T_n be the nth triangular number. Define $Q(n)$ to be the numerator of the quotient of T_n and $n!$ in simplest form. Given that the sum of all 306 prime numbers from 1 to 2019, inclusive, is 283,081, find the remainder when*

$$\sum_{k=1}^{2019} Q(k)$$

is divided by 1000.

Exercise 181. *(1994 AIME) The equation*

$$x^{10} + (13x - 1)^{10} = 0$$

has 10 complex roots $r_1, \overline{r_1}, r_2, \overline{r_2}, r_3, \overline{r_3}, r_4, \overline{r_4}, r_5, \overline{r_5}$, where the bar denotes complex conjugation. Find the value of

$$\frac{1}{r_1\overline{r_1}} + \frac{1}{r_2\overline{r_2}} + \frac{1}{r_3\overline{r_3}} + \frac{1}{r_4\overline{r_4}} + \frac{1}{r_5\overline{r_5}}$$

Exercise 182. *(2015 AIME II) There are $2^{10} = 1024$ possible 10-letter strings in which each letter is either an A or a B. Find the number of such strings that do not have more than 3 adjacent letters that are identical.*

Exercise 183. *We can generalize the concept of Fibonacci numbers to the sequences of n-acci numbers, with $n > 1$ a positive integer. The n-acci sequence starts with n 1's, and has each successive term equal to the sum of the n terms before it. For example, the 3-acci sequence begins $1, 1, 1, 3, 5, 9, 17, \ldots$ For how many values of $m \le 2019$ does an m-acci sequence contain a multiple of 13 less than or equal to $13m$?*

Exercise 184. *(1990 AIME) Find $ax^5 + by^5$ if the real numbers a, b, x, and y satisfy the equations*

$$ax + by = 3$$
$$ax^2 + by^2 = 7$$
$$ax^3 + by^3 = 16$$
$$ax^4 + by^4 = 42$$

Exercise 185. *Using generating functions, show that the closed form of F_n (the nth Fibonacci number, where $F_1 = F_2 = 1$) is*

$$F_n = \frac{1}{\sqrt{5}}\left(\left(\frac{1+\sqrt{5}}{2}\right)^n - \left(\frac{1-\sqrt{5}}{2}\right)^n\right)$$

198

Exercise 186. *(2011 Stanford Math Tournament) How many polynomials P of degree 4 satisfy $P(x^2) = P(x)P(-x)$?*

Exercise 187.

(a) *Prove that, in any triangle $\triangle ABC$, $\sin A + \sin B + \sin C \leq \dfrac{3\sqrt{3}}{2}$.*

(b) *What, if anything, can we conclude about the sums $\cos A + \cos B + \cos C$ and $\tan A + \tan B + \tan C$?*

Exercise 188. *(2009 USAMO P4) For $n \geq 2$ let $a_1, a_2, \ldots, a_n$ be positive real numbers such that $(a_1 + a_2 + \ldots + a_n)\left(\dfrac{1}{a_1} + \dfrac{1}{a_2} + \ldots + \dfrac{1}{a_n}\right) \leq \left(n + \dfrac{1}{2}\right)^2$ Prove that $\max(a_1, a_2, \ldots, a_n) \leq 4\min(a_1, a_2, \ldots, a_n)$.*

Exercise 189. *(1994 USAMO P4) Let $a_1, a_2, a_3, \ldots$ be a sequence of positive real numbers satisfying $\sum_{j=1}^{n} a_j \geq \sqrt{n}$ for all $n \geq 1$. Prove that, for all $n \geq 1$,*

$$\sum_{j=1}^{n} a_j^2 > \frac{1}{4}\left(1 + \frac{1}{2} + \cdots + \frac{1}{n}\right).$$

Exercise 190. *(2015 IMO P5) Let $\mathbb{R}$ be the set of real numbers. Determine all functions $f : \mathbb{R} \mapsto \mathbb{R}$ satisfying the equation $f(x + f(x + y)) + f(xy) = x + f(x + y) + yf(x)$ for all real numbers x and y.*

Exercise 191. *(2001 IMO P2) Let $a, b, c \in \mathbb{R}^+$. Prove that $\dfrac{a}{\sqrt{a^2 + 8bc}} + \dfrac{b}{\sqrt{b^2 + 8ca}} + \dfrac{c}{\sqrt{c^2 + 8ab}} \geq 1$.*

Extension Exercise 71. *(2010 Putnam B1) Is there an infinite sequence of real numbers $a_1, a_2, a_3, \ldots$ such that $a_1^m + a_2^m + a_3^m + \ldots = m$ for every positive integer m?*

Extension Exercise 72. *(2017 USAMO P6) Find the minimum possible value of*

$$\frac{a}{b^3 + 4} + \frac{b}{c^3 + 4} + \frac{c}{d^3 + 4} + \frac{d}{a^3 + 4}$$

given that a, b, c, d are nonnegative real numbers such that $a + b + c + d = 4$.

Extension Exercise 73. *(1973 USAMO P2) Let $\{X_n\}$ and $\{Y_n\}$ denote two sequences of integers defined as follows:*

$$X_0 = 1, X_1 = 1, X_{n+1} = X_n + 2X_{n-1}(n = 1, 2, 3, \ldots)$$
$$Y_0 = 1, Y_1 = 7, Y_{n+1} = 2Y_n + 3Y_{n-1}(n = 1, 2, 3, \ldots)$$

Thus, the first few terms of the sequences are:

$$X : 1, 1, 3, 5, 11, 21, \ldots$$
$$Y : 1, 7, 17, 55, 161, 487, \ldots$$

Prove that, except for the "1", there is no term which occurs in both sequences.

Extension Exercise 74. *(2017 Putnam A2) Let $Q_0(x) = 1$, $Q_1(x) = x$, and*

$$Q_n(x) = \frac{(Q_{n-1}(x))^2 - 1}{Q_{n-2}(x)}$$

for all $n \geq 2$. Show that, whenever n is a positive integer, $Q_n(x)$ is equal to a polynomial with integer coefficients.

Extension Exercise 75. *(2008 USAMO P6) At a certain mathematical conference, every pair of mathematicians are either friends or strangers. At mealtime, every participant eats in one of two large dining rooms. Each mathematician insists upon eating in a room which contains an even number of his or her friends. Prove that the number of ways that the mathematicians may be split between the two rooms is a power of two (i.e. is of the form 2^k for some positive integer k).*

Extension Exercise 76. *$\triangle ABC$ has side lengths a, b, and c. Let $n \geq 1$ be an integer. Prove that*

$$\sum_{cyc} \frac{a^n}{b + c} \geq \frac{(a + b + c)^{n-1}}{2 \cdot 3^{n-2}}$$

Part III

Geometry

Why Study Geometry?

Mathematical thinking and reasoning extends far beyond the most immediate applications that spring to mind (perhaps figuring change at the cash register or calculating a tip after a meal, filing taxes, or fundamental number sense in general). Just as important as a feel for the technical and more abstract aspects of math is the intuition for its visual and spatial components that are commonplace in the everyday world. Not only is geometry literally everywhere in the professional world - with construction workers, city planners, engineers, pilots, and countless others absolutely reliant on it for the foundation of the tasks they perform on a day-to-day basis to ensure the safety and well-being of the public - but it also manifests in places that might be completely unexpected to the untrained eye!

The basic constructs of geometry - angles, lines, polygons, curves, and area, to name a few - all serve a distinct purpose. Moreover, they teach extremely valuable lessons about how we perceive and conceive the world around us. Things may not be what they seem - they might be greater than the sum of their parts in that there might be some hidden element to them. We can clearly see this by studying geometry problems in which there are other constructions to be made that are not visible from the beginning. By drawing our own auxiliary lines, circles, and angle measures, we are effectively honing our creative capabilities, as well as our abilities to synthesize scattered information into one coherent, meaningful picture. Not to mention, each and every geometry problem is excellent practice in thinking critically and outside the box! As with many other disciplines in math - particularly the more abstract ones - it is not so much the content itself that is the most valuable of all in everyday life, but more the skills to be gleaned from it, and the intuition associated with it.

Chapter 7

Beginning Geometry

7.1 An Introduction to Geometric Terminology

In geometry is where mathematical vocabulary is more important at the fundamental level than in any other discipline. Following (next page) are several of the most crucial terms, though keep in mind that this list is far from all-encompassing:

Definition 7.1: Geometric Definitions

- A *point* is the most fundamental of geometric objects, which has a location but no dimensions. Their locations in a relative space (usually Cartesian space; see §6.2) are described using *coordinates*.

- A *line* is the one-dimensional analogue of a point, in that it is a representation of length without thickness or height. A line, unlike a *ray* or *line segment*, extends infinitely in both directions. A line segment connects two points to each other, and a ray is the infinite extension from one endpoint.

- A *vertex* is either a point defining a shape, or the intersection point of two line segments (or, in an algebraic context, the minimum/maximum point of the graph of a function).

- An *angle* is the amount of turn formed by two lines/line segments/rays that diverge from a common intersection point (the vertex). Angles are most commonly measured in *degrees* ($°$), but they are also measured in *radians* (abbreviated rad), where $180° = \pi$ rad (most commonly in trigonometric contexts). An angle measuring less than $180°$ is *acute* if it measures less than $90°$, *obtuse* if it measures more than $90°$, and *right* if it measures exactly $90°$.

- A *polygon* is a two-dimensional shape formed solely from straight lines. It cannot self-intersect; that is, none of its edges, or line segments connecting two of its adjacent vertices (points), may intersect another of its edges. Furthermore, a polygon is *convex* if all of its interior angles measure less than $180°$; otherwise, it is *concave*.

We can also classify different types of polygons as follows:

- A polygon is *equilateral* if all of its sides/edges have the same length.
- A polygon is *equiangular* if all of its interior angles have the same measure.
- A polygon is *regular* if it is both equilateral and equiangular.

A regular polygon with n sides ($n \geq 3$) is also called an *n-gon* (most commonly for large values of n, usually $n > 10$). Of a triangle ($n = 3$), we can apply the classifications *equilateral, isosceles,* or *scalene*. An equilateral triangle has all side lengths and angles equal to each other (namely

$60°$ for the angles, since all angles in a triangle must sum to $180°$); an isosceles triangle has two sides equal and the remaining side unequal (the same holds for the angles); a scalene triangle has all angles distinct and all side lengths distinct.

- A *circle* is the locus of points that are equidistant from the center of the circle. Namely, if r is the radius of the circle, then all points on the circle are at a distance r from the center.

    - A *diameter* of a circle is the maximum distance from one point on the circle to another (which can be achieved by drawing a *straight* line between the two points).

    - A *radius* of a circle is half of the diameter (from the center to any point on the circle).

    - A *chord* of a circle is a line segment connecting any two points that lie on the circle (not necessarily a diameter).

    - The *circumference* of a circle is the distance around the circle (it is effectively the circular analogue to perimeter), given by $2\pi r = \pi d$ where r is the radius and d is the diameter.

- The *area* of a closed region is the number of units contained within the region. That is, the number of units of area is the number of unit squares that can fit within the region (not necessarily whole).

- The *perimeter* of a closed region is the sum of all of its edge lengths; i.e. the total distance from one vertex all the way back around to the same vertex in a loop that visits each vertex exactly once.

7.1.1 Examples

Example 7.1.1.1. *Compute the area of the polygon below:*

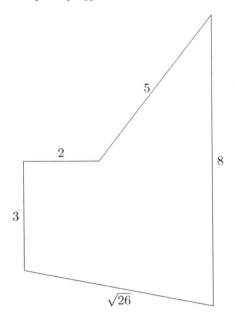

Solution 7.1.1.1. *Note that we can draw auxiliary lines beyond the line segment with length 2 to hit the line segment with length 8, and to form a right triangle with leg lengths 5 and 1, and hypotenuse length $\sqrt{26}$. We note that the figure consists of a rectangle with side lengths 3 and 5, a right triangle with leg lengths 3 and 4, and a right triangle with leg lengths 5 and 1. The total area is therefore $3 \cdot 5 + \dfrac{3 \cdot 4}{2} + \dfrac{5 \cdot 1}{2} = 15 + 6 + \dfrac{5}{2} = \dfrac{47}{2}$.*

Example 7.1.1.2. *A circle has an area that is numerically equal to π times its circumference. What is its diameter?*

Solution 7.1.1.2. *Let the circle have radius r. Then its area is πr^2, and its circumference is $2\pi r$. Setting $\pi r^2 = 2\pi r$ yields $r = 2\pi$, so $d = 2r = 4\pi$.*

Example 7.1.1.3. *At most how many acute angles can a convex octagon with integral angle measures have? (Note: The sum of the interior angle measures of an octagon is $1080°$; see §6.3.2.)*

Solution 7.1.1.3. *A convex polygon cannot have any angles that are $180°$ or greater, so the maximum angle measure is $179°$. If there are n $179°$ angles, that leaves $8 - n$ acute angles. To maximize the number of acute angles, we wish to set each of them equal to $89°$, then reduce them as necessary to reach $1080°$. Thus, we want to compute the smallest positive integer n such that $179n + 89(8 - n) \geq 1080$. Expanding and solving for n yields $90n \geq 368$, or $n \geq 5$. Thus, $8 - n \leq 3$, so a convex octagon with positive integral degree measures can have at most 3 acute angles.*

Example 7.1.1.4. *Show that the area of a rectangle with a constant perimeter is maximized if it is a square.*

Solution 7.1.1.4. *Let the rectangle's perimeter be p, and its side lengths a and b. Then $p = 2a + 2b \implies a + b = \dfrac{p}{2}$, and we want to maximize the value of ab. We have $a = \dfrac{p}{2} - b$, so $ab = b\left(\dfrac{p}{2} - b\right) = \dfrac{pb}{2} - b^2$. $\dfrac{pb}{2} - b^2$ attains its maximum value when $b = \dfrac{p}{4}$,[60] which implies $a = \dfrac{p}{4}$. Hence, $a = b$, and so the rectangle is a square.*

---

[60]This follows from the axis of symmetry formula. One may also use the derivative, but this is not necessary to complete the proof.

7.1.2   Exercises

Exercise 192. *What is the area of a regular polygon with side length 1, each of whose interior angles measure 120°? (Stuck? See §7.3.2 for the angle measure formulas of a regular polygon.)*

Exercise 193. *(2016 AMC 10A/12A) Find the area of the shaded region.*

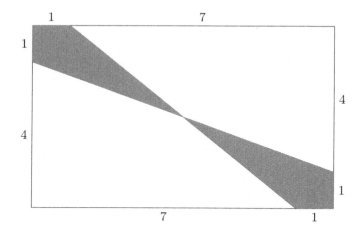

Exercise 194. *A rectangle has perimeter P and area A. What is the length of its diagonal, in terms of P and A?*

Exercise 195. *An equiangular hexagon has side lengths 1, 2, 2, 1, 2, 2, in clockwise order. What is its area?*

Exercise 196. *Find the area bounded by the graphs of $y = |x + 2|$ and $y = |2x + 6|$.*

Exercise 197. *(2002 AMC 12B) Let $f(x) = x^2 + 6x + 1$, and let R denote the set of points (x, y) in the coordinate plane such that*

$$f(x) + f(y) \le 0 \qquad and \qquad f(x) - f(y) \le 0$$

The area of R is closest to [what positive integer?]

Extension Exercise 77. *Circle Γ_1 has equation $(x - 2)(x - 4) + (y - 1)(y - 3) = 167$. Circles Γ_2 and Γ_3 have center $(4, 3)$ and radii $\sqrt{157}$ and $\sqrt{185}$ respectively. Compute the area of the quadrilateral with vertices at the points of intersection of the circles.*

Extension Exercise 78. *(2013 AIME I) Let $\triangle PQR$ be a triangle with $\angle P = 75°$ and $\angle Q = 60°$. A regular hexagon $ABCDEF$ with side length 1 is drawn inside $\triangle PQR$ so that side $\overline{AB}$ lies on $\overline{PQ}$, side $\overline{CD}$ lies on $\overline{QR}$, and one of the remaining vertices lies on $\overline{RP}$. There are positive integers $a, b, c,$ and d such that the area of $\triangle PQR$ can be expressed in the form $\frac{a + b\sqrt{c}}{d}$, where a and d are relatively prime, and c is not divisible by the square of any prime. Find $a + b + c + d$.*

7.2 Translating Algebra to Geometry: The Coordinate Plane

While geometry answers the question of how we visualize mathematical ideas, algebra formalizes them into rigorous equations based on axioms. We can bridge the gap between them with one familiar tool: the *Cartesian coordinate plane*.

The coordinate plane has an interesting history behind it, one which it symbolizes in its very existence. In the 17th century, René Descartes entertained the idea of connecting algebra and geometry, and in that moment discovered the solution: by graphing functions and equations according to their coordinates and how they relate one quantity to another. The Cartesian plane unites two measures: the $x-$ and $y-$ axes, and the quantities associated with them, much like Descartes himself had united two seemingly disparate fields of mathematics with his unforgettable footprint on the face of mathematics.

The Cartesian plane, or xy-plane (customarily abbreviated as $\mathbb{R}^2$, and in n dimensions as $\mathbb{R}^n$, especially in algebra- or analysis-related contexts), represents any relation graphically: it plots each point that a relation includes and connects them to form the graph of that relation. In its two-dimensional form, it relates the $x-$ coordinate of a relation to the corresponding $y-$ coordinate:

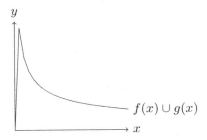

Figure 7.1: The two-dimensional Cartesian coordinate plane for $x, y \geq 0$.
Here we have plotted the union of the functions $f : [\frac{1}{4}, \infty) \mapsto (0, 2]$, $f(x) = \frac{1}{\sqrt{x}}$, and $g : [0, \frac{1}{4}] : [0, 2]$, $g(x) = 8x$.

Note that $(x, y) = (0, 0)$ (the point where the $x-$ and $y-$ axes intersect) is known as the *origin*. (This holds for any number of dimensions; the point whose coordinates are all zero is always the origin.) In addition, we split the plane into *quadrants*, as on the next page:

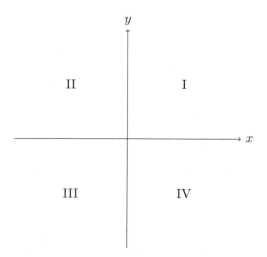

Figure 7.2: The two-dimensional Cartesian coordinate plane with quadrants I, II, III, IV labeled.

We can compactly define each of the ith quadrants Q_i as follows:

$$Q_1 := \{(x, y) \in \mathbb{R}^2 : x, y > 0\} \tag{7.1}$$

$$Q_2 := \{(x, y) \in \mathbb{R}^2 : x < 0, y > 0\} \tag{7.2}$$

$$Q_3 := \{(x, y) \in \mathbb{R}^2 : x, y < 0\} \tag{7.3}$$

$$Q_4 := \{(x, y) \in \mathbb{R}^2 : x > 0, y < 0\} \tag{7.4}$$

That is, Q_1 consists of all points whose x- and y-coordinates are both positive; Q_2 consists of those points whose x-coordinate is negative, but whose y-coordinate is positive; Q_3 consists of those points whose x-coordinate and y-coordinate are both negative; and Q_4 are consists of those points whose x-coordinate is positive but whose y-coordinate is negative.

Remark. Points that lie on one or both coordinate axes, including the origin, do not belong to any quadrant. We now study some properties of lines in the plane:

Definition 7.2: Relevant Definitions of Lines

A *line* in the xy-plane is a straight curve that has no endpoints and no thickness. (If it does not extend infinitely, and has exactly one endpoint, it is a *ray*; if it has two endpoints, it is a *line segment*.)

Any line $l \in L$, where $L = \{l := mx + b | m \in \mathbb{R} \cup \pm\infty, b \in \mathbb{R}\}$ is the set of all lines in $\mathbb{R}^2$, has an equation in the form $y = mx + b$, where m represents the *slope* of the line and b is the line's y-*intercept*. The slope of a line is the ratio of its "rise" to its "run"; i.e. the change in the y-value divided by the change in the x-value, more concisely written as $\dfrac{\Delta y}{\Delta x}$. That is, for any two points (x_1, y_1) and (x_2, y_2) on l, $m = \dfrac{y_2 - y_1}{x_2 - x_1}$, $x_1 \neq x_2$.[a] The y-intercept of l is $y = b$; the point with coordinates $(0, b)$ is where l hits the y-axis. (Similarly, we define the x-*intercept* of l as $x = -\dfrac{b}{m}$; the point $\left(-\dfrac{b}{m}, 0\right)$ is where l hits the x-axis.)

---

[a] If $x_1 = x_2$, then l is a vertical line and m is undefined, or in some cases set to $\pm\infty$.

In addition, we say that line $l_1 := m_1 x + b_1$ is *parallel* to line $l_2 := m_2 x + b_2$ iff $m_1 = m_2$; we say l_1 and l_2 are *perpendicular* or *orthogonal* iff $m_1 m_2 = -1$.[61]

There is one more formula of which competition math problems frequently make use, the *point-to-line distance formula*:

Theorem 7.1: Point-to-Line Distance Formula

Let $l := ax + by + c = 0$ (with a and b not both zero) be a line and $P = (x_0, y_0)$ be a point in the xy-plane. Then the minimal distance between l and P is $\dfrac{|ax_0 + by_0 + c|}{\sqrt{a^2 + b^2}}$. The closest point on l to P has coordinates $\left(\dfrac{b(bx_0 - ay_0) - ac}{a^2 + b^2}, \dfrac{a(-bx_0 + ay_0) - bc}{a^2 + b^2} \right)$.

There are multiple different approaches to proving the theorem; here is a geometric proof:

Proof. Note that this proof is only valid if the line is neither horizontal nor vertical.

From the point P drop a perpendicular to the line l, and let R be the foot of the perpendicular. Then draw a vertical line through P that intersects l at point S. For any point T on l, let U lie on l and orient V such that $\overline{TV}$ is horizontal and $\overline{UV}$ is vertical, thus forming a right triangle $\triangle TVU$. Then assume WLOG that $\overline{UV}$ will have length $|a|$ and $\overline{TV}$ will have length $|b|$.

We observe that $\triangle TVU$ and $\triangle PRS$ are similar,[62] so $\dfrac{PR}{PS} = \dfrac{TV}{TU}$. Hence, set $S = (x_0, m)$; it follows that $PS = y_0 - m$ and $PR = \dfrac{|y_0 - m| \cdot |b|}{\sqrt{a^2 + b^2}}$. S lies on l, so $m = -\dfrac{ax_0 + c}{b}$, which leads to $PR = \dfrac{|ax_0 + by_0 + c|}{\sqrt{a^2 + b^2}}$ as desired. $\square$

We may also take a strictly algebraic approach, using a few geometric facts:

Proof. Note that this proof is also invalid if the line is horizontal or vertical.

Observe that l has slope $-\dfrac{a}{b}$, so a line perpendicular to l will have slope $\dfrac{b}{a}$. Let (m, n) be the point of intersection of l and a perpendicular line $L \perp l$ such that L passes through (x_0, y_0). Hence, we can write $\dfrac{y_0 - n}{x_0 - m} = \dfrac{b}{a}$, and so $a(y_0 - n) - b(x_0 - m) = 0$. Squaring both sides and rearranging yields $a^2(y_0 - n)^2 + b^2(x_0 - m)^2 = 2ab(y_0 - n)(x_0 - m)$. It follows from this equation that $(a(x_0 - m) + b(y_0 - n))^2 = (a^2 + b^2)((x_0 - m)^2 + (y_0 - n)^2)$, and also that $(a(x_0 - m) + b(y_0 - n))^2 = (ax_0 + by_0 + c)^2$, since (m, n) lies on l, and so it satisfies the equation $am + bn + c = 0$. Thus, $(a^2 + b^2)((x_0 - m)^2 + (y_0 - n)^2) = (ax_0 + by_0 + c)^2$, which implies $(x_0 - m)^2 + (y_0 - n)^2 = \dfrac{(ax_0 + by_0 + c)^2}{a^2 + b^2}$, or $\sqrt{(x_0 - m)^2 + (y_0 - n)^2} = \dfrac{ax_0 + by_0 + c}{\sqrt{a^2 + b^2}}$, where the LHS is the distance from (x_0, y_0) to (m, n), the point of intersection of l with the perpendicular line L, and also at a minimal distance from (x_0, y_0). This completes the proof. $\square$

---

[61] The term *orthogonal* derives from the interpretation of lines as vectors, where two vectors are orthogonal if their dot product is zero.

[62] See §7.4, Working with Similarity.

7.2.1 Examples

Example 7.2.1.1. *Show that parallel lines with different y-intercepts never intersect.*

Solution 7.2.1.1. *Assume the contrary. Let the lines be defined by $l_1 : m_1 x + b_1$ and $l_2 : m_2 x + b_2$; then $m_1 = m_2$, but $b_1 \neq b_2$. For l_1 and l_2 to intersect, there must exist some $x \in \mathbb{R}$ such that $l_1(x) = l_2(x)$. However, this implies $m_1 x + b_1 = m_2 x + b_2$, which leads to $b_1 = b_2$, a contradiction. Hence, if $m_1 = m_2$ but $b_1 \neq b_2$, l_1 and l_2 cannot intersect.*

Example 7.2.1.2. *Give an example of two perpendicular lines that intersect the x-axis exactly 1 unit away from each other.*

Solution 7.2.1.2. *Let the lines be $l_1 := m_1 x + b_1$, $l_2 := m_2 x + b_2$. The respective x-intercepts are given by $x = -\dfrac{b_1}{m_1}$, $x = -\dfrac{b_2}{m_2}$. Assume WLOG that $1 + \dfrac{b_1}{m_1} = \dfrac{b_2}{m_2}$. In addition, we must have $m_1 m_2 = -1$, or $m_1 = -\dfrac{1}{m_2}$, so $1 + \dfrac{b_1}{m_1} = -b_2 m_1$. An example (of infinitely many) of an ordered triple satisfying this equation is $(b_1, b_2, m_1) = (-2, 1, 1)$, so that $l_1 := x - 2$, $l_2 = -x + 1$, which have x-intercepts of $x = 2$ and $x = 1$ respectively. (Note that m_1 uniquely determines m_2, under the perpendicularity condition.)*

Example 7.2.1.3. *What is the minimal distance from the line $y = 2x + 3$ to the point $(4, 5)$?*

Solution 7.2.1.3. *We can write $y = 2x + 3$ as $2x - y + 3 = 0$, then apply the formula:* $\dfrac{|2(4) + (-1)(5) + 3|}{\sqrt{2^2 + (-1)^2}} = \dfrac{6}{\sqrt{5}}$.

7.2.2 Exercises

Exercise 198.

(a) At what point do the graphs of the lines $y = 2x + 9$ and $y = -3x - 1$ intersect in the Cartesian plane?

(b) What is the equation of the line perpendicular to the line $y = 10x + 5$ passing through the point in (a)?

(c) The line in (b) passes through all quadrants except for one. What is this quadrant? Show algebraically that the line never passes through this quadrant.

(d)   (i) Give an example of a line that passes through two or fewer quadrants, or show that no such line exists.

    (ii) Give an example of a line that passes through all four quadrants, or show that no such line exists.

Exercise 199. Show that, if two lines intersect in the xy-plane, they must intersect at either zero, one, or infinitely many distinct points.

Exercise 200. Suppose that points $A = (x, y)$ and $B = (y, x)$ are equidistant from the origin, with AB also equal to this common distance. Given that $y > x$, what is $\dfrac{y}{x}$?

Exercise 201. *(2009 AIME I)* In parallelogram $ABCD$, point M is on $\overline{AB}$ so that $\frac{AM}{AB} = \frac{17}{1000}$ and point N is on $\overline{AD}$ so that $\frac{AN}{AD} = \frac{17}{2009}$. Let P be the point of intersection of $\overline{AC}$ and $\overline{MN}$. Find $\frac{AC}{AP}$.

Exercise 202. The x-coordinates of the points of intersection of the circles with radii 6 and centers $(-2, 3)$ and $(2, -3)$ can be written in the form $\pm\sqrt{\dfrac{a}{b}}$, where a and b are positive integers such that $\gcd(a, b) = 1$. Find $a + b$.

Extension Exercise 79. Prove that any three non-collinear points uniquely determine a circle in the plane.

Extension Exercise 80. *(2011 AIME I)* Let L be the line with slope $\frac{5}{12}$ that contains the point $A = (24, -1)$, and let M be the line perpendicular to line L that contains the point $B = (5, 6)$. The original coordinate axes are erased, and line L is made the x-axis and line M the y-axis. In the new coordinate system, point A is on the positive x-axis, and point B is on the positive y-axis. The point P with coordinates $(-14, 27)$ in the original system has coordinates (α, β) in the new coordinate system. Find $\alpha + \beta$. (Hint: Use the point-to-line distance formula.)

Extension Exercise 81. Try to generalize the point-to-line distance formula to dimensions ≥ 3! (Hint: interpret lines as vectors.)

7.3 Proving Formulas in Cartesian Coordinate Geometry

The following are all crucial formulas that serve as the crux of a wide berth of geometry problems in both the traditional curriculum and competition math. In this section we will both inspire motivation behind the formulas and prove them rigorously, rather than merely present them and utilize them in the standard computational manner.

7.3.1 Number of Diagonals in a Regular Polygon

> **Theorem 7.2: Number of Diagonals in a Regular n-Gon**
>
> First we must define a *diagonal* of a polygon:
>
> > **Definition 7.3: Diagonal of a Polygon**
> >
> > A diagonal of a polygon is any line segment that is not an edge of the polygon whose endpoints are both vertices of the polygon.
>
> Now we have the tools to present the definition:
>
> For all positive integers $n \geq 3$, a regular n-gon has $\dfrac{n(n-3)}{2}$ diagonals.

Proof. Let the polygon have vertices $\{V_1, V_2, V_3, \ldots, V_n\}$. WLOG consider the vertex V_1. There are $n-1$ edges emanating from this vertex that connect to the other vertices, but two of them connect to adjacent vertices, as there are two edges of the polygon with V_1 as an endpoint. Hence, there are $n(n-3)$ choices of non-adjacent vertices. However, we need to divide by 2 to account for the fact that order is irrelevant, so the final count is $\dfrac{n(n-3)}{2}$. $\qquad\square$

Another proof operates via induction:

Induction proof.

Base case: For $n = 3$, the statement is true (a triangle has no diagonals; the two line segments emanating from any given vertex are both edges of the triangle).

Inductive step: Assume that the statement is true for some $n \in \mathbb{N}$. Then we prove it for $n + 1$ as follows. Adding the vertex V_{n+1} to the polygon will produce additional diagonals $V_{n+1}V_2$, $V_{n+1}V_3, \ldots, V_{n+1}V_{n-1}$, as well as the diagonal V_1V_n (since V_n is no longer directly connected to V_1). This adds $n-1$ more diagonals, and so the total becomes $\dfrac{n(n-3)}{2} + (n-1) = \dfrac{n^2 - n - 2}{2} = \dfrac{(n+1)(n-2)}{2}$, which is consistent with the initial assumption. $\qquad\square$

7.3.2 Interior and Exterior Angle Measures

> **Theorem 7.3: Interior and Exterior Angle Measures of a Regular n-Gon**
>
> For all positive integers $n \geq 3$, an n-gon (not necessarily regular) has interior angle measures summing to $180(n-2)°$. (An equiangular n-gon has all angles equal to $\dfrac{180(n-2)}{n}°$.) The sum of the exterior angle measures of any polygon is always $360°$, with each angle measuring $\dfrac{360°}{n}$ in an equiangular n-gon.

Proof. Extend each edge of the polygon, which proves that the interior and exterior angle measures must sum to 180°. To prove that the interior angle measures sum to $180(n-2)°$, and thus that each interior angle measures $\dfrac{180(n-2)}{n}°$ in an equiangular polygon, draw each of the diagonals of the polygon emanating from a specific vertex. They will form $n-2$ triangles, each of whose angle measures sum to 180°. Hence, the sum of all angle measures in the polygon is $180(n-2)°$, as desired. It directly follows that the sum of all exterior angle measures must be 360°.[63] $\qquad\square$

7.3.3 The Triangle Inequality

> **Theorem 7.4: Triangle Inequality**
>
> Let x, y, z be positive real numbers. Assume WLOG that $x > y > z$. Then in order for x, y, and z to be the side lengths of a non-degenerate triangle, $x < y + z$ (and if y and z are the maximums of the side lengths, then $y < x + z$ and $z < x + y$, respectively). Equality holds if the triangle is *degenerate;* i.e. a straight line with zero area.
>
> In vector form, the inequality is written as $||x + y|| \leq ||x|| + ||y||$, where $||\cdot||$ denotes the *vector norm.*

Proof. The following is Euclid's proof in *Euclid's Elements, Book 1.*

Let $\triangle ABC$ have $AB = x$, $BC = y$, $AC = z$. Extend $\overline{AB}$ past B to a point D such that $BC = BD$. Then $\triangle BCD$ is an isosceles triangle, and $\angle ACD > \angle BCD$, so $AD > AC$. But since $AD = AB + BD = AB + BC$, it follows that $AB + BC > AC$. $\qquad\square$

Here is a more generalized proof using absolute value in place of length/distance:

Proof. We want to prove that $|x + y| \leq |x| + |y|$ for all $x, y \in \mathbb{R}$, with equality iff $x = y$. For $x = y$, equality follows immediately, so we assume that $x \neq y$. We invoke the following lemma:

> **Lemma 7.4: Definition of Absolute Value**
>
> For any $x, a \in \mathbb{R}$, $|x| \leq a$ iff $-a \leq x \leq a$.

We have $-|x| \leq x \leq |x|$ and $-|y| \leq y \leq |y|$ for $x, y \in \mathbb{R}$, so $-(|x| + |y|) \leq x + y \leq |x| + |y|$. By the lemma, we have $|x + y| \leq |x| + |y|$, as desired. $\qquad\square$

Generalization to n-Gons

We can generalize the Triangle Inequality to all n-gons; in addition, other inequalities, such as the Cauchy-Schwarz Inequality, have offshoots in proving this generalized form of the Triangle Inequality. Here is the statement of the Generalized Triangle Inequality:

> **Theorem 7.5: Generalized Triangle Inequality**
>
> Let the n-gon P_n have side lengths s_1, s_2, s_3, ..., s_n. Then
>
> $$2\max\{s_1, s_2, s_3, \ldots, s_n\} < \sum_{k=1}^{n} s_k$$
>
> Id est, the longest side length is less than the sum of the other side lengths.

---

[63]There are many more ways to prove the theorem; here are several of them: (link: Sum of interior angles of an n-sided polygon)

Proof. An equivalent statement of the Triangle Inequality ($n = 3$) is that, for all $a, b \in \mathbb{R}$, $|a + b| \leq |a| + |b|$ (with equality iff $a = b$). For $n \in \mathbb{N}$, the statement generalizes as follows:

$$\text{For all } x_1, x_2, x_3, \ldots, x_n \in \mathbb{R}, \left| \sum_{i=1}^{n} x_i \right| \leq \sum_{i=1}^{n} |x_i|.$$

We will prove this by induction.

Base cases ($n = 1, 2$): For $n = 1$, this is trivial. For $n = 2$, this is equivalent to the Triangle Inequality.

Inductive hypothesis: Assume the statement is true for some $n \in \mathbb{N}$; that is, assume

$$\left| \sum_{i=1}^{n} x_i \right| \leq \sum_{i=1}^{n} |x_i|.$$

We have

$$\left| \sum_{i=1}^{n+1} x_i \right| = \left| \sum_{i=1}^{n} x_i + x_{n+1} \right|.$$

By the Triangle Inequality,

$$\left| \sum_{i=1}^{n} x_i + x_{n+1} \right| \leq \left| \sum_{i=1}^{n} x_i \right| + |x_{n+1}|.$$

In addition,

$$\left| \sum_{i=1}^{n} x_i \right| + |x_{n+1}| \leq \sum_{i=1}^{n} |x_i| + |x_{n+1}| = \sum_{i=1}^{n+1} |x_i|.$$

Thus, it follows that

$$\sum_{i=1}^{n+1} |x_i| \leq \left| \sum_{i=1}^{n+1} x_i \right|,$$

which completes the proof. $\qquad\square$

Classifying Acute and Obtuse Triangles

We can additionally classify triangles as acute, right, or obtuse based solely on their side lengths:

> **Theorem 7.6: Classifying a Triangle as Acute or Obtuse**
>
> Let a triangle have side lengths a, b, and c, and assume WLOG that $c \geq b \geq a$. Iff $c^2 = a^2 + b^2$, the triangle is right (this is the converse of the Pythagorean Theorem); iff $c^2 > a^2 + b^2$, the triangle is obtuse; and iff $c^2 < a^2 + b^2$, the triangle is acute.

Proof. (Proving the converse of the Pythagorean Theorem is left as an exercise to the reader.)

Consider an acute triangle such as the following:

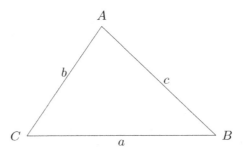

Figure 7.3: An acute triangle with side lengths a, b, c.

By drawing the perpendicular from A down to $\overline{BC}$, we can split $\overline{BC}$ into line segments of length t and $a - t$. Denote the foot of the perpendicular by M; then $AM = \sqrt{b^2 - t^2}$ by the Pythagorean Theorem. We have $c^2 = (b^2 - t^2) + (a - t)^2 = (b^2 - t^2) + (a^2 - 2at + t^2) = a^2 + b^2 - 2at < a^2 + b^2$, as desired.

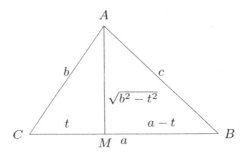

Figure 7.4: Figure 7.4 with an auxiliary altitude added.

Next, consider the following obtuse triangle:

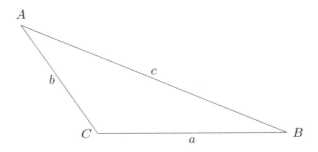

Figure 7.5: An obtuse triangle with side lengths a, b, c.

We can extend side $\overline{BC}$ to a point P such that $\overline{CP} = t$, and thus $\overline{AP} = \sqrt{b^2 - t^2}$ by the Pythagorean Theorem. It follows that $c^2 = (a + t)^2 + (b^2 - t^2) = (a^2 + 2at + t^2) + (b^2 - t^2) = a^2 + b^2 + 2at > a^2 + b^2$, as desired. Here is the final diagram:

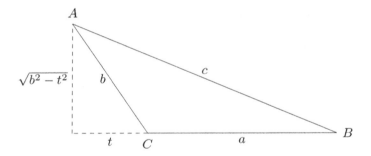

Figure 7.6: Figure 7.6 with auxiliary lines added.

This proves the theorem for all three cases. □

7.3.4 Area Formula for Regular Polygons

> **Theorem 7.7: Area Formula for an n-Gon**
>
> A regular polygon with n sides of length s has an area of $\dfrac{s^2 \cdot \cot \frac{\pi}{n}}{4}$.

Proof. Assume WLOG that $s = 1$. Then the desired area in terms of n is $\dfrac{1}{4} \cot \dfrac{\pi}{n}$. From the center of the polygon, draw n auxiliary lines to each vertex, dividing the polygon into n congruent triangles. Each of these triangles will be an isosceles triangle with vertex angle $\dfrac{360°}{n}$; the other two angles will have measure $\left(90 - \dfrac{180}{n}\right)°$:

Figure 7.7: A regular n-gon divided into n isosceles triangles. In the figure above, $n = 6$.

We can bisect the vertex angle, forming two right triangles with angle measures $90°$, $\dfrac{180°}{n}$, and $\left(90 - \dfrac{180}{n}\right)°$. The side opposite the $\dfrac{180°}{n}$ angle has length $\dfrac{1}{2}$, so by Law of Sines,[64] the side opposite the $\left(90 - \dfrac{180}{n}\right)°$ angle has length $\dfrac{\sqrt{1 - \sin^2 \frac{180°}{n}}}{2 \sin \frac{180°}{n}} = \dfrac{1}{2} \cdot \cot \dfrac{180°}{n}$. Thus, the area of each right triangle is $\dfrac{1}{8} \cdot \cot \dfrac{180°}{n}$, and so the area of each large isosceles triangle is twice that, or $\dfrac{1}{4} \cdot \cot \dfrac{180°}{n}$. Finally, we multiply the area of each isosceles triangle by n to obtain $\dfrac{n}{4} \cot \dfrac{\pi}{n}$, completing the proof. (Note that $\dfrac{180°}{n} = \dfrac{\pi}{n}$ radians, since $180° = \pi$ rad.) □

---

[64]If you are unfamiliar with the Law of Sines, please see §8.1.

7.3.5 Area-Finding Theorems: Heron's Formula, Shoelace Theorem, and Pick's Theorem

Rigorously speaking, how do we compute area? With less complicated shapes such as rectangles, one can multiply side lengths or subtract off extraneous area, but with more irregular shapes, it is often cumbersome to calculate area without the aid of area formulas, such as the following:

> **Theorem 7.8: Heron's Formula**
>
> The area of a triangle with side lengths a, b, and c, such that $\max(a, b, c)$ is strictly less than the sum of the other two side lengths, is equal to $\sqrt{s(s-a)(s-b)(s-c)}$ where $s = \dfrac{a+b+c}{2}$ is the *semi-perimeter* of the triangle.

We present a set of proofs from a variety of angles:

Algebraic proof. For this proof, we drop a perpendicular altitude down from vertex B to side b. Our initial diagram of the triangle upon dropping the perpendicular altitude is as follows:

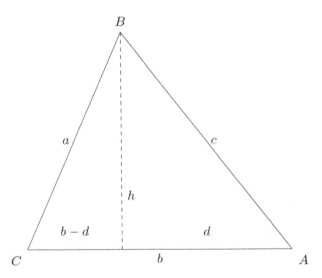

Figure 7.8: A triangle with side lengths a, b, and c split into two right triangles down its b-altitude.

It follows that $c^2 = h^2 + d^2$, and $a^2 = h^2 + (b-d)^2 = h^2 + b^2 - 2bd + d^2 = c^2 + b^2 - 2bd$. Subtracting b^2 from both sides of this equation yields $a^2 - b^2 = c^2 - 2bd$. Thus, $d = \dfrac{c^2 + b^2 - a^2}{2b}$.

We have $h^2 = c^2 - d^2$, so

$$h^2 = c^2 - \left(\frac{c^2 + b^2 - a^2}{2b}\right)^2$$

$$= \frac{(2bc - a^2 + b^2 + c^2)(2bc + a^2 - b^2 - c^2)}{4b^2}$$

$$= \frac{((b+c)^2 - a^2)(a^2 - (b-c)^2)}{4b^2}$$

$$= \frac{(b+c-a)(b+c+a)(a+b-c)(a+b+c)}{4b^2}$$

$$= \frac{2(s-a) \cdot 2(s-b) \cdot 2(s-c) \cdot 2s}{4b^2}$$

217

$$= \frac{4s(s-a)(s-b)(s-c)}{b^2}.$$

Therefore, we have

$$h = \frac{2\sqrt{s(s-a)(s-b)(s-c)}}{b}.$$

The area of the triangle is then $\frac{bh}{2} = \sqrt{s(s-a)(s-b)(s-c)}$, as desired. $\qquad\square$

Geometric proof. Note: The following proof is adapted from the work of Shannon Umberger and Howard Eves, originating in a loosely sketched form, from the latter's *An Introduction to the History of Mathematics* (Eves 1953).

Consider the following generic triangle with incenter I and incircle O:

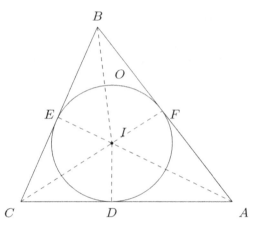

Figure 7.9: A generic triangle with incenter I, and incircle O with center I. Here the dashed lines represent the internal angle bisectors of the triangle.

Since $\overline{IB}$ bisects $\angle ABC$, $\angle IBF = \angle IBE$. Angles $\angle IFB$ and $\angle IEB$ are both right,[65] so they must be equal. It follows that $\angle BIF = \angle BIE$. In addition, we know that $IE = IF$ (as they are both inradii), and so $\triangle BIF$ is SAS congruent to $\triangle BIE$. Similarly, by SAS, triangles $\triangle CID$ and $\triangle CIE$ are congruent, as well as triangles $\triangle AID$ and $\triangle AIF$. Thus, $BE = BF$, $CE = CD$, and $AD = AF$.[66]

Now we note that the area of $\triangle ABC$ is the sum of the areas of triangles $\triangle IBC$, $\triangle IBA$, and $\triangle IAC$. The area of $\triangle IBC$ is equal to $\frac{BC \cdot IE}{2}$, the area of $\triangle IBA$ is equal to $\frac{AB \cdot IF}{2}$, and the area of $\triangle IAC$ is equal to $\frac{AC \cdot ID}{2}$. Summing these gives an expression for the area of $\triangle ABC$:

$$|\triangle ABC| = \frac{(AB \cdot IF) + (BC \cdot IE) + (CA \cdot ID)}{2}$$

Since $ID = IE = IF$, as all of them are inradii, we can rewrite the area of $\triangle ABC$ as $\frac{ID \cdot (AB + BC + CA)}{2}$. Furthermore, $AB + BC + CA$ is the perimeter of $\triangle ABC$, which we will denote by p; thus, $|\triangle ABC| = \frac{p \cdot ID}{2} = ID \cdot s$, where s is the semi-perimeter of $\triangle ABC$.

Next, let G lie on $\overline{AC}$ such that C lies between A and G, and $CG = BE$. Recall that $p = AB + BC + CA$, which equals $(AF + FB) + (BE + EC) + (CD + DA) = (AD + AF) + (BE + BF) + (CD + CE)$, which in

---

[65]The proof of this is left as an exercise.

[66]This is in fact a special case of Ceva's Theorem; see chapter 8.

turn simplifies to $2(AD + BE + CD)$,[67] using the fact that $AD = AF$, $BF = BE$, and $CE = CD$. Because $BE = CG$, $p = 2(AD + CG + CD)$. Substituting AG for $AD + DC + CG$, we obtain $p = 2AG$, or $s = AG$. Therefore, the area of $\triangle ABC$ is equal to $ID \cdot AG$.

The next step is to draw the perpendicular to $\overline{AI}$ through I, and to draw the perpendicular to $\overline{AG}$ through C; label their point of intersection as H. Also let $\overline{IH}$ and $\overline{AG}$ intersect at point J, then draw line segment $\overline{AH}$:

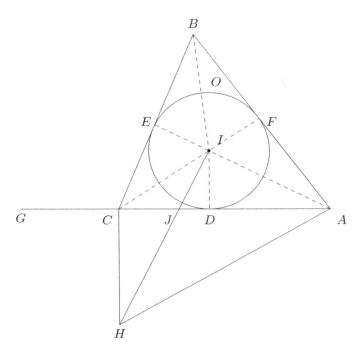

Figure 7.10: Figure 6.10 with points G, H, and J added, and with the corresponding auxiliary lines drawn.

Because $\overline{AI} \perp \overline{IH}$ and $\overline{AC} \perp \overline{CH}$, $\triangle AIH$ and $\triangle ACH$ are both right triangles, and have common hypotenuse $\overline{AH}$. It follows that quadrilateral $AICH$ is cyclic,[68] and so it, along with triangles $\triangle AIH$ and $\triangle ACH$, can be inscribed in circle ω with diameter $\overline{AH}$, as follows: It follows from quadrilateral $AICH$ being cyclic that $\angle AIC + \angle AHC = 180°$ (i.e. they are supplementary). Furthermore, we have $\angle BIE + \angle BIF + \angle CIE + \angle CID + \angle AID + \angle AIF = 360°$, and so $\angle BIE + \angle CID + \angle AID = 180°$. We also have $\angle CID + \angle AID = \angle AIC$, from which it follows that $\angle BIE + \angle AIC = 180°$ (so that, once again, they are supplementary). Since $\angle BIE$ and $\angle AHC$ are both supplementary to $\angle AIC$, $\angle BIE = \angle AHC$. This implies that $\angle BEI = \angle ACH = 90°$ by construction, and so $\triangle BIE$ and $\triangle AHC$ are AA-similar.

By substitution, $\dfrac{AC}{CG} = \dfrac{AC}{BE}$, and furthermore, $\dfrac{AC}{BE} = \dfrac{HC}{IE}$. Also, because $\angle IJD$ and $\angle HJC$ are vertical angles, they are equal. Moreover, $\angle IDJ = \angle HCJ = 90°$ by construction. Thus, $\triangle IJD$ and $\triangle HCJ$ are AA-similar as well. We then have $\dfrac{HC}{IE} = \dfrac{HC}{ID}$, and $\dfrac{HC}{ID} = \dfrac{CJ}{DJ}$ by similarity, so $\dfrac{AC}{CG} = \dfrac{CJ}{DJ}$ from the fact that $\dfrac{HC}{IE} = \dfrac{AC}{CG}$. Since $\dfrac{AC}{CG} = \dfrac{CJ}{DJ}$, $\dfrac{AC}{CG} + 1 = \dfrac{CJ}{DJ} + 1$; this equation can be re-written as $\dfrac{AC + CG}{CG} = \dfrac{CJ + DJ}{DJ}$. We have $AC + CG = AG$ and $CJ + DJ = CD$, so $\dfrac{AG}{CG} = \dfrac{CD}{DJ}$. Then we can write $\dfrac{AG}{CG} \cdot \dfrac{AG}{AG} = \dfrac{CD}{DJ} \cdot \dfrac{AD}{AD}$, which yields $\dfrac{AG^2}{CG \cdot AG} = \dfrac{CD \cdot AD}{DJ \cdot AD}$.

---

[67] Here we have chosen AD rather than AF to make the following step more clear.

[68] For a definition of a cyclic quadrilateral, see §8.6.

219

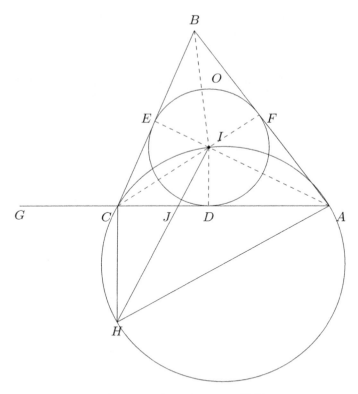

Figure 7.11: Figure 6.11 with circle ω drawn. ω has diameter $\overline{AH}$ and passes through A, I, C, and H.

Now since $\triangle AIJ$ is right, and D lies on the hypotenuse $\overline{AJ}$, the length of $\overline{ID}$ is the geometric mean of $\overline{JD}$ and $\overline{DA}$; i.e. $\sqrt{JD \cdot DA}$. Thus, $ID^2 = JD \cdot AD$; it follows that $\dfrac{AG^2}{CG \cdot AG} = \dfrac{CD \cdot AD}{ID^2}$. Then $AG^2 \cdot ID^2 = (CG \cdot AG) \cdot (CD \cdot AD)$, or $AG \cdot ID = \sqrt{CG \cdot AG \cdot CD \cdot AD}$. Recall that the area of $\triangle ABC$ is $AG \cdot ID$, which we have just shown is equal to $\sqrt{CG \cdot AG \cdot CD \cdot AD}$.

Finally, recall that $AG = s$. So $CG = AG - AC = s - b$, and similarly, $CD = s - c$ and $AD = s - a$. Therefore, the area of $\triangle ABC$ is $\sqrt{s(s-a)(s-b)(s-c)}$, which is what we originally set out to prove. (Whew, that was a long one!) $\qquad\square$

Trigonometric proof. This proof uses the Law of Cosines (LoC) as its crux; if you are unfamiliar with the LoC, please read through §7.2 before continuing with this proof.

Let $\triangle ABC$ have side lengths a, b, and c opposite vertices A, B, and C respectively, and angles α, β, γ associated to A, B, and C respectively:

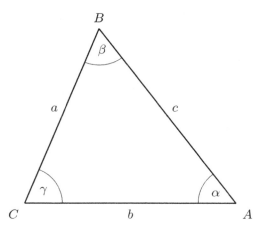

Figure 7.12: A generic triangle $\triangle ABC$ with vertices, sides, and angles labeled.

Applying the Law of Cosines on γ yields $\cos \gamma = \dfrac{a^2 + b^2 - c^2}{2ab}$ (setting $\gamma = C$ in the general formula). Then it follows that

$$\sin \gamma = \sqrt{1 - \cos^2 \gamma} = \frac{\sqrt{4a^2b^2 - (a^2 + b^2 - c^2)^2}}{2ab}$$

The A-altitude of $\triangle ABC$ has length $b \sin \gamma$, and so the area of $\triangle ABC$ is

$$\frac{1}{2} \cdot (\text{base}) \cdot (\text{height})$$

$$= \frac{1}{2} ab \cdot \sin \gamma$$

$$= \frac{1}{4} \sqrt{4a^2b^2 - (a^2 + b^2 - c^2)^2}$$

$$= \frac{1}{4} \sqrt{(2ab + (a^2 + b^2 - c^2))(2ab - (a^2 + b^2 - c^2))}$$

$$= \frac{1}{4} \sqrt{((a+b)^2 - c^2)(c^2 - (a-b)^2)}$$

$$= \sqrt{\frac{((a+b+c)(a+b-c)(c+a-b)(c-a+b)}{16}}$$

$$= \sqrt{\frac{a+b+c}{2} \cdot \frac{a+b-c}{2} \cdot \frac{c+a-b}{2} \cdot \frac{c-a+b}{2}}$$

$$= \sqrt{s(s-a)(s-b)(s-c)}$$

where we have made the substitution $s = \dfrac{a+b+c}{2}$. $\qquad\qquad\square$

You're probably wondering: "Well, that's nice and all, but will this ever show up on a contest?" The answer to that is a resounding *no;* you will (thankfully) not need to memorize these proofs for any contest! Nevertheless, it is good to assimilate the concepts and motivation behind the proofs so that you can get a better feel for the methods customarily used in Olympiad-style contest proofs and proofs in standard mathematical literature such as research papers.

There is another area-finding method that serves as a sort of generalization of Heron's Formula to polygons with 4 or more sides: the *Shoelace Theorem*. Unlike Heron's Formula, however, the Shoelace Theorem does not use the side lengths; it instead uses the coordinates of the vertices of the polygon in 2-dimensional Cartesian space.

Theorem 7.9: Shoelace Theorem

Let an n-sided polygon P have vertices $V_1, V_2, V_3, \ldots, V_n$ such that

$$V_1 = (x_1, y_1)$$

$$V_2 = (x_2, y_2)$$

$$V_3 = (x_3, y_3)$$

$$\vdots$$

$$V_n = (x_n, y_n)$$

Then

$$|P| = \frac{1}{2} \left| \left(\sum_{i=1}^{n-1} x_i y_{i+1} + x_n y_1 \right) - \left(\sum_{i=1}^{n-1} y_i x_{i+1} + y_n x_1 \right) \right|$$

where $|P|$ denotes the area of P.

Alternatively, using matrices, we can restate the theorem as follows:

$$|P| = \frac{1}{2} \left| \sum_{i=1}^{n-1} x_i (y_{i+1} - y_{i-1}) \right|$$

$$= \frac{1}{2} \left| \sum_{i=1}^{n} y_i (x_{i+1} - x_{i-1}) \right|$$

$$= \frac{1}{2} \left| \sum_{i=1}^{n} (x_{i+1} + x_i)(y_{i+1} - y_i) \right|$$

$$= \frac{1}{2} \left| \sum_{i=1}^{n} \det \begin{bmatrix} x_i & x_{i+1} \\ y_i & y_{i+1} \end{bmatrix} \right|$$

Here, $\det A$ denotes the *determinant* of the matrix A.

Proof. We first prove the Shoelace Theorem for a generic non-degenerate triangle in the xy-plane.

Let $\triangle ABC$ have coordinates $A = (0,0)$, $B = (x_1, y_1)$, $C = (x_2, y_2)$ (since any triangle can be translated such that one of the vertices is $(0,0)$, to simplify computations). Circumscribe a rectangle R around $\triangle ABC$ such that R has side lengths x_1 and y_2. Label the vertices of R, R_1, R_2, R_3, and R_4 such that R_1 coincides with $A = (0,0)$ and the vertices are in clockwise order (there is no requirement for how the vertices are oriented, but we can assume this particular setup *a priori*, as this changes nothing in terms of the actual computation process). Then the area of $\triangle ABC$ is the area of R, or $|x_2 y_1|$, minus the combined areas of triangles $\triangle R_1 R_2 B$, $\triangle B R_3 C$, and $\triangle C R_4 R_1$. The area of $\triangle R_1 R_2 B$ is $\left| \frac{x_1 y_1}{2} \right|$, the area of $\triangle B R_3 C$ is $\left| \frac{(x_2 - x_1)(y_1 - y_2)}{2} \right|$, and the area of $\triangle C R_4 R_1$ is $\left| \frac{x_2 y_2}{2} \right|$. Thus, the area of $\triangle ABC$ is $\left| \frac{x_2 y_1 - x_1 y_2}{2} \right|$, which corresponds with the statement of the Sholeace Theorem for $n = 3$, as desired.

Here is a diagram illustrating the Shoelace Theorem proof for $n = 3$:

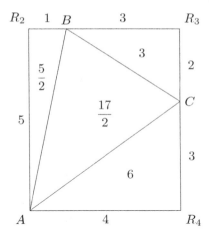

Figure 7.13: An example of the Shoelace Theorem proof in action. Here, $\triangle ABC$ has vertices $A = (2, 3)$, $B = (3, 8)$, and $C = (6, 6)$, which shows that the theorem holds even when A is not at the origin (as $\triangle ABC$ can be translated such that A moves to the origin). Indeed, by the Shoelace Theorem, the area of $\triangle ABC$ is $\frac{1}{2} |(2 \cdot 8 + 3 \cdot 6 + 6 \cdot 3) - (3 \cdot 3 + 8 \cdot 6 + 6 \cdot 2)| = \frac{17}{2}$.

We henceforth outline a proof of the theorem for all n. WLOG set $V_1 = (0, 0)$ and assign to each of the other vertices the coordinates $V_2 = (x_1, y_1)$, $V_3 = (x_2, y_2)$, ..., $V_n = (x_{n-1}, y_{n-1})$. For an n-gon, draw from one vertex (WLOG V_1) each of the $n - 3$ diagonals emanating from that vertex. These diagonals divide the polygon into $n - 2$ triangular regions. The areas of each of these triangles can be computed using the Shoelace Theorem for $n = 3$, and summing all of the areas yields the statement of the theorem for any $n \in \mathbb{N}$. $\square$

Extension Exercise 82. *The Shoelace Theorem can also be proven via induction; doing so is left as an exercise.*

Another method of polygon area-finding is *Pick's Theorem*, named for Georg Alexander Pick, which provides a simple formula for the area in terms of the numbers of *lattice points* within, and on the boundary of, the polygon. First we define a lattice point:

Definition 7.4: Lattice Points

A lattice point is a point in coordinate space whose coordinates are all integers.

Theorem 7.10: Pick's Theorem

The area A of a polygon P is given by $I + \frac{B}{2} - 1$, where I is the number of lattice points strictly within the interior of the polygon, and B is the number of lattice points that lie on the boundary/edges of the polygon.

Proof. Define *Pick's function* $F(P) = I + \frac{B}{2} - 1$ where I and B are defined as above. We want to show that area$(P) - F(P) = 0$ for all P. We will use a form of induction, starting with a triangle as the base case. Consider a rectangle in the coordinate plane whose sides are parallel to the axes; observe that Pick's Theorem holds for any such rectangle.[69] Then split that rectangle down either of its diagonals to form two congruent

---

[69] Proving this is left as an exercise.

right triangles, and since Pick's Theorem holds for the rectangle, it must hold for each the two triangles as well. (If it did not, then the sums of the interior points and boundary points would be inconsistent with the implications of Pick's Theorem for the rectangle.) Thus we have proven the base case ($n = 1$, where n is the number of non-overlapping triangles in the polygon).

We henceforth prove our inductive hypothesis: assuming Pick's Theorem holds for a general polygon P and a triangle T, then it also holds for $P + T$, which is the polygon formed by adjoining T to an edge of P. As P and T share an edge, all the boundary points they have an common become interior points, with the exception of the two endpoints of that shared edge, which are still boundary points. If c is the initial number of common boundary points, then of the polygon $P + T$, we have $I_{P+T} = I_P + I_T + (c - 2)$ and $B_{P+T} = B_P + B_T - 2c + 2$, where I_{P+T} denotes the number of interior points in the polygon $P + T$, and other notations are analogous.

It directly follows that $I_P + I_T = I_{P+T} - (c - 2)$ and $B_P + B_T = B_{P+T} + (2c - 2)$. We are assuming the theorem for P and T individually, so

$$F(P + T) = F(P) + F(T) = \left(I_P + \frac{B_P}{2} - 1\right) + \left(I_T + \frac{B_T}{2} - 1\right)$$

$$= (I_P + I_T) + \frac{B_P + B_T}{2} - 2$$

$$= I_{P+T} - (c - 2) + \frac{B_{P+T} + (2c - 2)}{2} - 2$$

$$= I_{P+T} + \frac{B_{P+T}}{2} - 1$$

hence proving the theorem for all polygons $P + T$ constructed from a polygon P consisting of n triangles and a triangle T. Thus the theorem is proven for all polygons P with n triangles for $n \geq 1$. $\qquad \square$

7.3.6 Examples

Example 7.3.6.1. *For some real value of x, the triangle with vertices at $(2, x)$, $(4, 3)$, and $(8, 4)$ has an area twice that of the triangle with vertices at $(x, 6)$, $(3, x)$, and $(3, 7)$. What is the sum of the possible values of x?*

Solution 7.3.6.1. *By the Shoelace Theorem, the first triangle (which we will call T_1) has area*

$$\frac{1}{2} \left|(2 \cdot 3 + 4 \cdot 4 + x \cdot 8) - (x \cdot 4 + 3 \cdot 8 + 4 \cdot 2)\right|$$

$$= \frac{1}{2} \left|(22 + 8x) - (32 + 4x)\right|$$

$$= \frac{1}{2} \left|4x - 10\right| = \left|2x - 5\right|$$

The second triangle, T_2, has area

$$\frac{1}{2} \left|(x^2 + 3 \cdot 7 + 3 \cdot 6) - (6 \cdot 3 + x \cdot 3 + x \cdot 7)\right|$$

$$= \frac{1}{2} \left|(x^2 + 39) - (10x + 18)\right|$$

$$= \frac{1}{2} \left|x^2 - 10x + 21\right|$$

Thus, $\left|2x - 5\right| = \left|x^2 - 10x + 21\right|$.

We now have four cases to check. If $2x - 4 = x^2 - 10x + 21$ (or, equivalently, if their negatives are both equal to each other), then the sum of the solutions for x is 12, by Vieta's sum formula. If $-(2x-4) = x^2 - 10x + 21$, then $x^2 - 8x + 17 = 0$; however, this produces complex solutions, which should be discarded.[70] Therefore, the desired sum of solutions is 12. (The specific values of x are $6 \pm \sqrt{10}$.)

Example 7.3.6.2. *Demonstrate algebraically that a right triangle with leg lengths a and b, and hypotenuse c, has area $\dfrac{ab}{2}$ using Heron's formula.*

Solution 7.3.6.2.

Proof. By the Pythagorean Theorem, we have $c^2 = a^2 + b^2$, or $c = \sqrt{a^2 + b^2}$ (since all side lengths must be positive). By Heron's, it follows that the area of the triangle is

$$\sqrt{\frac{a + b + \sqrt{a^2 + b^2}}{2} \cdot \frac{a + b - \sqrt{a^2 + b^2}}{2} \cdot \frac{\sqrt{a^2 + b^2} + (b - a)}{2} \cdot \frac{\sqrt{a^2 + b^2} - (b - a)}{2}}$$

$$= \sqrt{\frac{(a + b)^2 - (a^2 + b^2)}{4} \cdot \frac{(a^2 + b^2) - (b - a)^2}{4}}$$

$$= \sqrt{\frac{(2ab) \cdot (2ab)}{16}}$$

$$= \sqrt{\frac{4a^2 b^2}{16}} = \left| \frac{ab}{2} \right| = \frac{ab}{2}$$

(since $a, b > 0$) as desired. $\qquad \square$

Example 7.3.6.3. *Let $\triangle ABC$ have vertices $A = (0, 1)$, $B = (9, 2)$, $C = (12, 3)$. Given that there is exactly 1 lattice, non-vertex point on the boundary of $\triangle ABC$, how many lattice points lie strictly within the interior of $\triangle ABC$?*

Solution 7.3.6.3. *If you draw the triangle carefully enough, you won't even need Pick's Theorem for this! This example, however, serves to illustrate the extremes of Pick's Theorem, where it still works reliably. Here, we have $B = 4$ (including the vertices), and $A = 3$ by Shoelace, so $I = A + 1 - \dfrac{B}{2} = 2$. (Namely, the interior lattice points are $(7, 2)$ and $(8, 2)$. Clearly, the interior lattice points must have a y-coordinate of 2, and must lie between $x = 6$ and $x = 9$, exclusive.)*

Example 7.3.6.4. *$\triangle ABC$ is a lattice triangle; i.e. a triangle whose coordinates are all lattice points. If its area is 12, and the number of lattice points strictly within its interior is 7, provide a possibility for the set of coordinates of the triangle.*

Solution 7.3.6.4. *WLOG assume that $A = (0, 0)$, and set $B = (x_1, y_1)$, $C = (x_2, y_2)$. Then we must have $|x_1 y_2 - x_2 y_1| = 24$, by the Shoelace Theorem. Furthermore, by Pick's Theorem, the triangle has 12 lattice points on its boundary. From A to B, the number of lattice points on the edge $\overline{AB}$ is given by the number of positive integers k less than x_1 such that $k \cdot \dfrac{y_1}{k_1}$ is a positive integer. Excluding A, B, and C themselves, the total number of lattice points on the edges should be 9. An example of a triangle that works has coordinates $A = (0, 0)$, $B = (6, 6)$, $C = (6, 2)$ (this may require some trial-and-error).*

---

[70]Furthermore, even if the problem did not require x to be a real number, it is meaningless to take the absolute value of a complex number, unless we consider the absolute value signs as magnitude instead.

7.3.7 Exercises

Exercise 203. *Classify each of the following sets of triangle side lengths as corresponding to an equilateral, isosceles, scalene, or degenerate triangle. If non-degenerate, further classify them as corresponding to an acute, right, or obtuse triangle.*

(a) $\{3, 4, 6\}$

(b) $\{1, 2, 2\}$

(c) $\{7, 24, 25\}$

(d) $\{99, 100, 101\}$

(e) $\{20, 19, 50\}$

(f) $\{x, 1.5x, 2x\}$ *for some* $x \in \mathbb{N}$

Exercise 204. *(2014 Berkeley Math Tournament) Call two regular polygons supplementary if the sum of an internal angle from each polygon adds up to 180°. For instance, two squares are supplementary because the sum of the internal angles is $90° + 90° = 180°$. Find the other pair of supplementary polygons. Write your answer in the form (m, n) where m and n are the number of sides of the polygons and $m < n$.*

Exercise 205. *What is the area of the quadrilateral with vertices at $(0, 2)$, $(3, 5)$, $(8, 6)$, and $(9, 8)$ (not necessarily in that order)?*

Exercise 206.

(a) *A regular n-gon has 2019 times as many diagonals as edges. What is the value of n?*

(b) *For the value of n in (a), set up (but do not evaluate) an expression for the area of the polygon. Fully justify your answer with supporting work.*

Exercise 207. *(2017 ARML Individual Round #2) Trapezoid ARML has $AR \parallel ML$. Given that $AR = 4$, $RM = \sqrt{26}$, $ML = 12$, and $LA = \sqrt{42}$, compute AM.*

Exercise 208. *A rectangle is made out of unit squares as shown below. Find the total area in square units of the magenta shaded region.*

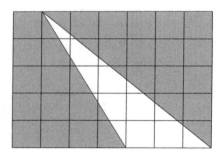

Exercise 209. *(2016 Berkeley Math Tournament) Let S be the set of all non-degenerate triangles with integer side lengths, such that two of the sides are 20 and 16. Suppose we pick a triangle, at random, from this set. What is the probability that it is acute?*

Exercise 210.

(a) *Is there an isosceles triangle with integer side lengths and area 12?*

(b) *Is there an isosceles triangle with integer side lengths and area 18?*

(c) *For what positive integers k does there exist an isosceles triangle with integer side lengths and area k?*

Extension Exercise 83. *(2015 ARML Individual Round #9) An (a, r, m, l)-trapezoid is a trapezoid with bases of length a and r, and other sides of length m and l. Compute the number of positive integer values of l such that there exists a $(20, 5, 15, l)$-trapezoid.*

Extension Exercise 84. *If a, b, and c are the side lengths of a triangle with $a \geq b \geq c$, prove that* $\min\left(\dfrac{a}{b}, \dfrac{b}{c}\right) \in [1, \phi)$ *where* $\phi = \dfrac{1 + \sqrt{5}}{2}$.

Extension Exercise 85. *Propose a generalization of Pick's Theorem for a polygon with n holes (where a hole of a polygon is defined to be a polygon with a positive area $\mathcal{A}$ whose union of interior and boundary points is contained strictly within the polygon).*

7.4 Working With Similarity

We say that two geometric objects (not necessarily polygons, *per se*) are *similar* if they have the same general shape (or one has the same shape asthe mirror image of the other). More precisely, we define similarity of two shapes as follows:

Definition 7.5: Similarity

Two shapes are similar iff one can be translated, rotated about a point, reflected about a point or line, or dilated (i.e. the coordinates of all points multiplied by a scalar factor).

For instance, all circles are similar to each other, as are all equilateral triangles and all squares.

In particular, shapes can be similar but not *congruent;* i.e. they may not have the exact same side lengths or angle measures. However, it is impossible for two shapes to be congruent, but not similar. As such, similarity is a type of congruence. We can further define and characterize congruence by the various types of congruence. Here are the familiar types of congruence of triangles, the same types you likely studied (or are studying) in geometry class:

Definition 7.6: Types of Congruence

- Side-side-side (SSS) congruence: If two triangles have all of the same side lengths, then they are congruent.

- Side-angle-side (SAS) congruence: If two triangles have two corresponding pairs of side lengths equal, and the angles spanning them are of equal measure, then the triangles are congruent. Note that the angle *must* be between the two equal pairs of side lengths; side-side-angle, or *SSA*, is *not* a valid type of congruence![a]

- Angle-side-angle (ASA) congruence: If within two triangles, two pairs of corresponding angles have equal measure, and a side length adjoining them in one triangle has equal length to the other, then the triangles are congruent.

- Angle-angle-angle (AAA) congruence: If each of the three pairs of angles in the triangles have equal measure, then the triangles are congruent.

- Hypotenuse-Leg (HL) congruence, also known as Right-Angle-Hypotenuse-Side (RHS) congruence: If two right triangles have equal hypotenuse lengths, and their shorter legs have equal length, then the triangles are congruent.

---

[a]The Law of Sines and Law of Cosines shed light on why this is; see §8.1, 8.2 for in-depth explanations of those laws.

Note that, for similarity, we need only test the AAA condition. Two polygons (not just triangles) are similar to each other iff all of their corresponding angles have equal measure.

7.4.1 Observing Triangle Similarity

In this section, we shift our focus to the observational part of the proof process. In particular, we will focus on triangle similarity as applied to competition problems.

Drawing Auxiliary Lines

Often, the similarity of two triangles does not pop out until we draw one (or several) auxiliary lines to divide a figure into multiple triangles. A prime example of this phenomenon is as follows:

Example 7.4.1.1. *Triangle* $\triangle ABC$ *with* $\angle B = 90°$ *has* $\overline{AB} = 10$. *Point D lies on* $\overline{AC}$ *such that* $\overline{BD}$ *is an altitude and* $AD = 6$. *Find* $BC + CD$.

Solution 7.4.1.1. *Triangle $\triangle ABD$ is right with $\angle BDA = 90°$, so $BD = 8$ (where we have drawn the auxiliary altitude $\overline{AD}$ to form the right triangle). Observe that triangle $\triangle BDA$ is similar to $\triangle BDC$ by ASA similarity, so $CD = 8 \cdot \dfrac{8}{6} = \dfrac{32}{3}$ and $BC = 10 \cdot \dfrac{8}{6} = \dfrac{40}{3}$. Thus, $BC + CD = 24$.*

Example 7.4.1.2. *(2017 AMC 8) In the right triangle ABC, $AC = 12$, $BC = 5$, and angle C is a right angle. A semicircle is inscribed in the triangle as shown. What is the radius of the semicircle?*

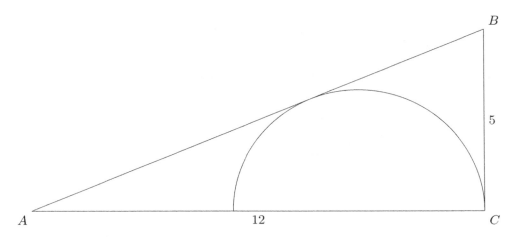

Figure 7.14: Diagram for Example 6.4.1.2.

Solution 7.4.1.2. *By the Pythagorean Theorem, $AB = 13$. Let O be the center of the semicircle, and let D be the intersection of the semicircle's arc with $\overline{AB}$. Then $\overline{OD}$ is a radius of the semicircle, and forms a right angle with $\overline{AB}$.*

We then observe that triangles $\triangle ODB$ and $\triangle OCB$ are both right triangles; in addition, they have the same hypotenuse. Furthermore, they have the same leg lengths, since $\overline{OD}$ and $\overline{OC}$ are both radii of the semicircle. As such, $\triangle ODB$ and $\triangle OCB$ are HL-congruent, and so $BD = 5$, which implies $DA = 8$. Finally, note that $\triangle ADO$ and $\triangle ABC$ are similar, so $\dfrac{8}{r} = \dfrac{12}{5}$, which yields $r = \dfrac{10}{3}$.

(Another solution, which does not involve similar triangles, is to reflect the figure about $\overline{AC}$ to form a circle inscribed in a triangle, then use the inradius formula, $A = rs$, to obtain the answer directly.)

Using the Side-Splitter Theorem

The *Side-Splitter Theorem* is a form of SSS similarity, in which a transversal line splits a triangle into two similar triangles:

> **Theorem 7.11: Side-Splitter Theorem**
>
> Let $\triangle ABC$ be a triangle, and let l be a line parallel to $\overline{BC}$ that intersects $\overline{AB}$ at D and $\overline{AC}$ at E. Then $\dfrac{AD}{AB} = \dfrac{AE}{AC}$.

Proof. First note that, since l is parallel to $\overline{BC}$, it contains neither B nor C. Furthermore, l does not pass through A since it intersects $\overline{AB}$ at a point D that does not coincide with A or B. This observation is crucial in determining that l must intersect $\overline{AC}$, as well as $\overline{AB}$.

Draw line segment $\overline{BE}$. Consider $\triangle ABE$, and denote the foot of the perpendicular from E to $\overline{AB}$ as P. The area of this triangle is given by $\dfrac{AB \cdot EP}{2}$, whereas the area of $\triangle ADE$ is equal to $\dfrac{AD \cdot EP}{2}$; hence,

$\dfrac{|\triangle ABE|}{|\triangle ADE|} = \dfrac{AB}{AD}$ (where $|\triangle ABC|$ denotes the area of triangle $\triangle ABC$). Similarly, we obtain $\dfrac{|\triangle ADE|}{|\triangle ADC|} = \dfrac{AE}{AC}$.

Area of disjoint polygons is additive (that is, the area of the polygon consisting of one polygon adjoined to another by a single common edge is equal to the sum of the areas of each of the polygons), so $|\triangle ABE| = |\triangle ADE| + |\triangle DEB|$ and $|\triangle ADC| = |\triangle ADE| + |\triangle DEC|$. Now observe that $|\triangle DEB| = |\triangle DEC|$ (as both triangles have the same height, which is the height of $\triangle ABC$ less the height of $\triangle ADE$, as well as the same base length, which is that of $\overline{DE}$), so it follows that $|\triangle ABE| = |\triangle ADC|$. Thus, we have $\dfrac{AD}{AB} = \dfrac{|\triangle ADE|}{|\triangle ABE|} = \dfrac{|\triangle ADE|}{|\triangle ADC|} = \dfrac{AE}{AC}$, which finishes the proof. $\qquad\square$

Example 7.4.1.3. *In $\triangle ABC$, line l parallel to $\overline{BC}$ splits $\overline{AB}$ into two line segments at point D, with $AD = 5$ and $DB = 10$. In addition, $BC = 20$. If l also hits the triangle at E, what is $BC - DE$?*

Solution 7.4.1.3. *Note that $AD = AD + DB = 15$, so by triangle similarity and the Side-Splitter Theorem, $BC = \dfrac{5}{15} \cdot BC = \dfrac{20}{3}$. Then $BC - DE = \dfrac{40}{3}$.*

Example 7.4.1.4. *Triangle $\triangle ABC$ has $AB = 9$, $BC = 10$, $CA = 11$. Line l hits $\triangle ABC$ at D and E with $l \parallel BC$. If $AE = 2BE$, compute the perimeter of $\triangle CDE$.*

Solution 7.4.1.4. *We have $AE = 6$ and $BE = 3$, since $AE + EB = AB$. It then follows that $CD = \dfrac{11}{3}$ and $DE = \dfrac{20}{3}$ from the Side-Splitter Theorem. To compute CE, note that the altitude of the triangle has length $6\sqrt{2}$ (which can be verified using Heron's formula), and so the foot of the perpendicular from E has length $2\sqrt{2}$. Then $CE^2 = \left(\dfrac{20}{3}\right)^2 + (2\sqrt{2})^2 = \dfrac{472}{9}$ by the Pythagorean Theorem, and so $CE = \dfrac{2\sqrt{118}}{3}$. It follows that the perimeter of $\triangle CDE$ is $\dfrac{31 + 2\sqrt{118}}{3}$.*

A Note on Angle Chasing and Similarity

Angle chasing, or using properties of triangles and other polygons to determine the angle measures formed by certain auxiliary lines and their intersections, is a very useful geometry technique in general, and especially in terms of detecting, and working with, similar polygons. For now, we will leave our study of angle-chasing at that, but in the Intermediate Geometry chapter (chapter 8), we will delve further into the specific techniques of angle-chasing and analyze which ones work best in which situations. In this section, however, we will provide one example in which angle chasing can be used to determine and prove angle-angle similarity.

Example 7.4.1.5. *(2001 AMC 12) In $\triangle ABC$, $\angle ABC = 45°$. Point D is on $\overline{BC}$ so that $2 \cdot BD = CD$ and $\angle DAB = 15°$. Find $\angle ACB$.*

Solution 7.4.1.5. *We can draw a diagram like the following:*

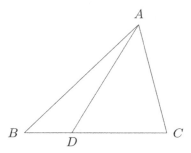

Figure 7.15: Drawing all points and the auxiliary line $\overline{AD}$.

Observe that $\angle BDA = 120°$, *and so* $\angle ADC = 60°$. *Draw the C-altitude, and have it intersect* $\overline{AD}$ *at E.* *Then* $\angle CED = 90°$, $\angle CDE = \angle ADC = 60°$, *and as a result,* $\angle DCE = 30°$.

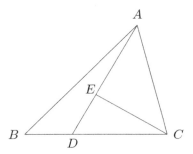

Figure 7.16: Drawing the altitude from C to hit $\overline{AD}$.

We have $\cos\angle CDE = \cos 60° = \dfrac{1}{2} = \dfrac{ED}{CD}$, *so* $ED = \dfrac{CD}{2}$. *In addition, observe that* $BD = \dfrac{CD}{2}$; *thus,* $ED = BD$. *Hence,* $\triangle BED$ *is isosceles (this is where the AA similarity comes in).*

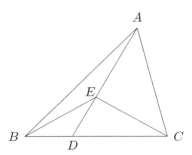

Figure 7.17: Drawing auxiliary line $\overline{BE}$ to form isosceles triangle $\triangle BED$.

From here, it follows that, since $\angle BDA = 120°$, $\angle DBE = \angle DEB = 30°$. *Furthermore,* $\angle ABE = \angle BAE = \angle DAB = 15°$, *so* $\triangle BEA$, *too, is isosceles (with vertex angle* $\angle BEA = 150°$).

Now, note that $\angle AEC = 90°$ *and* $\angle DCE = 30°$. *From the latter, it follows that* $\triangle BEC$ *is also isosceles, and from the former, a direct consequence of the previous fact is that* $\angle EAC = \angle ECA = 45°$ *(as* $AE = BE = EC$). *Finally,* $\angle ACB = \angle BCE + \angle ACE = 30° + 45° = 75°$.

7.4.2 Examples

Example 7.4.2.1. *(2014 MathCounts State Sprint) The side lengths of three similar regular pentagons are in the ratio* $2 : 5 : 7$. *The area of the smallest pentagon is* 40 in^2. *What is the area of the largest pentagon?*

Solution 7.4.2.1. *By definition, the side lengths of the pentagons are all in direct proportion. In addition, the areas of the pentagons are all in direct proportion with the square of their side lengths, so the area of the largest pentagon is* 40 $in^2 \cdot \dfrac{7^2}{2^2} = 490$ in^2.

Example 7.4.2.2. *(2013 MathCounts State Sprint) In isosceles trapezoid* $ABCD$, *shown below,* $AB = 4$ *units and* $CD = 10$ *units. Points* E *and* F *are on* CD *with* BE *parallel to* AD *and* AF *parallel to* BC. AF *and* BE *intersect at point* G. *What is the ratio of the area of triangle* EFG *to the area of trapezoid* $ABCD$? *Express your answer as a common fraction.*

Solution 7.4.2.2. *Because* $ABCD$ *is an isosceles trapezoid, let* $A = (3, h)$ *and* $B = (7, h)$ *for some* $h > 0$. *Both* $\overline{AF}$ *and* $\overline{BE}$ *intersect* G, *which has an x-coordinate of 5. Since* $\overline{BE}$ *and* $\overline{AF}$ *are parallel to* $\overline{AD}$ *and*

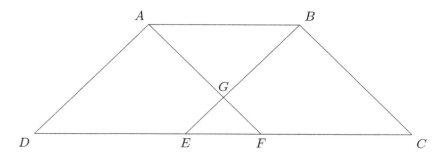

Figure 7.18: Diagram for Example 6.4.2.2.

$\overline{BC}$ respectively, the difference in the x-coordinates from E and B (and A and F) is 3. Then $EF = 2$, and by AA similarity of $\triangle AGB$ and $\triangle EGF$, G has a y-coordinate of $\dfrac{h}{3}$. Thus, the area of $\triangle EGF$ is $\dfrac{h}{3}$. The area of trapezoid $ABCD$ is $7h$, so the desired ratio is $\dfrac{1}{21}$.

Example 7.4.2.3. *Triangle $\triangle ABC$ has $AB = 13$, $BC = 14$, $CA = 15$. Line l is parallel to $\overline{BC}$ and hits $\triangle ABC$ at two points P and Q such that $PQ = 2$. What is the area of triangle $\triangle PQB$?*

Solution 7.4.2.3. *Triangles $\triangle ABC$ and $\triangle APQ$ are AA-similar, so all side lengths of $\triangle APQ$ are $\dfrac{1}{7}$ those of $\triangle ABC$. The altitude of $\triangle ABC$ minus that of $\triangle APQ$ is $12 - \dfrac{12}{7} = \dfrac{72}{7}$, so the area of $\triangle PQB$ is*
$$\dfrac{2 \cdot \dfrac{72}{7}}{2} = \dfrac{72}{7}.$$

Example 7.4.2.4. *(2017 AMC 10B/12B) The diameter $\overline{AB}$ of a circle of radius 2 is extended to a point D outside the circle so that $BD = 3$. Point E is chosen so that $ED = 5$ and line ED is perpendicular to line AD. Segment $\overline{AE}$ intersects the circle at a point C between A and E. What is the area of $\triangle ABC$?*

Solution 7.4.2.4. *Note: It is also possible to coordinate bash this, but we will be solving this problem via similarity methods.*

(On the following page is a diagram to represent the problem.)

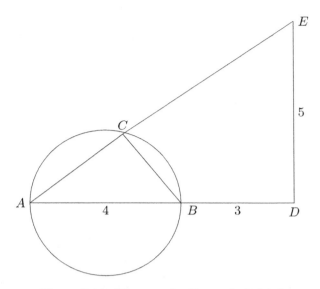

Figure 7.19: Diagram for Example 7.4.2.4.

Note that $\angle ACB$ and $\angle ADE$ are both right angles ($\angle ACB$ by construction, $\angle ADE$ by definition as per the problem statement). Thus, $\triangle ADE$ and $\triangle ABC$ are AA similar (since they share the angle $\angle EAD$). Since $AE = \sqrt{74}$, this means that the ratio of the side lengths of $\triangle ABC$ to those of $\triangle ADE$ is $\dfrac{4}{\sqrt{74}}$. Hence, $AC = \dfrac{28}{\sqrt{74}}$ and $CB = \dfrac{20}{\sqrt{74}}$. The area of $\triangle ABC$ is equal to $\dfrac{AC \cdot CB}{2} = \dfrac{140}{37}$.

7.4.3 Exercises

Exercise 211. *(2017 AMC 8) In the figure below, choose point D on $\overline{BC}$ so that $\triangle ACD$ and $\triangle ABD$ have equal perimeters. What is the area of $\triangle ABD$?*

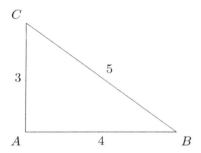

Figure 7.20: Diagram for this Exercise.

Exercise 212. *(2018 AMC 10A) Two circles of radius 5 are externally tangent to each other and are internally tangent to a circle of radius 13 at points A and B, as shown in the diagram. The distance AB can be written in the form $\frac{m}{n}$, where m and n are relatively prime positive integers. What is $m + n$?*

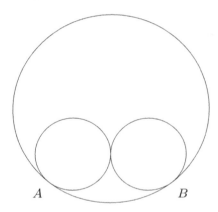

Figure 7.21: Diagram for this Exercise.

Exercise 213. *(2016 AMC 8) A semicircle is inscribed in an isosceles triangle with base 16 and height 15 so that the diameter of the semicircle is contained in the base of the triangle as shown. What is the radius of the semicircle?*

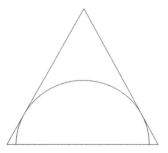

Figure 7.22: Diagram.

Exercise 214. *(2008 AIME I) Square AIME has sides of length 10 units. Isosceles triangle GEM has base EM, and the area common to triangle GEM and square AIME is 80 square units. Find the length of the altitude to EM in $\triangle GEM$.*

Exercise 215. *(2001 AIME I) In triangle $\triangle ABC$, angles A and B measure 60 degrees and 45 degrees, respectively. The bisector of angle A intersects $\overline{BC}$ at T, and $AT = 24$. Find the area of triangle ABC.*

Exercise 216. *In triangle $\triangle ABC$, let $AB = 13$, $AC = 14$, $BC = 15$. Point P_0 lies on $\overline{AC}$ such that $\overline{BP_0} \perp \overline{AC}$. Point P_1 lies on $\overline{BC}$ such that $\overline{P_0P_1} \perp \overline{BC}$, point P_2 lies on $\overline{AC}$ with $\overline{P_1P_2} \perp \overline{AC}$, and so forth. What is the value of the sum*

$$\sum_{i=0}^{\infty} [P_i P_{i+1} P_{i+2}]$$

(where $[A]$ denotes the area of A)?

Exercise 217. *(2011 AMC 10B) Let T_1 be a triangle with sides $2011, 2012$, and 2013. For $n \geq 1$, if $T_n = \triangle ABC$ and $D, E,$ and F are the points of tangency of the incircle of $\triangle ABC$ to the sides AB, BC and AC, respectively, then T_{n+1} is a triangle with side lengths AD, BE, and CF, if it exists. What is the perimeter of the last triangle in the sequence (T_n)?*

Exercise 218. *(2009 AIME I) In right $\triangle ABC$ with hypotenuse $\overline{AB}$, $AC = 12$, $BC = 35$, and $\overline{CD}$ is the altitude to $\overline{AB}$. Let ω be the circle having $\overline{CD}$ as a diameter. Let I be a point outside $\triangle ABC$ such that $\overline{AI}$ and $\overline{BI}$ are both tangent to circle ω. The ratio of the perimeter of $\triangle ABI$ to the length AB can be expressed in the form $\frac{m}{n}$, where m and n are relatively prime positive integers. Find $m + n$.*

Extension Exercise 86. *(2002 AIME II) The perimeter of triangle APM is 152, and the angle $\angle PAM$ is a right angle. A circle of radius 19 with center O on $\overline{AP}$ is drawn so that it is tangent to $\overline{AM}$ and $\overline{PM}$. Given that $OP = \frac{m}{n}$ where m and n are relatively prime positive integers, find $m + n$.*

Extension Exercise 87. *(2004 AIME II) Let ABCDE be a convex pentagon with $AB \parallel CE, BC \parallel AD, AC \parallel DE, \angle ABC = 120°, AB = 3, BC = 5$, and $DE = 15$. Given that the ratio between the area of triangle ABC and the area of triangle EBD is m/n, where m and n are relatively prime positive integers, find $m + n$.*

Extension Exercise 88. *(2013 AIME I) Triangle AB_0C_0 has side lengths $AB_0 = 12$, $B_0C_0 = 17$, and $C_0A = 25$. For each positive integer n, points B_n and C_n are located on $\overline{AB_{n-1}}$ and $\overline{AC_{n-1}}$, respectively, creating three similar triangles $\triangle AB_nC_n \sim \triangle B_{n-1}C_nC_{n-1} \sim \triangle AB_{n-1}C_{n-1}$. The area of the union of all triangles $B_{n-1}C_nB_n$ for $n \geq 1$ can be expressed as $\frac{p}{q}$, where p and q are relatively prime positive integers. Find q.*

7.5 Proving Angle Theorems

7.5.1 Central Angle Theorem

The fundamentals of angle chasing apply not only to closed polygonal shapes, but to circles as well (in fact, many of their most useful applications are in circle geometry). The theorem simplifies complicated problems involving circular arc angle measures with one rule of thumb:

Theorem 7.12: Central Angle Theorem

Consider circle O, with points P, A, and B lying on O. Then $\angle AOB = 2\angle APB$. (P can lie anywhere on the outer arc APB, and the theorem still holds.)

Here is a diagram to illustrate the theorem:

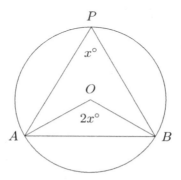

Figure 7.23: The Central Angle Theorem in action.

Proof. Draw line $\overline{OP}$. Then observe that $OP = OA = OB$ (since they are all radii of the circle), so triangles $\triangle AOP$, $\triangle AOB$, and $\triangle POB$ are all isosceles. In particular, let $\angle APO = \angle PAO = a°$ and $\angle BPO = \angle PBO = b°$. Then $\angle AOB = (180° - \angle AOP) + (180° - \angle BOP) = 360° - (\angle AOP + \angle BOP) = 360° - (180° - 2a°) - (180° - 2b°) = 2a° + 2b° = 2(a° + b°) = 2(\angle APB)$. $\qquad\square$

7.5.2 Exterior Angle Theorem

In any triangle, we can use the *Exterior Angle Theorem* to determine constraints on both the interior and exterior angle measures, even if several of them are unknown:

Theorem 7.13: Exterior Angle Theorem

In any triangle $\triangle ABC$ with $\angle A = a°$, $\angle B = b°$, $\angle C = c°$, an exterior angle is equal to the sum of the other two interior angles. (That is, the exterior angle of angle A has degree measure $(b + c)°$, and same for the other two exterior angles.)

Proof. In any triangle, $(a + b + c)° = 180°$, and any pair of interior and exterior angles sum to $180°$ as well. Assume WLOG that we are dealing with exterior angle A; then the exterior angle corresponding to angle A has measure $(180 - a)°$. We have $b° + c° = (180 - a)°$ as well; hence, the exterior angle of $\angle A$ has the same degree measure as the sum of interior angles $\angle B$ and $\angle C$ (this also holds, of course, for exterior angles of $\angle B$ and $\angle C$). $\qquad\square$

Noteworthily, this theorem is Proposition 1.16 in Euclid's *Elements,* in which he shows that its proof does not depend on the parallel postulate (which is his fifth postulate in the very same work, stating that, if a line segment intersects two straight lines forming two interior angles on the same side of the line that sum to less than 180°, then those two lines will eventually meet on that same side). This theorem was, indeed, quite controversial in Euclid's time despite the straightforwardness of the proof we have offered here, as Euclid's own proof was flawed in a crucial way (at least without an underlying set of axioms for Euclidean geometry). His proof went as follows:

Proof.

- First, let D be a point on $\overrightarrow{BC}$ beyond C.

- Let E be the midpoint of $\overline{AC}$.

- Then draw the ray $\overrightarrow{BE}$, and let F be a point on the extension of $\overrightarrow{BE}$ such that $BE = EF$.

- Finally, draw line segment $\overline{FC}$.

After all these constructions are completed, we observe that $\angle BAC = \angle ECF$ (vertical angles), and $\angle ECF < \angle ECD = \angle ACD$, so $\angle BAC < \angle ACD$. A similar result can be obtained for $\angle CBA$ (this is left as an exercise). $\qquad\square$

The key flaw lies in the assumption that F lies on the same side as angle $\angle ACD$. Without a rigorous axiomatic foundation, we cannot simply handwave this automatically. However, Euclid later set forth the foundations of Euclidean geometry, which allows for the proof to be completely valid.[71]

7.5.3 Angle Bisector Theorem

The Angle Bisector Theorem, the third of Euclid's Propositions in *Elements,* provides an extremely convenient way for us to extract similarity (and/or use coordinates), regardless of how messy angle measures can get (or if you aren't willing to use trigonometry):

Theorem 7.14: Angle Bisector Theorem

Let $\triangle ABC$ be a triangle with side lengths a, b, c. Then the bisectors of $\angle A$, $\angle B$, and $\angle C$ divide sides a, b, and c into the ratios $b : c$, $a : c$, and $a : b$ respectively.

Below is a diagram illustrating the theorem for the bisector of angle $\angle A$ in $\triangle ABC$:

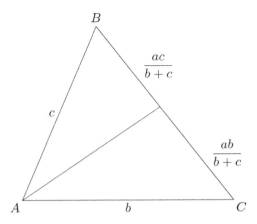

Figure 7.24: The Angle Bisector Theorem for angle $\angle A$.

Important remark! If we draw all three internal angle bisectors, their point of intersection is the *incenter* I; i.e. the center of the *incircle,* the largest circle that can be perfectly inscribed in the triangle. (The

---

[71]But not in spherical geometry.

circumcenter, or center of the *circumcircle* (the smallest circle that perfectly contains the triangle), is the intersection of the *perpendicular bisectors,* or lines perpendicular to each edge of the triangle that bisect them.)

Proof. Instead of proving this theorem as-is, we will take a look at the generalized form and prove that instead. □

Generalized Angle Bisector Theorem

The Angle Bisector Theorem can be generalized to all points of intersection with a side of a triangle of a line protruding from a vertex of the triangle. Specifically, the theorem statement is as follows:

> **Theorem 7.15: Generalized Angle Bisector Theorem**
>
> If D is the intersection of a line passing through A (strictly within the interior of the triangle) with the side $\overline{BC}$, then $\dfrac{BD}{DC} = \dfrac{AB \cdot \sin \angle DAB}{AC \cdot \sin \angle DAC}$.

Proof. (Note: The following proof makes extensive use of the Law of Sines. If you are unfamiliar with LoS, please read §8.1 first.)

Using Law of Sines on $\triangle ABD$ gives us

$$\frac{AB}{BD} = \frac{\sin \angle BDA}{\sin \angle BAD} \tag{7.5}$$

Similarly, LoS on $\triangle ACD$ yields

$$\frac{AC}{CD} = \frac{\sin \angle ADC}{\sin \angle DAC} \tag{7.6}$$

Henceforth, note that $\angle BDA + \angle CDA = 180°$, so $\sin \angle BDA = \sin \angle CDA$ (try visualizing the angles on the unit circle to see why this should be true, or use the addition formula for sin).

For the regular Angle Bisector Theorem, where $\overline{AD}$ bisects $\angle A$, we have $\angle BAD = \angle CAD$, so $\dfrac{AB}{BD} = \dfrac{AC}{CD}$. This rearranges to the desired form $\dfrac{BD}{DC} = \dfrac{AB}{AC}$, hence we are done.

If $\angle BAD \neq \angle DAC$, however, we can still write equations (7.5), (7.6) obtained from the Law of Sines in the following forms, respectively:

$$\frac{AB}{BD} \cdot \sin \angle BAD = \sin \angle BDA \tag{7.7}$$

$$\frac{AC}{CD} \cdot \sin \angle DAC = \sin \angle ADC \tag{7.8}$$

Regardless of whether $\angle BAD \neq \angle DAC$, they must still be supplementary, so $\sin \angle BAD = \sin \angle DAC$ still. Thus, we obtain

$$\frac{AB}{BD} \cdot \sin \angle BAD = \frac{AC}{CD} \cdot \sin \angle DAC \tag{7.9}$$

which rearranges to the desired form, completing the proof. □

Here is another proof that constructs an auxiliary point:

Alternate proof. In triangle $\triangle ABC$, let D lie on $\overline{BC}$ such that $\overline{AD}$ is *not* an altitude of $\triangle ABC$. (That is, $\overline{AD} \not\perp \overline{BC}$.) Let B_1 be the foot of the perpendicular from B intersecting $\overline{AD}$, and let C_1 be the foot of the perpendicular from C intersecting $\overline{AD}$. Then one and only one of B_1 and C_1 lie within the triangle; WLOG assume that B_1 does.

238

We have $\angle DB_1B = \angle DC_1C = 90°$, and $\angle B_1DB \cong \angle C_1DC$ (vertical angles), so triangles $\triangle DB_1B$ and $\triangle DC_1C$ are AA-similar. This implies that

$$\frac{BD}{CD} = \frac{BB_1}{CC_1} = \frac{AB \cdot \sin \angle BAD}{AC \cdot \sin \angle CAD} \tag{7.10}$$

Finally, $\sin BAD = \dfrac{BD}{AB}$ and $\sin DAC = \dfrac{CD}{AC}$; the desired result follows. $\qquad\square$

7.5.4 Examples

Example 7.5.4.1. *(2014 AMC 12A) Two circles intersect at points A and B. The minor arcs AB measure $30°$ on one circle and $60°$ on the other circle. What is the ratio of the area of the larger circle to the area of the smaller circle?*

Solution 7.5.4.1. *Let r, R be the radii of the smaller and larger circles respectively. Half the chord length is $r \sin 30° = R \sin 15°$. Then $r = R \cdot \dfrac{\sqrt{6} - \sqrt{2}}{2} \implies R^2 = \boxed{(2 + \sqrt{3})}\, r^2$.*

Example 7.5.4.2. *Triangle $\triangle ABC$ has $\angle A = 45°$, $\angle B = 60°$, $\angle C = 75°$, with $BC = 1$. $\overline{BC}$ is extended to a point D such that $\overline{AC}$ is the angle bisector of $\angle BAD$. What is $\dfrac{[ABC]}{[ABD]}$?*

Solution 7.5.4.2. *$\angle BAD$ will be right, so $\triangle ABD$ becomes a 30-60-90 right triangle. By Angle Bisector Theorem, it follows that the desired ratio is $\boxed{\dfrac{\sqrt{3} - 1}{2}}$ using the sine law.*

7.5.5   Exercises

Exercise 219. *(2016 AMC 12A) In* $\triangle ABC$, $AB = 6$, $BC = 7$, *and* $CA = 8$. *Point* D *lies on* $\overline{BC}$, *and* $\overline{AD}$ *bisects* $\angle BAC$. *Point* E *lies on* $\overline{AC}$, *and* $\overline{BE}$ *bisects* $\angle ABC$. *The bisectors intersect at* F. *What is the ratio* $AF : FD$?

Exercise 220. *(2018 AMC 12B) Side* $\overline{AB}$ *of* $\triangle ABC$ *has length* 10. *The bisector of angle* A *meets* $\overline{BC}$ *at* D, *and* $CD = 3$. *The set of all possible values of* AC *is an open interval* (m, n). *What is* $m + n$?

Exercise 221. *(2010 AMC 12A) Nondegenerate* $\triangle ABC$ *has integer side lengths,* $\overline{BD}$ *is an angle bisector,* $AD = 3$, *and* $DC = 8$. *What is the smallest possible value of the perimeter?*

Exercise 222. *(2017 AIME II) A triangle has vertices* $A(0,0)$, $B(12,0)$, *and* $C(8,10)$. *The probability that a randomly chosen point inside the triangle is closer to vertex* B *than to either vertex* A *or vertex* C *can be written as* $\frac{p}{q}$, *where* p *and* q *are relatively prime positive integers. Find* $p + q$.

Exercise 223. *Let* $\triangle ABC$ *be a triangle with* $AB = 3$, $BC = 4$, $CA = 5$. *Point* D *is on* $\overline{BC}$ *such that the circumcircles of* $\triangle ABD$ *and* $\triangle ADC$ *have areas that differ by* $\dfrac{37\pi}{16}$. *The angle bisector of* $\triangle ABD$ *hits* $\overline{AC}$ *at* E. *Compute the area of* $\triangle ADE$.

Exercise 224. *(1990 AIME) A triangle has vertices* $P = (-8, 5)$, $Q = (-15, -19)$, *and* $R = (1, -7)$. *The equation of the bisector of* $\angle P$ *can be written in the form* $ax + 2y + c = 0$. *Find* $a + c$.

Exercise 225. *(2011 AIME II) In triangle* ABC, $AB = \frac{20}{11}AC$. *The angle bisector of* $\angle A$ *intersects* BC *at point* D, *and point* M *is the midpoint of* AD. *Let* P *be the point of the intersection of* AC *and* BM. *The ratio of* CP *to* PA *can be expressed in the form* $\dfrac{m}{n}$, *where* m *and* n *are relatively prime positive integers. Find* $m + n$.

Exercise 226. *(2009 AIME I) Triangle* ABC *has* $AC = 450$ *and* $BC = 300$. *Points* K *and* L *are located on* $\overline{AC}$ *and* $\overline{AB}$ *respectively so that* $AK = CK$, *and* $\overline{CL}$ *is the angle bisector of angle* C. *Let* P *be the point of intersection of* $\overline{BK}$ *and* $\overline{CL}$, *and let* M *be the point on line* BK *for which* K *is the midpoint of* $\overline{PM}$. *If* $AM = 180$, *find* LP.

Exercise 227. *(2005 AIME II) In triangle* ABC, $AB = 13, BC = 15$, *and* $CA = 14$. *Point* D *is on* $\overline{BC}$ *with* $CD = 6$. *Point* E *is on* $\overline{BC}$ *such that* $\angle BAE \cong \angle CAD$. *Given that* $BE = \frac{p}{q}$ *where* p *and* q *are relatively prime positive integers, find* q.

Extension Exercise 89. *Triangle* $\triangle ABC$ *has circumcircle* Ω. *The internal angle bisectors of* $\angle A$, $\angle B$, *and* $\angle C$ *are all extended to intersect* Ω *at points* D, E, F *respectively. What is the area of* $\triangle DEF$?

Extension Exercise 90. *Let* $\triangle ABC$ *have side lengths* a, b, *and* c *opposite vertices* A, B, *and* C, *respectively. Compute the length of each internal angle bisector of* $\triangle ABC$ *in terms of* a, b, *and* c.

Extension Exercise 91. *(1992 USAMO P4) Chords* AA', BB', *and* CC' *of a sphere meet at an interior point* P *but are not contained in the same plane. The sphere through* A, B, C, *and* P *is tangent to the sphere through* A', B', C', *and* P. *Prove that* $AA' = BB' = CC'$.

Extension Exercise 92. *(1959 IMO P5b) An arbitrary point* M *is selected in the interior of the segment* AB. *The squares* $AMCD$ *and* $MBEF$ *are constructed on the same side of* AB, *with the segments* AM *and* MB *as their respective bases. The circles about these squares, with respective centers* P *and* Q, *intersect at* M *and also at another point* N. *Let* N' *denote the point of intersection of the straight lines* AF *and* BC. *Prove that the straight lines* MN *pass through a fixed point* S *independent of the choice of* M.

7.6 Introduction to 3D Geometry

Just as interesting as two-dimensional Cartesian geometry, if not more so, is its generalization to higher dimensions. In three dimensions and higher, visualization becomes much more of a challenge, but we can also think and imagine beyond the boundaries and confines of the world around us. We will begin this section by studying properties of three-dimensional geometric figures, and how they relate to their two-dimensional equivalents.

7.6.1 Shedding Light on Three-Dimensional Area/Volume Formulas

This section serves as an analogue in three dimensions to the analysis of area formulas in two dimensions.[72] Indeed, these can theoretically be extended to n dimensions for any n with some tweaking, but in this section we will focus purely on three-dimensional objects and the derivations of their volumes.

Theorem 7.16: Three-Dimensional Area/Volume Formulas

- The volume of a rectangular prism with length l, width w, and height h is lwh. (Thus, the volume of a cube with side length s is s^3.) Its surface area is given by $2(lw + lh + wh)$ (and the surface area of a cube with side length s is $6s^2$).

- More generally, the volume of a parallelepiped (a figure whose faces are all parallelograms) is given by the area of the base multiplied with the height. If we express the volume in terms of dimension vectors a, b, and c, we obtain the formula $|(a \times b) \cdot c|$ where $\times$ denotes the cross product and $\cdot$ denotes the dot product. We can also write this as $||a \times b|| \cdot ||c|| \cdot \cos\theta$, where θ is the angle between c and the z-axis.

- The volume of a cylinder with base radius r and height h is $\pi r^2 h$, and the volume of a cone with the same dimensions is $\dfrac{\pi r^2 h}{3}$. In general, if A is the base area, we can write these as Ah and $\dfrac{1}{3}Ah$ for rectangular figures, namely prisms and pyramids. The surface area of a cylinder with base radius r and height h is $2\pi r(h + r)$, and the surface area of a cone with those dimensions is $\pi r(r + \sqrt{h^2 + r^2})$.

- The volume of a sphere with radius r is $\dfrac{4}{3}\pi r^3$, and its surface area is $4\pi r^2$. In general, the volume and surface area of an ellipsoid with semi-axis lengths r_1, r_2, r_3 is $\dfrac{4}{3}\pi r_1 r_2 r_3$.

At first glance, many of these formulas may seem arbitrary, and without the proper context, it can be quite difficult to understand why they should be true on an intuitive level. Fear not! There is always intuitive reasoning behind these formulas - and more rigorous reasoning to go along with it as well.

Proof.

- Extending the definition of area as the number of unit blocks that can fit inside of a two-dimensional region, we can apply the concept to three dimensions, by defining *volume* as the number of unit cubes that can fit inside a three-dimensional region. Thus, if A blocks can fit inside an $l \times w$ region, then $V = ah$ blocks can fit inside of an $l \times w \times h$ region (which can be seen by duplicating the $l \times w \times 1$ three-dimensional volume h times onto itself). As for the surface area, by definition, to obtain the surface area, we sum the areas of each face of the figure. A rectangular prism has two faces with dimensions l and w, two faces with dimensions l and h, and two faces with dimensions w and h. The areas of those types of faces are lw, lh, and wh, so the desired sum is $2(lw + lh + wh)$. For a cube, the side lengths are all equal to s, so the volume is s^3 and the surface area is $6s^2$ (applying our definitions).

---

[72]See §7 for the proofs of several of the most prominent/common area formulae.

- By definition, the area of the base of the parallelepiped is $\left| \vec{b} \times \vec{c} \right|$, since the general area of a parallelogram is $bc\sin\theta$ (where, in this case, $\theta = 90°$. We then multiply by the height a to obtain the desried result, since $\vec{a}$ is perpendicular to the plane of the other two vectors.

- The cylinder's volume is a direct corollary of the circle's area (which we will not prove here). Its surface area results from summing the areas of the two circular bases (namely $2\pi r^2$), and the circumference multiplied with the height (namely $2\pi rh$). The cone's volume can be derived from the volume of revolution - namely the volume when the cone is rotated about the x-axis with its apex at the origin. Taking the integral from 0 to h of πy^2 with respect to x, and with $y = x \cdot \dfrac{r}{h}$, yields the desired result after some computation. (The details are outside the scope of this section, but this is the general motivating idea.) Finally, the surface area of the cone is a consequence of unrolling the cone into a circular sector with radius $\sqrt{r^2 + h^2}$ and arc length $2\pi r$.

- This is a proof that typically involves calculus on some level: the most common proof involves a horizontal slice with height z and length x. Setting the sphere's radius equal to r, we obtain $x = \sqrt{r^2 - z^2}$ by the Pythagorean Theorem. Then $V = \int_{-r}^{r} \pi(r^2 - z^2)\,dz$, and we can evaluate using integral calculus methods to obtain the desired result. To prove the surface area formula, we can take a similar approach, using the definition of arc length as

$$S = \int_a^b \sqrt{\left(\frac{dx}{dt}\right) + \left(\frac{dy}{dt}\right)^2}\,dt$$

and realizing that the sphere is the solid of revolution of a half-circle about the x-axis. Hence, parametrizing $x(t) = r\cos t$ and $y(t) = r\sin t$ yields the desired result after some substitution and subsequent computation.

These are the derivations of some of the most commonly-used three-dimensional volume and surface area formulas at this level of study. $\qquad\square$

7.6.2 Example: Coordinates in Three-Dimensional Space

Just as we can assign coordinates to each point in a figure in 2-dimensional space (or, colloquially, "coordinate bash"), so we can perform this process in 3-dimensional space to extract vital data regarding lengths, areas, and especially volumes that might otherwise be inaccessible without much more complicated methods.

The following problem is a prime example of 3-dimensional coord-bashing:

Example 7.6.2.1. *(2019 AMC 12B) Square pyramid $ABCDE$ has base $ABCD$, which measures 3 cm on a side, and altitude AE perpendicular to the base, which measures 6 cm. Point P lies on BE, one third of the way from B to E; point Q lies on DE, one third of the way from D to E; and point R lies on CE, two thirds of the way from C to E. What is the area, in square centimeters, of $\triangle PQR$?*

Solution 7.6.2.1. *Let $A = (0,0,0)$, so that the rest of the coordinates follow suit: $B = (0,3,0), C = (3,3,0), D = (3,0,0), E = (0,0,6)$ without loss of generality. Then $P = (0,2,2)$, $Q = (2,0,2)$, $R = (1,1,4)$. We have $PQ = \sqrt{8}, QR = \sqrt{6}, RP = \sqrt{6}$ by Distance Formula, and subsequently $[\triangle PQR] = \boxed{2\sqrt{2}}$.*

7.6.3 Examples

Example 7.6.3.1. *Let $ABCDE$ be a pyramid, where the base $ABCD$ is a square of side length 15. The total surface area of pyramid $ABCDE$ (including all five faces) is 585.*

Let M, N, P, and Q be the midpoints of $\overline{AE}$, $\overline{BE}$, $\overline{CE}$, and $\overline{DE}$, respectively. Find the total surface area of frustum $ABCDMNPQ$.

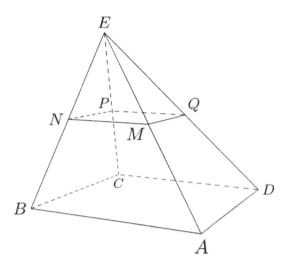

Solution 7.6.3.1. *The area of square $ABCD$ is $15^2 = 225$, so the lateral surface area of pyramid $ABCDE$ is $585 - 225 = 360$.*

The dimensions of pyramid $MNPQE$ are exactly half that of pyramid $ABCDE$, so the lateral surface area of pyramid $MNPQE$ is $\dfrac{360}{2^2} = 90$. Therefore, the lateral surface area of frustum $ABCDMNPQ$ is $360 - 90 = 270$.

The area of square $MNPQ$ is $\dfrac{225}{2^2} = 56\dfrac{1}{4}$, so the total surface area of frustum $ABCDMNPQ$ is $270 + 225 + 56\dfrac{1}{4} = \boxed{551\dfrac{1}{4}}$.

Example 7.6.3.2. *Let $ABCDEFGH$ be a cube of side length 10, as shown. Let P and Q be points on $\overline{AB}$ and $\overline{AE}$, respectively, such that $AP = 4$ and $AQ = 2$. The plane through C, P, and Q intersects $\overline{DH}$ at R. Find DR.*

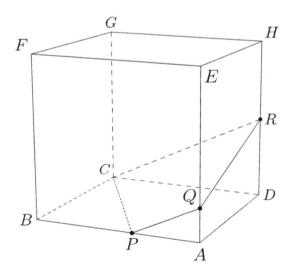

Solution 7.6.3.2. *Let the plane through C, P, and Q intersect $\overline{AD}$ at S.*

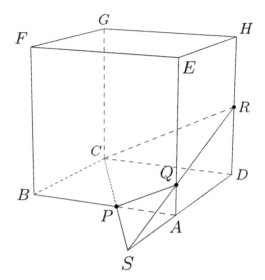

Triangles SAP and SDC are similar by AA Similarity because $\overline{AP} \parallel \overline{CD}$, so $\dfrac{SA}{SD} = \dfrac{AP}{CD}$. Similarly, triangles SAQ and SDR are similar, so $\dfrac{SA}{SD} = \dfrac{AQ}{RD}$. Combining these gives $\dfrac{AQ}{RD} = \dfrac{AP}{CD}$, so $RD = \dfrac{CD \cdot AQ}{AP} = \dfrac{10 \cdot 2}{4} = \boxed{5}$.

Example 7.6.3.3. *A right circular cone and a right circular cylinder each have a radius of 12 and a height of 5. Let A and B be the lateral surface areas of the cone and cylinder, respectively. Find $\dfrac{A}{B}$.*

Solution 7.6.3.3. *If h is the height and r is the radius, then the lateral surface area of a right circular cone is $\pi r\sqrt{r^2 + h^2}$, while the lateral area of a right circular cylinder is $2\pi rh$. Hence, for $r = 12$ and $h = 5$, the ratio of the lateral surface area of a right circular cone to the lateral surface area of a right circular cylinder is $\dfrac{\pi r\sqrt{r^2 + h^2}}{2\pi rh} = \dfrac{\sqrt{r^2 + h^2}}{2h} = \dfrac{\sqrt{12^2 + 5^2}}{2 \cdot 5} = \boxed{\dfrac{13}{10}}$.*

7.6.4 Exercises

Exercise 228. *A cube has a surface area of 96 square feet. A sphere is inscribed within that cube; then, another cube is inscribed within the sphere. Find the surface area, in square feet, of the smaller cube.*

Exercise 229. *Determine the sum of the edge lengths of a cube whose volume is numerically equal to six times its surface area (ignoring units).*

Exercise 230. *(2000 AMC 12) The point $P = (1, 2, 3)$ is reflected in the xy-plane, then its image Q is rotated by $180°$ about the x-axis to produce R, and finally, R is translated by 5 units in the positive-y direction to produce S. What are the coordinates of S?*

Exercise 231. *(2014 AMC 10A) Four cubes with edge lengths 1, 2, 3, and 4 are stacked [at the lower-left corner]. What is the length of the portion of $\overline{XY}$ contained in the cube with edge length 3 [where X is the lower-left, uppermost point of the smallest cube and Y is the upper-right, lowermost point of the largest cube]?*

Exercise 232. *Three points, P_1, P_2, and P_3, lie on the surface of a cube with side length 1. Find the maximal area of $\triangle P_1 P_2 P_3$ and prove that your construction is optimal.*

Exercise 233. *(2008 AMC 12B) A pyramid has a square base $ABCD$ and vertex E. The area of square $ABCD$ is 196, and the areas of $\triangle ABE$ and $\triangle CDE$ are 105 and 91, respectively. What is the volume of the pyramid?*

Exercise 234. *(2009 ARML Individual Round #9) A cylinder with radius r and height h has volume 1 and total surface area 12. Compute $\dfrac{1}{r} + \dfrac{1}{h}$.*

Extension Exercise 93. *(2013 AMC 12A) Six spheres of radius 1 are positioned so that their centers are at the vertices of a regular hexagon of side length 2. The six spheres are internally tangent to a larger sphere whose center is the center of the hexagon. An eighth sphere is externally tangent to the six smaller spheres and internally tangent to the larger sphere. What is the radius of this eighth sphere?*

Extension Exercise 94. *(2018 Berkeley Math Tournament) A plane cuts a sphere of radius 1 into two pieces, one of which has three times the surface area of the other. What is the area of the disk that the sphere cuts out of the plane?*

Extension Exercise 95. *(2018 AMC 12B) Ajay is standing at point A near Pontianak, Indonesia, $0°$ latitude and $110°$ E longitude. Billy is standing at point B near Big Baldy Mountain, Idaho, USA, $45°$ N latitude and $115°$ W longitude. Assume that Earth is a perfect sphere with center C. What is the degree measure of $\angle ACB$?*

Extension Exercise 96. *Compute the number of lattice points lying within the interior of the hypersphere in $\mathbb{R}^4$ with radius $2\sqrt{2}$.*

Extension Exercise 97. *Let H_n denote the n-dimensional unit hypercube, and denote by $H_n(m)$ the number of m-dimensional faces contained within H_n. Show that*

$$\sum_{m=0}^{n} H_n(m) = 3^n$$

for all positive integers n.

Chapter 7 Review Exercises

Exercise 235. *The perimeter of a sector of a circle is equal to 2.018 times its radius. In radians, what is the angle measure of the sector's arc?*

Exercise 236. *(2009 AMC 12A) One dimension of a cube is increased by 1, another is decreased by 1, and the third is left unchanged. The volume of the new rectangular solid is 5 less than that of the cube. What was the volume of the cube?*

Exercise 237. *In rectangle $ABCD$ with $AB = 4$ and $BC = 3$, M is the midpoint of $\overline{CD}$. Lines $\overline{AC}$ and $\overline{BM}$ form four distinct regions in the interior of the rectangle: R_1, R_2, R_3, and R_4, whose areas are in decreasing order from R_1 to R_4. What is $[R_1] - [R_2] + [R_3] - [R_4]$? ($[A]$ denotes the area of A.)*

Exercise 238. *Rectangle $MATH$ has $MA = 20$ and $AT = 18$. Point P is on diagonal $\overline{MT}$ such that $TP = 3PM$, point R is on diagonal $\overline{AH}$ such that $AR = 7RH$, and points O and F are the centroids of triangles $\triangle MAP$ and $\triangle TOR$. What is the area of quadrilateral $PROF$?*

Exercise 239. *In $\triangle ABC$, $AB = 10$, $BC = 17$, $CA = 21$. Let M be the midpoint of $\overline{AB}$. The perpendicular bisector of $\overline{AB}$ passing through M cuts $\overline{CA}$ at D. Compute the ratio of the area of $\triangle AMD$ to the area of $\triangle ABC$.*

Exercise 240. *What is the tangent of the acute angle formed by $y = 20x + 18$ and $y = 4x + 2$?*

Exercise 241. *(2018 AMC 12B) Square $ABCD$ has side length 30. Point P lies inside the square so that $AP = 12$ and $BP = 26$. The centroids of $\triangle ABP$, $\triangle BCP$, $\triangle CDP$, and $\triangle DAP$ are the vertices of a convex quadrilateral. What is the area of that quadrilateral? (Note: The centroid of a polygon is the point whose coordinates are the averages of the respective coordinates of the vertices.)*

Exercise 242. *(2002 AMC 12A) Triangle ABC is a right triangle with $\angle ACB$ as its right angle, $m\angle ABC = 60°$, and $AB = 10$. Let P be randomly chosen inside ABC, and extend $\overline{BP}$ to meet $\overline{AC}$ at D. What is the probability that $BD > 5\sqrt{2}$?*

Exercise 243. *A right triangle has integer side lengths, and its longer leg length is 1 less than its hypotenuse length. Let r be its inradius, and let R be its circumradius. If $1 < |R - r| < 100$, find the sum of the possible values for the triangle's hypotenuse.*

Exercise 244. *(2002 AMC 12A) In triangle $\triangle ABC$, side $\overline{AC}$ and the perpendicular bisector of $\overline{BC}$ meet in point D, and $\overline{BD}$ bisects $\angle ABC$. If $AD = 9$ and $DC = 7$, what is the area of triangle $\triangle ABD$?*

Extension Exercise 98. *(2013 AIME I) A rectangular box has width 12 inches, length 16 inches, and height $\frac{m}{n}$ inches, where m and n are relatively prime positive integers. Three faces of the box meet at a corner of the box. The center points of those three faces are the vertices of a triangle with an area of 30 square inches. Find $m + n$.*

Extension Exercise 99.

   (a) *Does a tetrahedron exist with edge lengths 1, 2, 2, 3, 4, 5? If so, construct an example of such a tetrahedron. If not, prove that no such tetrahedron exists.*

   (b) *(2014 Caltech-Harvey Mudd Math Competition) What's the greatest pyramid volume one can form using edges of length 2, 3, 3, 4, 5, 5, respectively?*

Extension Exercise 100. *Recall that, in a triangle, the intersection of the angle bisectors determines the incenter. Is it necessarily true that the intersection of the angle bisector planes in a tetrahedron determines the center of the inscribed sphere?*

Extension Exercise 101. *(1972 USAMO P2) A given tetrahedron $ABCD$ is isosceles, that is, $AB = CD$, $AC = BD$, $AD = BC$. Show that the faces of the tetrahedron are acute-angled triangles.*

Extension Exercise 102. *(2010 ARML Individual Round #8) In square $ABCD$ with diagonal 1, E is on AB and F is on BC with $\angle BCE = \angle BAF = 30°$. If CE and AF intersect at G, compute the distance between the incenters of triangles AGE and CGF.*

Extension Exercise 103. *Right triangle $\triangle ABC$ with $\angle B = 90°$ and hypotenuse $\overline{AC}$ has square $ACDE$ with $\overline{AC}$ as an edge, and not overlapping with the interior of $\triangle ABC$. If $(AD + BE)(AD - BE) = 24$, and $\triangle ABC$ has integer side lengths, compute the area of pentagon $ABCDE$.*

Chapter 8

Intermediate Geometry

8.1 Law of Sines

A beautiful theorem that relates the side lengths of any triangle to its interior angle measures, the *Law of Sines* provides a clear-cut, guaranteed way out of a situation in which it may seem impossible to determine either the side lengths or angles of a triangle given only one of the two.

> **Theorem 8.1: Law of Sines**
>
> For $\triangle ABC$ with circumradius R,
> $$\frac{a}{\sin A} = \frac{b}{\sin B} = \frac{c}{\sin C} = 2R.$$

Remark. The formulation of the theorem which includes the $2R$ is sometimes known as the *Extended Law of Sines*. Here we will simply refer to it as the Law of Sines, full stop. (There's no point in abridging it, after all!)

Proof. WLOG we prove
$$\frac{a}{\sin A} = 2R$$
since A is nothing special.

Note that we can make A any point on the circumcircle of $\triangle ABC$ and $\sin A$ stays the same, by the Inscribed Angle Theorem. Thus we let A make $\triangle ABC$ right at $\angle B$. Then note that $\frac{a}{\sin A} = 2R$, as desired. (This is by the right triangle definition of sine.) $\square$

Granted, Law of Sines is most effective when either all three sides or all three angles are known, but it can also work with two sides and an angle given, if the angle is opposite one of the known sides. Otherwise, if we only know, say, $\angle A$ and $\angle B$, and side length c, it is impossible to use Law of Sines to determine all the angles and side lengths.

Note that, in particular, in a valid configuration, the (extended) Law of Sines can be used as a way to determine the circumradius of a triangle (with an alternate, more direct formula being $R = \dfrac{abc}{4[ABC]}$, where $[ABC]$ denotes the area of $\triangle ABC$).

Indeed, we can relate the formula for the circumradius back to the Law of Sines. Recall that the area of any triangle $\triangle ABC$ is given by $\dfrac{ab\sin\theta}{2}$, where θ is the angle between side lengths a and b. Then substitute $\sin\theta = \sin C = \dfrac{c}{2R}$ into the formula to obtain $[ABC] = \dfrac{abc}{4R}$, as desired. (We can also prove Heron's formula using this same general method.)

The Ambiguous Case

As aforementioned, the Law of Sines does have some limitations, depending on which side lengths and angles have been given. In particular, if we only know two of the side lengths a and c, then a triangle constructed from the acute angle A could either be acute or obtuse. More specifically, we require $a < c$ and $a > c\sin A = h$, where h is the length of the B-altitude. In this case, we may have an acute triangle with $C = \arcsin\left(\dfrac{c}{a}\sin A\right)$ and $C' = \pi - C$ for an obtuse triangle. Here, it is possible to compute B and b, but this is the case where we know both the angle A and the side length a. Without a (say, instead, side length b), it would be impossible to conclusively determine any more information about the triangle. Even in this case, the Law of Sines produces an ambiguous result.

8.1.1 Examples

Example 8.1.1.1. *Consider $\triangle ABC$ with $a = 4$, $b = 2\sqrt{6}$, and $c = 2\sqrt{3} + 2$. Find $\angle A, \angle B, \angle C$.*

Solution 8.1.1.1. *Notice that $a : b : c = \dfrac{\sqrt{2}}{2} : \dfrac{\sqrt{3}}{2} : \dfrac{\sqrt{2}+\sqrt{6}}{4} = \sin 45° : \sin 60° : \sin 75°$. Thus, by the Law of Sines, $\angle A = 45°$, $B = 60°$, and $C = 75°$.*

Example 8.1.1.2. *In the above problem, find the circumradius of the given triangle.*

Solution 8.1.1.2. *By the Extended Law of Sines, $R = \dfrac{a}{2\sin A} = \dfrac{4}{\sqrt{2}} = 2\sqrt{2}$.*

8.1.2 Exercises

Exercise 245. *If $\triangle ABC$ is a 13-14-15 triangle, compute $\sin A + \sin B + \sin C$.*

Exercise 246. *Using the Law of Sines, show that the area of $\triangle ABC$ is $\dfrac{abc}{4R}$, where a, b, and c are the side lengths and R is the circumradius.*

Exercise 247. *(2014 AMC 10A) In rectangle $ABCD$, $\overline{AB} = 20$ and $\overline{BC} = 10$. Let E be a point on $\overline{CD}$ such that $\angle CBE = 15°$. What is $\overline{AE}$?*

Exercise 248. *Prove that the area of $\triangle ABC$ is $\dfrac{ab\sin C}{2}$.*

Exercise 249. *Show that, in $\triangle ABC$, $\sin A$, $\sin B$, and $\sin C$ must obey the Triangle Inequality.*

Extension Exercise 104. *(2003 AIME I) Triangle ABC is isosceles with $AC = BC$ and $\angle ACB = 106°$. Point M is in the interior of the triangle so that $\angle MAC = 7°$ and $\angle MCA = 23°$. Find the number of degrees in $\angle CMB$.*

Extension Exercise 105. *For $\triangle ABC$, prove the following:*

(a) $a = b\cos C + c\cos B$

(b) $a(\sin B - \sin C) + b(\sin C - \sin A) + c(\sin A - \sin B) = 0$

Extension Exercise 106. *(2001 HMMT February Geometry Test) Equilateral triangle $\triangle ABC$ with side length 1 is drawn. A square is drawn such that its vertex at A is opposite to its vertex at the midpoint of $\overline{BC}$. Find the area enclosed within the intersection of the insides of the triangle and square.*

8.2 Law of Cosines

The *Law of Cosines* is similar to the Law of Sines in that it reveals information about a triangle, not given in a problem statement, that would otherwise be unknown. However, we *require* two known side lengths and a different angle, as opposed to in the Law of Sines, where this is actually the sole inconclusive scenario! In this way, think of the Law of Sines and Law of Cosines as polar opposites. If one cannot be used, then surely the other will work. Thus, there is no triangle for which neither rule will solve the problem.

Theorem 8.2: Law of Cosines (LoC)

In $\triangle ABC$,
$$a^2 + b^2 - 2ab\cos C = c^2.$$

Cyclic variants hold.

Corollary 8.2: Isosceles Case

WLOG assume $a = b$. Then we have $a^2 + b^2 = 2ab$, or $\cos C = \dfrac{2a^2 - c^2}{2a^2}$.

Proof. We'll just prove it for an altitude inside the triangle. (The proof for obtuse is nearly identical, and the proof for right is trivial.)

If the altitude is inside the triangle, then note that $BH = a\sin C$, because $\sin C = \dfrac{h}{a}$, and note that $AH = b - a\cos C$ because $b - a\cos C = b - a \cdot \dfrac{CH}{a} = b - CH$, and $b - CH = AH$. By the Pythagorean Theorem, we see that $c^2 = (a\sin C)^2 + (b - a\cos C)^2 = a^2\sin^2 C + b^2 - 2ab\cos C + a^2\cos^2 C$. Since $\sin^2 C + \cos^2 C = 1$, factoring gives us $c^2 = a^2(\sin^2 C + \cos^2 C) + b^2 - 2ab\cos C$. Substituting yields $c^2 = a^2 + b^2 - 2ab\cos C$, as desired. $\qquad\square$

It is actually possible to prove the law of cosines directly using the law of sines!

Proof by LoS. The angles in $\triangle ABC$ sum to $180° = \pi$ radians, so we have that $\dfrac{c}{\sin C} = \dfrac{b}{\sin B} = \dfrac{a}{\sin A}$, and $A + B + C = \pi$. From the last equation, we obtain the system of equations $\dfrac{c}{\sin C} = \dfrac{b}{\sin(A + B)}$, $\dfrac{c}{\sin C} = \dfrac{a}{\sin A}$.

From here, we apply the addition formula for sin:

$$\implies c(\sin A\cos C + \sin C\cos A) = b\sin C, \quad c\sin A = a\sin C$$

Dividing through by $\cos C$ yields

$$c(\sin A + \tan C\cos A) = b\tan C$$

$$\frac{c\sin A}{\cos C} = a\tan C$$

$$\frac{c^2\sin^2 A}{\cos^2 C} = a^2\tan^2 C$$

Hence, we obtain $\dfrac{c\sin A}{b - c\cos A} = \tan C$.

Subsequently substituting this into the second equation, and applying the identity $1 + \tan^2 C = \sec^2 C$, yields

$$c^2 \sin^2 A \left(1 + \frac{c^2 \sin^2 A}{(b - c \cos A)^2} \right) = a^2 \frac{c^2 \sin^2 A}{(b - c \cos A)^2}$$

Multiplying through by $(b - c \cos A)^2$ yields $(b - c \cos A)^2 + c^2 \sin^2 A = a^2$, or $b^2 - 2bc \cos A + c^2 \cos^2 A + c^2 \sin^2 A = a^2$. Finally, from the Pythagorean Theorem, the Law of Cosines follows. $\square$

Much like the Law of Sines, we can also apply the Law of Cosines to compute the angle measures if we know what all three side lengths are. Specifically, we have that $\angle C = \arccos \left(\dfrac{a^2 + b^2 - c^2}{2ab} \right)$ from algebraic manipulation.

8.2.1 Examples

Example 8.2.1.1.
(2014 ARML Individual #4) In triangle ABC, $a = 12$, $b = 17$, and $c = 13$. Compute $b \cos C - c \cos B$.

Solution 8.2.1.1. *Applying LoC on $\angle C$, $c^2 = a^2 + b^2 - 2ab \cos C$. Doing the same on $\angle B$ gives us $b^2 = a^2 + c^2 - 2ac \cos B$. Subtracting the first equation from the second gives us $b^2 - c^2 = (a^2 + c^2 - 2ac \cos B) - (a^2 + b^2 - 2ab \cos C) = c^2 - b^2 + 2ab \cos C - 2ac \cos B$. It follows that $b \cos C - c \cos B = \dfrac{2(b^2 - c^2)}{2a} = \dfrac{b^2 - c^2}{a} = \dfrac{17^2 - 13^2}{12} = \boxed{10}$.*

Example 8.2.1.2. *Triangle ABC is inscribed in equilateral triangle PQR, as shown. If $PC = 3$, $BP = CQ = 2$, and $\angle ACB = 90°$, then compute AR.*

Solution 8.2.1.2. *The side length of equilateral triangle PQR is $PQ = PC + CQ = 3 + 2 = 5$. Let $x = AR$. Then $AQ = QR - AR = 5 - x$. By the law of cosines on triangle BCP,*

$$\begin{aligned} BC^2 &= BP^2 + PC^2 - 2BP \cdot PC \cos \angle BPC \\ &= 2^2 + 3^2 - 2 \cdot 2 \cdot 3 \cos 60° \\ &= 7. \end{aligned}$$

By the law of cosines on triangle ACQ, we have

$$\begin{aligned} AC^2 &= AQ^2 + CQ^2 - 2AQ \cdot CQ \cos \angle AQC \\ &= (5 - x)^2 + 2^2 - 2 \cdot (5 - x) \cdot 2 \cos 60° \\ &= (5 - x)^2 - 2(5 - x) + 4 \qquad\qquad = x^2 - 8x + 19. \end{aligned}$$

By the law of cosines on triangle ABR, we have

$$\begin{aligned} AB^2 &= AR^2 + BR^2 - 2AR \cdot BR \cos \angle ARB \\ &= x^2 + 3^2 - 2x \cdot 3 \cos 60° \\ &= x^2 - 3x + 9. \end{aligned}$$

But by Pythagoras on right triangle ABC, we have $AC^2 + BC^2 = AB^2$, so $(x^2 - 8x + 19) + 7 = x^2 - 3x + 9$. Solving for x, we find $AR = x = \boxed{\dfrac{17}{5}}$.

8.2.2 Exercises

Exercise 250. *(2017 AMC 12B) Let ABC be an equilateral triangle. Extend side $\overline{AB}$ beyond B to a point B' so that $BB' = 3AB$. Similarly, extend side $\overline{BC}$ beyond C to a point C' so that $CC' = 3BC$, and extend side $\overline{CA}$ beyond A to a point A' so that $AA' = 3CA$. What is the ratio of the area of $\triangle A'B'C'$ to the area of $\triangle ABC$?*

Exercise 251. *A triangle has side lengths 4 and 8, and it has an area of $3\sqrt{15}$. Find the possible lengths of the third side.*

Exercise 252. *(2006 AMC 12B) Isosceles $\triangle ABC$ has a right angle at C. Point P is inside $\triangle ABC$, such that $PA = 11$, $PB = 7$, and $PC = 6$. Legs $\overline{AC}$ and $\overline{BC}$ have length $s = \sqrt{a + b\sqrt{2}}$, where a and b are positive integers. What is $a + b$?*

Exercise 253. *Derive the Triangle Inequality from the Law of Cosines. (Hint: Use the fact that $\cos$ is bounded.)*

Exercise 254. *$\triangle ABC$ has perimeter 19. We know that $\angle A = 60°$, and $BC = 9$. What is the area of $\triangle ABC$?*

Extension Exercise 107. *(2013 AIME I) A paper equilateral triangle ABC has side length 12. The paper triangle is folded so that vertex A touches a point on side $\overline{BC}$ a distance 9 from point B. The length of the line segment along which the triangle is folded can be written as $\frac{m\sqrt{p}}{n}$, where m, n, and p are positive integers, m and n are relatively prime, and p is not divisible by the square of any prime. Find $m + n + p$.*

Extension Exercise 108. *(2013 HMMT November Team Round) Consider triangle $\triangle ABC$ with side lengths $AB = 4$, $BC = 7$, and $AC = 8$. Let M be the midpoint of segment $\overline{AB}$, and let N be the point on the interior of segment $\overline{AC}$ that also lies on the circumcircle of triangle $\triangle MBC$. Compute BN.*

Extension Exercise 109. *(2018 AIME I) In $\triangle ABC$, $AB = AC = 10$ and $BC = 12$. Point D lies strictly between A and B on $\overline{AB}$ and point E lies strictly between A and C on $\overline{AC}$ so that $AD = DE = EC$. Then AD can be expressed in the form $\frac{p}{q}$, where p and q are relatively prime positive integers. Find $p + q$.*

Extension Exercise 110.

(a) *(2014 HMMT February Algebra Test) Given that a, b, and c are complex numbers satisfying*

$$a^2 + ab + b^2 = 1 + i$$

$$b^2 + bc + c^2 = -2$$

$$c^2 + ca + a^2 = 1$$

compute $(ab + bc + ca)^2$.

(b) *Can you generalize this to any system of the form*

$$a^2 + ab + b^2 = x$$

$$b^2 + bc + c^2 = y$$

$$c^2 + ca + a^2 = z$$

8.3   Strategies for Angle Chasing

For this section, we will not be focusing as much on the proofs as the technique: the craft of angle chasing so as to solve angle problems with the greatest efficiency and the least wasted time/effort/scratch paper space. In essence, angle chasing is exactly what it sounds like: the method of scouting angle measures based on what is already given in the problem/accompanying diagram. In a way, angle chasing is very similar to a Sudoku puzzle: the more information we reveal about the angles (often by turning to alternative, creative means such as drawing auxiliary lines and/or figures such as circumcircles that are not given), the more progress we can make from there. (Continuing the analogy of the Sudoku puzzle, the more numbers we fill in, the easier it becomes to fill in the rest of the numbers and solve the puzzle. Just like in Sudoku, we can make pencil marks - or assumptions about our angles - that lead us down a rabbit hole that eventually lends itself to *reductio ad absurdum*.)

In general, we can greatly facilitate our efforts in angle chasing by actively seeking patterns in our diagram (which we should draw neatly, cleanly, nice and large, with everything labeled, and as close to scale as possible to begin with). To be more specific, we make note of any right angles, triangle similarity (and, at that, what *type* of triangle similarity), altitudes, triangle centers, and points of tangency. It might also considerably help to extend lines past points and draw auxiliary lines from other points to meet the extended lines at new points. This may make a diagram more conductive to similarity or tangency applications, thereby paving the way for angle chasing.

8.3.1   Helpful Theorems

In any triangle $\triangle ABC$, if we extend $\overline{BC}$ past C to D, then $\angle ACD = \angle ABC + \angle BAC$. This turns out to be equivalent to the Exterior Angle Theorem! As a refresher in case you are unfamiliar, here is the theorem:

Theorem 8.3: Exterior Angle Theorem

In any triangle $\triangle ABC$, the measure exterior angle of a triangle is greater than that of either of the two interior angles of the triangle that are not adjacent to the exterior angle.

Proof. We know that the sum of the angles in any triangle is $180°$. Let the interior angle adjacent to the exterior angle have measure $a°$; then the exterior angle has measure $(180 - a)°$, which is also the sum of the two other interior angles. But then this forces both interior angles to be strictly less than $(180 - a)°$, as otherwise, the triangle would be degenerate. Hence the theorem is proved. $\square$

Remark. This theorem is *invalid* in spherical geometry, where the angles in a triangle may sum to more than $180°$.

Equally helpful (and arguably the single most common construction in all of angle chasing), are the theorems pertaining to subtended arcs, in which we use the circumcircle of a triangle to derive some of its angles. The inscribed angle theorem (which states that an inscribed angle is half the central angle subtending the arc), in particular, has the potential to help tremendously with angle chasing problems that involve a circumscribing circle.

Symmetry is also a majorly helpful property to exploit in an angle chasing problem! If you find that a figure is symmetric about some point, it may further benefit to reflect about that point. In general, reflection can lead to some major breakthroughs in that we can reflect about an angle bisector, median, or altitude (among several other lines) to obtain other angle measures. (Perhaps the possible candidates for the reflected points could be our Sudoku pencil marks.)

8.3.2 Examples

The meat of this section lies in the examples, as angle chasing is a honed skill and not a memorized one. With this skill in particular, practice makes better!

Example 8.3.2.1. *(2011 AMC 10B) Rectangle $ABCD$ has $AB = 6$ and $BC = 3$. Point M is chosen on side AB so that $\angle AMD = \angle CMD$. What is the degree measure of $\angle AMD$?*

Solution 8.3.2.1. *We are given that $\angle AMD \cong \angle CMD$. Since $\angle AMD, \angle CDM$ are alternate interior angles, and $\overline{AB} \parallel \overline{DC}$, it follows that $\angle AMD \cong \angle CDM \implies \angle CMD \cong \angle CDM$. $\overline{DC} \cong \overline{MC}$, so $CM = 6$. Furthermore, $\triangle BMC$ is a 30-60-90 triangle with $\angle BMC = 30°$. Thus, $\angle AMD = \boxed{75°}$.*

Example 8.3.2.2. *(2006 ARML) If $ABCDE$ is a regular pentagon and $MNCD$ is a square (with M and N inside $ABCDE$), compute the value of $m\angle AMN - m\angle EAM$ in degrees.*

Solution 8.3.2.2. *Let $\angle AMN = x, \angle EAM = y$. If we draw a line $\overline{PA} \parallel \overline{MN}$, then it follows from observing that $\angle PAM = \angle AMN = x$ that $\angle PAE = x - y$. Where Q is the point past A on $\overline{PA}$ such that $PA = AQ$, we have $\angle QAB = x - y$, or $180 = 2(x - y) + 108 \implies x - y = \boxed{36°}$.*

Example 8.3.2.3. *(2017 CMIMC Geometry Test) Let $\triangle ABC$ be a triangle with $\angle BAC = 117°$. The angle bisector of $\angle ABC$ intersects side $\overline{AC}$ at D. Suppose $\triangle ABD \sim \triangle ACB$. Compute the measure of $\angle ABC$, in degrees.*

Solution 8.3.2.3. *By AA similarity of triangles $\triangle ABD$ and $\triangle ACB$, we have $\angle ABD = \angle ACB$, and so $\angle ABC = 2\angle ACB$. Then $\angle ACB + 2\angle ACB + 117° = 180° \implies \angle ACB = 21°$, i.e. $\angle ABC = \boxed{42°}$.*

Example 8.3.2.4. *Triangle $\triangle ABC$ has $\angle A = 60°$, $\angle B = 90°$, $\angle C = 30°$. The internal angle bisectors of $\angle A$ and $\angle B$ intersect at point D. What is the measure of angle $\angle ADC$?*

Solution 8.3.2.4. *We have $\angle ABD = 45°, \angle BAD = 30°$, so $\angle ABD = 105°$. Furthermore, $\angle BCD = 15°$, and $\angle CAD = 30°$; then if we call the intersection of the bisector of $\angle B$ with $\overline{AC}$ point E, $\angle AED = 75°$. As a result, $\angle CED = 105°$, meaning that $\angle CPE = 60°$ and $\angle APC = \boxed{135°}$.*

8.3.3 Exercises

Exercise 255. *(2017 CMIMC Geometry Test) Cyclic quadrilateral ABCD satisfies $\angle ABD = 70°$, $\angle ADB = 50°$, and $BC = CD$. Suppose $\overline{AB}$ intersects $\overline{CD}$ at point P, while $\overline{AD}$ intersects $\overline{BC}$ at point Q. Compute $\angle APQ - \angle AQP$.*

Exercise 256. *(2008 AIME I) Let ABCD be an isosceles trapezoid with $\overline{AD} \parallel \overline{BC}$ whose angle at the longer base $\overline{AD}$ is $\frac{\pi}{3}$. The diagonals have length $10\sqrt{21}$, and point E is at distances $10\sqrt{7}$ and $30\sqrt{7}$ from vertices A and D, respectively. Let F be the foot of the altitude from C to $\overline{AD}$. The distance EF can be expressed in the form $m\sqrt{n}$, where m and n are positive integers and n is not divisible by the square of any prime. Find $m + n$.*

Exercise 257. *Let $\triangle ABC$ be an isosceles right triangle with $\angle B = 90°$. Points D and E lie in the plane of $\triangle ABC$ such that ACDE is a square that does not overlap with $\triangle ABC$. Let O be the center of ACDE. Compute $\angle OBC$ in degrees.*

Exercise 258. *(2018 AMC 12A) In $\triangle PAT$, $\angle P = 36°$, $\angle A = 56°$, and $PA = 10$. Points U and G lie on sides $\overline{TP}$ and $\overline{TA}$, respectively, so that $PU = AG = 1$. Let M and N be the midpoints of segments $\overline{PA}$ and $\overline{UG}$, respectively. What is the degree measure of the acute angle formed by lines MN and PA?*

Extension Exercise 111. *(2013 HMMT February Guts Round) Consider triangle $\triangle ABC$ with $\angle A = 2\angle B$. The angle bisectors from A and C intersect at D, and the angle bisector from C intersects $\overline{AB}$ at E. If $\frac{DE}{DC} = \frac{1}{3}$, compute $\frac{AB}{AC}$.*

Extension Exercise 112. *(2011 USAJMO P5) Points A, B, C, D, E lie on a circle ω and point P lies outside the circle. The given points are such that*

  (i) lines PB and PD are tangent to ω,

  (ii) P, A, C are collinear, and

  (iii) $\overline{DE} \parallel \overline{AC}$.

Prove that $\overline{BE}$ bisects $\overline{AC}$.

8.4 Proving Triangle Theorems: Part 1

8.4.1 Thales' Theorem

Much of geometry revolves around inscribing and circumscribing figures about others - it is this type of construction that leads to some of the most special, grandiose, and meaningful mathematical art forms (figuratively and literally). A special case of the inscribed angle theorem, *Thales' Theorem* can be profoundly useful in working with points on a circle, especially when concerning the circumcircle of a triangle (in particular, a right triangle):

> **Theorem 8.4: Thales' Theorem**
>
> Let A, B, and C be distinct points lying on a circle with center O. If line $\overline{AC}$ is a diameter of O, then $\angle ABC = 90°$.

Proof. We can take (at least) three different approaches, the first of which is to observe that $\overline{OA}$, $\overline{OB}$, and $\overline{OC}$ are all radii of the circle, and subsequently $\triangle OBA$ and $\triangle OBC$ are isosceles. In turn, $\angle OBC = \angle OCB$ and $\angle OAB = \angle OBA$. Note that the internal angles of $\angle ABC$ are $\angle OAB$, $\angle OBC$, and $\angle OAB + \angle OBC$. Since the sum of the interior angles in a triangle is $180°$, it follows that $\angle OAB + \angle OBC = 90° \implies \angle B = 90°$, as desired.

Alternatively, we can let D be the point diametrically opposed from B. Since $\overline{AB}$ and $\overline{CD}$ are parallel, as are $\overline{AD}$ and $\overline{BC}$, $ABCD$ is a parallelogram. In addition, since $\overline{AC}$ and $\overline{BD}$ have the same length, and are both diameters of the circle, $ABCD$ is a rectangle, which has all right angles.

Finally, we can use trigonometry to prove the theorem. WLOG let $O = (0,0)$, and let the radius of the circle be 1. Then $A = (-1, 0)$ and $C = (1, 0)$, with $B = (\cos\theta, \sin\theta)$ for some $0 \leq \theta < 2\pi$. We prove that $\overline{AB}$ and $\overline{BC}$ are perpendicular; i.e. the product of their slopes is equal to -1. With an application of the identity $\sin^2\theta + \cos^2\theta = 1$, this is straightforward algebra (details left to the reader). $\qquad\square$

We can show that the converse also holds:

> **Theorem 8.5: Converse of Thales' Theorem**
>
> The circumcenter of a right triangle is the midpoint of its hypotenuse. (Equivalently, the hypotenuse of a right triangle is the diameter of its circumcircle.)

Proof. WLOG let $\angle B = 90°$. Then let l_1 be a line passing through A and parallel to $\overline{BC}$, and let l_2 be a line passing through C and parallel to $\overline{AB}$. Let l_1 and l_2 intersect at P. By construction, $ABCD$ is a parallelogram and its adjacent angles are supplementary. Because $\angle B = 90°$, all the other angles must be right as well, proving that $ABCD$ is a rectangle, and that the intersection of its diagonals is equidistant from A, B, and C - and hence is the circumcenter, with $\overline{AC}$ as the diameter of the circumcircle. $\qquad\square$

This is in fact a very useful result in not only proof-based geometry, but in computational problems as well. We can place some arbitrary points on a circle, determine the possible choices of points for which the resulting triangle is right, and then construct the circum-diameter.

8.4.2 Ceva's Theorem

In our study of notions such as the incircle and circumcircle, we often invoke the idea of concurrence inside the triangle. That is, a great part of the foundations we have established for these important geometrical constructs in fact relies on line segments in a triangle all passing through a common point (this common point, in special cases, being the incenter, cirucmcenter, centroid, etc.) *Ceva's Theorem* tells us a crucial fact regarding this point of concurrence and relates it to the distances between points lying on the triangle and its vertices. Equipped with this theorem, along with Menelaus' Theorem (see the immediately following section), we can firmly cement yet another perfectly beautiful property of the almighty triangle.

As a refresher, we can define a *cevian* of triangle $\triangle ABC$ as follows:[73]

Definition 8.1: Cevian of a Triangle

A *cevian* of a triangle is a line segment from a vertex of the triangle to the opposite side length of the triangle.

Note that medians, angle bisectors, and altitudes are all special types of cevians.

We can now state and prove Ceva's Theorem:

Theorem 8.6: Ceva's Theorem

In $\triangle ABC$, let O be a point lying strictly in the interior of $\triangle ABC$. Then extend $\overline{AO}, \overline{BO}, \overline{CO}$ to meet $\overline{BC}, \overline{AC}$, and $\overline{AB}$ at D, E, and F respectively. Then $\dfrac{AF}{FB} \cdot \dfrac{BD}{DC} \cdot \dfrac{CE}{EA} = 1$ (using signed lengths).[a]

---

[a]Length can also be negative under this definition; the length $\overline{AB}$ is said to be positive if B is to the right of A, and vice versa. In this way, order matters!

Stated another way, the product is 1 iff the line segments $\overline{AD}, \overline{BE}$, and $\overline{CF}$ are concurrent at a point O in the triangle (or all three are parallel). (This is the converse of the theorem.)

Proof. Suppose that $\overline{AD}, \overline{BE}, \overline{CF}$ do concur at $O \in \triangle ABC$. Then we notice that $\triangle ABD, \triangle ADC$ have the same altitude to $\overline{BC}$, and respective bases $\overline{BD}$ and $\overline{DC}$. We then have that $\dfrac{BD}{DC} = \dfrac{[ABD]}{[ADC]}$. The same follows for $\triangle OBD, \triangle ODC$, such that

$$\frac{BD}{DC} = \frac{[ABD]}{[ADC]} = \frac{[OBD]}{[ODC]} = \frac{[ABD] - [OBD]}{[ADC] - [ODC]} = \frac{[ABO]}{[AOC]}$$

Similarly, we have $\dfrac{CE}{EA} = \dfrac{[BCO]}{[BOA]}$ and $\dfrac{AF}{FB} = \dfrac{[CAO]}{[COB]}$, so it follows that $\dfrac{AF}{FB} \cdot \dfrac{BD}{DC} \cdot \dfrac{CE}{EA} = 1$.

To conclude, suppose D, E, F satisfy Ceva's Theorem, and suppose $\overline{AD}$ and $\overline{BE}$ intersect at O. Further suppose that $\overline{CO}$ intersects $\overline{AB}$ at P. P must satisfy the criterion, so $\dfrac{AP}{PB} = \dfrac{AF}{FB}$, which implies $P = F$ and so that $\overline{CF}$ concurs with $\overline{AD}$ and $\overline{BC}$. $\qquad \square$

Remark. We have placed Routh's Theorem after this section, since the proof of Ceva's Theorem using Routh's Theorem is immediate and likely of little interest to the reader (simply use the definitions of the notation).

8.4.3 Menelaus' Theorem

In a similar vein as Ceva's Theorem, *Menelaus' Theorem* posits that the product of the same lengths from Ceva's Theorem is -1, rather than $+1$ as in that theorem, only with a different construction:

Theorem 8.7: Menelaus' Theorem

Let $\triangle ABC$ has transversal line l passing through $\overline{BC}, \overline{AC}$, and $\overline{AB}$ respectively at points D, E, F

---

[73]Indeed, we call these *cevians* in Giovanni Ceva's honor!

distinct from A, B, C, we have $\dfrac{AF}{FB} \cdot \dfrac{BD}{DC} \cdot \dfrac{CE}{EA} = -1$ (again using signed lengths).

The converse is also true: if the product is -1, then D, E, F are collinear.

Proof. First note (as a point of justification) that $\dfrac{AF}{FB} \cdot \dfrac{BD}{DC} \cdot \dfrac{CE}{EA} < 0$, since either all three fractions are negative (the case where the transversal line l completely misses $\triangle ABC$), or one is negative and the others are positive (the case where l hits two of the sides of $\triangle ABC$).

Construct the perpendiculars from A, B, C to the line DEF with lengths a, b, c respectively. Then using similar triangles, we have that $\left|\dfrac{AF}{FB}\right| = \left|\dfrac{a}{b}\right|$, $\left|\dfrac{BD}{DC}\right| = \left|\dfrac{b}{c}\right|$, and $\left|\dfrac{CE}{EA}\right| = \left|\dfrac{c}{a}\right|$. So then it follows that the product is -1 (the vertical bars denote magnitude - i.e. *signed* length - not absolute value). $\square$

We can check the converse of Menelaus' Theorem as well, using similar logic as we used to verify Ceva's Thoerem.

8.4.4 Routh's Theorem

A somewhat lesser-known, yet just as helpful (albeit hard to memorize) theorem is *Routh's Theorem*, which generalizes Ceva's and Menelaus' Theorems:

Theorem 8.8: Routh's Theorem

In $\triangle ABC$, if points D, E, F lie on $\overline{BC}$, $\overline{AC}$, and $\overline{AB}$ respectively, and we have $\dfrac{CD}{BD} = x$, $\dfrac{AE}{CE} = y$, and $\dfrac{BF}{AF} = z$, then the cevians $\overline{AD}, \overline{BE}, \overline{CF}$ form a triangle with area of $\dfrac{(xyz - 1)^2}{(xy + y + 1)(yz + z + 1)(zx + x + 1)} \cdot [ABC]$.

Corollary 8.8: Special Cases

For $x = y = z = 1$, the medians concur, yielding Ceva's Theorem. For $x = y = z = 2$, this is the "one-seventh triangle."

For the sake of convenience in the proof, we let P denote the intersection of $\overline{AD}$ and $\overline{BC}$, Q denote the intersection of $\overline{BE}$ and $\overline{CF}$, and R denote the intersection of $\overline{CF}$ and $\overline{AD}$.

Proof. WLOG assume that $[ABC] = 1$. Then we can apply Menelaus' Theorem on $\triangle ABD$ and line FRC (with F, H, C collinear). We have that $\dfrac{AF}{FB} \cdot \dfrac{BC}{CD} \cdot \dfrac{DR}{RA} = 1$, so $\dfrac{DR}{RA} = \dfrac{BF}{FA} \cdot \dfrac{DC}{CB} = \dfrac{zx}{z + 1}$. This makes $[ARC] = \dfrac{AR}{AD} \cdot [ADC] = \dfrac{AR}{AD} \cdot \dfrac{DC}{BC} \cdot [ABC] = \dfrac{x}{zx + x + 1}$. We can repeat this process for $[BPA]$ and $[CQB]$, and combine this with the observation that $[PQR] = [ABC] - [ARC] - [BPA] - [CQB]$ to obtain $[PQR] = 1 - \dfrac{x}{zx + x + 1} - \dfrac{y}{xy + y + 1} - \dfrac{z}{yz + z + 1} = \dfrac{(xyz - 1)^2}{(xz + x + 1)(yx + y + 1)(zy + z + 1)}$, as desired. $\square$

8.4.5 Examples

Example 8.4.5.1. *The area of a circle with center O is 100π. $\overline{AB}$ is a diameter of the circle. C is a point on the circle such that $AC = BC + 4$. Compute $AC + BC$.*

Solution 8.4.5.1. *By Thales' Theorem, $\overline{AB}$ is the hypotenuse of right $\triangle ACB$, with right angle at $\angle C$. Hence, by the Pythagorean Theorem, $AC^2 + BC^2 = AB^2 \implies (BC + 4)^2 + BC^2 = 400 \implies BC = 12, AC = 16 \implies \boxed{28}$.*

Example 8.4.5.2. *Let $\triangle ABC$ have $AB = 13$, $AC = 14$, $BC = 15$. If $\frac{AD}{DB} = \frac{8}{5}$ and $\frac{CE}{EA} = \frac{1}{6}$, F is the intersection point of $\overline{BE}$ and $\overline{CD}$, and G is the intersection point of $\overline{BC}$ and $\overline{AF}$, compute $BG \cdot GC$.*

Solution 8.4.5.2. *This is an application of Ceva's Theorem; the product of the ratios of the segments around the triangle (keeping direction consistent) should be 1, so our answer is $\boxed{\dfrac{15}{4}}$.*

8.4.6 Exercises

Exercise 259. *Prove that every triangle has an orthocenter (which need not lie within the interior of the triangle).*

Exercise 260. *Suppose $\triangle ABC$ has area 2019. Let D, E, and F be points on $\overline{AB}$, $\overline{BC}$, and $\overline{CA}$, respectively, such that $AD = 3BD$, $BE = 3CE$, and $CF = 3AF$. Compute the area of $\triangle DEF$.*

Exercise 261. *Show that Routh's Theorem implies that the medians of a triangle are concurrent.*

Exercise 262. *Triangles $\triangle ABC$ and $\triangle DEF$ are similar (with AB corresponding to $\overline{DE}$, $\overline{BC}$ corresponding to $\overline{EF}$, and $\overline{CA}$ corresponding to $\overline{FA}$). Show that the ratio of correspondent medians is the same as the ratio of correspondent sides.*

Exercise 263. *(2011 Stanford Math Tournament) Let $\triangle ABC$ be any triangle, and D, E, F be points on $\overline{BC}$, $\overline{CA}$, $\overline{AB}$ such that $CD = 2BD$, $AE = 2CE$ and $BF = 2AF$. $\overline{AD}$ and $\overline{BE}$ intersect at X, $\overline{BE}$ and $\overline{CF}$ intersect at Y, and $\overline{CF}$ and $\overline{AD}$ intersect at Z. Find $\dfrac{Area(\triangle ABC)}{Area(\triangle XYZ)}$.*

Exercise 264. *Prove that the external angle bisectors of a triangle intersect their opposite sides at three collinear points.*

Extension Exercise 113. *(1996 USAMO P5) Let ABC be a triangle, and M an interior point such that $\angle MAB = 10°$, $\angle MBA = 20°$, $\angle MAC = 40°$ and $\angle MCA = 30°$. Prove that the triangle is isosceles.*

8.5 Power of a Point

Among a veritable sea of theorems relating to circles, one in particular stands out as well-known, versatile, and bottomlessly deep all at once: the *power of a point theorem*. The theorem statement is actually three-fold, in that there are three different possible cases to consider. In each of these cases, there is a setup that differs ever-so-slightly from the others, yet has its own interesting nuances that set it apart.

As a pre-requisite to understanding the theorem, we will define the *power* of a point in the plane of a circle:

Definition 8.2: Power with Respect to a Circle

The *power* of a point P with respect to a circle with center O and radius r is equal to $d^2 - r^2$, where $d = OP$. (Note that $d^2 - r^2$ can be negative, in which case P lies strictly inside the circle.)

All points with positive power lie strictly outside the circle, all points with negative power lie strictly inside the circle, and all points with zero power lie strictly on the boundary of the circle. (Equivalently, for P outside the circle, we can define the power of P as the product of the distances from P to the intersection points of a ray from P with the circle. If the intersection point is at a point of tangency, this distance is then squared. This is in fact the statement of the pivotal Power of a Point Theorem, as we shall soon see.)

Rather than using the traditional theorem-proof format, we will present each of the three cases of the Power of a Point theorem as individual theorems, and prove each in its respective section.

8.5.1 Case 1: The Intersecting Chords Theorem

If two chords of a circle intersect at a point in the circle, we can apply the first variation of Power of a Point to determine the lengths of the chords, as well as the lengths of the line segments from the circle to the point of intersection, as in the following diagram:

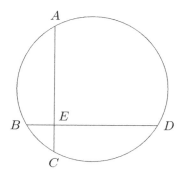

Figure 8.1: Case 1 of the Power of a Point Theorem.

Theorem 8.9: Power of a Point Case 1: Intersecting Chords

In the diagram above, we have $AE \cdot CE = BE \cdot DE$.

Proof. This case of the theorem is a consequence of triangle similarity (namely angle-angle similarity). We have that $\angle ADE = \angle BCE$, as inscribed angles both subtended by $\overline{AB}$; $\angle DAE = \angle CBE$, as inscribed angles both subtended by $\overline{CD}$; and $\angle AED = \angle BEC$ as vertical angles.

Furthermore, triangles $\triangle AED$ and $\triangle BEC$ are similar, such that we obtain $\dfrac{AE}{ED} = \dfrac{BE}{EC} \implies AE \cdot CE = BE \cdot DE$. $\qquad \square$

8.5.2 Case 2: Tangent-Secant Theorem

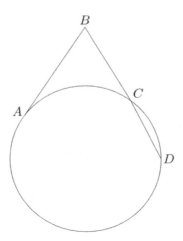

Figure 8.2: Case 2 of the Power of a Point Theorem.

Theorem 8.10: Power of a Point Case 2: Tangent-Secant

In the diagram above, $AB^2 = BC \cdot BD$.

Proof. Notice that $\triangle BAD$ is similar to $\triangle BCA$ (not to scale in the figure). Thus, $\dfrac{BA}{BD} = \dfrac{BC}{BA}$ which yields the desired conclusion. $\square$

8.5.3 Case 3: Secant-Secant Theorem

If two secants of a circle intersect at a point outside the circle, we can apply the third variation of Power of a Point to determine the distances between the secants' intersection point and each point of intersection of a secant and the circle, as in the following diagram:

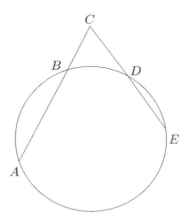

Figure 8.3: Case 3 of the Power of a Point Theorem.

Theorem 8.11: Power of a Point Case 3: Secant-Secant

In the above diagram, $CA \cdot CB = CD \cdot CE$.

Proof. This follows from the similarity of triangles $\triangle CBE$ and $\triangle CDA$ (diagram not to scale). The two triangles share angles $\angle ACE$ and $\angle BAD \equiv \angle BED$ (as both are inscribed angles over $\overline{BD}$. Hence, it follows that $\dfrac{CB}{CE} = \dfrac{CD}{CA} \implies CB \cdot CA = CE \cdot CD$ as desired. $\qquad\square$

8.5.4 The General Formulation

We can actually extend Power of a Point such that it makes a considerably stronger statement regarding a choice of point P outside a circle O, and any line passing through P:

Theorem 8.12: Generalized Power of a Point Theorem

Consider circle O, and a point P outside O. Then we can draw a line through P that intersects O at two points. Then the product of the distances from P to the first point of intersection, and from P to the second point of intersection, is constant regardless of the choice of line through P.

The following is a significant result that we leave to you to engage with in the process of self-discovery:

Corollary 8.12: Equal Tangent Lengths

Both tangents from P to the circle O are equal in length.

8.5.5 Examples

Example 8.5.5.1. *(ARML) In a circle, chords $\overline{AB}$ and $\overline{CD}$ intersect at R. If $AR : BR = 1 : 4$ and $CR : DR = 4 : 9$, find the ratio $AB : CD$.*

Solution 8.5.5.1. *Power of a Point: in this form, it is just the Intersecting Chords Theorem. Here, we have $AR \cdot BR = CR \cdot DR$. Let $AR = x$, $CR = y$; then $BR = 4x$ and $DR = 9y$. Hence, $4x^2 = 36y^2 \implies x^2 = 9y^2$. Then $\dfrac{x}{y} = 3$ and $\dfrac{AB}{CD} = \dfrac{5x}{13y} = \boxed{\dfrac{15}{13}}$.*

Example 8.5.5.2. *Let A and B be the points of tangency to a circle from P. Draw a line through P that intersects the circle at C and D. Prove that $AC \cdot BD = BC \cdot AD$.*

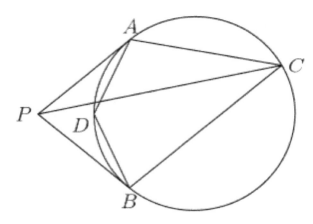

Solution 8.5.5.2. *We have* $\angle ACP \cong \angle DAP$ *because these angles are both inscribed in arc* AD. *Since we also have* $\angle APC \cong \angle DPA$, $\triangle PAC \sim \triangle PDA$. *Thus,* $\dfrac{AC}{AD} = \dfrac{PC}{PA}$.

Similarly, $\triangle PBC \sim \triangle PDB$, *so* $\dfrac{CB}{BD} = \dfrac{PC}{PB}$. *Since* $\overline{PA}$ *and* $\overline{PB}$ *are tangents from the same point* P *to the same circle,* $PA = PB$. *Therefore,* $\dfrac{AC}{AD} = \dfrac{CB}{BD}$; *it follows that* $AC \cdot BD = BC \cdot AD$, *as desired.*

8.5.6 Exercises

Exercise 265. *(1971 CMO P1) DEB is a chord of a circle such that $DE = 3$ and $EB = 5$. Let O be the center of the circle. Join $\overline{OE}$ and extend $\overline{OE}$ to cut the circle at C. Given $EC = 1$, find the radius of the circle.*

Exercise 266. *(1995 AHSME) In the figure, $\overline{AB}$ and $\overline{CD}$ are diameters of the circle with center O, $\overline{AB} \perp \overline{CD}$, and chord $\overline{DF}$ intersects $\overline{AB}$ at E. If $DE = 6$ and $EF = 2$, then the area of the circle is*

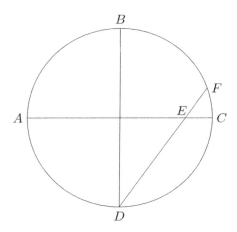

Figure 8.4: Diagram for this Exercise.

Exercise 267. *(2017 AMC 12A) Quadrilateral $ABCD$ is inscribed in circle O and has side lengths $AB = 3, BC = 2, CD = 6$, and $DA = 8$. Let X and Y be points on $\overline{BD}$ such that $\frac{DX}{BD} = \frac{1}{4}$ and $\frac{BY}{BD} = \frac{11}{36}$. Let E be the intersection of line AX and the line through Y parallel to $\overline{AD}$. Let F be the intersection of line CX and the line through E parallel to $\overline{AC}$. Let G be the point on circle O other than C that lies on line CX. What is $XF \cdot XG$?*

Exercise 268. *(1992 AHSME) A circle of radius r has chords $\overline{AB}$ of length 10 and $\overline{CD}$ of length 7. When $\overline{AB}$ and $\overline{CD}$ are extended through B and C, respectively, they intersect at P, which is outside of the circle. If $\angle APD = 60°$ and $BP = 8$, then compute r^2.*

Extension Exercise 114. *(2016 AIME I) Circles ω_1 and ω_2 intersect at points X and Y. Line ℓ is tangent to ω_1 and ω_2 at A and B, respectively, with line AB closer to point X than to Y. Circle ω passes through A and B intersecting ω_1 again at $D \neq A$ and intersecting ω_2 again at $C \neq B$. The three points C, Y, D are collinear, $XC = 67$, $XY = 47$, and $XD = 37$. Find AB^2.*

Extension Exercise 115. *(1998 USAMO P2) Let $\mathcal{C}_1$ and $\mathcal{C}_2$ be concentric circles, with $\mathcal{C}_2$ in the interior of $\mathcal{C}_1$. From a point A on $\mathcal{C}_1$ one draws the tangent AB to $\mathcal{C}_2$ ($B \in \mathcal{C}_2$). Let C be the second point of intersection of AB and $\mathcal{C}_1$, and let D be the midpoint of AB. A line passing through A intersects $\mathcal{C}_2$ at E and F in such a way that the perpendicular bisectors of DE and CF intersect at a point M on AB. Find, with proof, the ratio AM/MC.*

8.6 Cyclic Quadrilaterals

Good news: The definition for this section can be summed up in a single sentence! A *cyclic quadrilateral* is a quadrilateral that can be inscribed in a circle. Even better news: this seemingly very simple construct, below its surface, has a veritable wealth of uniquely defining properties that are nice to prove and even nicer to work with computationally!

Essentially, beyond the very basic, intuitive definition, a cyclic quadrilateral is such that every vertex of the quadrilateral lies on the circumscribing circle. The circle's center, just as with a triangle, is the circumcenter of the quadrilateral, and its radius is the circumradius of the quadrilateral. (Usually, we assume that a cyclic quadrilateral is convex; i.e. all of its angles have measure less than 180°.) We also have that the sides of the quadrilateral are chords of its circumcircle, by definition.

8.6.1 Properties of Cyclic Quadrilaterals

Though we have just listed a few properties, these are mostly accessible from either the definition or from taking a look at a few examples of cyclic quadrilaterals. Here, we will be examining the more nitty-gritty aspects of cyclic quadrilaterals in terms of rigorous geometry.

Clearly, the sum of the angles in a cyclic quadrilateral is 360°, as this is true for *any* quadrilateral. However, what distinguishes a cyclic quadrilateral from all other quadrilaterals with respect to angle measures is that both its opposite angles must be supplementary. (This is, in fact, Proposition 22 in Book 3 of *Elements* by Euclid.) Equivalently, we can say that $ABCD$ is cyclic iff each exterior angle is congruent to the opposite interior angle.

In fact, we have a trigonometric corollary from this that is somewhat vestigial (since it is harder to remember by comparison), but can nevertheless by useful when reverse-engineering a trig problem:

Corollary 8.12: Trigonometric Cyclic Condition

Quadrilateral $ABCD$ is cyclic iff $\tan\frac{\alpha}{2}\tan\frac{\gamma}{2} = \tan\frac{\beta}{2}\tan\frac{\delta}{2}$, where $\alpha, \beta, \gamma, \delta$ denote the angle measures opposite vertices A, B, C, D respectively.

This follows as a result of the tangent multiplication identity.

A necessary and sufficient condition for quadrilateral $ABCD$ to be cyclic is that the angle between a side and a diagonal must be congruent to that between the opposite side and the other diagonal. Id est, $\angle ACB = \angle ADB$. This follows as a consequence of the previous property.

8.6.2 Brahmagupta's Formula

Brahmagupta's formula is an analogue of Heron's Formula for *cyclic* quadrilaterals. Note that this formula will not work for non-cyclic quadrilaterals; this is due to the

Theorem 8.13: Brahmagupta's Formula

For a cyclic quadrilateral $ABCD$ with side lengths a, b, c, d and semi-perimeter $s = \dfrac{a+b+c+d}{2}$,

$$[ABCD] = \sqrt{(s-a)(s-b)(s-c)(s-d)}$$

Notice how strikingly similar this looks to Heron's Formula! We can justify this remarkable resemblance by observing that a triangle is really just a degenerate quadrilateral; that is, since it has three side lengths, it is a quadrilateral with one side length of zero. From this point of view, as d approaches zero, s approaches $\frac{a+b+c}{2}$, and the area $[ABCD]$ approaches $\sqrt{s(s-a)(s-b)(s-c)}$ as all triangles are cyclic (they can be inscribed in a circumcircle, regardless of their side lengths). For $d = 0$, the two formulae are one and the same. (We can prove this using similar triangles, and subsequently by applying Heron's Formula twice.)

As such, the area of a cyclic quadrilateral is maximal, amongst all quadrilaterals with side lengths a, b, c, d. Proving this relies on the fact that the formula can be generalized to non-cyclic quadrilaterals as well, with the area formula $[ABCD] = \sqrt{(s-a)(s-b)(s-c)(s-d) - \frac{1}{2}abcd(1 + \cos(\alpha + \gamma))}$. (The proof for this generalized formula - known as *Bretschneider's formula* - is quite similar to the proof of Brahmagupta's formula, and so its derivation is left to the reader.)

8.6.3   Ptolemy's Theorem

Ptolemy's Theorem, named in honor of astronomer Ptolemy,[74] Interestingly, Ptolemy originally used this theorem not for purely geometrical purposes, but rather as a means of aiding him in constructing the *table of chords:* a table of values of the sin function that included values in increments as small as $\frac{1}{2}$ of a degree![75]

Theorem 8.14: Ptolemy's Theorem

In cyclic quadrilateral $ABCD$ with vertices in that order, $AC \cdot BD = AB \cdot CD + BC \cdot AD$.

That is, the product of the diagonals is the sum of the pairs of opposite edge lengths.

Proof. Note that, on the chord BC, we have $\angle BAC = \angle BDC$, as both are inscribed angles. Furthermore, on chord AB, we have $\angle ACB = \angle ADB$ by the same logic. Let E be a point on AC such that $\angle ABE = \angle CBD$; this implies $\angle CBE = \angle ABD$.

We have, then, that $\triangle ABE$ and $\triangle DBC$ are similar, and so $\triangle ABD$ is also similar with $\triangle EBC$. It follows that $\frac{AE}{AB} = \frac{CD}{BD}$, and $\frac{CE}{BC} = \frac{DA}{BD}$; i.e. $AE \cdot BD = AB \cdot CD$ and $CE \cdot BD = BC \cdot AD$. Summing these two equations yields $AE \cdot BD + CE \cdot BD = AB \cdot CD + BC \cdot AD$; factoring yields $BD(AE + CE) = AB \cdot CD + BC \cdot AD$. Finally, we have that $AE + EC = AC$, so $AC \cdot BD = AB \cdot CD + BC \cdot AD$, as claimed. $\square$

Remark. Ptolemy's Theorem does not hold for non-cyclic quadrilaterals; instead, Ptolemy's Inequality, which states that $AB \cdot CD + BC \cdot DA \geq AC \cdot BD$ (with equality iff $ABCD$ is indeed cyclic), holds.

8.6.4   The Butterfly Theorem

The *butterfly theorem* - so named for how it generates the likeness of a butterfly - is not so much related to cyclic quadrilaterals as it is to inscribed quadrilaterals; however, one of the proofs does involve a cyclic quadrilateral construction, so we feel it is appropriate to discuss the theorem here.

Theorem 8.15: Butterfly Theorem

Let $\overline{PQ}$ be a chord of circle O, and let M be its midpoint. Draw two more chords $\overline{AB}$ and $\overline{CD}$ through M; then $\overline{AD}$ and $\overline{BC}$ intersect $\overline{PQ}$ at X and Y, respectively. Then M is the midpoint of $\overline{XY}$.

---

[74] A little background: Ptolemy proposed the geocentric model of the Solar System - and the Universe - that was later discredited by Copernicus and Galileo in his wake. He had also authored the sole surviving text from his time period that is considered to be an authoritative astronomical treatise: the *Almagest*.

[75] It should be noted that, for such tiny increments, Ptolemy used *approximations* based on Aristarchus' Inequality.

Proof. Drop perpendiculars $\overline{XX_1}$ and $\overline{XX_2}$ from X onto the lines $\overline{AM}$ and $\overline{DM}$ respectively. Similarly, drop perpendiculars $\overline{YY_1}$ and $\overline{YY_2}$ from Y onto $\overline{BM}$ and $\overline{CM}$ respectively. Since $\triangle MXX_1 \sim \triangle MYY_1$, $\frac{MX}{MY} = \frac{XX_1}{YY_1}$, $\triangle MXX_2 \sim \triangle MYY_2$, $\frac{MX}{MY} = \frac{XX_2}{YY_2}$, $\triangle AXX_1 \sim \triangle\triangle CYY_2 \implies \frac{XX_1}{YY_2} = \frac{AX}{CY}$, and $\triangle DXX_2 \sim \triangle BYY_1 \implies \frac{XX_2}{YY_1} = \frac{DX}{BY}$, by the intersecting chords theorem, we have that

$$\frac{MX^2}{MY^2} = \frac{XX_1 \cdot XX_2}{YY_1 \cdot YY_2} = \frac{AX \cdot DX}{BY \cdot CY}$$

$$= \frac{PX \cdot QX}{PY \cdot QY} = \frac{(PM+XM)(PM-XM)}{(PM+MY)(PM-MY)} = \frac{PM^2 - MX^2}{PM^2 - MY^2}$$

because $PM = MQ$. Thus, $MX = MY$, and M is the midpoint of $\overline{XY}$, as desired. $\square$

8.6.5 The Japanese Theorem

A powerful (but admittedly oddly-named)[76] tool, the *Japanese theorem,* in general terms, states that the incenters of the triangles formed from a *triangulation*[77] of a cyclic quadrilateral are the vertices of a rectangle.

Theorem 8.16: The Japanese Theorem

Let $ABCD$ be a cyclic quadrilateral, and let I_i ($1 \le i \le 4$) be the incenters of $\triangle ABC$, $\triangle ABD$, $\triangle ACD$, $\triangle BCD$. Then $I_1 I_2 I_3 I_4$ is a rectangle.

Proof. We construct the parallelogram tangent to each of the I_i such that its sides are parallel to the diagonals of the cyclic quadrilateral $ABCD$. Then it follows that this parallelogram is a rhombus, which shows that the sums of the inradii of the circles that are tangent to each diagonal are equal (and so that $I_1 I_2 I_3 I_4$ has two pairs of parallel sides with equal lengths, hence confirming that it is a rectangle). $\square$

Generalization to Cyclic Polygons

The Japanese Theorem applies not only to cyclic quadrilaterals, but to all polygons that can be inscribed in a circle (cyclic polygons).

Theorem 8.17: Generalized Japanese Theorem

Under triangulation of a cyclic polygon, the sum of the inradii of the triangles is invariant.

Proof. This is a corollary of the quadrilateral case that follows from induction. (Loosely speaking, we induct upon the set of triangulating partitions of the polygon with 1 fewer side.) $\square$

---

[76] After all, if this is the Japanese theorem, whatever happened to the Greek Theorem or the American Theorem? Considering this is far from the *only* theorem hailing from Japan, the name just seems ill-fitting... But there is actually an explanation for it! Initially, in 1900, the theorem was known as the Chinese theorem (the irony!) But when the generalization came about later, in a 1906 article simply entitled "Japanese Mathematics," the resulting name stuck ever since.

[77] A *triangulation* of a polygon is a splitting of the polygon into triangles whose union is the polygon.

8.6.6   Examples

Example 8.6.6.1.

(a) Cyclic quadrilateral $ABCD$ has side lengths $AB = 6$, $BC = 7$, $CD = 7$, $DA = 6$. What is the area of $ABCD$?

(b) (Purple Comet HS Spring 2014) Let $ABCD$ be a cyclic quadrilateral with $AB = 1$, $BC = 2$, $CD = 3$, $DA = 4$. Find the square of the area of quadrilateral $ABCD$.

Solution 8.6.6.1.

(a) Notice that $ABCD$ is just a rectangle, so its area is $\boxed{42}$. Alternatively, use Brahmagupta's Formula.

(b) Brahmagupta's Formula yields $\boxed{24}$. (A good sanity check for this is to recognize that the area must be bounded between $4 \cdot 1 = 4$ and $2 \cdot 3 = 6$.)

Example 8.6.6.2. Let $ABCD$ be a cyclic quadrilateral. Show that $\cos A + \cos B + \cos C + \cos D = 0$.

Solution 8.6.6.2. In any cyclic quadrilateral, the two pairs of supplementary angles ($\angle A, \angle C$ and $\angle B, \angle D$) must both sum to $180°$. Then $C = 180 - A$ and $D = 180 - B$; indeed, we have $\cos A + \cos(180 - A) = \cos A - \cos A = 0$ for both pairs. Thus, the equation must hold for all cyclic quadrilaterals $ABCD$.

8.6.7 Exercises

Exercise 269. *(2004 AMC 10B) In triangle $\triangle ABC$ we have $AB = 7$, $AC = 8$, $BC = 9$. Point D is on the circumscribed circle of the triangle so that $\overline{AD}$ bisects angle $\angle BAC$. What is the value of $\dfrac{AD}{CD}$?*

Exercise 270. *It's time for reverse cyclic quads!*

(2000 AIME II) A circle is inscribed in quadrilateral $ABCD$, tangent to $\overline{AB}$ at P and to $\overline{CD}$ at Q. Given that $AP = 19$, $PB = 26$, $CQ = 37$, and $QD = 23$, find the square of the radius of the circle.

Exercise 271. *(2018 AMC 12A) Triangle ABC is an isosceles right triangle with $AB = AC = 3$. Let M be the midpoint of hypotenuse $\overline{BC}$. Points I and E lie on sides $\overline{AC}$ and $\overline{AB}$, respectively, so that $AI > AE$ and $AIME$ is a cyclic quadrilateral. Given that triangle EMI has area 2, the length CI can be written as $\frac{a-\sqrt{b}}{c}$, where a, b, and c are positive integers and b is not divisible by the square of any prime. What is the value of $a + b + c$?*

Exercise 272. *(2013 AIME II) A hexagon that is inscribed in a circle has side lengths 22, 22, 20, 22, 22, and 20 in that order. The radius of the circle can be written as $p + \sqrt{q}$, where p and q are positive integers. Find $p + q$.*

Exercise 273. *(2014 AMC 12B) Let $ABCDE$ be a pentagon inscribed in a circle such that $AB = CD = 3$, $BC = DE = 10$, and $AE = 14$. The sum of the lengths of all diagonals of $ABCDE$ is equal to $\frac{m}{n}$, where m and n are relatively prime positive integers. What is $m + n$?*

Extension Exercise 116. *(APMO 2007) Let $\triangle ABC$ be an acute-angled triangle with $\angle BAC = 60°$ and $AB > AC$. Let I be the incenter and H the orthocenter of the triangle $\triangle ABC$. Prove that $2\angle AHI = 3\angle ABC$.*

Extension Exercise 117. *(2018 AIME I) David found four sticks of different lengths that can be used to form three non-congruent convex cyclic quadrilaterals, A, B, C, which can each be inscribed in a circle with radius 1. Let φ_A denote the measure of the acute angle made by the diagonals of quadrilateral A, and define φ_B and φ_C similarly. Suppose that $\sin \varphi_A = \frac{2}{3}$, $\sin \varphi_B = \frac{3}{5}$, and $\sin \varphi_C = \frac{6}{7}$. All three quadrilaterals have the same area K, which can be written in the form $\dfrac{m}{n}$, where m and n are relatively prime positive integers. Find $m + n$.*

Extension Exercise 118. *Suppose that $ABCD$ is not a cyclic quadrilateral.*

   (a) *Provide a construction of such a quadrilateral (that is, specify its side and diagonal lengths).*

   (b) *Prove an inequality pertaining to this quadrilateral, and show that Ptolemy's Theorem does not hold for it.*

Extension Exercise 119.

   (a) *From Ptolemy's Theorem, derive the addition and subtraction formulas for $\sin$ and $\cos$.*

   (b) *Can we also accomplish the converse? How so?*

8.7   Mass Points

This particular topic - *mass points* - is inseparably connected with physics. At its core, it relates to the idea of the *center of mass* of a system, which is essentially the point where all masses balance out; i.e. the center of gravity, or the mean of the coordinates multiplied with the masses of each point. Mass points work much the same way: by assigning weights, or masses, to points in the plane, we can determine the center of mass of those points.

Mass points can also greatly simplify the *proofs* of fundamental geometric theorems - not just the computations that they power. They are particularly helpful in solving problems involving points that divide line segments into ratios (with bisectors, trisectors, etc. being special cases of such points).

8.7.1   Definitions

We define a *mass point* as a point in the plane equipped with a *mass*. Formally, a mass point is an ordered pair (m, P), where m is the mass of the point P and $P \in \mathbb{R}^2$ is a point in the Cartesian coordinate plane. (Two mass points (m_1, P_1) and (m_2, P_2) are equal, or *coincide*, iff $m_1 = m_2$ and $P_1 = P_2$.)

Addition of mass points invokes the idea of center of mass in that it effectively "balances" the sum between the two points being added. That is, if point $A = mP$ and point $B = nQ$, then $C = A + B$ has mass $m + n$ and coordinate R, where $\dfrac{RQ}{RP} = \dfrac{m}{n}$. That is, the sum point should "lean" more towards the "lighter" points than the "heavier" ones, in proportion with their masses:

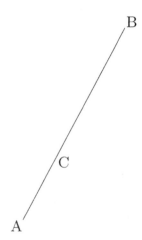

Figure 8.5: A demonstration of addition of mass points. Here, we have $A = (1, P)$, $B = (2, Q)$, $C = (3, R)$, where R is $\dfrac{1}{3}$ of the way from A to B.

Mass point addition is closed, associative, and commutative:

Proof. First, we show mass point addition is closed. Note that adding two mass points $A = (m, P)$, $B = (n, Q)$ will produce another mass point, $C = \left(m + n, \dfrac{nP + mQ}{m + n} \right)$. Hence the set of mass points is closed under addition.

Next we prove associativity. For any three mass points $A = (a, P), B = (b, Q), C = (c, R)$, we have $(A + B) + C = A + (B + C)$ (which we leave to the reader to verify manually). Finally, we prove commutativity. We have that $A + B = B + A$ (where A, B are defined as above), since $m + n = n + m$ and $mQ + nP = nP + mQ$. □

We define scalar multiplication $k(m, P) = (km, P)$ such that the mass is multiplied by k, but the coordinate of the point remains constant.

8.7.2   Assigning Masses

The most common type of problem in which we apply mass points is that which involves concurrent cevians of a triangle; i.e. in determining points such as the incenter, circumcenter, and centroid (as the cevian is a generalization of the angle bisector, perpendicular bisector, and median, among other, similar points).

In assigning mass points to the points on a triangle with concurrent cevians, we assign each point a whole number mass, with the vertices being assigned masses in accordance with the problem statement, and with each intermediary point (on an edge, or the point of concurrence) having a mass that is equal to the sum of the masses of the points it is directly connected to. (Note that each point will always have a unique mass, since if the vertices have masses a, b, c, the point of concurrence will have mass $a + b + c$. This is because the points on the edges will have masses $a + b$, $b + c$, and $a + c$, and the sum of one of these with the sum of the corresponding vertex will always be equal to $a + b + c$.)

In the method of *splitting mass points*, we consider a triangle that is split by a transversal line. We then assign one specific vertex two masses; one corresponding to the vertex on one side, and one corresponding to the vertex on the other side. We can use proportionality/similarity to determine the masses of the vertex which is split, then sum individually on each side, as opposed to using the sum value associated with the vertex.

8.7.3   Examples

For this section, the bulk of our concentration will be on exercises, since mass points is a topic that is perhaps confusing if explained solely from the passive perspective of the written word, and not practiced actively by the reader.

Example 8.7.3.1. *(2013 AMC 10B) In triangle ABC, medians AD and CE intersect at P, $PE = 1.5$, $PD = 2$, and $DE = 2.5$. What is the area of $AEDC$?*

Solution 8.7.3.1. *Assign to point B a mass of 1. Then A also has mass 1, since E is the midpoint of $\overline{AB}$. Similarly, Cs mass is 1, and D, E have mass 2. Then $AP = 2PD$, and $PD = 2$, so $AP = 4$, and $AD = 6$. Similarly, $CP = 2PE = 3$, since $PE = \dfrac{3}{2}$; hence, $CE = \dfrac{9}{2}$. $\triangle PED$ is subsequently a 3-4-5 right triangle with $\angle DPE = 90°$. Thus, $[AEDC] = \dfrac{6 \cdot \frac{9}{2}}{2} = \boxed{\dfrac{27}{2}}$.*

Example 8.7.3.2. *Isosceles $\triangle MAT$ has $MA = AT = 20$ and $TM = 24$. The perpendicular bisectors of $\overline{MA}$ and $\overline{AT}$ intersect at H. Line $\overline{AH}$ intersects $\overline{MT}$ at S. Compute the ratio $\dfrac{AH}{HS}$.*

Solution 8.7.3.2. *WLOG assume $A = (0,0), B = (12, 16), C = (24, 0)$. By symmetry, the x-coordinates of H and S are both 12. Then set the mass of A equal to 1, which allows for the calculation of the masses of the midpoints of $\overline{AM}$ and $\overline{AT}$. Extend the perpendicular bisectors to meet $\overline{MT}$ in order to determine the mass of H, which leads to the answer $\boxed{\dfrac{25}{7}}$.*

8.7.4 Exercises

Exercise 274. *(2004 AMC 10B) In $\triangle ABC$ points D and E lie on BC and AC, respectively. If AD and BE intersect at T so that $\dfrac{AT}{DT} = 3$ and $\dfrac{BT}{ET} = 4$, what is $\dfrac{CD}{BD}$?*

Exercise 275. *In triangle $\triangle ABC$, $AB = 7$, $BC = 8$, $CA = 9$. Point D is the midpoint of $\overline{BC}$. Point E trisects $\overline{CA}$, and is closer to C. The angle bisector of $\angle A$ hits $\overline{BC}$ at point F. What is the area of $\triangle DEF$?*

Exercise 276. *$\triangle ABC$ has $AB = 13$, $BC = 15$, $CA = 14$. Let $\overline{BD}$ be an altitude of $\triangle ABC$. The bisector of $\angle C$ hits $\overline{AB}$ at E and $\overline{BD}$ at F. Compute $\dfrac{CE}{EF}$.*

Exercise 277. *(1975 AHSME) In $\triangle ABC$, M is the midpoint of side $\overline{BC}$, $AB = 12$ and $AC = 16$. Points E and F are taken on $\overline{AC}$ and $\overline{AB}$, respectively, and lines $\overline{EF}$ and $\overline{AM}$ intersect at G. If $AE = 2AF$ then compute $\dfrac{EG}{GF}$.*

Exercise 278. *In $\triangle ABC$ with $AB = 7$, $BC = 3$, and $CA = 5$, angle bisectors $\overline{AD}$ and $\overline{BE}$ intersect at point P. Compute $\dfrac{BP}{PE}$.*

Exercise 279. *Three similar problems, one from MathCounts and two from the 2016 AMC 10 tests:*

(a) *(2012 MathCounts State Sprint) In rectangle $ABCD$, point M is the midpoint of side $\overline{BC}$, and point N lies on $\overline{CD}$ such that $DN : NC = 1 : 4$. Segment $\overline{BN}$ intersects $\overline{AM}$ and $\overline{AC}$ at points R and S, respectively. If $NS : SR : RB = x : y : z$, where x, y and z are positive integers, what is the minimum possible value of $x + y + z$?*

(b) *(2016 AMC 10A) In rectangle $ABCD$, $AB = 6$ and $BC = 3$. Point E between B and C, and point F between E and C are such that $BE = EF = FC$. Segments $\overline{AE}$ and $\overline{AF}$ intersect $\overline{BD}$ at P and Q, respectively. The ratio $BP : PQ : QD$ can be written as $r : s : t$ where the greatest common factor of r, s, and t is 1. What is $r + s + t$?*

(c) *(2016 AMC 10B) Rectangle $ABCD$ has $AB = 5$ and $BC = 4$. Point E lies on $\overline{AB}$ so that $EB = 1$, point G lies on $\overline{BC}$ so that $CG = 1$. and point F lies on $\overline{CD}$ so that $DF = 2$. Segments $\overline{AG}$ and $\overline{AC}$ intersect $\overline{EF}$ at Q and P, respectively. What is the value of $\frac{PQ}{EF}$?*

Note: These, like many other exercises in this part, can be coordinate bashed, but we recommend using synthetic methods in favor of bashing.

Exercise 280. *(2001 AIME I) Triangle ABC has $AB = 21$, $AC = 22$ and $BC = 20$. Points D and E are located on $\overline{AB}$ and $\overline{AC}$, respectively, such that $\overline{DE}$ is parallel to $\overline{BC}$ and contains the center of the inscribed circle of triangle ABC. Then $DE = m/n$, where m and n are relatively prime positive integers. Find $m + n$.*

Extension Exercise 120. *(1989 AIME) Point P is inside $\triangle ABC$. Line segments APD, BPE, and CPF are drawn with D on $\overline{BC}$, E on $\overline{AC}$, and F on $\overline{AB}$. Given that $AP = 6$, $BP = 9$, $PD = 6$, $PE = 3$, and $CF = 20$, find the area of $\triangle ABC$.*

8.8 3D Geometry, Part 2

8.8.1 Working with Planes

We will continue our foray into the expansive world of three dimensions by, ironically enough, taking a look at two-dimensional figures once again. Namely, we will be taking a look at *planes* - no different than our very familiar 2-dimensional xy-plane $\mathbb{R}^2$. We can also have the xz-plane, the yz-plane, and an infinitude of other planes that pass through points, just as lines do within planes.

To be precise:

> **Definition 8.3: Two-Dimensional Plane**
>
> A *plane* in two dimensions is a flat, two-dimensional surface extending infinitely in both directions.

In three-dimensional space $\mathbb{R}^3$, a plane has general equation of the form $ax + by + cz = d$, where not all of a, b, c are zero. This is similar to the equation of a line in $\mathbb{R}^2$, namely $ax + by = c$ (i.e., in slope-intercept form, $y = mx + b$). We can also think of the plane as a flattened prism (or, more generally, a flattened parallelepiped).

A crucial theorem/property regarding planes is that, if we know three points that lie on a plane, we can uniquely determine the equation of the plane (i.e. three points uniquely determine a plane, much like two points uniquely determine a line). That is, let $ax + by + cz = d$ be the equation of a plane passing through some points A, B, C. Then if $A = (x_1, y_1, z_1)$ (with B and C defined analogously), we can substitute the x_i, y_i, z_i into the equation of the plane to obtain a system of three equations with one free variable d and three unknowns.

Furthermore, recall the point-to-line distance formula, namely $D((x_0, y_0), l) = \dfrac{|ax_0 + by_0 + c|}{\sqrt{a^2 + b^2}}$ (where $l :=$ $ax + by = c$). We can similarly extend this to the point-to-plane and line-to plane distance formulae:

> **Theorem 8.18: Point-to-Plane and Line-to-Plane Distance Formulae**
>
> - Point-to-plane: The shortest distance from a point to a plane is the length of the perpendicular to the plane through the point, or $\dfrac{|ax_0 + by_0 + cz_0 + d|}{\sqrt{a^2 + b^2 + c^2}}$ (where the point has coordinates (x_0, y_0, z_0) and the plane has equation $ax + by + cz + d = 0$, with $d = -Ax_0 - By_0 - Cz_0$).
>
> - Line-to-plane: If the line and plane are parallel, simply pick any point on the line and apply the point-to-plane formula. Otherwise, their distance is clearly 0.

Proof. We prove the point-to-plane formula here (the line-to-plane formula is trivial). Let n be the normalized unit normal vector; i.e. N with its components divided by its norm (see §9.6 for a rigorous explanation of the norm. For now, all we need to know is that the norm is equal to the square root of the sum of the squares of the components of the vector), denoted $n = \dfrac{N}{||N||}$. Then

$$n = \frac{(a, b, c)}{\sqrt{a^2 + b^2 + c^2}}$$

(see the immediately following section for some context as to why this is).

Let v be a vector from $Q = (x_0, y_0, z_0)$ to $P = (x_1, y_1, z_1)$, so that $v = (x_1 - x_0, y_1 - y_0, z_1 - z_0)$. Since the distance is the projection of v (with $||v|| = 1$ onto n, it is the dot product $|v \cdot n|$, which simplifies to the desired value of $\dfrac{|ax_0 + by_0 + cz_0 + d|}{\sqrt{a^2 + b^2 + c^2}}$ after we eliminate the x_1, y_1, z_1 terms. $\square$

8.8.2 A (Brief) Introduction to Vectors

We can also define a plane in $\mathbb{R}^3$ as that which is uniquely determine by a point and a *vector* perpendicular to the plane. But what is a vector?

Simply put, a *vector* is a quantity with both direction and magnitude (i.e. length). It determines the position of one point in space with respect to another. We will not dwell on the properties of vectors here; we will save this for our discussion of vector spaces (§9.6). However, we will say here that many physical quantities - such as velocity and acceleration - can be thought of as vectors, and that they are opposed to scalar quantities (which are only numbers and do not have a directional component).

Using this (rough) idea of a vector, we can define the *normal vector* in reference to a plane in $\mathbb{R}^3$:

Definition 8.4: Normal Vector

A *normal vector* with respect to a plane is a vector perpendicular to the plane.

This definition applies not only to planes, but to lines as well, and also to quantities other than vectors (such as the *normal line* of a curve). "Normal," in this context, essentially means "perpendicular."

Once we compute the normal vector of a plane, we can rewrite our formula for the plane in terms of the normal vector:

Theorem 8.19: Plane Equation in Terms of the Normal Vector

If $P := ax + by + cz + d = 0$ and $N = (a, b, c)$ is the normal vector of P, with point $Q = (x_0, y_0, z_0)$, then P's equation can be written as $A(x - x_0) + B(y - y_0) + C(z - z_0) = 0$ with $d = -Ax_0 - By_0 - Cz_0$.

8.8.3 Euler's Formula for Polyhedra

We conclude this chapter with a formula that may seem quite familiar, but whose proof may not be as clear as it seems: Euler's formula for polyhedra. This is the formula $F + V - E = 2$, where F, V, and E are the numbers of faces, vertices, and edges of a polyhedron, respectively. (Note that this formula applies not only to 3-dimensional polyhedra, but also to n-dimensional polyhedra for any $n \geq 3$. For $n = 2$, this does not hold; consider a square, which has $F = 1$, $V = 4$, and $E = 4$.)

This is yet another prime example of a theorem with a surprisingly large number of proofs; here, we will not go over all known proofs (numbering at least 20), but we will focus on the two that are perhaps most well-known and which are most accessible without specialized knowledge in other disciplines.

Seeing as to how this theorem in fact dates in an earlier form to Descartes in 1630 (but was popularized by Euler in 1752 after he formally published it[78]), it seems appropriate to discuss Descartes' original proof, which takes advantage of the fact that every polyhedron has a planar graph (a graph that can be drawn on the plane such that its edges intersect only at its endpoints; that is, no edges cross each other - the graph is not self-intersecting.[79]) which can be embedded with all edges forming straight line segments.

Proof. In the planar graph of the polyhedron, observe that each edge belongs to two faces, and that the sum of the angles in a k-gon is $180(k-2)° = \pi(k-2)$ radians, so the sum of the angles in the entire polyhedron is $\pi(2E - 2F) = 2\pi(E - F)$.

---

[78]Albeit having a faulty proof; this was not corrected until 1794 by Legendre, who used spherical angles to provide a correct proof.

[79]See §11.3 for more information regarding these and other types of graphs, as well as an overarching discussion of graph theory.

Similarly, note that each interior vertex is surrounded by triangles, each of which adds 2π to the sum of angles. On the outer face, the vertices add $2(\pi - \theta(v))$, where $\theta(v)$ denotes the exterior angle at vertex v. Summing over all $v \in \mathcal{V}$ (where $\mathcal{V}$ denotes the set of vertices of the polyhedron) yields $2\pi \cdot V - 2(2\pi) = 2\pi \cdot V - 4\pi$, since the sum of exterior angles of any polygon is $360° = 2\pi$ radians.

Finally, set the resulting two expressions equal to each other, and divide through by 2π, to obtain $V - 2 = E - F \implies V + F - E = 2$ as desired. $\square$

8.8.4 Examples

Example 8.8.4.1. *(1996 AHSME) Triangle PAB and square $ABCD$ are in perpendicular planes. Given that $PA = 3, PB = 4$ and $AB = 5$, what is PD?*

Solution 8.8.4.1. *As the planes are perpendicular, $\triangle PAD$ is right and $PD = \boxed{\sqrt{34}}$.*

Example 8.8.4.2. *(1998 AIME) Eight spheres of radius 100 are placed on a flat surface so that each sphere is tangent to two others and their centers are the vertices of a regular octagon. A ninth sphere is placed on the flat surface so that it is tangent to each of the other eight spheres. The radius of this last sphere is $a + b\sqrt{c}$, where a, b, and c are positive integers, and c is not divisible by the square of any prime. Find $a + b + c$.*

Solution 8.8.4.2. *The figures are spheres - not circles! - and so setting up auxiliary lines between the centers of circles perpendicular to the flat surface will yield a trapezoid and a corresponding right triangle. By the Pythagorean Theorem, letting the length of the segment connecting the centers be r and the horizontal distance between the smaller circles center and the radius of the larger circle be x, we obtain $x^2 + (r - 100)^2 = (r + 100)^2 \implies x = 20\sqrt{r}$, and so $2x = 40\sqrt{r}$.*

Now drawing another right triangle between two diametrically-opposed circle with hypotenuse $40\sqrt{r}$ and one leg length of 200 yields the other leg length of $200 + 200\sqrt{2}$. Pythagorean Theorem again yields $r = 100 + 50\sqrt{2} \implies \boxed{152}$.

Example 8.8.4.3. *(1967 IMO P2) Prove that a tetrahedron with just one edge length greater than 1 has volume less than or equal to $\frac{1}{8}$.*

Solution 8.8.4.3. *Call the tetrahedron $ABCD$. Assume that all edges except for AB have length no greater than 1. We can then maximize the volume by taking planes ACD, BCD to be perpendicular. If $AC, AD < 1$, we can keep increasing the length of the altitude from A to CD, which implies $AC = AD = 1$. Similarly, $BC = BD = 1$.*

To maximize the volume, we adjust CD as well, such that the angle between ACD and ABD should be as close as possible to a right angle. $\angle CMD$ would then become a right angle. When $\angle CMD = 90°$, $CD > 1$, so the height of D above AC becomes maximal when $CD = 1$ exactly.

Thus, $\triangle BCD$ is equilateral, as is $\triangle ACD$ with altitude $\frac{\sqrt{3}}{2}$ (this is also the altitude of A to BCD). Hence, the volume is $\frac{1}{8}$, as desired.

8.8.5 Exercises

Exercise 281. *(1996 AHSME) On a $4 \times 4 \times 3$ rectangular parallelepiped, vertices A, B, and C are adjacent to vertex D. What is the perpendicular distance from D to the plane containing A, B, and C?*

Exercise 282. *(2012 AIME I) Cube $ABCDEFGH$ [with E above A, F above B, and likewise for G and H] has edge length 1 and is cut by a plane passing through vertex D and the midpoints M and N of $\overline{AB}$ and $\overline{CG}$ respectively. The plane divides the cube into two solids. The volume of the larger of the two solids can be written in the form $\frac{p}{q}$, where p and q are relatively prime positive integers. Find $p + q$.*

Exercise 283. *(2000 AIME II) The points A, B and C lie on the surface of a sphere with center O and radius 20. It is given that $AB = 13$, $BC = 14$, $CA = 15$, and that the distance from O to $\triangle ABC$ is $\frac{m\sqrt{n}}{k}$, where m, n, and k are positive integers, m and k are relatively prime, and n is not divisible by the square of any prime. Find $m + n + k$.*

Exercise 284. *(2016 HMMT November Guts Round) Create a cube C_1 with edge length 1. Take the centers of the faces and connect them to form an octahedron O_1. Take the centers of the octahedron's faces and connect them to form a new cube C_2. Continue this process infinitely. Find the sum of all the surface areas of the cubes and octahedrons.*

Exercise 285. *Let P_1 be the plane passing through the points $(4, 5, 3)$, $(2, 0, 1)$, $(1, 9, 4)$. Let P_2 pass through the points $(20, 19)$ and $(-4, -5)$ with arbitrary z-coordinate (i.e. the line $y = x - 1$ in $\mathbb{R}^2$ with $z \in \mathbb{R}$). Either compute the minimum distance between P_1 and P_2, or show that P_1 and P_2 overlap (and determine the locus of overlap).*

Exercise 286. *(2009 AMC 12A) A regular octahedron has side length 1. A plane parallel to two of its opposite faces cuts the octahedron into the two congruent solids. The polygon formed by the intersection of the plane and the octahedron has area $\frac{a\sqrt{b}}{c}$, where a, b, and c are positive integers, a and c are relatively prime, and b is not divisible by the square of any prime. What is $a + b + c$?*

Exercise 287. *(2011 AIME I) In triangle ABC, $BC = 23$, $CA = 27$, and $AB = 30$. Points V and W are on $\overline{AC}$ with V on $\overline{AW}$, points X and Y are on $\overline{BC}$ with X on $\overline{CY}$, and points Z and U are on $\overline{AB}$ with Z on $\overline{BU}$. In addition, the points are positioned so that $\overline{UV} \parallel \overline{BC}$, $\overline{WX} \parallel \overline{AB}$, and $\overline{YZ} \parallel \overline{CA}$. Right angle folds are then made along $\overline{UV}$, $\overline{WX}$, and $\overline{YZ}$. The resulting figure is placed on a level floor to make a table with triangular legs. Let h be the maximum possible height of a table constructed from triangle ABC whose top is parallel to the floor. Then h can be written in the form $\frac{k\sqrt{m}}{n}$, where k and n are relatively prime positive integers and m is a positive integer that is not divisible by the square of any prime. Find $k + m + n$.*

Extension Exercise 121. *(1960 IMO P5) Consider the cube $ABCDA'B'C'D'$ (with face $ABCD$ directly above face $A'B'C'D'$).*

    (a) Find the locus of the midpoints of the segments XY, where X is any point of AC and Y is any point of $B'D'$;

    (b) Find the locus of points Z which lie on the segment XY of part a) with $ZY = 2XZ$.

Extension Exercise 122. *(1966 IMO P3) Prove that the sum of the distances of the vertices of a regular tetrahedron from the center of its circumscribed sphere is less than the sum of the distances of these vertices from any other point in space.*

Extension Exercise 123. *(2002 Putnam A2) Given any five points on a sphere, show that some four of them must lie on a closed hemisphere.*

This is not an exercise, but perhaps as a catalyst for thought and experimentation: how can we prove that the shortest path between any two points on a sphere is the minor arc of the sphere's great circle?

Chapter 8 Review Exercises

Exercise 288. *A right triangle has hypotenuse 15, and leg lengths 9 and 12. What is the product of the radii of its incircle and circumcircle?*

Exercise 289. *(2015 Berkeley Math Tournament) Let $\mathcal{C}$ be the sphere $x^2 + y^2 + (z-1)^2 = 1$. Point P on $\mathcal{C}$ is $(0, 0, 2)$. Let $Q = (14, 5, 0)$. If PQ intersects $\mathcal{C}$ again at Q_0, then find the length PQ_0.*

Exercise 290. *(2018 HMMT February Geometry Round) How many noncongruent triangles are there with one side of length 20, one side of length 17, and one $60°$ angle?*

Exercise 291. *Two line segments with a common endpoint have lengths a and b. If they span an angle of $\theta = \cos^{-1}\left(\frac{a}{b}\right)$, show that the triangle formed by drawing the line segment from the endpoints of the line segments cannot be a right triangle.*

Exercise 292. *$\triangle ABC$ has $AB = 13$, $BC = 14$, $CA = 15$. $\overline{AD}$ is an altitude of the triangle. What is the maximum x-distance between some two points on the circumcircles of $\triangle ABD$ and $\triangle ADC$?*

Exercise 293. *(2010 AIME II) Triangle ABC with right angle at C, $\angle BAC < 45°$ and $AB = 4$. Point P on $\overline{AB}$ is chosen such that $\angle APC = 2\angle ACP$ and $CP = 1$. The ratio $\frac{AP}{BP}$ can be represented in the form $p + q\sqrt{r}$, where p, q, r are positive integers and r is not divisible by the square of any prime. Find $p + q + r$.*

Exercise 294. *(2018 AMC 12B) Let $ABCDEF$ be a regular hexagon with side length 1. Denote by X, Y, and Z the midpoints of sides $\overline{AB}$, $\overline{CD}$, and $\overline{EF}$, respectively. What is the area of the convex hexagon whose interior is the intersection of the interiors of $\triangle ACE$ and $\triangle XYZ$?*

Exercise 295. *(2018 AMC 12B) In $\triangle ABC$ with side lengths $AB = 13$, $AC = 12$, and $BC = 5$, let O and I denote the circumcenter and incenter, respectively. A circle with center M is tangent to the legs AC and BC and to the circumcircle of $\triangle ABC$. What is the area of $\triangle MOI$?*

Exercise 296. *(2014 Caltech-Harvey Mudd Math Competition) Find the area of the cyclic quadrilateral with side lengths given by the solutions to $x^4 - 10x^3 + 34x^2 - 45x + 19 = 0$.*

Exercise 297. *(1999 USAMO P2) Let $ABCD$ be a cyclic quadrilateral. Prove that*

$$|AB - CD| + |AD - BC| \geq 2|AC - BD|.$$

Exercise 298. *(2014 Stanford Math Tournament) In cyclic quadrilateral $ABCD$, $AB \cong AD$. If $AC = 6$ and $\frac{AB}{BD} = \frac{3}{5}$, find the maximum possible area of $ABCD$.*

Exercise 299. *(2017 HMMT February Geometry Test) Let $\triangle ABC$ be a triangle with circumradius $R = 17$ and inradius $r = 7$. Find the maximum possible value of $\sin\frac{A}{2}$.*

Exercise 300. *(2015 AMC 10A/12A) Let S be a square of side length 1. Two points are chosen independently at random on the sides of S. The probability that the straight-line distance between the points is at least $\frac{1}{2}$ is $\frac{a - b\pi}{c}$, where a, b, and c are positive integers and $\gcd(a, b, c) = 1$. What is $a + b + c$?*

Extension Exercise 124. *(2008 USAMO P2) Let ABC be an acute, scalene triangle, and let M, N, and P be the midpoints of $\overline{BC}$, $\overline{CA}$, and $\overline{AB}$, respectively. Let the perpendicular bisectors of $\overline{AB}$ and $\overline{AC}$ intersect ray AM in points D and E respectively, and let lines BD and CE intersect in point F, inside of triangle ABC. Prove that points A, N, F, and P all lie on one circle.*

Extension Exercise 125. *(2012 Putnam A1) Let $d_1, d_2, \ldots, d_{12}$ be real numbers in the open interval $(1, 12)$. Show that there exist distinct indices i, j, k such that d_i, d_j, d_k are the side lengths of an acute triangle.*

Extension Exercise 126. *(2015 AIME I) A block of wood has the shape of a right circular cylinder with radius 6 and height 8, and its entire surface has been painted blue. Points A and B are chosen on the edge of one of the circular faces of the cylinder so that arc AB on that face measures $120°$. The block is then sliced in half along the plane that passes through point A, point B, and the center of the cylinder, revealing a flat, unpainted face on each half. The area of one of these unpainted faces is $a\pi + b\sqrt{c}$, where a, b, and c are integers and c is not divisible by the square of any prime. Find $a + b + c$.*

Extension Exercise 127. *(1998 Putnam A1) A cone has circular base radius 1, and vertex at height 3 directly above the center of the circle. A cube has four vertices in the base and four on the sloping sides. What length is a side of the cube?*

Extension Exercise 128. *(1999 USAMO P6) Let $ABCD$ be an isosceles trapezoid with $AB \parallel CD$. The inscribed circle ω of triangle BCD meets CD at E. Let F be a point on the (internal) angle bisector of $\angle DAC$ such that $EF \perp CD$. Let the circumscribed circle of triangle ACF meet line CD at C and G. Prove that the triangle AFG is isosceles.*

Extension Exercise 129. *(1995 IMO Shortlist G3) $\triangle ABC$ is a triangle. The incircle touches $\overline{BC}$, $\overline{CA}$, $\overline{AB}$ at D, E, F respectively. X is a point inside the triangle such that the incircle of $\triangle XBC$ touches $\overline{BC}$ at D. It touches $\overline{CX}$ at Y and $\overline{XB}$ at Z. Show that $EFZY$ is cyclic.*

Chapter 9

Advanced Geometry

9.1 Proving Triangle Theorems: Part 2

9.1.1 Collinearity: Pascal's Theorem

The great Blaise Pascal is known in most circles mostly, if not solely, for Pascal's Triangle and its inextricable link with the Binomial Theorem, but as it turns out, his mathematical work extends so much deeper. A paragon of his devotion to the field - and to his reverence for those giants upon whose shoulders he stood - is *Pascal's Theorem*, named in his honor:

> **Theorem 9.1: Pascal's Theorem**
>
> Given any six points (not necessarily distinct) lying on circle O, namely A, C, E, B, F, D **in that order**, $\overline{AB}$ and $\overline{DE}$ will intersect at P, $\overline{AF}$ and $\overline{CD}$ will intersect at Q, and $\overline{BC}$ and $\overline{EF}$ will intersect at R. Then P, Q, and R are collinear.

Remark. Pascal's Theorem seems most natural in the projective plane, where lines at infinity are defined such that Pascal's Theorem has no limitations due to lines being extended to meet at some point. However, it still holds perfectly well in the standard Euclidean plane.

Several different proofs exist for Pascal's Theorem, ranging in simplicity of the techniques used from law of sines and similarity, to the isogonal conjugate (see following section) and conic sections. For the purposes of this section, we will prove the theorem through repeated applications of Menelaus' Theorem.

Proof. Let $\overline{CD}$ and $\overline{EF}$ intersect at point G; let $\overline{AB}$ and $\overline{EF}$ intersect at H; and let $\overline{AB}$ and $\overline{CD}$ intersect at I. Applying Menelaus' Theorem on $\triangle GHI$ and line QDE, we obtain $\dfrac{HP}{IP} \cdot \dfrac{ID}{GD} \cdot \dfrac{GE}{HE} = 1$. Repeating the process on $\triangle GHI$ and line AQF yields $\dfrac{HA}{IA} \cdot \dfrac{IP}{GP} \cdot \dfrac{GF}{HF} = 1$, and on $\triangle GHI$ and BCR yields $\dfrac{HB}{IB} \cdot \dfrac{IC}{GC} \cdot \dfrac{GR}{HR} = 1$.

Multiplying all of these together and applying Power of a Point yields $\dfrac{IP}{GP} \cdot \dfrac{HQ}{IQ} \cdot \dfrac{GR}{HR} = 1$. Hence, by one final application of Menelaus, P, Q, R are collinear, as desired. $\qquad \square$

It is worth noting that Pascal had first proven this theorem at sixteen years of age; while his original proof has never been recovered, it is likely that he had based it on Menelaus' Theorem, which had been around for several centuries beforehand. In this way, Pascal was the Fermat of geometry (except with a much less difficult and much less tantalizing proof)!

9.1.2 Isogonal and Isotomic Conjugates

In this section, we cover two of the most important and versatile constructions in high-level geometry problems: the *isogonal* and *isotomic* conjugates of a point P in a triangle $\triangle ABC$ (and not lying on any of the edges). Their constructions come from theorems which we have already proven; now is the time to put them to good use!

Theorem 9.2: Construction of the Isogonal Conjugate

Let $\triangle ABC$ have incenter I, and let $P \in \triangle ABC$. Then $\overline{AP}$, $\overline{BP}$, and $\overline{CP}$, when reflected about $\overline{AI}$, $\overline{BI}$, and $\overline{CI}$ respectively, concur at a point Q, which is the *isogonal conjugate* of P.

We denote the isogonal conjugate of P by P^*.

Corollary 9.2: Involution of the Isogonal Conjugate

The isogonal conjugate operator is an involution; i.e. $(P^*)^* = P$.

Furthermore, each excenter of $\triangle ABC$ is its own isogonal conjugate. If we denote O as the circumcenter, H as the orthocenter, I as the incenter, G as the centroid, and S as the symmedian point (the point of concurrence of the lines consisting of points that are the isogonal conjugates of the points on the median with respect to each vertex; i.e. the concurrence point of the reflections of each median about each internal angle bisector),[80] then $O^* = H$ (and $H^* = O$) and $G^* = S$, $S^* = G$.

Remark. How can we be sure that the symmedian point exists? As an exercise, try proving its existence for all triangles $\triangle ABC$.

O and H are isogonal conjugates. Let Γ be the circumcircle of $\triangle ABC$. Let $F \neq A$ be the intersection of the A-altitude AH_A with Γ, and let $E \neq A$ be the intersection of the bisector of $\angle A$ with Γ. Furthermore, let M_A be the midpoint of $\overline{BC}$.

Observe that O, M_A, and E are collinear, and $\angle AEO = \angle EAO$. We also have $\angle FAE = \angle AEO$, so $\angle FAE = \angle EAO$. Applying this to all three vertices A, B, C yields the desired result. $\qquad \square$

G and S are isogonal conjugates. Since the centroid is, by definition, the intersection point of the medians of the triangle, and each symmedian is the isogonal line of each median, we have that the intersection of the symmedians is the isogonal conjugate of the centroid, which happens to be the symmedian point by definition. $\qquad \square$

Finally, as an interesting property that vastly simplifies problems with isogonal conjugates, if $x : y : z$ are the trilinear coordinates of a point $P \in \triangle ABC$ (not on an edge), then $\dfrac{1}{x} : \dfrac{1}{y} : \dfrac{1}{z}$ are the coordinates of P^* (see §9.3.3). We will leave this as an exercise once you have read through that section.

We finally define the *isotomic conjugate* of a point $P \in \triangle ABC$. Denoted by P',[81] the isotomic conjugate yields many similar properties as the isogonal conjugate when manipulated:

Theorem 9.3: Construction of the Isotomic Conjugate

Extend $\overline{AP}$, $\overline{BP}$, $\overline{CP}$ to meet $\overline{BC}$, $\overline{AC}$, $\overline{AB}$ at A', B', C' respectively. Let D, E, F be the midpoints of $\overline{BC}$, $\overline{AC}$, $\overline{AB}$ respectively. Then reflect A', B' and C' over D, E, F respectively to obtain points A'', B'', C''. The isotomic lines $\overline{AA''}, \overline{BB''}, \overline{CC''}$ intersect at the isotomic conjugate P' of P.

Remark. Note that the isotomic conjugate of the centroid is itself.

---

[80] We can similarly define the *symmedial triangle*, which has vertices at the points of intersection of the symmedians with the edges of the triangle.

[81] There is no standard notation for the isotomic conjugate of P; P' is just what we have chosen.

In addition, if P has trilinear coordinates $x : y : z$, we can compute the trilinear coordinates (see §9.3.3) of the isotomic conjugate P' to be $(a^{-2}x^{-1}, b^{-2}y^{-1}, c^{-2}z^{-1})$, where a, b, c are the side lengths of $\triangle ABC$ opposite vertices A, B, C respectively.

9.1.3 Miquel's Theorem and the Miquel Point

Miquel's Theorem represents a leap forward with the idea of the circumcircle of a triangle by considering three arbitrary points on the edges of the triangle, and all *three* circumcircles that they spawn:

Theorem 9.4: Miquel's Theorem

In $\triangle ABC$, let A', B', C' be points lying strictly on $\overline{BC}, \overline{AC}, \overline{AB}$ (not coinciding with any vertex). Then the circumcircles of the triangles $\triangle AB'C', \triangle A'BC', \triangle A'B'C$ will intersect at the Miquel point M. Furthermore, $\angle MA'B = \angle MB'C = \angle MC'A$, and $\angle MA'C = \angle MB'A = \angle MC'B$.

Believe it or not, this proof is very short and sweet by comparison to the theorem statement!

Proof. Cyclic quadrilaterals are our best friends! Here, we let the circumcircles of $\triangle A'B'C$ and $\triangle AB'C'$ intersect at $M \neq A$. Then observe that $\angle A'MC' = 2\pi - \angle B'MA' - \angle C'MB' = 2\pi - (\pi - \angle C) - (\pi - \angle A) = \angle A + \angle C = \pi - \angle B$, which shows that quadrilateral $BA'MC'$ is cyclic; Miquel's Theorem then follows. $\square$

Note that, if A', B', C' are collinear, the theorem still holds, albeit as a special case:

Corollary 9.4: Collinearity Case

If A', B', C' are collinear, then the Miquel point M lies on the circumcircle of $\triangle ABC$. (The converse holds as well.)

In addition, we can generalize Miquel's Theorem in the standard three-circle case to the quadrilateral theorem, the five-circle theorem, and, of special mention, the *six-circle theorem*:

Theorem 9.5: Generalization: Miquel's Six-Circle Theorem

Given points A, B, C, D lying on a circle, and four circles passing through each of the four pairs of *adjacent* points, there exists a sixth circle passing through the points of intersection of the four circles.

Proof. This proof (sketch) utilizes the idea of *cross ratios*.

Consider $A - B$ as a vector from A to B. In polar coordinates, $A - B = r_{AB} \cdot e^{i\theta_{AB}}$, where r_{AB} denotes the real cross ratio and θ_{AB} denotes the difference of the angles of the vectors A and B. Then note that $\dfrac{A - C}{B - C} = \dfrac{r_{AC}}{r_{BC}} \cdot e^{i(\theta_{AC} - \theta_{BC})}$; extending this to the entire cross ratio of the four points on the circle yields

$$\frac{(A - C)(B - D)}{(A - D)(B - C)} = \frac{r_{AC}r_{BD}}{r_{AD}r_{BC}} \cdot e^{i(\theta_{AC} + \theta_{BD} - \theta_{AD} - \theta_{BC})}$$

which is a real number iff $\theta_{AC} + \theta_{BD} - \theta_{AD} - \theta_{BC}$ is a multiple of π, or in other words, $\angle ACB \equiv \angle ADB \pmod{\pi}$.

In our formulation of the first (largest) circle, we must have

$$\frac{(A - C)(B - D)}{(A - D)(B - C)} \in \mathbb{R}$$

along with the other four cross ratios. Observe that, if all five cross ratios are real (i.e. the four points corresponding to them have the property that $\angle ACB \equiv \angle ADB \pmod{\pi}$), then the last cross ratio (representing the circle containing all four intersection points) must also be real, and thus they are concyclic. $\square$

We can even generalize the theorem to three dimensions, in which case the four spheres passing through a vertex of a tetrahedron and arbitrary points on the edges of the tetrahedron all intersect at a common point in 3-space. (We will not prove this here, however.)

9.1.4 Examples

Example 9.1.4.1. *(2016 AIME I) In $\triangle ABC$ let I be the center of the inscribed circle, and let the bisector of $\angle ACB$ intersect $\overline{AB}$ at L. The line through C and L intersects the circumscribed circle of $\triangle ABC$ at the two points C and D. If $LI = 2$ and $LD = 3$, then $IC = \frac{p}{q}$, where p and q are relatively prime positive integers. Find $p + q$.*

Solution 9.1.4.1. *WLOG assume that $\triangle ABC$ is isosceles with vertex angle $\angle C$. Then L is the midpoint of $\overline{AB}$, and $\angle CLB = \angle CLA = 90°$. Drawing the perpendicular to $\overline{BC}$ from I to E reveals that $IE = 2$. Let $BD = x$; then $\triangle BLD \sim \triangle CEI$, so $\dfrac{CI}{x} = \dfrac{2}{3} \implies CI = \dfrac{2x}{3}$. Also, $\triangle CBD \sim \triangle CEI$, so $\dfrac{IE}{DB} = \dfrac{CI}{CD}$. Thus, $\dfrac{2}{x} = \dfrac{\frac{2x}{3}}{\frac{2x}{3} + 5} \implies x = 5, -3 \implies x = 5$. Thence $CI = \dfrac{10}{3} \implies \boxed{13}$.*

Example 9.1.4.2. *(2012 Caltech-Harvey Mudd Math Competition) In triangle $\triangle ABC$, the angle bisector from A and the perpendicular bisector of $\overline{BC}$ meet at point D, the angle bisector from B and the perpendicular bisector of $\overline{AC}$ meet at point E, and the perpendicular bisectors of $\overline{BC}$ and $\overline{AC}$ meet at point F. Given that $\angle ADF = 5°$, $\angle BEF = 10°$, and $AC = 3$, find the length of DF.*

Solution 9.1.4.2. *$\triangle ABC$ has circumcenter F, so D is the midpoint of arc BC, and E is the midpoint of arc AC as a direct result. Then all radii of the circumcircle are equal, hence $FA = FB = FC = FD = FE$. $\triangle ADF, \triangle BEF$ are isosceles, so $\angle ADF = \angle DAF = 5°$ and $\angle BEF = EBF = 10°$.*

Henceforth, let $\angle FBC = \angle FCB = x°, \angle FCA = \angle FAC = y°, \angle FAB = \angle FBA = z°$. Then $x + y + z = 90°, y + 5° = z - 5°, x + 10° = z - 10° \implies x = 20°, y = 30°, z = 40°$; then $\angle ABC = 60°$. Law of Sines finishes the problem, with $DF = \boxed{\sqrt{3}}$.

9.1.5 Exercises

Exercise 301. *(2012 Online Math Open) Let $\triangle ABC$ be a triangle with circumcircle ω. Let the bisector of $\angle ABC$ meet segment $\overline{AC}$ at D and circle ω at $M \neq B$. The circumcircle of $\triangle BDC$ meets line $\overline{AB}$ at $E \neq B$, and $\overline{CE}$ meets ω at $P \neq C$. The bisector of $\angle PMC$ meets segment $\overline{AC}$ at $Q \neq C$. Given that $PQ = MC$, determine the degree measure of $\angle ABC$.*

Exercise 302. *(2014 Online Math Open) Let $\triangle ABC$ be a triangle with $AB = 26$, $AC = 28$, $BC = 30$. Let X, Y, and Z be the midpoints of arcs BC, CA, AB (not containing the opposite vertices) respectively on the circumcircle of $\triangle ABC$. Let P be the midpoint of arc BC containing point A. Suppose lines $\overline{BP}$ and $\overline{XZ}$ meet at M, while lines $\overline{CP}$ and $\overline{XY}$ meet at N. Find the square of the distance from X to $\overline{MN}$.*

Exercise 303. *Let O be the circumcenter of $\triangle ABC$. O is tangent to the sides of $\triangle ABC$ at points X, Y, and Z. Let the orthic triangle of $\triangle ABC$ be $\triangle DEF$. Prove that the isogonal conjugate of O with respect to $\triangle DEF$ is the orthocenter of $\triangle XYZ$.*

Extension Exercise 130.

   (a) What are the trilinear coordinates of the Miquel point of a triangle?

   (b) Determine when the Miquel point coincides with the circumcenter of the triangle.

(If you're stuck, it may help to read through §9.3 first.)

Extension Exercise 131. *(2006 IMO P1) Let $\triangle ABC$ be a triangle with incenter I. A point P in the interior of the triangle satisfies $\angle PBA + \angle PCA = \angle PBC + \angle PCB$. Show that $AP \geq AI$, and that equality holds if and only if $P = I$.*

Extension Exercise 132. *(2012 IMO P5) Let $\triangle ABC$ be a triangle with $\angle BCA = 90°$, and let D be the foot of the altitude from C. Let X be a point in the interior of the segment CD. Let K be the point on the segment $\overline{AX}$ such that $BK = BC$. Similarly, let L be the point on the segment $\overline{BX}$ such that $AL = AC$. Let M be the point of intersection of $\overline{AL}$ and $\overline{BK}$. Show that $MK = ML$.*

Extension Exercise 133. *(2006 USAMO P6) Let $ABCD$ be a quadrilateral, and let E and F be points on sides AD and BC, respectively, such that $AE/ED = BF/FC$. Ray FE meets rays BA and CD at S and T respectively. Prove that the circumcircles of triangles SAE, SBF, TCF, and TDE pass through a common point.*

9.2 (Some) Advanced Synthetic Geometry

9.2.1 Stewart's Theorem

In most situations, we are not given all three sides of a triangle, or otherwise lack the necessary information to apply other formulas. However, using Stewart's Theorem, we can relate the side lengths of any triangle to its cevian length.

Theorem 9.6: Stewart's Theorem

Let $\triangle ABC$ have side lengths a, b, c opposite vertices A, B, C respectively. Draw cevian $\overline{AD}$ such that $BD = m$, $DC = n$, and $AD = d$. Then we have $mb^2 + nc^2 = amn + ad^2$.

We present two different proofs of this theorem:

Using the Pythagorean Theorem. Drop the altitude from A down to P on $\overline{BC}$; then $DP^2 + PA^2 = AD^2$ since $\angle DPA = 90°$ by virtue of P being the foot of the altitude $\overline{AP}$. Then write b, c, d in terms of AP to reduce the resulting expression algebraically to Stewart's Theorem. (We don't want to bore you with computations; besides, it is more helpful to you, the reader, to work through them for the sake of self-discovery!) □

Using the Law of Cosines. Let θ be the angle spanned by $\overline{BD}$ and $\overline{DA}$, and let $\theta' = 180 - \theta$ be the complement of θ (and so $\cos\theta' = -\cos\theta$). By the Law of Cosines,

$$m^2 + d^2 - 2md\cos\theta$$

$$n^2 + d^2 + 2nd\cos\theta$$

If we add n times the first equation to m times the second equation, this will eliminate the $\cos\theta$ term and yield

$$mb^2 + nc^2 = nm^2 + n^2m + d^2(m+n)$$
$$= (m+n)(mn + d^2) = a(mn + d^2)$$

which is the desired equation. □

A common mnemonic device for Stewart's Theorem is phonetic: "A man and his dad put a bomb in the sink." (Don't recite this in public!) This arises from rewriting the equation as $man + dad = bmb + cnc$.

Note that, when the cevian coincides with the median of $\triangle ABC$, then Stewart's Theorem reduces to the special case of *Apollonius' Theorem* (proving this is left as an end-of-section exercise). Moreover, when $\triangle ABC$ is isosceles, Stewart's Theorem simplifies to the equation $b^2 = d^2 + mn$. (Why?)

9.2.2 The Radical Axis

The *radical axis* of two (non-concentric) circles has a definition that is remarkably intuitive: it is the locus of points - namely the line - for which the tangents drawn to both circles have the same length. From this single, highly-understandable definition, we can extract a wealth of truly interesting and - well, radical - properties of the radical axis!

General Properties

First things first: the radical axis is always a straight line, regardless of where the circles are positioned in the plane. Furthermore, it is perpendicular to the line connecting the centers of the circles. If the circles intersect, then their radical axis coincides with the line passing through the intersection points of the circles. Similarly, if the circles are (either internally or externally tangent), then the radical axis is just the common tangent.

Let P be a point on the radical axis. Then there exists a unique circle with center P intersecting both circles at right angles. In addition, P has the same power with respect to both circles; i.e.

$$R^2 = d_1^2 - r_1^2 = d_2^2 - r_2^2$$

where R denotes the radius of the circle with center P, d_1 and d_2 denote the distances from P to the centers of the first and the second circle, and r_1 and r_2 denote the radii of the first and second circles respectively.

If we have *three* circles ω_1, ω_2, ω_3, no two of which are concentric, then the (extended) *radical axis theorem* applies:

Theorem 9.7: Radical Axis Theorem

This theorem consists of two major statements that apply to the two-circle case (with circles ω_1 and ω_2, and respective centers O_1 and O_2), and one additional statement that applies to the analogous three-circle case (statement (e)).

(a) The radical axis is perpendicular to the line connecting the centers of the circles (henceforth line l).

(b) If the two circles intersect at two points, their radical axis is the line through the two intersection points.

(c) (Radical Axis Concurrence) The three radical axes of each pair of circles are either parallel,[a] or intersect at the *radical center*.

---
[a]In which case, they still concur at a point at infinity in a projective geometry.

Proof.

(a) For part (a), we invoke the following two lemmas:

Lemma 9.7: Difference of Powers of P and P' (9.2a)

Let P be a point in the plane and let P' be the foot of the perpendicular from P to l. Then we have $\mathrm{pow}(P, O_1) - \mathrm{pow}(P, O_2) = \mathrm{pow}(P', O_1) - \mathrm{pow}(P', O_2)$.

Proof. This is left to the reader as an application of the Pythagorean Theorem. $\square$

Lemma 9.7: Equality of Powers (9.2b)

On line l, there exists a unique point P such that $\mathrm{pow}(P, O_1) = \mathrm{pow}(P, O_2)$.

Proof. Using proof by contradiction, we can show that P is between O_1 and O_2 (coupled with the fact that $O_1 O_2 > r_1 + r_2$). This, combined with the fact that $O_1 P + P O_2$, completes the proof (the reader is left to fill in the details). $\square$

By Lemma 9.2a, every point in $\mathbb{R}^2$ can be mapped linearly to a line on l. It also follows from Lemma 9.2b that this point is unique. Hence, the radical axis must be a line perpendicular to l.

(b) We can also prove part (b) using Lemmas 9.2a and 9.2b, but here is a different approach:

Denote the intersection points of the circles ω_1, ω_2 by X and Y. Then by Power of a Point, with respect to both circles the powers of P are given by $PX \cdot PY$, which implies that $\mathrm{pow}(P, \omega_1) = \mathrm{pow}(P, \omega_2)$.

Hence, if P lies on XY, then these powers are equal.

Next, suppose P does not lie on XY. Then PY would not intersect X, and PY intersects ω_1 and ω_2 at points M and N respectively, with $M \neq N$, $M \neq X$, and $N \neq Y$. Since $PM \neq PN$, $PY \cdot PM \neq PY \cdot PN$; by Power of a Point, $\text{pow}(P, \omega_1) \neq \text{pow}(P, \omega_2)$ as desired.

(c) This is actually a deceptively simple result! This follows immediately from the transitive property of equality, as by definition, the radical axis of two of the circles (assume ω_1, ω_2) is such that the tangents a, b from those circles to any point on the radical axis have equal length. Similarly, ω_2 and ω_3 have tangents b, c of equal length. Then $a = b$, $b = c \implies a = b = c$ at the intersection point of those two radical axes. Finally, the radical axis for circles ω_1, ω_3 must pass through that intersection point, since $a = c$ at that point (which is then the radical center).

In particular, the radical center proves an extremely useful construction in applied problems. $\qquad \square$

We also have two other sub-statements following from (b):

Corollary 9.7: 1 or 0 Intersection Points

This particular corollary of statement (b) in Theorem 9.2 is two-fold:

1. If the two circles intersect at one point, then their radical axis is their common internal tangent.

2. If the two circles are disjoint, and provided one does not fully contain the other, then their radical axis is perpendicular to l through point A, which is the unique point on l such that $\text{pow}(A, O_1) = \text{pow}(A, O_2)$.

Relationship with the Determinant

By representing the circles and lines in *trilinear coordinates* (see §9.3.3 for further reference), we can derive a relationship between the radical axis and the matrix determinant. This relationship is often unseen or overlooked, but it wields the immense power to pin down the radical center with little need for concurrence or pairwise radical axis determination.

Theorem 9.8: Radical Center as a Determinant

Let $X = x : y : z$ be the trilinear coordinates of a point X in the interior of $\triangle ABC$ with side lengths $a = BC$, $b = AC$, $c = AB$. Then the circles have trilinear equations given by

$$(dx + ey + fz)(ax + by + cz) + g(ayz + bzx + cxy) = 0$$

$$(hx + iy + jz)(ax + by + cz) + k(ayz + bzx + cxy) = 0$$

$$(lx + my + nz)(ax + by + cz) + p(ayz + bzx + cxy) = 0$$

Then the radical center has trilinear coordinates

$$\det \begin{bmatrix} g & k & p \\ e & i & m \\ f & j & n \end{bmatrix} : \det \begin{bmatrix} g & k & p \\ f & j & n \\ d & h & l \end{bmatrix} : \det \begin{bmatrix} g & k & p \\ d & h & l \\ e & i & m \end{bmatrix}$$

9.2.3 Spiral Similarity

Definition 9.1: Spiral Similarity

A spiral similarity is the combination of a homothety and a rotation about the same point.

Right now, this probably seems like a quite pointless transformation. However, we will be building up to the Fundamental Theorem of Spiral Similarity, a lemma that can be used to simplify many problems.

Theorem 9.9: Center of Spiral Similarity

For any A, B, C, D that does not form a parallelogram, there is a unique spiral similarity that sends $A \to B$ and $C \to D$.

Proof. We use the complex plane to prove this. Let the spiral similarity be $f(z) = z_o + \alpha(z - z_o)$. (This can easily be proven to be the general form of a spiral similarity.) Notice this amounts to solving $z_o + \alpha(a - z_o) = c$ and $z_o + \alpha(b - z_o) = d$. This yields $\alpha = \frac{c-d}{a-b}$ and $z_0 = \frac{ad-bc}{a+d-b-c}$. $\square$

Remark. This theorem is only here to ensure the validity of our lemma, as it is torn apart if there is not exactly one center of spiral similarity.

And now for the lemma itself.

Theorem 9.10: The Fundamental Theorem of Spiral Similarity

Consider A, B, C, D such that AC and BD are not parallel. Let AC intersect BD at X. Then let the circumcircles of $\triangle ABX$ and $\triangle CDX$ intersect at O. Then the center of the spiral that sends $A \to C$ and $B \to D$ is O.

Proof. Because the orientation matters in spiral similarity, we use directed angles mod $180°$. (This means lines, not rays!) This means four points A, B, C, D are concyclic if and only if $\angle(AB, BC) = \angle(AD, DC)$.

Notice that $\angle(OA, AC) = \angle(OA, AX) = \angle(OB, BX) = \angle(OB, BD)$ and $\angle(OC, CA) = \angle(OC, CX) = \angle(OD, DX) = \angle(OD, DB)$ by Inscribed Angle. It then follows from Angle-Angle Similarity that $\triangle AOC \sim \triangle BOD$ and the two triangles have the same orientation. Thus, there is a spiral similarity centered at O that sends $A \to C$ and $B \to D$. $\square$

Theorem 9.11: Pairs of Spiral Similarities

If O is the center of the spiral similarity from $A \to C$, $B \to D$, then O also is the center of the spiral similarity from $A \to B$, $C \to D$.

Proof. We use directed angles once more due to orientation.

Notice $\angle(AO, OB) = \angle(CO, OD)$ as similitudes preserve angles. Also, $r = \frac{OC}{OA} = \frac{OD}{OB}$. Rearranging yields $\frac{OB}{OA} = \frac{OD}{OC}$. Then the similitude of angle $\angle(AO, OB) = \angle(CO, OD)$ and dilation of $\frac{OB}{OA} = \frac{OD}{OC}$ centered at O sends $A \to B, C \to D$. $\square$

9.2.4 Examples

Example 9.2.4.1. *Consider $\triangle ABC$ and its medial triangle $\triangle A'B'C'$. Prove that both triangles share the same centroid G and that a spiral similarity of $\triangle ABC$ about G yields $\triangle A'B'C'$.*

Solution 9.2.4.1. *The two triangles are clearly similar by a ratio of $\frac{1}{2}$ with identical orientation, so there must be a center of spiral similarity. Then notice by similar triangles, AA' passes the midpoint of $B'C'$, so the medians of $\triangle ABC$ are also the medians of $\triangle A'B'C'$. Thus a spiral similarity about G with ratio $\frac{1}{2}$ and angle $180°$ will yields $\triangle A'B'C'$.*

Example 9.2.4.2. *(ARML) In triangle $\triangle ABC$, $AB = AC = 17$. Point D is on $\overline{BC}$ such that $CD = BD + 8$. If $AD = 16$, what is BC?*

Solution 9.2.4.2. *Let $BD = x, CD = x + 8$. By Stewart's Theorem, let $a = 2x + 8, b = c = 17, d = 16, m = x, n = x + 8$. Then $man + dad = bmb + cnc \implies x(2x + 8)(x + 8) + 256(2x + 8) = 289x + 289(x + 8) = 289(2x + 8) \implies x(x + 8) + 256 = 289 \implies x^2 + 8x - 33 = 0 \implies x = \dfrac{-8 \pm 3\sqrt{22}}{2} \implies x = \dfrac{3\sqrt{22} - 8}{2}$. Hence, $BC = 2x + 8 = \boxed{3\sqrt{22}}$.*

9.2.5 Exercises

Exercise 304. *Prove that any $\triangle ABC \sim \triangle DEF$ with the same orientation has a center O such that some similitude about O sends $\triangle ABC \to \triangle DEF$ given that none of the 15 lines formed by the 6 points A, B, C, D, E, F are parallel.*

Exercise 305. *(2015 AIME I) Point B lies on line segment $\overline{AC}$ with $AB = 16$ and $BC = 4$. Points D and E lie on the same side of line AC forming equilateral triangles $\triangle ABD$ and $\triangle BCE$. Let M be the midpoint of $\overline{AE}$, and N be the midpoint of $\overline{CD}$. The area of $\triangle BMN$ is x. Find x^2.*

Exercise 306. *(2017 AIME II) Circle C_0 has radius 1, and the point A_0 is a point on the circle. Circle C_1 has radius $r < 1$ and is internally tangent to C_0 at point A_0. Point A_1 lies on circle C_1 so that A_1 is located $90°$ counterclockwise from A_0 on C_1. Circle C_2 has radius r^2 and is internally tangent to C_1 at point A_1. In this way a sequence of circles $C_1, C_2, C_3, \ldots$ and a sequence of points on the circles $A_1, A_2, A_3, \ldots$ are constructed, where circle C_n has radius r^n and is internally tangent to circle C_{n-1} at point A_{n-1}, and point A_n lies on C_n $90°$ counterclockwise from point A_{n-1}, as shown in the figure below. There is one point B inside all of these circles. When $r = \frac{11}{60}$, the distance from the center C_0 to B is $\frac{m}{n}$, where m and n are relatively prime positive integers. Find $m + n$.*

Exercise 307. *Prove that two directly similar triangles are related by a spiral similarity, and that two triangles related by a spiral similarity are directly similar.*

Exercise 308. *Consider $\triangle ABO$ and $\triangle CDO$ such that $\triangle ABO \sim \triangle CDO$ with the same orientation. Then let AC intersect BD at X. Prove that $\angle AXB = \angle COD$.*

Exercise 309. *(Fundamental Theorem of Similarity) Given $A_1 A_2 \ldots A_n \sim B_1 B_2 \ldots B_n$ with the same orientation, prove that for all $M_1, M_2 \ldots M_n$ such that $\frac{A_i M_i}{B_i M_i} = r$ where r is constant for all $1 \leq i \leq n$, $M_1 M_2 \ldots M_n \sim A_1 A_2 \ldots A_n \sim B_1 B_2 \ldots B_n$.*

Exercise 310. *(2014 IMO P4) Points P and Q lie on side BC of acute-angled $\triangle ABC$ so that $\angle PAB = \angle BCA$ and $\angle CAQ = \angle ABC$. Points M and N lie on lines AP and AQ, respectively, such that P is the midpoint of AM, and Q is the midpoint of AN. Prove that lines BM and CN intersect on the circumcircle of $\triangle ABC$.*

Exercise 311. *Consider directly similar $\triangle ABC \sim \triangle DEF$. Let X, Y, Z be the intersection of AD and BE, BE and CF, and CF and AD, respectively. Prove that $\triangle XYZ$ is isosceles.*

Extension Exercise 134. *(1978 IMO P4) In the triangle $\triangle ABC$, $AB = AC$. A circle is tangent internally to the circumcircle of the triangle and also to $\overline{AB}$, $\overline{AC}$ at P, Q respectively. Prove that the midpoint of $\overline{PQ}$ is the center of the incircle of the triangle.*

Extension Exercise 135. *(1995 IMO P1) Let A, B, C, D be four distinct points on a line, in that order. The circles with diameters AC and BD intersect at X and Y. The line XY meets BC at Z. Let P be a point on the line XY other than Z. The line CP intersects the circle with diameter AC at C and M, and the line BP intersects the circle with diameter BD at B and N. Prove that the lines AM, DN, XY are concurrent.*

Extension Exercise 136. *(2003 USAMO P4) Let ABC be a triangle. A circle passing through A and B intersects segments AC and BC at D and E, respectively. Lines AB and DE intersect at F, while lines BD and CF intersect at M. Prove that $MF = MC$ if and only if $MB \cdot MD = MC^2$.*

Extension Exercise 137. *(1981 IMO P5) Three congruent circles have a common point O and lie inside a given triangle. Each circle touches a pair of sides of the triangle. Prove that the incenter and the circumcenter of the triangle and the point O are collinear.*

9.3 Polar, Cylindrical, and Barycentric Coordinates

9.3.1 Polar Coordinates: Connections with Complex Numbers

In working with real numbers, we ccan safely and conveniently restrict ourselves to the standard, familiar tow-dimensional Cartesisan plane $\mathbb{R}^2$, with no ill effect. However, in working with the complex numbers, it is far more efficient and less messy to use the *polar coordinate system* to manipulate expressions, make substitutions, use complex formulae, and derive important properties of the complex number system.

Instead of each quantity being assigned x- and y-coordinates corresponding to its position in space, it is assigned r- and θ-coordinates, with $r \geq 0$ representing the radius (i.e. the scale factor of homothety) and $0 \leq \theta < 2\pi$ representing the angle of rotation in radians (with $0°$ being the rightmost part of the circle, and $90° = \dfrac{\pi}{2}$ being the uppermost part). The larger the value of r, the greater the distance from the point to the origin. The angle θ starts at $0°$ from the counter-clockwise reference direction, and continues until $2\pi = 360°$ (assuming uniqueness of polar coordinates).

A critical construction of the plane is the conversion of points between Cartesian and polar coordinates. We make the following substitutions:
$$x = r\cos\theta, y = r\sin\theta$$

which also paints a picture of why, in a right triangle, we define the sine of an angle as the ratio of the opposite to the hypotenuse, and the cosine of that angle as the ratio of the adjacent to the hypotenuse. Thus, $r = \sqrt{x^2 + y^2}$ (which is conducive to a Pythagorean Theorem-based interpretation), and $\theta = \text{atan2}(x, y)$, where we define the function atan2 as follows:

$$\text{atan2}(x, y) = \begin{cases} \arctan\left(\frac{y}{x}\right) & x > 0 \\ \arctan\left(\frac{y}{x}\right) + \pi & x < 0, y \geq 0 \\ \arctan\left(\frac{y}{x}\right) - \pi & x < 0, y < 0 \\ \dfrac{\pi}{2} & x = 0, y > 0 \\ -\dfrac{\pi}{2} & x = 0, y < 0 \\ \text{undefined} & x = y = 0 \end{cases}$$

9.3.2 Cylindrical Coordinates

In similar fashion to how we can re-write Cartesian coordinates in 2-space as polar coordinates, so we can generalize this to 3-space through *cylindrical coordinates*. The cylindrical coordinates of a point $P = (x, y, z) \in \mathbb{R}^3$ are of the form (ρ, θ, z), where ρ is the distance from P to the z-axis, θ is the angle (or *azimuth*) between the reference direction and the line from P to the projection of P on the plane, and z is the distance from P to the chosen plane.

The conversions between Cartesian coordinates and cylindrical coordinates are given by

$$x = \rho\cos\theta$$

$$y = \rho\sin\theta$$

$$z = z$$

Note that z is unchanged (assuming WLOG that P is suspended above the xy-plane), and much like with polar coordinates, we have $\rho = \sqrt{x^2 + y^2}$ (using the identity $\sin^2\theta + \cos^2\theta = 1$ for all θ).

9.3.3 The Barycentric Coordinate System

For the previous two sections, we primarily went over polar and cylindrical coordinates for purposes of completeness, and leave the reader to ponder their implications with regard to applications in contest problems.

(Hint: Using the identity $\sin^2 x + \cos^2 x = 1$ will get you very far!) Here, we will spend much more time on *barycentric coordinates,* as their use is a fundamental component of several challenging contest problems, and moreover presents a wonderfully efficient and versatile "bashing" technique.

Think back to how we solved problems using mass points. We can further extend this line of thought to any point P in the interior of a triangle $\triangle ABC$. If we consider the triangle as a convex set of points, with a spectrum from 0 to 1 representing the proximity of P to the respective vertex, we can set $A = (1, 0, 0)$, $B = (0, 1, 0)$, and $C = (0, 0, 1)$, and furthermore any point that lies on an edge of the triangle (suppose $\overline{AB}$ has a barycentric coordinate of 0 corresponding the other vertex (since it is a function of the form $tA + (1-t)B$, where $t \in [0, 1]$ is a parameter). In general, we define barycentric coordinates in an n-dimensional simplex as follows:

Definition 9.2: Barycentric Coordinates in n-Space

In $\mathbb{R}^n$, the n-simplex (analogue of the cube in $\mathbb{R}^3$) has barycentric coordinates $(1, 0, 0, \ldots, 0)$, $(0, 1, 0, \ldots, 0)$, $(0, 0, 1, \ldots, 0)$, $\ldots$, $(0, 0, 0, \ldots 1)$ where each tuple has $n + 1$ coordinates. Any point P lying in the interior of the n-simplex has (homogenized) coordinates $(t_1, t_2, t_3, \ldots, t_{n+1})$, with $t_1 + t_2 + t_3 + \ldots + t_{n+1} = 1$ and t_i is proportional to the distance of P from its respective vertex.

Remark. Barycentric coordinates are homogeneous (such that $(t_1, t_2, t_3) = (\alpha t_1, \alpha t_2, \alpha t_3)$ for all $\alpha > 0$).

In particular, we will focus on barycentric coordinates in triangles. The following is a table of common barycentric coordinates (not necessarily homogeneous) for points relative to a triangle $\triangle ABC$. (Here a, b, c are the side lengths of $\triangle ABC$ and $s = \dfrac{a+b+c}{2}$ is the semi-perimeter.)

point	bary coordinates
circumcenter	$(a^2(b^2 + c^2 - a^2), b^2(a^2 + c^2 - b^2), c^2(a^2 + b^2 - c^2))$
incenter	(a, b, c)
A-excenter	$(-a, b, c)$
B-excenter	$(a, -b, c)$
C-excenter	$(a, b, -c)$
orthocenter	$((a^2 + b^2 - c^2)(a^2 + c^2 - b^2), (b^2 + c^2 - a^2)(a^2 + b^2 - c^2), (a^2 + c^2 - b^2)(b^2 + c^2 - a^2))$
centroid	$(1, 1, 1)$
symmedian point	(a^2, b^2, c^2)

In particular, note that the centroid has equal coordinates (this makes sense, since by definition, it is equidistant from all each vertices); when homogenized, its coordinates become $\left(\dfrac{1}{3}, \dfrac{1}{3}, \dfrac{1}{3}\right)$). Also note the convolutedness of the circumcenter and orthocenter formulae! (Even these formulae are still somewhat elegant in that they have a symmetry to them; however, we still don't recommend memorizing them. If dealing with a circumcenter or orthocenter in a bary-bashing problem, either try to re-derive it using properties of barycentric coordinates, or find an alternate approach.)

Trilinear Coordinates

Much like trilinear coordinates, trilinear coordinates are defined relative to the sides of the triangle in which a point P lies in the interior. However, we are now considering the least distances to each *side* of the triangle (perhaps using the point-to-line distance formula as an aid in this process), as opposed to the proximity to each vertex.

To be precise:

Definition 9.3: Trilinear Coordinates

Let P be a point in $\mathbb{R}^3$. If P lies outside $\triangle ABC$, then the trilinear coordinates of P are (x, y, z), where $x : y$ is the ratio of the lengths of the perpendiculars to $\overline{BC}$ and $\overline{AC}$ respectively, $y : z$ is defined likewise for $\overline{AC}$ and $\overline{AB}$, and $z : x$ is defined likewise for $\overline{AB}$ and $\overline{BC}$.

In particular, if P lies in the interior of $\triangle ABC$, then the trilinear coordinates of P, (x, y, z), are given by (a', b', c'), where a' is the length of the perpendicular from P to $\overline{BC}$, and b', c' are defined analogously. (These coordinates are known as *exact* or *normalized* trilinear coordinates.)

The Art of Bary Bashing

We now discuss a favorite tool of Olympiad bashers everywhere: *bary bashing!* Though this section does not place the focus on mathematical rigor that the other sections do, it is arguably just as important to master the *technique,* as well as the motivation, behind problem solving. Without technique (both in terms of the thought process and the grindy, nitty-gritty computation), one may find oneself severely hindered in reaching the final answer or desired conclusion despite knowing the general solution path.

In this section, it is instrumental to make the connection that barycentric coordinates yield an immediate relationship with area ratios: namely, the areas of the regions formed by the point P and the pairs of vertices of $\triangle ABC$ should be in the ratios specified by the respective barycentric coordinates of P. That is, if $P = (t_1, t_2, t_3)$, then $[PAB] : [PAC] : [PBC] = t_1 : t_2 : t_3$, with $t_1 + t_2 + t_3 = 1$ (homogenized) or $t_1 + t_2 + t_3 = [ABC]$ (non-homogenized). Thus, we can interpret the bary coordinates of P as

$$P = \left(\frac{[PBC]}{[ABC]}, \frac{[PAC]}{[ABC]}, \frac{[PAB]}{[ABC]} \right)$$

A line in the barycentric coordinate system has an equation of the form $\alpha x + \beta y + \gamma z$, where $P = (x, y, z)$ and $\alpha + \beta + \gamma = [ABC]$ (non-homogenized) or $\alpha + \beta + \gamma = 1$ (homogenized). In particular, if the line passes through a vertex (say A), then it is of the form $y = kz$ for some $k \in \mathbb{R}$.

We can also re-state the Distance Formula in $\mathbb{R}^n$ in terms of barycentric coordinates:

Theorem 9.12: Barycentric Distance Formula

Let $P = (x_1, y_1, z_1)$ and $Q = (x_2, y_2, z_2)$ be points in the plane of a triangle $\triangle ABC$ such that $|P - Q| = (x, y, z)$. Then $PQ^2 = -a^2 yz - b^2 xz - c^2 xy$.

Furthermore, we can state the general barycentric equation of a circle:

Theorem 9.13: Barycentric Circle Equation

For $\alpha + \beta + \gamma = 1$, the equation of a circle is given by

$$-a^2 yz - b^2 xz - c^2 xy + (\alpha x + \beta y + \gamma z)(x + y + z) = 0$$

Proof. If the circle has center (d, e, f) and radius r, then

$$-a^2 (y - e)(z - f) - b^2 (x - d)(z - f) - c^2 (x - d)(y - e)$$

Expanding and simplifying yields

$$-a^2 yz - b^2 xz - c^2 xy + C_1 x + C_2 y + C_3 z = C$$

where C_1, C_2, C_3, and C are constants. Because $\alpha + \beta + \gamma = 1$, however, this further re-writes to the desired form, from setting $\alpha = C_1 - C$, $\beta = C_2 - C$, $\gamma = C_3 - C$. $\qquad \square$

In particular, this yields the equation of the circumcircle of $\triangle ABC$.

Though these are the fundamentals of bary-bashing, we still have more ground to cover! Recall that, in Cartesian coordinates with respect to the Euclidean plane, we can use a matrix determinant to apply Shoelace Theorem and obtain the area of any solid in $\mathbb{R}^3$. Here, the basic principle is the same, except arguably much simpler:

Theorem 9.14: Barycentric Shoelace Theorem

In barycentric coordinates, the area of the triangle $\triangle PQR$ with respect to $\triangle ABC$, and with $P = (x_1, y_1, z_1)$, $Q = (x_2, y_2, z_2)$, $R = (x_3, y_3, z_3)$ is given by

$$[ABC] \cdot \det \begin{bmatrix} x_1 & y_1 & z_1 \\ x_2 & y_2 & z_2 \\ x_3 & y_3 & z_3 \end{bmatrix}$$

Proof. Let point X lie outside of the plane of $\triangle ABC$. Then let $X = (0,0,0)$, $A = (1,0,0)$, $B = (0,1,0)$, $C = (0,0,1)$ (up to translation). Let $\mathcal{P}(A, B, C)$ be the parallelepiped spanned by the vectors $\overrightarrow{XA}$, $\overrightarrow{XB}$, $\overrightarrow{XC}$ (and define $\mathcal{P}(PQR)$ similarly). By the definition of the determinant in terms of volume, we have

$$\frac{V(\mathcal{P}(PQR))}{V(\mathcal{P}(ABC))} = \det \begin{bmatrix} x_1 & y_1 & z_1 \\ x_2 & y_2 & z_2 \\ x_3 & y_3 & z_3 \end{bmatrix}$$

where $V(A)$ denotes the volume of A. Since $V(\mathcal{P}(ABC)) = 6V(XABC) = 6 \cdot \dfrac{h[ABC]}{3} = 2[ABC]$ (and similarly for $V(\mathcal{P}(PQR))$), we have the determinant of the matrix equal to $\dfrac{[PQR]}{[ABC]}$, as desired. $\qquad \square$

Corollary 9.14: Zero Determinant

Iff the determinant is zero, then A, B, C are collinear (hence forming a degenerate triangle with area zero).

As a prime example of the effectiveness of bary-bashing, we can prove Ceva's and Menelaus' Theorems in terms of barycentric coordinates:

Proofs using barycentric coordinates.

- Ceva's Theorem: As point D lies on $\overline{BC}$, it has barycentric coordinates $(0, d, 1-d)$ for $d \in (0,1)$. Then $\overline{AD}$ has equation $z = y \cdot \dfrac{1-d}{d}$. Similarly, $E = (1-e, 0, e)$, $F = (f, 1-f, 0)$, and so $\overline{BE}$ and $\overline{CF}$ have respective equations $x = z \cdot \dfrac{1-e}{e}$ and $y = x \cdot \dfrac{1-f}{f}$.

  Multiplying the three equations together and dividing by xyz yields $1 = \dfrac{1-d}{d} \cdot \dfrac{1-e}{e} \cdot \dfrac{1-f}{f}$ which reduces to Ceva's theorem, as desired.

- Menelaus' Theorem: Let P, Q, R be points on the extensions of $\overline{BC}$, $\overline{AC}$, $\overline{AB}$ respectively such that $BP \cdot CQ \cdot AR = -QA \cdot RB \cdot PC$ (directed segments). Assign P, Q, R the bary coordinates of $P = (0, P, 1-P)$, $Q = (1-Q, 0, Q)$, $R = (R, 1-R, 0)$. Note that this is equivalent to $\dfrac{CP}{PB} = \dfrac{1-P}{P}$,

$\frac{QA}{QC} = \frac{1-Q}{Q}, \frac{BR}{AR} = \frac{1-R}{R}$. Line $\overline{RP}$ has barycentric equation ensuing from the determinant

$$\det \begin{bmatrix} x & 0 & R \\ y & P & 1-R \\ z & 1-P & 0 \end{bmatrix} = 0$$

which implies that $z \cdot PR = -x \cdot (R-1)(P-1) + Y \cdot R(1-P)$. We also have $PQR = (P-1)(Q-1)(R-1)$, and in conjunction with $\frac{CP}{PB} = \frac{1-P}{P}$, this yields $P = \frac{BP}{CP} \cdot (1-P)$, $Q = \frac{QC}{QA} \cdot (1-Q)$, and $R = \frac{AR}{BR} \cdot (1-R)$. Substituting these into the last equation yields $QA \cdot RB \cdot PC = -AR \cdot BP \cdot CQ$.

$\square$

9.3.4 Examples

Example 9.3.4.1. *(2009 AMC 12B) A region S in the complex plane is defined by*

$$S = \{x + iy : -1 \le x \le 1, -1 \le y \le 1\}.$$

A complex number $z = x + iy$ is chosen uniformly at random from S. What is the probability that $\left(\frac{3}{4} + \frac{3}{4}i\right) z$ is also in S?

Solution 9.3.4.1. *We can compute $\left(\frac{3}{4} + \frac{3}{4}i\right) z = \frac{3}{4}(x-y) + \frac{3}{4}(x+y)i$ to be in S iff $-1 \le \frac{3(x-y)}{4} \le 1$ and $-1 \le \frac{3(x+y)}{4} \le 1 \implies |x-y| \le \frac{4}{3}, |x+y| \le \frac{4}{3}$.*

Let T be the set of $x + iy$ satisfying these conditions. Then we want the probability that $\frac{[S \cap T]}{[S]}$, i.e. $\frac{[S \cap T]}{4}$. We can compute the area of intersection graphically to get $\boxed{\frac{7}{9}}$.

Example 9.3.4.2. *(2018 AIME II) Suppose that x, y, and z are complex numbers such that $xy = -80 - 320i$, $yz = 60$, and $zx = -96 + 24i$, where $i = \sqrt{-1}$. Then there are real numbers a and b such that $x + y + z = a + bi$. Find $a^2 + b^2$.*

Solution 9.3.4.2. *We have $|xy| = 80\sqrt{17}, |yz| = 60, |xz| = 24\sqrt{17}$, so $|xyz| = 240\sqrt{34}$. Then $|x| = 40\sqrt{34}, |y| = 10\sqrt{2}, |z| = 3\sqrt{2}$. Thus, $\arg y = \frac{\pi}{4}, \arg z = -\frac{\pi}{4}$. Converting to rectangular form from polar form yields $y = 10 + 10i, z = 3 - 3i \implies x = -20 - 12i \implies x + y + z = -7 - 5i \implies \boxed{74}$.*

9.3.5 Exercises

Exercise 312. *(2013 AIME II) In $\triangle ABC$, $AC = BC$, and point D is on $\overline{BC}$ so that $CD = 3 \cdot BD$. Let E be the midpoint of $\overline{AD}$. Given that $CE = \sqrt{7}$ and $BE = 3$, the area of $\triangle ABC$ can be expressed in the form $m\sqrt{n}$, where m and n are positive integers and n is not divisible by the square of any prime. Find $m + n$.*

Exercise 313. *(1992 AIME) In triangle ABC, A', B', and C' are on the sides BC, AC, and AB, respectively. Given that AA', BB', and CC' are concurrent at the point O, and that $\dfrac{AO}{OA'} + \dfrac{BO}{OB'} + \dfrac{CO}{OC'} = 92$, find $\dfrac{AO}{OA'} \cdot \dfrac{BO}{OB'} \cdot \dfrac{CO}{OC'}$.*

Exercise 314. *(1999 AIME) Let $\mathcal{T}$ be the set of ordered triples (x, y, z) of nonnegative real numbers that lie in the plane $x + y + z = 1$. Let us say that (x, y, z) supports (a, b, c) when exactly two of the following are true: $x \geq a, y \geq b, z \geq c$. Let $\mathcal{S}$ consist of those triples in $\mathcal{T}$ that support $\left(\frac{1}{2}, \frac{1}{3}, \frac{1}{6}\right)$. The area of $\mathcal{S}$ divided by the area of $\mathcal{T}$ is m/n, where m and n are relatively prime positive integers. Find $m + n$.*

Exercise 315. *(1986 AIME) In $\triangle ABC$, $AB = 425$, $BC = 450$, and $AC = 510$. An interior point P is then drawn, and segments are drawn through P parallel to the sides of the triangle. If these three segments are of an equal length d, find d.*

Exercise 316. *(2012 ARML Team Round #7) Given noncollinear points A, B, C, segment $\overline{AB}$ is trisected by points D and E, and F is the midpoint of segment $\overline{AC}$. $\overline{DF}$ and $\overline{BF}$ intersect $\overline{CE}$ at G and H, respectively. If $[DEG] = 18$, compute $[FGH]$. [Author's note: $[A]$ denotes the area of A.]*

Extension Exercise 138. *(2006 Putnam A1) Find the volume of the region of points (x, y, z) such that*

$$(x^2 + y^2 + z^2 + 8)^2 \leq 36(x^2 + y^2)$$

Extension Exercise 139. *(2015 USAMO P2) Quadrilateral $APBQ$ is inscribed in circle ω with $\angle P = \angle Q = 90°$ and $AP = AQ < BP$. Let X be a variable point on segment $\overline{PQ}$. Line AX meets ω again at S (other than A). Point T lies on arc AQB of ω such that $\overline{XT}$ is perpendicular to $\overline{AX}$. Let M denote the midpoint of chord $\overline{ST}$. As X varies on segment $\overline{PQ}$, show that M moves along a circle.*

9.4 Inversion

9.4.1 Introduction

Inversion is a transformation that is taken with respect to a circle. It is an **involution,** meaning that if you invert twice, you get the original result. The idea of inversion directly and indirectly reveals a wealth of information and significantly expands your abilities. However, it's still a low-priority subject compared to fundamentals.

> **Definition 9.4: Inversion**
>
> To invert a point P about circle ω with radius r, we take the unique point P' on ray $\overrightarrow{OP}$ such that $OP \cdot OP' = r^2$.

9.4.2 Preliminary Exercise

Exercise 317.

(a) *Invert P about circle ω to get P'. If P' is outside of ω and the tangent points from P' to ω are X, Y, prove P, X', Y' are collinear.*

(b) *Consider circle ω with diameter AB. What do you get when you invert line AB about ω?*

(c) *What about inverting segment AB?*

(d) *Prove that if you invert about a unit circle centered at the origin in the Cartesian plane, $P = (x, y)$ goes to $P' = (\frac{x}{x^2+y^2} + \frac{y}{x^2+y^2})$.*

(e) *Prove that if you invert about a unit circle centered at the origin in the Complex plane, z goes to $\overline{z}^{-1}$.*

9.4.3 The Point at Infinity

Where does O go when you invert about circle ω with center O? Which point goes to O? To make life easier (and to make lines into generalized circles), we let O go to the point at infinity. Then every line must pass the point at infinity.

Now we can think of lines as circles passing through the point at infinity. This means that for any three points, a generalized circle is uniquely determined.

9.4.4 Generalized Circles

Inversion is a powerful tool that can turn collinearity problems into concyclic problems, and vice versa. What does that mean?

It means that usually lines go to circles, sometimes circles go to circles, and sometime circles go to lines. We'll take a look at all 3 cases (the case where a line goes to a line is obvious and is omitted), and we'll see when each happens.

> **Theorem 9.15: Circle to Circle**
>
> Consider circle ω with center O and circle Γ not passing through O. Then Γ inverts to another circle.

Proof. Draw line $\overline{OXY}$ such that $\overline{XY}$ is a diameter of Γ. Then draw arbitrary ray $\overrightarrow{OR}$, with R on Γ. Let S be the other intersection of $\overrightarrow{OR}$ with Γ. Then let R, S get inverted to R', S'. By the definition of inversion, $OR \cdot OR' = OS \cdot OS' = r^2$. Rearranging yields $OR' = \frac{r^2}{OR \cdot OS} \cdot OS$. But by Power of a Point,

$OR \cdot OS = OX \cdot OY$. This implies that $OR' = \frac{r^2}{OX \cdot OY} \cdot OS$, so R' is a dilation of S about O by some constant scale factor. Thus, as S traces Γ, R' traces a dilation of Γ, which is a circle. $\square$

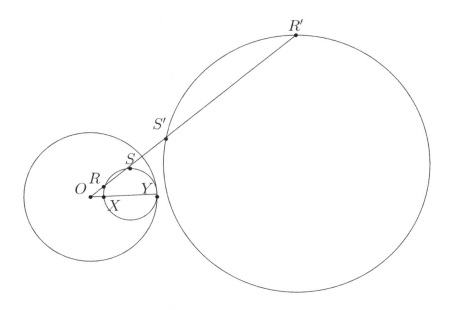

Figure 9.1: A visual representation of the circle-to-circle proof.

Theorem 9.16: Circle to Line

Consider circle ω with center O and circle Γ passing through O. Then Γ inverts to a line.

Proof. Let OP be a diameter of Γ. Then let the inversion of P about ω be Q. Then pick any point on Γ and let its inversion about ω be Y. By the definition of inversion, $OP \cdot OQ = OX \cdot OY = r^2$. Rearranging yields $\frac{OP}{OX} = \frac{OY}{OQ}$, implying $\triangle OPX \sim \triangle OYQ$. By Inscribed Angle, $\angle OXP = 90°$, implying $\angle OQY = 90°$. So the locus of Y is the locus of points such that $OQ \perp QY$, which is a line. $\square$

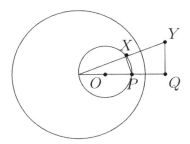

Figure 9.2: A visual representation of the circle-to-circle proof.

Theorem 9.17: Line to Circle

Consider circle ω with center O. Then the inversion of a line about ω yields a circle passing through O.

Proof. Let the line be l. Drop a perpendicular from O to l, and let the foot be Q. Then pick some other point Y on l. Let the inverse of Q be P, and the inverse of Y be X. Then notice $\triangle OXP \triangle OQY$. (This was proved in the proof of Circle to Line.) Thus, X is the locus of points such that $\angle OXP = 90$, which is also known as a circle with diameter OP. $\qquad\square$

As a side note: The proofs demonstrate that changing the radius of the circle of inversion merely changes the result by a dilation. This means that we can invert about a point and leave out the radius of the circle. Perhaps more powerful is the fact that we can choose the radius of the circle of inversion in a specific way to make the problem easier to solve (such as by cleverly invoking a lemma or reducing the problem to a more common configuration).

9.4.5 Check-up Exercises

Exercise 318. *Given the normal construction tools and the ability to dilate by any factor, construct the inversion of a circle not passing through the center of the circle of inversion.*

Exercise 319. *With the same tools, construct the inversion of a circle passing through the center of the circle of inversion.*

Exercise 320. *With the same tools, construct the inversion of a line.*

9.4.6 Poles and Polars

Poles and polars give name to some of the most tremendous powers of inversion, and make it so we can apply them easily. One of the most important results of poles and polars is La Hire's.

Definition 9.5: Poles and Polars

The pole of a point P with respect to circle ω is the point Q that results from an inversion of P about ω.

The polar of P with respect to ω is the line l passing through its pole Q such that $l \perp PQ$.

Theorem 9.18: La Hire's

If P lies on the polar of Q, then Q lies on the polar of P.

Proof. By Power of a Point, P, P, Q, Q are concyclic. Since P is on the polar of Q, $\angle PQQ = 90°$. By the Inscribed Angle Theorem, $\angle PP'Q = \angle PQ'Q = 90°$. Thus Q is on the polar of P. $\qquad\square$

9.4.7 Examples

Example 9.4.7.1. *Prove that the polar of P with respect to circle ω with center O is the result of an inversion of the circle with diameter OP.*

Solution 9.4.7.1. *This is a direct application of the Line to Circle Theorem. Let Γ be the circle with diameter OP. Notice that as Γ is traced out, a perpendicular line through the pole is traced out. (This is called the polar of P.)*

Example 9.4.7.2. *Two circles ω and Γ with centers O, O' that intersect at X, Y are called orthogonal if and only if $OX \perp O'X$ and $OY \perp O'Y$. Prove that if ω and Γ are orthogonal, then an inversion about ω preserves Γ.*

Solution 9.4.7.2. *Let a line passing through O intersect Γ at P, Q. But by Power of a Point, $OX^2 = OP \cdot OQ$, implying that Q is the polar of P. As P traces out Γ, so will Q.*

Example 9.4.7.3. *Consider $\triangle ABC$, and let its incircle touch BC, CA, AB at D, E, F, respectively. Prove that an inversion of the circumcircle of $\triangle ABC$ about the incircle of $\triangle ABC$ yields the nine-point circle of $\triangle DEF$.*

Solution 9.4.7.3. *Since circles go to circles, we only need to prove three points of the circumcircle of $\triangle ABC$ belong on the nine-point circle of $\triangle DEF$. The easiest points to do this with are A, B, C. Notice that the inverse of A is the midpoint of EF as AI perpendicularly bisects EF. (This is because AE, AF are tangents to the incircle.)*

Analogously, the inverse of B and C are the midpoints of CA and AB, respectively. Since the midpoints of $\triangle DEF$ are on the nine-point circle of $\triangle DEF$, we are done.

Example 9.4.7.4. *Consider $\triangle PAB$ with circumcenter X. Then consider an inversion about some circle ω with center P. If A, B, X are the poles of A, B, X with respect to ω, prove that X is the result of reflecting P about AB.*

Solution 9.4.7.4. *By the definition of inversion, $PA \cdot PA' = PX \cdot PX'$. Applying Power of a Point yields that $AA'X'X$ and $BB'X'X$ are cyclic quadrilaterals. Then notice that $\triangle PAX \sim \triangle PX'A'$.*

Since X is the center of a circle, $AX = PX$. By similiarity, $X'A' = PA'$, so $\angle PXA = \angle APX = \angle PX'A$. Similarly, $\angle BPX = PXB'$. This implies that $\angle AX'B' = \angle PA'B'$. Since P, X, X' are collinear, X' is the reflection of P about AB, as desired.

9.4.8 Exercises

Exercise 321. *(2013 Berkeley Math Tournament) From a point A construct tangents to a circle centered at point O, intersecting the circle at P and Q respectively. Let M be the midpoint of PQ. If K and L are points on circle O such that K, L, and A are collinear, prove $\angle MKO = \angle MLO$.*

Exercise 322. *Circles C_1, C_2, C_3, and C_4 lie in $\mathbb{R}^2$ such that C_2 and C_4 are both tangent to C_1 and C_3. Prove that the points of tangency are collinear or concyclic.*

Exercise 323. *Consider scalene $\triangle ABC$ with incenter I. Let the A excircle of $\triangle ABC$ intersect the circumcircle of $\triangle ABC$ at X, Y. Let XY intersect BC at Z. Then choose M, N on the A excircle of $\triangle ABC$ such that ZM, ZN are tangent to the A excircle of $\triangle ABC$. Prove I, M, N are collinear.*

Exercise 324. *Consider $\triangle ABC$ with point D on BC. Let M, N be the circumcenters of $\triangle ABD$ and $\triangle ACD$, respectively. Let the circumcircles of $\triangle ACD$ and $\triangle MND$ intersect at $H \neq D$. Prove A, H, M are collinear.*

Exercise 325. *Consider $\triangle ABC$ with orthocenter H and circumcenter O. Let X, Y, Z be the midpoints of AH, BH, CH and let D, E, F be the midpoints of BC, CA, AB. Prove that XD, YE, ZF are concurrent.*

Exercise 326. *(1993 USAMO P2) Let $ABCD$ be a convex quadrilateral such that diagonals AC and BD intersect at right angles, and let E be their intersection. Prove that the reflections of E across AB, BC, CD, DA are concyclic.*

Exercise 327. *(2003 IMO Shortlist G3) Let $\triangle ABC$ be a triangle and let P be a point in its interior. Denote by D, E, F the feet of the perpendiculars from P to the lines $\overline{BC}$, $\overline{CA}$, $\overline{AB}$, respectively. Suppose that $AP^2 + PD^2 = BP^2 + PE^2 = CP^2 + PF^2$. Denote by I_A, I_B, I_C the excenters of the triangle $\triangle ABC$. Prove that P is the circumcenter of the triangle $I_A I_B I_C$.*

Exercise 328. *A hexahedron is any polyhedron with six faces. Prove that if seven of the vertices of a hexahedron lie on a sphere, then so does the eighth vertex.*

Exercise 329. *Prove Feuerbach's Theorem:*

Theorem 9.19: Feuerbach's Theorem

The nine-point circle of triangle $\triangle ABC$ is tangent to the incircle, and all three excircles, of $\triangle ABC$.

Extension Exercise 140. *(1996 IMO P2) Let P be a point inside $\triangle ABC$ such that $\angle APB - \angle C = \angle APC - \angle B$. Let D and E be the incenters of $\triangle APB$ and $\triangle APC$ respectively. Show that AP, BD, and CE meet at a point.*

Extension Exercise 141. *(2002 IMO Shortlist G1) Let B be a point on a circle S_1, and let A be a point distinct from B on the tangent at B to S_1. Let C be a point not on S_1 such that the line segment $\overline{AC}$ meets S_1 at two distinct points. Let S_2 be the circle touching $\overline{AC}$ at C and touching S_1 at a point D on the opposite side of $\overline{AC}$ from B. Prove that the circumcenter of triangle $\triangle BCD$ lies on the circumcircle of triangle $\triangle ABC$.*

Extension Exercise 142. *(2003 IMO Shortlist G4) Let Γ_1, Γ_2, Γ_3, Γ_4 be distinct circles such that Γ_1, Γ_3 are externally tangent at P, and Γ_2, Γ_4 are also externally tangent at P. Suppose that, for $1 \leq i \leq 4$, C_i and C_{i+1} meet at A, B, C, and D, respectively, and that none of these points coincides with P. Prove that*
$$\frac{AB \cdot CD}{AD \cdot CD} = \left(\frac{PB}{PD}\right)^2.$$

Extension Exercise 143. *(1985 IMO P5) A circle with center O passes through the vertices A and C of the triangle ABC and intersects the segments AB and BC again at distinct points K and N respectively. Let M be the point of intersection of the circumcircles of triangles ABC and KBN (apart from B). Prove that $\angle OMB = 90°$.*

Extension Exercise 144. *(2007 USAMO P6) Let ABC be an acute triangle with ω, Ω, and R being its incircle, circumcircle, and circumradius, respectively. Circle ω_A is tangent internally to Ω at A and tangent externally to ω. Circle Ω_A is tangent internally to Ω at A and tangent internally to ω. Let P_A and Q_A denote the centers of ω_A and Ω_A, respectively. Define points P_B, Q_B, P_C, Q_C analogously. Prove that*

$$8P_AQ_A \cdot P_BQ_B \cdot P_CQ_C \leq R^3,$$

with equality if and only if triangle ABC is equilateral.

9.5 The Nine Facts of Euclidean Geometry

As a sort of summative sendoff for our explorative look into contest-based geometry, we will go over each of nine crucial, fundamental, and highly novel facts regarding Euclidean geometry as a whole. In particular, Fact 5 of these (bolded) is of special mention!

> **Theorem 9.20: The 9 Facts of Euclidean Geometry**
>
> 1. (Euler Line) In any triangle, the orthocenter H, centroid G, and circumcenter O are collinear, and moreover, $HG = 2GO$.
>
> 2. (Nine-point circle; see §9.1.6) The midpoints, feet of the altitudes, and the midpoints of the segments adjoining the vertices with the orthocenter, all lie on a circle with center at the midpoint of the orthocenter and the circumcenter.
>
> 3. (Addendum to fact 2)
>
>   (a) The diameter of the nine-point circle is equal to the circumradius.
>
>   (b) The center of the homothety taking the circumcircle to the nine-point circle is the orthocenter.
>
>   (c) The center of the inverse homothety taking the circumcircle to the nine-point circle is the centroid.
>
> 4. (Reflecting the orthocenter) The reflections of the orthocenter with respect to the sides of the triangle lie on the circumcircle.
>
> 5. **(Incenter-Excenter Lemma) $\triangle ABC$ has incenter I, A-excenter A_I, and L the midpoint of the arc BC. Then L is the center of the circle through I, I_A, B, and C.**
>
> 6. (Bisectors of a triangle) In $\triangle ABC$, the internal and external angle bisectors of $\angle A$ and the perpendicular bisector of $\overline{BC}$ form a triangle with vertices on the circumcircle of $\triangle ABC$.
>
> 7. (Isogonal conjugates) In the plane of $\triangle ABC$, if P and Q are points such that $\angle PAB = \angle CAQ$ and $\angle PBC = \angle ABQ$ (directed angles), then $\angle PCA = \angle BCQ$. Then P and Q are isogonal conjugates with respect to $\triangle ABC$.
>
> 8. (Symmedians) In $\triangle ABC$, the isogonal line of the median through A passes through the intersection of the lines through B and C tangent to the circumcircle.
>
> 9. (Addendum to fact 8) If AM is a symmedian with M on $\overline{BC}$, then
>
>   (a) Let P be a point on $\overline{AM}$. The ratio of the distances from P to AB and AC is $\dfrac{AB}{AC}$.
>
>   (b) $\dfrac{BM}{MC} = \dfrac{AB^2}{AC^2}$.

Remark. These facts are so designated from a 2011 MOP handout by Carlos Shine. Credit goes to him - and the many others who have built upon his original work - for immortalizing some of the most important cornerstones of Euclidean geometry!

In particular, we focus on the proofs of the first five facts; facts 6-9 are either provable by pure angle chasing, are corollaries of the first 5 facts, or are somewhat vestigial in terms of their use on an Olympiad.

Proof.

1. We extend $\overline{CG}$ to H such that $2CG = GH$. It suffices to show that H is the orthocenter of $\triangle ABC$.

Extend $\overline{AH}$ to meet $\overline{BD}$ at point P; then let $\overline{AG}$ intersect $\overline{BD}$ at Q. We then have $QGGA = \frac{1}{2}$, $\frac{QG}{GA} = \frac{CG}{GH}$. Thus, $AH \parallel QC$, but $QC \perp BD$, and so $\overline{AP}$ is an altitude. Extend $\overline{DH}$ to be perpendicular to $\overline{AB}$; this shows that H is the orthocenter, as desired.

2. Let D, E, F be the feet of the A-, B-, and C-altitudes respectively; let G, H, I be the midpoints of the sides opposite A, B, C respectively; and let J, K, L be the midpoints of the line segments from the orthocenter to a vertex. We know that the line connecting the midpoints of two sides of a triangle is parallel to the opposite side, so $GH \parallel AC$. In addition, $AC \parallel LK$ (and so $GH \parallel LK$); this is because L and K are the midpoints of line segments connecting the orthocenter to their respective vertices.

From here, we can form another triangle $\triangle BC\gamma$, where γ is the incenter. Then the line $\overline{B\gamma} \parallel \overline{HK}$. We also have $\overline{LG} \parallel \overline{B\gamma}$ (from), so $\overline{LG} \parallel \overline{HK}$. Henceforth, draw parallelogram $GHKL$, and since the perpendicular line forms the altitude of a triangle, we have that $GHKL$ is in fact a rectangle. Furthermore, this rectangle is a cyclic quadrilateral (in fact, we know this *a priori*). Similarly, we can draw rectangle $GJKI$, so J, G, D, L, K, H are concyclic. The circle on which they all lie has diameter $\overline{IJ}$, whose midpoint O is the circumcenter.

Finally, since F is the foot of the C-altitude, $\triangle JFI$ is right, and F lies on the circle. E, as well, lies on the circle, as does D by similar logic. Therefore, we have established that all nine points are concyclic, and so that the nine-point circle exists for all triangles $\triangle ABC$.

3. (a) If we let H be the orthocenter of $\triangle ABC$, and N the center of the nine-point circle of $\triangle ABC$, then observe that $\triangle ABC$, $\triangle ABH$, $\triangle BCH$, and $\triangle CAH$ all have the same nine-point circle (due to symmetry). Then their circumradii are all equal, and so $NA^2 + NB^2 + NC^2 + NH^2 = 3R^2$, where R is the circumradius of $\triangle ABC$. For any point P lying on the nine-point circle, we additionally have $PA^2 + PB^2 + PC^2 + PD^2 = 4R^2$; it follows that the radius of the nine-point circle is half that of the circumcircle of $\triangle ABC$.

   (b) It suffices to show that the nine-point center N bisects the segment from H to O (i.e. $ON = NH$). This follows immediately from (a).

   (c) Use the fact that N trisects the segment from the orthocenter H to the circumcenter; then the desired result follows.

4. (Assume WLOG that $\triangle ABC$ is acute.) Let the altitude from A hit $\overline{BC}$ at H_A. Let the extension of $\overline{AH_A}$ intersect the circumcircle at K_A. We show that $HH_A = H_A K_A$.

Let BH_B be the B-altitude (and, similarly, let CH_C be the C-altitude). Since $\triangle CAH_A$ and $\triangle CBH_B$ share angle $\angle C$, their other acute angles are equal, namely $\angle CAH_A = \angle CBH_B$. Arc CK_A subtends the inscribed angles $\angle CAK_A$ and $\angle CBK_A$, so $\angle CAH_A = \angle CBK_A$. It follows that $\angle CBH_B = \angle CBK_A$. Hence, $\overline{BH_A}$ is both an altitude and an angle bisector of $\triangle BHK_A$. It is also a B-median, which proves that $HH_A = H_A K_A$.

Henceforth, observe that the points A, B, C, K_A, K_B, K_C partition the circumcircle into three pairs of arcs, with each pair around a vertex being equal. The angle measure of each arc adjacent to A is $90 - \angle BAC$, and similarly for B and C. The circle passing through B, C, H is the reflection of the circumcircle in $\overline{BC}$, and so has equal radius. Then let S be the reflection of A, such that $AH_A = H_A S$. Vertices B and C are similar, so we have that the circumcircles of $\triangle ABC, \triangle ABH, \triangle BCH, \triangle CAH$ are all congruent. The circumcircles of $\triangle AK_B K_C$, $\triangle BK_C K_A$, and $\triangle CK_A K_B$ coincide with the circumcircle of $\triangle ABC$.

5. **Let $A = \angle BAC$, $B = \angle CAB$, and $C = \angle ACB$. Note that $\angle BIC = \left(90 + \frac{A}{2}\right)^\circ$, or $\angle BI_A C = \left(90 - \frac{A}{2}\right)^\circ$, as $BICI_A$ is a cyclic quadrilateral. Then $\angle BDC = (180 - A)^\circ$ as well; as**

$\angle BDC = 2\angle BI_AC$, L is the center of the circumcircle of $BICI_A$.

We can also take this a step further to show that $BD = BI_A$. Any point on the circumcircle of $BICI_A$ would satisfy the necessary angle relations, and observe that D is the midpoint of BC, so BD and BI_A are both radii of the circumcircle. Hence, $BD = BI_A$.

For facts 6-9, we nevertheless encourage you to find the link between them and the first 5 facts through your own exploration and independent thought. □

(Note: This section will have neither examples nor exercises, since it is meant to sum up everything we have studied thus far. To compensate, the Chapter 9 Review Exercises section is considerably lengthier than usual.)

Chapter 9 Review Exercises

Exercise 330. *Triangle $\triangle ABC$ has $AB = 10$, $BC = 12$, $CA = 14$. The angle bisector of $\angle BAC$ hits $\overline{BC}$ at D. Non-concentric circles C_1 and C_2 respectively have $\overline{BD}$ and $\overline{CD}$ as radii, with D being the center of C_1. What is the distance between the points of intersection of C_1 and C_2?*

Exercise 331. *(2003 AIME I) Point B is on $\overline{AC}$ with $AB = 9$ and $BC = 21$. Point D is not on $\overline{AC}$ so that $AD = CD$, and AD and BD are integers. Let s be the sum of all possible perimeters of $\triangle ACD$. Find s.*

Exercise 332. *$\triangle ABC$ has $AB = 4$, $BC = 5$, $CA = 6$. As point D on the incircle of $\triangle ABC$ varies, what is the maximum possible area of $\triangle BAD$?*

Exercise 333. *Prove that two circles cannot intersect at more than two points.*

Exercise 334. *In $\triangle ABC$, point D lies on $\overline{BC}$ such that $CD = 6$ and $BD = 3$. Given that $AC = 11$ and $AB = 7$, compute AD.*

Exercise 335. *Prove Apollonius' Theorem:*

> **Theorem 9.21: Apollonius' Theorem**
>
> *In $\triangle ABC$, if $\overline{AD}$ is a median, then $|AB|^2 + |AC|^2 = 2(|AD|^2 + |BD|^2)$.*

Exercise 336. *(1991 USAMO P1) In triangle ABC, angle A is twice angle B, angle C is obtuse, and the three side lengths a, b, c are integers. Determine, with proof, the minimum possible perimeter.*

Exercise 337. *(1984 IMO P4) Let $ABCD$ be a convex quadrilateral with the line CD being tangent to the circle on diameter AB. Prove that the line AB is tangent to the circle on diameter CD if and only if the lines BC and AD are parallel.*

Exercise 338. *(1989 USAMO P4) Let ABC be an acute-angled triangle whose side lengths satisfy the inequalities $AB < AC < BC$. If point I is the center of the inscribed circle of triangle ABC and point O is the center of the circumscribed circle, prove that line IO intersects segments AB and BC.*

Exercise 339. *(1968 IMO P1) Prove that there is one and only one triangle whose side lengths are consecutive integers, and one of whose angles is twice as large as another. Find this triangle.*

Exercise 340. *Does there exist a triangle whose side lengths and angle measures are both in arithmetic progression? What about geometric progression?*

Exercise 341. *(2012 Stanford Math Tournament) In quadrilateral $ABCD$, $m\angle ABD \cong m\angle BCD$ and $\angle ADB = \angle ABD + \angle BDC$. If $AB = 8$ and $AD = 5$, find BC.*

Exercise 342. *(2006 Romanian NMO Grade 8 P3) Let $ABCDA_1B_1C_1D_1$ be a cube and P a variable point on the side $[AB]$. The perpendicular plane on AB which passes through P intersects the line AC' in Q. Let M and N be the midpoints of the segments $A'P$ and BQ respectively.*

(a) Prove that the lines MN and BC' are perpendicular if and only if P is the midpoint of AB.

(b) Find the minimal value of the angle between the lines MN and BC'.

Exercise 343. *(2008 iTest) Points C and D lie on opposite sides of line $\overline{AB}$. Let M and N be the centroids of $\triangle ABC$ and $\triangle ABD$ respectively. If $AB = 841, BC = 840, AC = 41, AD = 609$, and $BD = 580$, find the sum of the numerator and denominator of the value of MN when expressed as a fraction in lowest terms.*

Exercise 344. *(1978 USAMO P4)*

(a) Prove that if the six dihedral (i.e. angles between pairs of faces) of a given tetrahedron are congruent, then the tetrahedron is regular.

(b) Is a tetrahedron necessarily regular if five dihedral angles are congruent?

Exercise 345. *(1994 AIME) Given a point P on a triangular piece of paper ABC, consider the creases that are formed in the paper when $A, B,$ and C are folded onto P. Let us call P a fold point of $\triangle ABC$ if these creases, which number three unless P is one of the vertices, do not intersect. Suppose that $AB = 36, AC = 72,$ and $\angle B = 90°$. Then the area of the set of all fold points of $\triangle ABC$ can be written in the form $q\pi - r\sqrt{s}$, where $q, r,$ and s are positive integers and s is not divisible by the square of any prime. What is $q + r + s$?*

Extension Exercise 145. *(2008 Putnam B1) What is the maximum number of rational points that can be on a circle in $\mathbb{R}^2$ whose center is not a rational point? (A rational point is a point both of whose coordinates are rational numbers.)*

Extension Exercise 146. *(1973 IMO Shortlist Bulgaria 1) A tetrahedron $ABCD$ is inscribed in the sphere S. Find the locus of points P, situated in S, such that*

$$\frac{AP}{PA_1} + \frac{BP}{PB_1} + \frac{CP}{PC_1} + \frac{DP}{PD_1} = 4,$$

where A_1, B_1, C_1, D_1 are the other intersection points of AP, BP, CP, DP with S.

Extension Exercise 147. *(2015 Putnam B4) Let T be the set of all triples (a, b, c) of positive integers for which there exist triangles with side lengths a, b, c. Express*

$$\sum_{(a,b,c)\in T} \frac{2^a}{3^b \cdot 5^c}$$

as a rational number in lowest terms.

Extension Exercise 148. *Points A, B, and C lie in the first quadrant of the Cartesian coordinate plane such that A lies on the line $x + y = 6$, B lies on the x-axis, and C is the point $(3, 4)$. Compute the minimum possible value of $AB + BC + CA$.*

Extension Exercise 149. *(2015 IMO P4) Triangle ABC has circumcircle Ω and circumcenter O. A circle Γ with center A intersects the segment BC at points D and E, such that $B, D, E,$ and C are all different and lie on line BC in this order. Let F and G be the points of intersection of Γ and Ω, such that $A, F, B, C,$ and G lie on Ω in this order. Let K be the second point of intersection of the circumcircle of triangle BDF and the segment AB. Let L be the second point of intersection of the circumcircle of triangle CGE and the segment CA.*

Suppose that the lines FK and GL are different and intersect at the point X. Prove that X lies on the line AO.

Extension Exercise 150. *(2016 HMIC P2) Let $\triangle ABC$ be an acute triangle with circumcenter O, orthocenter H, and circumcircle Ω. Let M be the midpoint of $\overline{AH}$ and N the midpoint of $\overline{BH}$. Assume the points M, N, O, H are distinct and lie on a circle ω. Prove that the circles ω and Ω are internally tangent to each other.*

Extension Exercise 151. *(1973 IMO P1) Point O lies on line g; $\overrightarrow{OP_1}, \overrightarrow{OP_2}, \cdots, \overrightarrow{OP_n}$ are unit vectors such that points $P_1, P_2, \cdots, P_n$ all lie in a plane containing g and on one side of g. Prove that if n is odd,*

$$\left|\overrightarrow{OP_1} + \overrightarrow{OP_2} + \cdots + \overrightarrow{OP_n}\right| \geq 1.$$

Here $\left|\overrightarrow{OM}\right|$ denotes the length of vector $\overrightarrow{OM}$.

Extension Exercise 152. *(2004 USAMO P1) Let $ABCD$ be a quadrilateral circumscribed about a circle, whose interior and exterior angles are at least 60 degrees. Prove that*

$$\frac{1}{3}|AB^3 - AD^3| \leq |BC^3 - CD^3| \leq 3|AB^3 - AD^3|.$$

When does equality hold?

Part IV

Solutions to Exercises

Part 1: The Nature of Mathematical Arguments

Chapter 1: Defining "Proof" and Proof Terminology

Section 1.1: Logical Notation

Exercise 1. *For further demonstrative purposes, both of the following statements are false. Why?*

$$\forall b \in \mathbb{R}, b^2 > 0.$$

$$\exists c \in \mathbb{R} \ s.t. \ c^2 = -1.$$

Solution 1.

(a) *It is not true that for all real numbers b, $b^2 > 0$ (though $b^2 \geq 0$ would be true). A counterexample (and in fact the only one) is $b = 0$.*

(b) *There is no real number c such that $c^2 = -1$ (the solutions to this equation are $c = \pm i$, neither of which is a real number).*

Section 1.2: What is a Statement?

Exercise 2. *Which of the following are, or are not, statements? Justify your answers.*

1. *I like cheesecake but not ice cream.*

2. *59 is a prime number.*

3. *Is 6 a perfect number?*

4. *The expanded form of $(2k+1)^2$ is $4k^2 + 4k + 1$.*

5. *$15 + 5 = 30$.*

Solution 2.

1. *This is **not** a statement, since it does not present a mathematically defensible thesis (its thesis is purely subjective - a matter of opinion not based in any mathematical axioms).*

2. *This is a statement, since it is mathematically verifiable (and a true one, at that).*

3. *This is **not** a statement, as it is a question (it does not present any thesis).*

4. *This is a statement, since it does present a thesis (which happens to be true).*

5. *This is a statement, despite it being false (not all statements have to be true).*

Section 1.3: Statements and Their Various Forms

Exercise 3. *Write the inverse, converse, and contrapositive of the following statements, then classify each of statements 1-5 as true or false with justification.*

1. *If the sky is blue, then 4 is a composite number.*

2. *1 is a composite number if a cube has 8 vertices.*

3. *If $x^2 + y^2 = 0$, and x and y are real numbers, then $x = y = 0$.*

4. *If x and y are positive integers less than or equal to 10 with $xy = 25$, then $x + y = 10$.*

5. *$2^4 = 4^2$ if 90 is less than 3^4.*

6. *If you have a good diet, you won't get sick.*

7. *If you study for the test, you will get an A.*

Solution 3. 1. **Inverse:** *If the sky is not blue, then 4 is not a composite number.*
   Converse: *If 4 is a composite number, then the sky is blue.*
   Contrapositive: *If the sky is not blue, then 4 is not a composite number.*
   The original statement is true, since it is of the form $T \implies T$ ($4 = 2^2$ is composite).

2. **Inverse:** *If a cube does not have 8 vertices, then 1 is not a composite number.*
   Converse: *If 1 is a composite number, then a cube has 8 vertices.*
   Contrapositive: *If 1 is not a composite number, then a cube does not have 8 vertices.*
   The original statement is false, since the second part of the statement does not follow from the first. Be careful with this one - the false part of the statement ("1 is a composite number") comes first, so the statement is actually of the form $T \implies F$, not $F \implies T$.

3. **Inverse:** *If $x^2 + y^2 \neq 0$ or one of x or y is not real, then $x \neq 0$ or $y \neq 0$.*
   Converse: *If $x^2 + y^2 = 0$ and $x, y \in \mathbb{R}$, then $x = y = 0$.*
   Contrapositive: *If $x \neq 0$ or $y \neq 0$, then $x^2 + y^2 \neq 0$ or one of x or y is not real.*
   The original statement is true by the Trivial Inequality (see §4.3).

4. **Inverse:** *If one of x or y is not a positive integer or $xy \neq 25$, then $x + y \neq 10$.*
   Converse: *If $x + y = 10$, then $x, y \in \mathbb{Z}^+$ and $xy = 25$.*
   Contrapositive: *If $x + y \neq 10$, then one of x or y is not a positive integer or $xy \neq 25$.*
   The statement is true; the only possible ordered pair (x, y) is $(5, 5)$.

5. **Inverse:** *If $90 \geq 3^4$, then $2^4 \neq 4^2$.*
   Converse: *If $2^4 = 4^2$, then 90 is less than 3^4.*
   Contrapositive: *If $2^4 \neq 4^2$, then $90 \geq 3^4$.*
   The statement is true, since $90 \not< 3^4$, so the "then" part of the statement is irrelevant.

6. **Inverse:** *If you don't have a good diet, then you will get sick.*
   Converse: *If you won't get sick, then you have a good diet.*
   Contrapositive: *If you will get sick, then you don't have a good diet.*

7. **Inverse:** *If you don't study for the test, then you won't get an A.*
   Converse: *If you will get an A, then you study for the test.*
   Contrapositive: *If you won't get an A, then you don't study for the test.*

Extension Exercise 1. *Let A be a statement in the form "if p, then q." Let B be the contrapositive of the inverse of A, let C be the converse of the contrapositive of B, and let D be the inverse of the converse of C. What can we infer about B, C, and D?*

Extension Solution 1. *B is the statement "if q, then p"; C is the converse of the inverse of A, which is "if not q, then not p"; and D is just itself A, which is "if p, then q." Hence, if A is true, then we cannot conclude anything about B; we know that C is true (as it is the contrapositive of A), and we know that D is false. We reverse these values for when A is false.*

Section 1.4: The Truth Table

Exercise 4. *Suppose we define an operator $+$ such that $p + q$ is true when $(pNORq) \oplus (p \lor (p \iff q))$ is false.*

(a) *Draw the truth table for $+$ given the four possibilities for (p, q).*

(b) *Draw the truth table for $(p + q) + (p + q)$. Your table should only have three rows, including the labels.*

(c) *Say an operator $\circ$ commutes if $p \circ q = q \circ p$ (i.e. returns the same truth values for all possible ordered Boolean pairs (p, q)). Does $+$ commute? Justify your answer.*

Solution 4.

(a)

p	q	$+$
T	T	F
T	F	F
F	T	T
F	F	T

(b)

$p + q$	$(p + q) + (p + q)$
T	F
F	T

(c) *We claim that $+$ does not commute. Observe that $T + F = F$, but $F + T = T$.*

Extension Exercise 2. *Say an operator $\circ$ has a left identity L if $L \circ q = q$ whether q is true or false. Similarly, say $\circ$ has a right identity R if $p \circ R = p$ whether p is true or false.*

(a) *Under what conditions will an operator have one identity, but not the other?*

(b) *Are there operators that commute, but do not have either a left or right identity? If so, what are they?*

Extension Solution 2.

(a) *An operator will have a left, but not right, identity if it is the converse non-implication (see below), q, or $\to$. Similarly, it will only have a right identity if it is the material non-implication (see below), p, or $\leftarrow$.*

(b) *Yes; these are T, F, NOR, NAND. It is up to the reader to verify these do indeed work (and are the only such operators).*

Extension Exercise 3. *Consider table 1.5 (the extended truth table) [in §1.4]. There are 14 columns in the table; however, there are $2^4 = 16$ possible operators from 4 possible Boolean pairs of input variables (p, q).*

(a) *Briefly describe the two missing operators. Do they commute, and do they have a left identity, right identity, or both?*

(b) *If we introduced a third Boolean input variable r, how many operators (columns) would be in a complete truth table?*

(c) *If we introduced more input variables v_i so that there are then V input variables, how many operators (columns) would be in a complete truth table?*

Extension Solution 3.

(a) *The missing operators are the converse non-implication (F/F/T/F), and the material non-implication (F/T/F/F). The converse non-implication does not commute, but has a left (and not right) identity of F. The material non-implication also does not commute, but has a right identity of F (and no left identity).*

(b) *There would be $2^{2^3} = 2^8 = \boxed{256}$ columns, as each of the $2^3 = 8$ choices of Boolean variable ordered triples would dictate its own truth operator which has 2 values.*

(c) *In general, there would be 2^{2^V} columns, by the same logic as above.*

Chapter 2: Arguments and Their Forms

Exercise 5. *Prove that the following argument is valid.*

1. $\neg p \rightarrow r \wedge \neg s$

2. $t \rightarrow s$

3. $u \rightarrow \neg p$

4. $\neg w$

5. $u \vee w$

Therefore, $\neg t$.

Solution 5.

Step	Statement	Reason
A	$\neg w$	5
B	u	6, A, Disruptive Syllogism
C	$\neg p$	B, 3
D	$r \wedge \neg s$	C, 1
E	$\neg s$	D, Specialization
F	$\neg t$	E, 2 via contrapositive

...and we are done!

Chapter 3: Types of Mathematical Proof

Section 3.1: Direct Proof (Proof by Construction)

Exercise 6. *Prove or disprove that $x^2 + y^2$ is a multiple of 8 iff x and y are both even.*

Solution 6. *Since x and y are even, both can be written in the form $2k$ for some integer k. To avoid confusion, let $x = 2k_1$ and $y = 2k_2$ for some $k_1, k_2 \in \mathbb{Z}$. Then $x^2 + y^2 = 4k_1^2 + 4k_2^2 = 4(k_1^2 + k_2^2)$. This guarantees that $x^2 + y^2$ is a multiple of 4 if x and y are even, but it does not guarantee that it is a multiple of 8 (since $k_1^2 + k_2^2$ could be odd; namely, when exactly one of k_1 or k_2 is odd). $x^2 + y^2$ would be a multiple of 8, however, if x and y were multiples of 4 (proving this alternate statement is left as an exercise to the reader).*

Section 3.4: Proof by Contradiction

Exercise 7. *Show that $\sqrt{p}$ is irrational for any prime number p.*

Solution 7. *Much like in Example 2.3.2, assume the contrary, and set $\sqrt{p} = \dfrac{a}{b} \implies pb^2 = a^2$ (where, as before, $\gcd(a, b) = 1$). If $\gcd(a, b) = 1$, then a and b have no common factors other than 1, and so p is a factor of a^2. This in turn implies that p is a factor of a (repeat the previous mini-contradiction argument in the example above). If we make the substitution $a = pk$, then $pb^2 = p^2 k^2 \implies b^2 = pk^2$, which implies that p divides b^2 (p is a prime, and so $\gcd(a, p) = 1$, regardless of the value of a). However, this leads to a contradiction, as a^2 and b^2 should be relatively prime (here, p is a common factor). Hence $\dfrac{a^2}{b^2}$ is not in simplest form, and $\sqrt{p}$ is irrational, as desired.*

Chapters 1-3 Review Exercises

Exercise 8. *Classify each of the following as true, false, inconclusive, or not a statement:*

(a) $\forall a \in \mathbb{R} \backslash \mathbb{Q}, \ a^2 \in \mathbb{Q}$.

(b) A real number has an imaginary part of 1.

(c) π is better than e.

(d) I live at $(1, 2)$ on the Cartesian plane. The nearest gas station is at $(2, 3)$. The gas station is not further than 1 mile from my house.

(e) $A > B$, and $C > B$. Then $A > C$.

Solution 8.

(a) False; $\pi \in \mathbb{R} \backslash \mathbb{Q}$ and $\pi^2 \notin \mathbb{Q}$.

(b) False; all real numbers have an imaginary part of 0. ($\mathbb{R}$ is a subset of $\mathbb{C}$, the set of complex numbers; for all $z \in \mathbb{C}$, $z = a + bi$ for some $a, b \in \mathbb{R}$. For a real number, we must have $b = 0$.)

(c) Not a statement; there is no mathematical axiom, or anything that follows from such an axiom, that says which of π or e is better.

(d) Inconclusive; the units of the distances on the Cartesian plane are not given. For all I know, they could be in inches or lightyears :P

(e) Inconclusive; take $(A, B, C) = (1, 0, 2)$ as a counterexample. However, the statement does hold for some ordered triples (A, B, C).

Exercise 9. *Given the following two premises:*

- *"All poisonous objects are bad-tasting."*

- *"Broccoli is bad-tasting."*

Why is the following an example of a faulty conclusion? "Broccoli is a poisonous object." How could we change the premises so that we can make a conclusion regarding broccoli?

Solution 9. *From the two generic statements $A \implies B$ and $C \implies B$ it does not follow that $C \implies A$ (nor that $A \implies C$, for that matter; we cannot conclude anything about the relationship between A and C, in this case whether broccoli is poisonous). We could fix the premises by taking the converse of the second premise: "All bad-tasting objects are broccoli," which, although nonsensical, leads to the logically sound conclusion in this context of broccoli being poisonous.*

Exercise 10. *Show that the sum of the squares of an odd integer and an integer divisible by 4 leaves a remainder of 1 when divided by 8.*

Solution 10. *First, note that odd integers are either ± 1 or $\pm 3 \pmod 8$, both of which, when squared, are congruent to $1 \pmod 8$. If an integer is divisible by 4, it is congruent to 0 or $4 \pmod 8$, both of which are congruent to $0 \pmod 8$ when squared. Thus, the sum of the squares of an odd integer and an integer divisible by 4 is congruent to $1 + 0 \equiv 1 \pmod 8$; in other words, the sum leaves a remainder of 1 when divided by 8 as desired.*

Exercise 11. *List the inverse, converse, and contrapositive of each of the following statements, and classify each as true or false with explanation.*

(a) *"If I am sick, I won't be able to go to school."*

(b) *"If a regular octahedron has an equal number of faces and vertices, then a cube with side length s has surface area given by $6s^2$."*

(c) *"If the statement '$\neg(p \wedge (q \oplus \neg p)) \vee (q \iff p)$' is true, then either p and q are both true or both false."*

Solution 11.

(a) (i) *Inverse: "If I am not sick, I can go to school." This is false, since there are plenty of other reasons a student cannot show up to school - perhaps there is a family emergency or some other event that takes place on a school day.*

(ii) *Converse: "If I can't go to school, I'm sick." This is false; although it is most likely that the reason one cannot go to school is because one is sick, this of course does not discount other reasons, such as those mentioned above.*

(iii) *Contrapositive: "If I can go to school, I am not sick." This is true - being sick is a sufficient condition for one not to be able to go to school. Otherwise put, just because one is healthy does not mean that one will be able to go to school, but being sick automatically prevents one from being able to attend (after all, one would pass on one's illness to other unsuspecting students and teachers!)*

(b) *First, we evaluate the truth of both parts of the original statement. The first part ("a regular octahedron has an equal number of faces and vertices") is false; a regular octahedron has 8 faces and 6 vertices. The second part ("a cube with side length s has surface area $6s^2$") is true.*

(i) *Inverse: "If an octahedron has a differing number of faces and vertices, then a cube with side length s does not have surface area $6s^2$." This is false, since it is a statement of the form $T \implies F$.*

(ii) *Converse: "If a cube with side length s has surface area $6s^2$, then an octahedron has an equal number of faces and vertices." This is also false, since it is of the form $T \implies F$.*

(iii) *Contrapositive: "If a cube with side length s does not have volume $6s^2$, then an octahedron has a differing number of faces and vertices." This is true, since it is of the form $F \implies T$. If the first part of an "implies" statement is false, the statement as a whole is necessarily true.*

316

(c) Let A be the Boolean value of "$\neg(p \wedge (q \oplus \neg p)) \vee (q \implies p)$." Note that A is true when either $(p \wedge (q \oplus \not{p}))$ or $q \implies p$ is true, else it is false. It is considerably easier to examine when it is false, since both parts must be false. $(p \wedge (q \oplus \not{p}))$ is false when $p \wedge (q \oplus \not{p})$ is true, which means that p must be true. In addition, $q \oplus \not{p}$ is true, but then $q \oplus F$ is true, which forces $q = T$. However, $q \implies p$ is true when both p and q are true, so A is always true.

  (i) *Inverse:* "If A is false, then exactly one of p or q is true" (i.e. $p \oplus q$ is true). This is true, since A being false is irrelevant.

  (ii) *Converse:* "If p and q have the same truth value, then A is true." This is true and follows immediately from the above.

  (iii) *Contrapositive:* "If $p \oplus q$ is true, then A is false." This is false (A can never be false).

Exercise 12. *Prove by induction that $2^n > n^3$ for all $n \geq 10$.*

Solution 12. *The base case is $n = 10$, for which $2^{10} = 1024 > 1000 = 10^3$. Hence the base case holds.*

Let $S(n)$ be the statement that $2^n > n^3$. We show that if $S(n)$ holds, then $S(n+1)$ must also hold. $S(n+1)$ is the statement $2^{n+1} > (n+1)^3$; we have $\dfrac{(n+1)^3}{n^3} = \left(\dfrac{n+1}{n}\right)^3$, which we want to show is less than 2 (since $\dfrac{2^{n+1}}{2^n} = 2$ for all n). $\left(\dfrac{n+1}{n}\right)^3$ is decreasing for all $n \geq 10$ (since $\dfrac{n+1}{n} < \dfrac{n}{n-1}$ for all $n \geq 10$); and since $\left(\dfrac{11}{10}\right)^3 = \dfrac{1331}{1000} < 2$, for all $n \geq 10$, $\left(\dfrac{n+1}{n}\right)^3 < 2$ and so $S(n+1)$ holds whenever $S(n)$ holds, proving that $2^n > n^3$ for all $n \geq 10$, $n \in \mathbb{N}$.

Exercise 13. *Prove that $\displaystyle\sum_{k=1}^{n} 2^k = 2^{n+1} - 2$ for all $n \geq 1$ by induction.*

Solution 13. *Base case: $n = 1$. We have $\displaystyle\sum_{k=1}^{1} 2^k = 2^2 - 2$, which is true.*

Inductive hypothesis: Let $S(n)$ be the statement. We wish to show $S(n+1)$ assuming $S(n)$. $S(n+1)$ is the statement that $\displaystyle\sum_{k=1}^{n+1} 2^k = 2^{n+2} - 2$. The LHS simplifies to $\displaystyle\sum_{k=1}^{n} 2^k + 2^{n+1}$, and indeed, $2^{n+2} - 2 = \left(2^{n+1} - 2\right) + 2^{n+1}$ because $2^{n+2} = 2^{n+1} + 2^{n+1}$ for all $n \in \mathbb{N}$. Hence the statement is proven for all $n \in \mathbb{N}$.

Exercise 14. *Prove that $1 \cdot 2 \cdot 3 + 2 \cdot 3 \cdot 4 + 3 \cdot 4 \cdot 5 + \ldots + n \cdot (n+1) \cdot (n+2) = \dfrac{n(n+1)(n+2)(n+3)}{4}$ for all $n \in \mathbb{N}$.*

Solution 14. *Let $S(n)$ be the statement. Base case: $S(1)$. $1 \cdot 2 \cdot 3 = \dfrac{1 \cdot 2 \cdot 3 \cdot 4}{4}$, so $S(1)$ is true.*

Inductive hypothesis: If $S(n)$ is true, we show $S(n+1)$ is also true: $S(n+1)$ is the statement $1 \cdot 2 \cdot 3 + 2 \cdot 3 \cdot 4 + 3 \cdot 4 \cdot 5 + \ldots + (n+1)(n+2)(n+3) = \dfrac{(n+1)(n+2)(n+3)(n+4)}{4}$. From subtracting $S(n)$ from both sides, it follows that $(n+1)(n+2)(n+3) = \dfrac{4(n+1)(n+2)(n+3)}{4} = (n+1)(n+2)(n+3)$, which is true.

Extension Exercise 4. *Using **strong** induction, prove the following:*

Factorial Representation Theorem. Every positive integer N can be written in the form $N = a_1 \cdot 1! + a_2 \cdot 2! + a_3 \cdot 3! + \ldots + a_k \cdot k!$, where k is a positive integer and $a_i \in \{0, 1, \ldots, i\}$ for all $i \in \mathbb{N}$.

Extension Solution 4. *For the base case, $n = 1$, we have $1 = 1 \cdot 1!$, hence $P(1)$ holds.*

Let $k \geq 1$ be given, and suppose that $P(n)$ holds for all $n \leq k$. We show that $n = k + 1$ also has such a representation.

Let r be the largest positive integer with $r! \leq k + 1$, id est $r! \leq k + 1 \leq (r + 1)!$. If $r! = k + 1$, then we are done. Otherwise, if $r! < k + 1$, let $k = k + 1 - r!$. As $k + 1 \geq r! \geq 1$, we must have $k \geq k \geq 1$. Applying the inductive hypothesis to k allows us to write k as a finite sum of $i! \cdot d_i$ terms, with $0 \leq d_i \leq i$.

By adding $r!$, we then obtain $k + 1 = k + r!$, and it suffices to show that this representation still has $0 \leq d_i \leq i$ for all digits d_i. If $r!$ is not in the representation of k, this immediately follows. Otherwise, add $r!$ to get another representation with d_i replaced by $d_i + 1$ satisfying the condition.

Furthermore, when k has a representation involving $r!$, with the digits maximized, then $k + 1 = k + r! \geq d_r r! + r_! = (r + 1)!$. Then $k + 1 \geq (r + 1)!$, which is a contradiction. Hence the representation of $k + 1$ can always be achieved in the desired form, completing the inductive step.

Extension Exercise 5. *Let $f : \mathbb{N} \to \mathbb{N}$ be an increasing function with $f(1) = 1$ such that $f(n + 1) - f(n) \geq f(n) - f(n - 1)$ for all $n \in \mathbb{N}$. Show that*

$$\sum_{k=1}^{n} f(k) \leq f\left(\sum_{k=1}^{n} k\right)$$

Extension Solution 5. *Base case: for $n = 1$, we have $1 \leq 1$, which is true.*

Inductive step: Assume that the statement holds for some $k + 1$, and show it also holds for all positive integers in $[1, k]$. Then use the linearity of the function to reach the desired conclusion.

Part 2: Algebra

Chapter 4: Beginning Algebra

Section 4.1: The Axioms of Algebraic Equations

Exercise 15. *For (a) through (c), use only the seven axioms in the subsection (listed in Theorem 4.1).*

(a) *Show that $x = 1$ is the solution to $3x + 4 = 7$.*

(b) *Show that $x = \dfrac{c - b + d}{a}$ is the solution to $ax + b - d = c$.*

(c) *Show that $x = \pm a$ is the solution to $x^2 - a^2 = 0$.*

(d) *When is the Reflexive Property necessary to prove a solution to an equation? What about the Symmetric Property?*

Solution 15.

(a) *By the addition property of equality, we have $(3x + 4) - 4 = 7 - 4 = 3$, so $3x = 3$. Then by the multiplication property of equality with $b = \dfrac{1}{3}$, it follows that $x = 1$.*

(b) *Using the addition property of equality shows that $ax = c - b + d$, and applying the multiplication property with $\dfrac{1}{a}$ yields $x = \dfrac{c - b + d}{a}$, as desired.*

(c) *We have $x^2 = a^2$ by the addition property, and the multiplication property says that $x \cdot x = a \cdot a \implies x = a, -a$ by inspection.*

(d) *The Reflexive Property is useful whenever we wish to show similarity in geometric figures, which requires equality of angles. The Symmetric property is useful in an algebraic context, in which we may conclude that the order of the variables in an equation does not change the desired conclusion. For instance, $x = 2$ and $2 = x$ express the same basic concept.*

Extension Exercise 6. *Define a binary operator ξ such that $a \xi b = \left| a^2 - b^2 \right|$ for all real numbers a and b. What can we conclude about the associativity and commutativity of ξ?*

Extension Solution 6. *As $(a \xi b) \xi c = \left| \left| a^2 - b^2 \right| - c^2 \right|$ and $a \xi (b \xi c) = \left| a^2 - \left| b^2 - c^2 \right| \right|$, we can observe that ξ is not associative (a counterexample being $(a, b, c) = (1, 2, 3)$). However, it is commutative, as $a \xi b = b \xi a$ for all a, b.*

Section 4.2: An Introduction to Proofs Involving Inequalities

Exercise 16. *Let $x \in \mathbb{R}$. Prove that $-1 \le 2 - 3 \cos \left(1 - x^2 \right) \le 5$.*

Solution 16. *Recall that $\forall a \in \mathbb{R}, -1 \le \cos a \le 1$. Thus, the minimum value of $\cos \left(1 - x^2 \right)$ is -1, and the maximum value of $\cos \left(1 - x^2 \right)$ is 1. Plugging in $\cos \left(1 - x^2 \right) = -1$, we find that $2 - 3 \cos \left(1 - x^2 \right) = 2 - 3(-1) = 5$, so the upper bound of $2 - 3 \cos \left(1 - x^2 \right)$ is 5. Plugging in $\cos \left(1 - x^2 \right) = 1$, $2 - 3 \cos \left(1 - x^2 \right) = 2 - 3 \cdot 1 = -1$, so the lower bound of $2 - 3 \cos \left(1 - x^2 \right)$ is -1. Thus, we have successfully proven that $-1 \le 2 - 3 \cos \left(1 - x^2 \right) \le 5$.*

Exercise 17. *Let x, y, and z be real numbers. Show that $xy + xz + yz \leq x^2 + y^2 + z^2$.*

Solution 17. *By AM-GM, since $x, y, z \in \mathbb{R}$, $x^2 + y^2 \geq 2xy$, $y^2 + z^2 \geq 2yz$, and $x^2 + z^2 \geq 2xz$. Adding these three inequalities together, we obtain that $2x^2 + 2y^2 + 2z^2 \geq 2xy + 2yz + 2xz$. We can divide by 2 and rearrange the inequality to get it in the desired form: $xy + xz + yz \leq x^2 + y^2 + z^2$.*

Exercise 18. *Let x, y, and z be real numbers. Show that $(0,0,0)$ is the only solution to the equation $10xy + 12yz + 2xz = 2x^2 + 29y^2 + 10z^2$.*

Solution 18. *We can combine like terms to obtain the equivalent form $2x^2 - 10xy + 29y^2 - 12yz + 10z^2 - 2xz = 0$; this, in turn, can be written as $(x^2 - 10xy + 25y^2) + (4y^2 - 12yz + 9z^2) + (z^2 - 2xz + x^2) = 0$ (with this factorization motivated by the coefficients of the degree-1 terms). Factoring, we obtain $(x - 5y)^2 + (2y - 3z)^2 + (z - x)^2 = 0$, which implies $x - 5y = 2y - 3z = z - x = 0$ (since $\forall x \in \mathbb{R}, x^2 \geq 0$ by the Trivial Inequality). This forces $(x, y, z) = (0, 0, 0)$ (since $x = z = 5y$, but $2y = 3z$).*

Section 4.3: Introduction to Functions

Exercise 19. *For each of the following, either provide a proof if the statement is true, or provide a counterexample if the statement is false.*

(a) *All bijective functions are strictly monotonic.*

(b) *All strictly monotonic functions are bijective.*

(c) *All bijective continuous functions are strictly monotonic.*

(d) *All strictly monotonic continuous functions are bijective.*

Solution 19.

(a) *False. Consider $f(x) = x$ if $x \in (-\infty, -1) \cup (0, \infty)$, $f(x) = -x - 1$ if $x \in [-1, 0]$.*

(b) *False. $f(x) = e^x$ is a counterexample (this is strictly monotonic, but fails the surjectivity test - there is no x-value in the reals for which $f(x) = 0$).*

(c) *True. It suffices to show that a continuous, injective map is strictly monotonic, since bijectivity requires both injectivity and surjectivity. Assume the contrary; using the intermediate value theorem, let there be $x_1 < x_2 < x_3$ such that $f(x_1) \geq f(x_2) \leq f(x_3)$. If there is equality, we are done. Otherwise, assume $f(x_1) > f(x_2) < f(x_3)$, and also $f(x_1) \neq f(x_3)$ (if not, then we are done). Hence apply IVT on x_1, x_2 if $f(x_1) < f(x_3)$, and x_2, x_3 if $f(x_1) > f(x_3)$.*

(d) *True. This is a corollary of the Inverse Function Theorem.*

Exercise 20. *Prove that*

$$\lim_{x \to 1} \frac{x^n - 1}{x - 1} = n$$

for all $n \in \mathbb{N}$.

Solution 20. *Let $n \in \mathbb{N}$. Since $\dfrac{x^n - 1}{x - 1} = 1 + x + \ldots + x^{n-1}$ by the geometric series formula, $\dfrac{x^n - 1}{x - 1}$ has a hole and not an asymptote at $x = 1$. Thus, we can plug in $x = 1$ to $1 + x + \ldots + x^{n-1}$ in order to calculate $\lim_{x \to 1} \dfrac{x^n - 1}{x - 1}$; the sum evaluates to n as desired.*

Extension Exercise 7. *Prove that all linear functions $f(x) = ax + b$ $(a \neq 0)$ are continuous at all $x \in \mathbb{R}$.*

Extension Solution 7. *Let $\epsilon > 0$, $\epsilon \in \mathbb{R}$, and let $\delta = \dfrac{\epsilon}{a}$. Given that $|x - c| < \delta$, it follows that $|f(x) - f(c)| < \epsilon$ through some manipulation. This satisfies the definition of continuity.*

Extension Exercise 8. *Prove that*

$$\lim_{x \to \infty} x = \infty$$

Extension Solution 8. *Let $\epsilon > 0$, with $f(x) = x$. Then note that $f(x)$ is a linear function $f(x) = 1x + 0$, so this is a corollary of the above result (as well as the continuity of f).*

Extension Exercise 9. *Let $f : \mathbb{R} \to \mathbb{R}$ and $g : \mathbb{R} \to \mathbb{R}$ be continuous functions. Show that $f(g(x))$ and $g(f(x))$ are also continuous.*

Extension Solution 9. *Let f be continuous at $g(a)$. Then for all $\epsilon > 0$, there exists δ such that*

$$|g(x) - g(a)| < \delta \implies |f(g(x)) - f(g(a))| < \epsilon$$

As g is continuous at a, there is some δ_1 such that

$$|x - a| < \delta_1 \implies |g(x) - g(a)| < \delta$$

Take $\epsilon = \delta$, which completes the proof.

Section 4.4: Introduction to the Binomial Theorem

Exercise 21. *In the expansion of $(1 - x)^n$, where n is a positive integer, the x^{n-2} term has a coefficient of 55. Find n.*

Solution 21. *By the Binomial Theorem, the coefficient of the x^{n-2} term is $\binom{n}{n-2} = \dfrac{n(n-1)}{2}$. Thus,*
$\dfrac{n(n-1)}{2} = 55$, *and $n = \boxed{11}$.*

Exercise 22. *There are 10 fruit snacks in a pouch: 5 strawberry-flavored, and 5 grape-flavored. If we must pick 5 random snacks, and at least two must be strawberry-flavored, in how many ways can we pick the snacks? (The specific snacks picked matter, but not the order in which they are picked.)*

Solution 22. *There are $\binom{5}{2}\binom{5}{3} = 100$ ways to pick 2 strawberry and 3 grape snacks with respect to the snacks picked, $\binom{5}{3}\binom{5}{2} = 100$ ways to pick 3 strawberry snacks, $\binom{5}{4}\binom{5}{1} = 25$ ways to pick 4 strawberry, and finally 1 way to pick all strawberry. The total is $\boxed{226}$ ways.*

Exercise 23. *Show that*

$$\sum_{k=0}^{n} \binom{n}{k} = 2^n$$

for all $n \in \mathbb{N}$.

Solution 23. *We can prove this statement via an application of the Binomial Theorem. Recall that, for all $n \in \mathbb{N}$, $(x+y)^n = \displaystyle\sum_{k=0}^{n} \binom{n}{k} x^k y^{n-k}$. Let $x = y = 1$. Then $(x+y)^n = (1+1)^n = 2^n$, which is equal to*

$$\sum_{k=0}^{n} \binom{n}{k} 1^k 1^{n-k} = \sum_{k=0}^{n} \binom{n}{k}. \text{ Thus, } \sum_{k=0}^{n} \binom{n}{k} = 2^n \text{ as desired.}$$

Exercise 24. *Show that*

$$\binom{2n}{n} = \sum_{k=0}^{n} \binom{n}{k}^2$$

for all $n \in \mathbb{N}$.

Solution 24. *We have $\binom{n}{k}^2 = \binom{n}{k}\binom{n}{n-k}$; hence, the summation can be written as*

$$\binom{n}{0}\binom{n}{n} + \binom{n}{1}\binom{n}{n-1} + \ldots + \binom{n}{n}\binom{n}{0}$$

Effectively, we are choosing n people from a group of $2n$ people (upon dividing said group into equal halves).

Extension Exercise 10. *Prove that*

$$\sum_{k=0}^{n}(-1)^k \cdot \binom{n}{k} = 0$$

for all $n \in \mathbb{N}$ using the Binomial Theorem.

Extension Solution 10. *The binomial coefficients are precisely the coefficients of the expansion of $(x+1)^n$. By plugging in $x = -1$, we see that the odd-powered coefficients become negative, and the entire expression cancels out to zero.*

Extension Exercise 11. *Using the Binomial Theorem, prove that*

$$\sum_{k=0}^{n}\binom{n}{k}2^k = 3^n$$

Extension Solution 11. *We can prove this by expanding $(1+2)^n$ using the Binomial Theorem. With $t = 2$ substituted into the more general form*

$$(1+t)^n = \sum_{k=0}^{n}\binom{n}{k}t^n$$

the result follows.

Extension Exercise 12. *Show that, for all positive integers n, $\binom{4n}{2} = \binom{3n}{2} + \binom{2n}{2} + \dfrac{3n^2 + n}{2}$.*

Extension Solution 12. *Using the definition of the binomial coefficient and expanding out the factorials yields the result after some not particularly difficult computation.*

Section 4.5: A Brief Introduction to Studying the Roots of Polynomials

Exercise 25. *How can we relate Vieta's Formulas to the vertex of a parabola (a quadratic polynomial)?*

Solution 25. *Let $ax^2 + bx + c$ be the quadratic polynomial under consideration, and let r and s be its roots. From Vieta's Formulas, we know that $r + s = -\dfrac{b}{a}$. We also know that the vertex of the parabola is $-\dfrac{b}{2a}$, which is equal to $\boxed{\dfrac{r+s}{2}}$.*

Exercise 26. *If a and b are the roots of $x^2 + 3x + 9$, compute $a^2 + b^2$ and $a^4 + b^4$.*

Solution 26. *From Vieta's Formulas, $a+b = -3$ and $ab = 9$. Then $a^2 + b^2 = (a+b)^2 - 2ab = (-3)^2 - 2\cdot 9 = 9 - 18 = \boxed{-9}$. We can find $a^4 + b^4$ in a similar manner: $a^4 + b^4 = (a^2+b^2)^2 - 2a^2b^2 = (-9)^2 - 2(ab)^2 = 9^2 - 2\cdot 9^2 = \boxed{-81}$.*

Exercise 27. *(2006 AMC 10B) Let a and b be the roots of the equation $x^2 - mx + 2 = 0$. Suppose that $a + \dfrac{1}{b}$ and $b + \dfrac{1}{a}$ are the roots of the equation $x^2 - px + q = 0$. What is q?*

Solution 27. *From Vieta's Formulas, $ab = 2$. In addition, $q = \left(a + \dfrac{1}{b}\right)\left(b + \dfrac{1}{a}\right) = ab + \dfrac{a}{a} + \dfrac{b}{b} + \dfrac{1}{ab} = 2 + 1 + 1 + \dfrac{1}{2} = \boxed{\dfrac{9}{2}}$.*

Exercise 28. *Let $P(x) = (x-3)(x-9)$. Let there exist a function $G(x)$ where $G(1) = P(x)$, $G(2) = P(P(x))$, $G(3) = P(P(P(x)))$, and so forth. What is the sum of the roots of $G(10)$?*

Solution 28. *Expanding $P(x)$ yields the sum of roots of 12. Computing $P(P(x))$ and further iterations yields the pattern $6 \cdot 2^x$ by engineer's induction. Our answer is $6 \cdot 2^{10} = \boxed{6144}$.*

Exercise 29. *If $\sin x$ and $\cos x$ are the roots of $-20x^2 + bx + 8$ for some constant $b > 0$ and some real number $x \in [0, \pi]$, compute the value of $\tan x$.*

Solution 29. *Observe that $\sin^2 x + \cos^2 x = 1$ for all x, so $(\sin x + \cos x)^2 - 2\sin x \cos x = 1 \implies$ $\left(\dfrac{b}{20}\right)^2 - 2\left(-\dfrac{8}{20}\right) = 1$ by Vieta's formulas. Thus, $b = 4\sqrt{5}$. Then the roots of $-20x^2 + 4x\sqrt{5} + 8$ are $x = \dfrac{-4\sqrt{5} \pm 12\sqrt{5}}{-40} = \dfrac{2\sqrt{5}}{5}, -\dfrac{\sqrt{5}}{5}$. For $x \in [0, \pi]$, $\sin x \geq 0$, so we have $\sin x = \dfrac{2\sqrt{5}}{5}$ and $\cos x = -\dfrac{\sqrt{5}}{5}$, which implies $\tan x = \dfrac{\sin x}{\cos x} = \boxed{-2}$.*

Exercise 30. *(1983 AHSME) If $\tan \alpha$ and $\tan \beta$ are the roots of $x^2 - px + q = 0$, and $\cot \alpha$ and $\cot \beta$ are the roots of $x^2 - rx + s = 0$, find the value of rs.*

Solution 30. *Recall that $\cot \alpha = \dfrac{1}{\tan \alpha}$ and $\cot \beta = \dfrac{1}{\tan \beta}$. From Vieta's formulas, $rs = (\cot \alpha + \cot \beta)(\cot \alpha \cot \beta) =$ $\left(\dfrac{1}{\tan \alpha} + \dfrac{1}{\tan \beta}\right)\left(\dfrac{1}{\tan \alpha} \cdot \dfrac{1}{\tan \beta}\right) = \left(\dfrac{\tan \alpha + \tan \beta}{\tan \alpha \tan \beta}\right)\left(\dfrac{1}{\tan \alpha \tan \beta}\right) = \dfrac{\tan \alpha + \tan \beta}{(\tan \alpha \tan \beta)^2}$. Vieta's formulas also inform us that $\tan \alpha + \tan \beta = p$ and $\tan \alpha \tan \beta = q$. It follows that $rs = \dfrac{\tan \alpha + \tan \beta}{(\tan \alpha \tan \beta)^2} = \boxed{\dfrac{p}{q^2}}$.*

Exercise 31. *(2014 Caltech-Harvey Mudd Math Competition) If a complex number z satisfies $z + \dfrac{1}{z} = 1$, then what is $z^{96} + \dfrac{1}{z^{96}}$?*

Solution 31. *Notice that $\left(z + \dfrac{1}{z}\right)^2 = z^2 + \dfrac{1}{z^2} + 2 = 1$, so $z^2 + \dfrac{1}{z^2} = -1$. In addition, $\left(z^2 + \dfrac{1}{z^2}\right)^2 = z^4 + \dfrac{1}{z^4} + 2 = 1$, so $z^4 + \dfrac{1}{z^4} = -1$. Similarly, we can find that $z^{2^n} + \dfrac{1}{z^{2^n}} = -1$ for any positive integer n. To find $z^{96} + \dfrac{1}{z^{96}}$, we will first compute $\left(z^{32} + \dfrac{1}{z^{32}}\right)^3 = z^{96} + \dfrac{1}{z^{96}} + 3\left(z^{32} + \dfrac{1}{z^{32}}\right)$. That is, $-1 = z^{96} + \dfrac{1}{z^{96}} + 3(-1)$, so $z^{96} + \dfrac{1}{z^{96}} = \boxed{2}$.*

Extension Exercise 13. *How can we derive formulas for $a^n + b^n$ where n is any nonzero integer?*

Extension Solution 13. *If n is even, we can continue to multiply by $a^2 + b^2$ and subtract any constant terms left over. Otherwise, if n is odd, use the sum of cubes identity and repeat as with the even exponents.*

Section 4.6: Introductory Sequences and Series

Exercise 32. *(2016 MathCounts Chapter Sprint) A bug crawls a distance of n miles at a speed of $n + 1$ miles per hour one day and then crawls $2n + 1$ miles at a speed of $n^2 + n$ miles per hour the next day. If the total time for the trip was 6 hours, what was the bug's average speed?*

Solution 32. *Using the formula $d = rt$, it follows that $\dfrac{n}{n+1} + \dfrac{2n+1}{n^2+n} = 6 \implies \dfrac{n^2 + 2n + 1}{n^2 + n} = 6 \implies$ $n^2 + 2n + 1 = 6n^2 + 6n \implies 5n^2 + 4n - 1 = 0 \implies n = -1, \dfrac{1}{5}$. Clearly, $n > 0$, so the bug's average speed in miles per hour is $3n + 1 = \dfrac{8}{5}$ divided by 6, or $\boxed{\dfrac{4}{15}}$ miles per hour.*

Exercise 33. *(2015 ARML Relay Round) Given that $1 - r + r^2 - r^3 + \ldots = s$ and $1 + r^2 + r^4 + r^6 + \ldots = 4s$, compute s.*

Solution 33. *From the infinite geometric series formula, $s = \dfrac{1}{1 - (-r)} = \dfrac{1}{1 + r}$ and $4s = \dfrac{1}{1 - r^2} = \dfrac{1}{1-r} \cdot \dfrac{1}{1+r}$. Thus, $4 \cdot \dfrac{1}{1+r} = \dfrac{1}{1-r} \cdot \dfrac{1}{1+r}$, or $4 = \dfrac{1}{1-r}$. Solving for r, we find that $r = \dfrac{3}{4}$, so $s = \boxed{\dfrac{4}{7}}$.*

Exercise 34. *Can a non-constant finite sequence with length ≥ 3 be both arithmetic and geometric? Justify your answer.*

Solution 34. *We claim that no such sequence exists. Assume that there exists an arithmetic sequence (a_n) with length ≥ 3 that is also geometric. Let $(a_n) = \{a, a+d, a+2d, a+3d, \ldots\}$ have the geometric subsequence $(b_n)_{n=1}^3 = \{a, a+d, a+2d\}$. However, note that, for (b_n) to be geometric, we must have $r = \dfrac{a+d}{a} = \dfrac{a+2d}{a+d}$, or $(a+d)^2 = a(a+2d) = a^2 + 2ad$. However, it follows from this equation that $d^2 = 0$, or $d = 0$, which implies that (b_n) is constant, and so that (a_n) is constant as well (since if the common difference of (a_n) were non-zero, any subsequence of (a_n) would also have a non-zero common difference). This is a contradiction; hence, no sequence with length ≥ 3 can be both arithmetic and geometric.*

Exercise 35. *(2000 AIME I) A sequence of numbers $x_1, x_2, x_3, \ldots, x_{100}$ has the property that, for every integer k between 1 and 100, inclusive, the number x_k is k less than the sum of the other 99 numbers. Given that $x_{50} = \dfrac{m}{n}$, where m and n are relatively prime positive integers, find $m + n$.*

Solution 35. *Let the sum of the numbers be S. Then note that $S = x_1 + x_2 + x_3 + \ldots + x_{100} = (x_2 + x_3 + \ldots + x_{100}) + (x_1 + x_3 + \ldots + x_{100}) + (x_1 + x_2 + x_4 + \ldots + x_{100}) + \ldots + (x_1 + x_2 + x_3 + \ldots + x_{100}) - (1 + 2 + 3 + \ldots + 100) = 99S - 5050 \implies S = \dfrac{5050}{98} = \dfrac{2525}{49}$. Then $x_{50} = (S - x_{50}) - 50 \implies 2x_{50} = S - 50 = \dfrac{75}{49} \implies x_{50} = \dfrac{75}{98}$. Finally, $m + n = 75 + 98 = \boxed{173}$.*

Extension Exercise 14. *(a) Suppose that the infinite sequence (a_n) has subsequences $(b_n) = (a_{2k})_{k \in \mathbb{N}}$ and $(c_n) = (a_{2k-1})_{k \in \mathbb{N}}$ with (b_n) arithmetic and (c_n) geometric. Propose and prove a formula for the sum of the first n terms of (a_n) in terms of the initial term a of (a_n), the common difference d of (b_n), and the common ratio r of (c_n).*

(b) Let $(a_n) = \{1, 2, 3, 6, 7, 14, 15, \ldots\}$ be the infinite sequence such that, for all odd n, $a_{n+1} = 2a_n$, for all even n, $a_{n+1} - a_n = 1$. Propose and prove a formula for the sum of the first n terms of (a_n).

Extension Solution 14. *If $n = 2m$ is even, then the sum is just $3 + 9 + 21 + 45 + \ldots = 3(1 + 3 + 7 + 15 + \ldots) = 3 \cdot \sum_{k=1}^m (2^k - 1)$. If $n = 2m - 1$ is odd, subtract the $2m^{th}$ term, namely $2 \cdot (2^m - 1)$, from the above formula.*

Extension Exercise 15. *Given that the roots of the polynomial $P(x) = x^3 + kx^2 + 20x - 18$ are in arithmetic progression, what are all possible values of k?*

Extension Solution 15. *Note that the roots sum to $-k$ by Vieta's formulas, so the middle root must be $-\dfrac{k}{3}$. In addition, their product is 18, also by Vieta's. As such, let the roots be $-\dfrac{k}{3} - d$, $-\dfrac{k}{3}$, and $-\dfrac{k}{3} + d$. Furthermore, we have $\left(-\dfrac{k}{3} - d\right)\left(-\dfrac{k}{3}\right) + \left(-\dfrac{k}{3} - d\right)\left(-\dfrac{k}{3} + d\right) + \left(-\dfrac{k}{3}\right)\left(-\dfrac{k}{3} + d\right) = 20$, which simplifies to $\left(-\dfrac{k}{3}\right)\left(-\dfrac{2k}{3}\right) + \dfrac{k^2}{9} - d^2 = 20 \implies \dfrac{k^2}{3} - d^2 = 20$. Furthermore, $\dfrac{k^2}{9} - d^2 = -\dfrac{54}{k}$, so $\dfrac{2k^2}{9} = \dfrac{54}{k} + 20 \implies k^3 = 243 + 90k \implies k^3 - 90k - 243 = 0$. Observe that $k = -3$ is a solution by inspection, and then factor using long division to obtain the other two values, namely $k = \dfrac{3 \pm 3\sqrt{37}}{2}$.*

Chapter 4 Review Exercises

Exercise 36. *Prove that the positive solution to the equation $20x^2 + 18 = 2018$ is $x = 10$.*

Solution 36. *Remember from the Fundamental Theorem of Algebra that a quadratic polynomial has exactly two complex roots. The equation in question simplifies to $x^2 = 100$, which has two solutions: $x = 10$ and $x = -10$. Thus, $x = 10$ is the positive solution to $20x^2 + 18 = 2018$.*

Exercise 37. *Consider the relation $f : \mathbb{R} \backslash \{-1, 2\} \to \mathbb{R} \backslash \{0\}$, $f(x) = \dfrac{2}{x^2 - x - 2}$.*

(a) Is f a function? Why or why not?

(b) Prove that f is continuous for all $x \in \mathbb{R}$, $x \neq -1, 2$. What type(s) of discontinuities does f have at $x = -1, 2$?

(c) Define $f_1(x) = \{f(x) : x < -1\}$, $f_2(x) = \{f(x) : -1 < x < 2\}$, $f_3(x) = \{f(x) : x > 2\}$. For each of f_1, f_2, f_3, is it injective, surjective, bijective, some combination thereof, or not a function?

Solution 37.

(a) Yes, as it passes the VLT (i.e. no x-value produces multiple values for the output of f).

(b) We can use the δ-ε strategy to prove continuity for all $x \neq -1, 2$ (the details are left as an exercise in applying the technique).

At $x = -1, 2$, f has a jump discontinuity.

(c) f_1 yields an injective, but not surjective (or bijective) function, due to the horizontal asymptote at the x-axis. f_2 yields a non-injective (fails the HLT) and non-surjective function (vertical asymptotes at $x = -1, 2$), and f_3 yields a similar outcome to f_1.

Exercise 38. *Consider the binomial expansion of $(x + y)^{19}$.*

(a) How many coefficients are in the expansion?

(b) What is the sum of the coefficients in the expansion?

(c) Show that the sum of the coefficients of the terms with odd powers of x is equal to the sum of the coefficients of the terms with odd powers of y, and compute this common sum.

Solution 38. *(a) There are $19 + 1 = \boxed{20}$ coefficients/terms in the expansion.*

(b) In Exercise 22, we showed that $\sum_{k=0}^{n} \binom{n}{k} = 2^n$ for all $n \in \mathbb{N}$. Since the coefficients of $(x + y)^{19}$ are $\binom{19}{0}, \binom{19}{1}, \ldots, \binom{19}{19}$, the sum of the coefficients is $2^{19} = \boxed{524288}$.

(c) Since the degree measure of $(x + y)^{19}$ is odd, and x and y are always raised to integer powers that sum to 19, x has an even degree measure while y has an odd one and vice versa. That is, the coefficients for which x is raised to an odd power are $\binom{19}{0}, \binom{19}{2}, \ldots, \binom{19}{18}$. These coefficients are equal to the coefficients for which y is raised to an odd power: $\binom{19}{19}, \binom{19}{17}, \ldots, \binom{19}{1}$, as we know that $\binom{n}{k} = \binom{n}{n-k}$. This common sum found is equal to $\frac{1}{2} \cdot 2^{19} = 2^{18} = \boxed{262144}$.

Exercise 39. *(2006 AMC 10A) How many non-similar triangles have angles whose degree measures are distinct positive integers in arithmetic progression?*

Solution 39. *The angle measures in any triangle sum to $180°$. If the angle measures are in arithmetic progression, then the middle angle must be $60°$ (this forces the other two angles to have measures $(60 - d)°$ and $(60 + d)°$ for some $d > 0$; the sum of n elements in arithmetic progression is n times the middle element (if n is even, n times the median)). Thus, any d from 1 to 59, inclusive, works. There are $\boxed{59}$ non-similar triangles whose degree measures are distinct positive integers in arithmetic progression.*

Exercise 40. *(a) Using Euclid's Division Algorithm, compute $\gcd(2016, 4200)$.*

(b) (1959 IMO) Prove that $\dfrac{21n + 4}{14n + 3}$ is irreducible for every natural number n.

Solution 40.

(a) $4200 - 2 \cdot 2016 = 168$, $2016 - 12 \cdot 168 = 0$, so $\gcd(2016, 4200) = \boxed{168}$.

(b) *Using the Euclidean Algorithm, we have* $(21n + 4) - (14n + 3) = 7n + 1$, $(14n + 3) - 2(7n + 1) = 1$, *and clearly, this shows that* $\gcd(21n + 4, 14n + 3) = 1$. *Hence the fraction* $\dfrac{21n + 4}{14n + 3}$ *is irreducible.*

Exercise 41. *Let* r_1 *and* r_2, *with* $r_1 \geq r_2$, *be the roots of* $2x^2 + 3x + 1$.

(a) *Compute* $r_1^2 + r_2^2$.

(b) *Compute* $r_1^2 r_2 + r_1 r_2^2$.

(c) *Compute* $r_1^3 + r_2^3$.

(d) *Without computing* r_1 *and* r_2, *compute* $r_1^4 - r_2^4$.

Solution 41. (a) *From Vieta's Formulas, we have that* $r_1 + r_2 = -\dfrac{3}{2}$ *and* $r_1 r_2 = \dfrac{1}{2}$. *Then* $r_1^2 + r_2^2 =$

$$(r_1 + r_2)^2 - 2r_1 r_2 = \left(-\frac{3}{2}\right)^2 - 2 \cdot \frac{1}{2} = \frac{9}{4} - 1 = \boxed{\frac{5}{4}}.$$

(b) *We have that* $r_1^2 r_2 + r_1 r_2^2 = r_1 r_2 (r_1 + r_2) = \dfrac{1}{2}\left(-\dfrac{3}{2}\right) = \boxed{-\dfrac{3}{4}}$.

(c) *We have that* $r_1^3 + r_2^3 = (r_1 + r_2)(r_1^2 - r_1 r_2 + r_2^2) = -\dfrac{3}{2}\left(\dfrac{5}{4} - \dfrac{1}{2}\right) = \boxed{-\dfrac{3}{8}}$.

(d) *We have that* $r_1^4 - r_2^4 = (r_1^2 + r_2^2)(r_1^2 - r_2^2) = (r_1^2 + r_2^2)(r_1 + r_2)(r_1 - r_2)$. *All of the quantities in the factorization of* $r_1^4 - r_2^4$ *are known except for* $(r_1 - r_2)$. *However, we can find* $(r_1 - r_2)^2 = r_1^2 - 2r_1 r_2 + r_2^2 = \dfrac{5}{4} = 2 \cdot \dfrac{1}{2} = \dfrac{1}{4}$. *Since* $r_1 \geq r_2$, $(r_1 - r_2) = \dfrac{1}{2}$. *Plugging this into the factorization found above, we find that* $r_1^4 - r_2^4 = \dfrac{5}{4}\left(-\dfrac{3}{4}\right)\dfrac{1}{2} = \boxed{-\dfrac{15}{32}}$.

Exercise 42.

(a) *Show that* $\dfrac{4x^2 - 16x + 17}{9x^2 - 30x + 26} > 0$ *for all* $x \in \mathbb{R}$.

(b) *Solve the inequality* $\dfrac{4x^2 - 16x + 17}{9x^2 - 30x + 26} \geq 1$ *for* $x \in \mathbb{R}$.

Solution 42.

(a) *We are essentially showing that either* $4x^2 - 16x + 17 > 0$ *and* $9x^2 - 30x + 26 > 0$, *or* $4x^2 - 16x + 17 < 0$ *and* $9x^2 - 30x + 26 < 0$. *In the first case,* $4(x^2 - 4x + 4) + 1 > 0 \implies 4(x - 2)^2 + 1 > 0$, *which is true by Trivial Inequality. Furthermore,* $9x^2 - 30x + 26 = (9x^2 - 30x + 25) + 1 = 9\left(x - \dfrac{5}{3}\right)^2 + 1 > 0$. *By Trivial Inequality, the second case is impossible, so we have proven the inequality.*

(b) *We have* $4x^2 - 16x + 17 \geq 9x^2 - 30x + 26 \implies 5x^2 - 14x + 9 \leq 0 \implies x \leq 1, x \geq \dfrac{9}{5}$. *Hence, the inequality is satisfied when* $x \in \left[1, \dfrac{9}{5}\right]$.

Extension Exercise 16. *Let* $f : \mathbb{R} \to \mathbb{R}$ *be a monotonic function. Can* f *have infinitely many discontinuities? An uncountably infinite number of discontinuities? (Hint: First prove that the set of rational numbers* $\mathbb{Q}$ *is countable.)*

Extension Solution 16. *The answer to the first part is yes; consider $f(x) = 2^{\lfloor \log_2 x \rfloor}$. The answer to the second part is no; this follows from the countable infinitude of $\mathbb{Q}$, the set of rational numbers. Specifically, let $f(x^-), f(x^+)$ denote the left- and right-hand limits of f at x respectively. Let S be the set of points x for which f is not continuous at x. For any $x \in S$, there exists $g(x) \in \mathbb{Q}$ such that $f(x^-) < g(x) < f(x^+)$. For $x_1 < x_2$, we have $f(x_1^+) < f(x_2^-)$. It follows that if $x_1 \neq x_2$, then $g(x_1) \neq g(x_2)$. Hence the map between S and $\mathbb{Q}$ is injective. As $\mathbb{Q}$ is countable, so is S.*

To prove $\mathbb{Q}$ is countably infinite, arrange them in the form

$$\frac{0}{1}, \frac{1}{1}, \frac{-1}{1}, \frac{1}{2}, \frac{-1}{2}, \frac{2}{1}, \frac{-2}{1}, \dots$$

with a bijection from $\mathbb{Q}$ to $\mathbb{N}$ (which we know to be countably infinite, with cardinality $\aleph_0$). (This is also known as a diagonal argument.)

Extension Exercise 17.

(a) *Prove Pascal's Identity for all $n, k \in \mathbb{N}$, $n > k$ (i.e. prove the equation $\binom{n-1}{k-1} + \binom{n-1}{k} = \binom{n}{k}$ for all $n, k \in \mathbb{N}$, $n > k$).*

(b) *Prove the Hockey Stick Identity for all $n, k \in \mathbb{N}$, $n > k$:*

$$\sum_{i=k}^{n} \binom{i}{k} = \binom{n+1}{k+1}.$$

Extension Solution 17.

(a) *Assume $k \leq n$ (else the result is trivialized). Then*

$$\binom{n-1}{k-1} + \binom{n-1}{k} = \frac{(n-1)!}{(k-1)!(n-k)!} + \frac{(n-1)!}{k!(n-k+1)!}$$

$$= (n-1)! \cdot \left(\frac{k}{k!(n-k)!} + \frac{n-k}{k!(n-k)!} \right)$$

$$= \frac{n!}{k!(n-k)!} = \binom{n}{k}$$

by definition.

(b) *Using induction on n, the base case is $n = r$, which leads to $\sum_{i=r}^{n} \binom{i}{r} = \binom{n+1}{r+1}$. Next suppose for some $k \in \mathbb{N}, k > r$, that $\sum_{i=r}^{k} \binom{i}{r} = \binom{k+1}{r+1}$. Then*

$$\sum_{i=r}^{k+1} \binom{i}{r} = \binom{k+1}{r+1} + \binom{k+1}{r} = \binom{k+2}{r+1}$$

which is true by Pascal's Identity.

Extension Exercise 18. *Show that if $f : \mathbb{N} \to \mathbb{N}$ is a multiplicative function, its values for all natural numbers are uniquely determined by $f(p^k)$ for prime p, $k \in \mathbb{N}$.*

Extension Solution 18. *Each positive integer has a prime factorization with prime factors p_i. Since f is multiplicative, we can uniquely determine its value solely from these prime factors, and each exponent of a prime factor is synonymous with raising the value of $f(p_i)$ to the corresponding power (again by multiplicativity of f).*

Chapter 5: Intermediate Algebra

Section 5.1: The Rearrangement Inequality

Exercise 43. *Prove* $a^4 + b^4 + c^4 \geq a^2bc + b^2ca + c^2ab$.

Solution 43. *Note that each term of the RHS has degree 4, hence this is a direct application of Rearrangement.*

Exercise 44. *Let a, b and c be positive real numbers. Prove that*

$$a^a b^b c^c \geq (abc)^{\left(\frac{a+b+c}{3}\right)}.$$

Solution 44. *Write the RHS as* $a^{\frac{a+b+c}{3}} b^{\frac{a+b+c}{3}} c^{\frac{a+b+c}{3}}$*, allowing us to apply Rearrangement, since* $3\left(\dfrac{a+b+c}{3}\right) = a+b+c$.

Exercise 45. *Let* $a, b, c > 0$. *Prove that if* $a^2 + b^2 + c^2 = 1$, *then*

$$\frac{1}{a^2} + \frac{1}{b^2} + \frac{1}{c^2} \geq 3 + \frac{2(a^3 + b^3 + c^3)}{abc}.$$

Solution 45.

Exercise 46. *Let* $y_1, y_2 \cdots y_8$ *be a permutation of* $1, 2 \cdots 8$. *Find*

$$\min\left(\sum_{i=1}^{8} (y_i + i)^2\right)$$

Solution 46.

Section 5.2: The AM-GM Inequality

Exercise 47.

(a) *A rectangular fence is constructed against a wall using 60 meters of wire. Show that the maximal area of the fence is 400 square meters.*

(b) *If the fence need not be rectangular, find the maximal area of the fence.*

Solution 47.

(a) *There are 3 sides that can be used for the fence (not counting the wall). By AM-GM, we should make each side the same length - 20 meters - yielding a square with area 400 square meters.*

(b) *The fence may be semi-circular with semi-circumference* $60 = \pi r \implies r = \dfrac{60}{\pi} \implies \dfrac{1}{2}\pi r^2 = \boxed{\dfrac{1800}{\pi}}$.

Exercise 48. *If a and b are real numbers such that* $a, b > 0$ *and* $a + b = 12$, *what is the maximum value of* ab^2?

Solution 48. *This is* $4 \cdot 8^2 = \boxed{256}$ *using the idea of weighted means in AM-GM.*

Exercise 49. *For a, b, and c positive real numbers, show that* $\dfrac{a^2}{bc} + \dfrac{b^2}{ca} + \dfrac{c^2}{ab} \geq 3$.

Solution 49. *Hint: We have equality at* $\dfrac{a^2}{bc} = 1$. *How can we apply AM-GM in this way?*

Exercise 50. *Let* $\pi = \{y_1, y_2, y_3, \ldots, y_n\}$ *be a permutation of the set* $S = \{x_1, x_2, x_3, \ldots, x_n\}$. *Show that* $\dfrac{x_1}{y_1} + \dfrac{x_2}{y_2} + \dfrac{x_3}{y_3} + \ldots + \dfrac{x_n}{y_n} \geq n$.

Solution 50. *Hint: This problem is quite similar to the one above, except with permutations that do not change the equality case.*

Extension Exercise 19. *Prove the AM-GM Inequality by induction.*

Extension Solution 19. *Hint: Assume it holds for n variables, with $n = 1$ the trivial base case. Then consider $(n+1)\alpha = \sum_i x_i$.*

Extension Exercise 20. *Let n be an integer greater than 1. Prove that $\left(\dfrac{n+1}{2}\right)^n > n!$*

Extension Solution 20. *Rearrangement Inequality in combination with AM-GM will suffice to prove this one (mainly left as an exercise to the reader; trying small cases is your best friend here!)*

Section 5.3: The Cauchy-Schwarz Inequality

Exercise 51. *Prove that $(x_1 + x_2 + \ldots + x_n)^2 \le n(x_1^2 + x_2^2 + \ldots + x_n^2)$ for all real numbers $x_1, x_2, x_3, \ldots, x_n$.*

Solution 51. *(This is pretty much a direct application of Cauchy, intended as a demonstration of its power.)*

Exercise 52. *(2013 AMC 12B) Let $a, b,$ and c be real numbers such that*

$$a + b + c = 2, \text{ and}$$

$$a^2 + b^2 + c^2 = 12.$$

What is the difference between the maximum and minimum possible values of c?

Solution 52. *Note that $a + b = 2 - c$. Applying Cauchy-Schwarz, we obtain $a^2 + b^2 \ge \dfrac{(2-c)^2}{2}$, or $\dfrac{(2-c)^2}{2} + c^2 \le 12$. Solving for c and taking the difference yields $\boxed{\dfrac{16}{3}}$.*

Exercise 53. *If x, y, and z are real numbers such that $x^2 + y^2 + z^2 = 1$, and $x + 2y + 3z$ is as large as possible, compute $x + y + z$.*

Solution 53. *By Cauchy, we assign "weights" to x, y, z in accordance with their coefficients. These weights are $\dfrac{1}{6}, \dfrac{1}{3}, \dfrac{1}{2}$, and scaling accordingly, $(x, y, z) = \left(\dfrac{1}{\sqrt{14}}, \dfrac{2}{\sqrt{14}}, \dfrac{3}{\sqrt{14}}\right) \implies x + y + z = \boxed{\dfrac{\sqrt{14}}{2}}$.*

Section 5.4: Binomial Theorem, Part 2: Generalizing to Real Coefficients

Exercise 54. *(2000 AIME I) In the expansion of $(ax + b)^{2000}$, where a and b are relatively prime positive integers, the coefficients of x^2 and x^3 are equal. Find $a + b$.*

Solution 54. *From the Binomial Theorem, we know that the coefficients of the x^2 and x^3 terms are $\dbinom{2000}{2}a^2 b^{1998}$ and $\dbinom{2000}{3}a^3 b^{1997}$, respectively. Expanding out the binomial coefficients and cancelling, we have that $\dfrac{2000 \cdot 1999}{2 \cdot 1}a^2 b^{1998} = \dfrac{2000 \cdot 1999 \cdot 1998}{3 \cdot 2 \cdot 1}a^3 b^{1997}$. Doing more cancelling, $b = \dfrac{1998}{3}a = 666a$. Since a and b are relatively prime, $a = 1$ and $b = 666$; thus, $a + b = \boxed{667}$.*

Exercise 55. *How many terms are in the expansion of $(5 \log_5 x + \log_{25} y)^{125}$, after combining like terms?*

Solution 55. *Notice that $5 \log_5 x = \log_5(x^5)$ and $\log_{25} y = \dfrac{\log_5 y}{\log_5 25} = \dfrac{1}{2}\log_5 y = \log_5 \sqrt{y}$. Then the expression inside the parentheses evaluates to $5 \log_5 x + \log_{25} y = \log_5(x^5) + \log_5 \sqrt{y} = \log_5(x^5 \sqrt{y})$, which has one term. Raising that to the 125th power, we get $(\log_5(x^5 \sqrt{y}))^{125}$, which has only $\boxed{1}$ term.*

Exercise 56. *(2002 AIME I) The Binomial Expansion is valid for exponents that are not integers. That is, for all real numbers x, y and r with $|x| > |y|$,*

$$(x+y)^r = x^r + rx^{r-1}y + \frac{r(r-1)}{2}x^{r-2}y^2 + \frac{r(r-1)(r-2)}{3!}x^{r-3}y^3 \cdots$$

What are the first three digits to the right of the decimal point in the decimal representation of $\left(10^{2002}+1\right)^{\frac{10}{7}}$?

Solution 56. *Using the given equation, and the fact that $\dfrac{2002}{7} = 286$, yields*

$$\left(10^{2002}+1\right)^{\frac{10}{7}} = 10^{2860} + \frac{10}{7}10^{858} + \frac{15}{40}10^{-1144} + \cdots$$

Essentailly, this boilds down to finding the last three digits after the decimal in the term $\dfrac{10}{7}10^{858}$. As $\dfrac{1}{7}$ has a period of 6, this is equivalent to the last three digits after the decimal in $\dfrac{10}{7}$, or $\boxed{428}$.

Exercise 57. *(1995 AIME) Let $f(n)$ be the integer closest to $\sqrt[4]{n}$. Find $\displaystyle\sum_{k=1}^{1995} \frac{1}{f(k)}$.*

Solution 57. *We can use casework up to $6^4 = 1296$, $1785 < 6.5^4 < 1786$, and $7^4 = 2401$. But a solution with the Binomial Theorem involves observing that $\left(k - \frac{1}{2}\right)^4 \leq n < \left(k + \frac{1}{2}\right)^4$, then setting $f(n) = k$ (which has $\lfloor \left(k + \frac{1}{2}\right)^4 - \left(k - \frac{1}{2}\right)^4 \rfloor$ satisfactory values). Using the Binomial Theorem, $\left(k + \frac{1}{2}\right)^4 - \left(k - \frac{1}{2}\right)^4 = 4k^3 + k$. The sum of all $\frac{1}{k}$ is then $4k^2 + 1$, and from $k = 1$ to $k = 6$, we have $\sum_{k=1}^{6} 4k^2 + 1 = 370$. However, we still have $1995 - 1785 = 210$ terms unaccounted for, and for all of these, $f(n) = 7$, so the desired sum is $370 + 30 = \boxed{400}$.*

Extension Exercise 21. *Can we derive Bernoulli's Inequality from the Binomial Theorem? Briefly explain how you might approach such a derivation.*

Extension Solution 21. *Using the Binomial Theorem, write $(1+x)^r = 1 + rx + \binom{r}{2}x^2 + \ldots + \binom{r}{r}x^r > 1 + rx$ (for all $x > 0$). Then for $x = 0$, $(1+x)^r = 1 + rx = 1$, and for $-1 \leq x < 0$, let $y = -x$, so that $0 < y \leq 1$.*

Extension Exercise 22. *In the expansion of $(1 + x)^n$ $(n > 2)$, prove that if the coefficients of three consecutive terms are in arithmetic progression, then $n + 2$ is a perfect square.*

Extension Solution 22. *This requires that $\binom{n}{k-1}, \binom{n}{k}, \binom{n}{k+1}$ are in A.P., or $\dfrac{n}{n-k+1}, 1, \dfrac{n-k}{k+1}$ after scaling upward. In turn, $k(k+1)$, $(k+1)(n-k+1)$, $(n-k)(n-k+1)$ are in A.P., and so $k(k+1) + (n-k)(n-k+1) = 2(k+1)(n-k+1) \implies (2k-n)^2 = n+2$, as desired.*

Section 5.5: Rational Root Theorem

Exercise 58.

(a) **Using the RRT,** *show that $\sqrt{2}$ is irrational.*

(b) **Using the RRT,** *show that $\sqrt{p}$ is irrational for all prime p.*

(c) **Using the RRT,** *show that $\sqrt{2} + \sqrt{3}$ is irrational.*

Solution 58.

(a) *For contradiction, assume $\sqrt{2}$ were rational. Note that $\sqrt{2}$ is the root of $x^2 - 2$. But by RRT, $x^2 - 2$ may only have rational roots $\pm\frac{1}{2}, 1$, so $\sqrt{2}$ cannot be rational - contradiction.*

(b) *Repeat the same process as above, with $P(x) = x^2 - p$ (with $\sqrt{p}$ as a root of P in this case).*

(c) *Here, we try to write $\sqrt{2} + \sqrt{3}$ in a form that is conducive to it being a root of a polynomial. We have $(\sqrt{2} + \sqrt{3})^2 = 5 + 2\sqrt{6}$, and $(\sqrt{2} + \sqrt{3})^4 = 49 + 20\sqrt{6}$. Then $\sqrt{2} + \sqrt{3}$ is a root of $x^4 - 10x^2 + 1$. By RRT, we again see that $\sqrt{2} + \sqrt{3}$ is not included in the list of viable candidates for rational roots, hence it must be irrational.*

Exercise 59. *Let S_n be the set of all monic polynomials (polynomials with leading coefficient 1) with degree n that have no rational roots. Show that, for all polynomials $p(x) \in S_n$, $n \leq 3$, $p(x)$ is irreducible over $\mathbb{Q}$.*

Solution 59. *If $p(x)$ were reducible over $\mathbb{Q}$, then it would have a linear factor in $\mathbb{Q}$ (with a rational root). But since $p(x) \in S_n$ (and thus has no rational roots), this is a contradiction, as any polynomial with degree ≤ 3 must contain a linear factor (as the product of two quadratics is a quartic, which has degree 4).*

Exercise 60. *How many rational roots does*

$$P(x) = \sum_{k=0}^{2019} x^k \cdot (-1)^{2020-k}$$

have?

Solution 60. *We claim the answer is $\boxed{1}$. Observe that the leading coefficient of $P(x)$ is -1 and the constant term is $+1$, so the only possible rational roots are ± 1. Testing both reveals that only $+1$ works.*

Extension Exercise 23. *Let $\mathbb{Z}_n$ be the group of integers modulo n; for instance, $\mathbb{Z}_5 = \{0, 1, 2, 3, 4\}$. Let $\mathbb{Z}_n[x]$ be the field of polynomials with integer coefficients modulo n; for instance, $\mathbb{Z}_{10}[x] = \{a_n x^n + a_{n-1} x^{n-1} + \ldots + a_1 x + a_0, a_i \in \mathbb{Z}_{10}, \forall i \in [0, n]\}$. For which polynomials $f(x) \in \mathbb{Z}_2[x]$ does f have at least one root in $\mathbb{Z}_2$?*

Extension Solution 23. *If $f(x) \in \mathbb{Z}_2[x]$, then its coefficients are all either 0 or 1. Then we require that it has at least one root that is either 0 or 1, which means that it has a factor of either x or $x - 1$. In the former case, all coefficients except for a_0 must be 1, and $a_0 = 0$. In the latter case, the sum of the coefficients must be 0.*

Section 5.6: Continuing Sequences, Series, and Recursion

Exercise 61. *What is the partial fraction decomposition of $\dfrac{1}{n(n + k)}$, in terms of k?*

Solution 61. *Note that we want to express this in the form of $\dfrac{K}{n} + \dfrac{J}{n + k}$ such that $\dfrac{K}{n} + \dfrac{J}{n + k} = \dfrac{K(n + k) + J(n)}{n(n + k)} = \dfrac{1}{n(n + k)}$. Co-efficient matching yields the system of equations*

$$K + J = 0$$

$$Kk = 1.$$

Thus, $K = \dfrac{1}{k}$, implying that $J = -\dfrac{1}{k}$. Thus, our partial fraction decomposition is $\boxed{\dfrac{1}{kn} - \dfrac{1}{k(n + k)}}$.

Exercise 62. *Find the partial fraction decomposition of $\dfrac{1}{n(n + 1)(n + 2)}$.*

Solution 62. *Note that* $\dfrac{1}{n(n+1)(n+2)} = \dfrac{1}{n(n+1)} \cdot \dfrac{1}{n+2}$. *Using the formula we found in the previous exercise,* $\dfrac{1}{n(n+1)} = \dfrac{1}{n} - \dfrac{1}{n+1}$. *Thus,* $\dfrac{1}{n(n+1)(n+2)} = \left(\dfrac{1}{n} - \dfrac{1}{n+1} \right) \dfrac{1}{n+2} = \dfrac{1}{n(n+2)} - \dfrac{1}{(n+1)(n+2)}$. *Again, via the formula, we find that* $\dfrac{1}{n(n+2)} = \dfrac{1}{2n} - \dfrac{1}{2(n+2)}$. *In addition, we have that* $\dfrac{1}{(n+1)(n+2)} = \dfrac{1}{n+1} - \dfrac{1}{n+2}$. *Thus,* $\dfrac{1}{n(n+1)(n+2)} = \dfrac{1}{n(n+2)} - \dfrac{1}{(n+1)(n+2)} = \dfrac{1}{2n} - \dfrac{1}{2(n+2)} - \left(\dfrac{1}{n+1} - \dfrac{1}{n+2} \right) = \boxed{\dfrac{1}{2n} - \dfrac{1}{n+1} + \dfrac{3}{2(n+2)}}$.

Exercise 63. *A fair coin is flipped until it lands on tails. What is the expected number of times the coin is flipped?*

Solution 63. *The probability that the first flip is the first tails is* $\dfrac{1}{2}$, *the probability that the second flip is the first tails is* $\dfrac{1}{4}$, *and in general, the probability that the* n^{th} *flip is the first tails is* $\dfrac{1}{2^n}$. *Thus, we want to compute the sum* $1 \cdot \dfrac{1}{2} + 2 \cdot \dfrac{1}{4} + 3 \cdot \dfrac{1}{8} + \ldots = \dfrac{1}{2} + \dfrac{2}{4} + \dfrac{3}{8} + \ldots = \left(\dfrac{1}{2} + \dfrac{1}{4} + \dfrac{1}{8} + \ldots \right) + \left(\dfrac{1}{4} + \dfrac{1}{8} + \ldots \right) + \ldots = 1 + \dfrac{1}{2} + \dfrac{1}{4} + \ldots = \boxed{2}$.

Exercise 64. *(1993 ARML Team Round) The Fibonacci numbers* F_a, F_b, *and* F_c *form an arithmetic sequence. If* $a + b + c = 2000$, *compute* a.

Solution 64. *The recursive, additive nature of the Fibonacci sequence motivates us to try values close to 666 or 667. Indeed,* $a = \boxed{665}$, $b = 667$, $c = 668$ *works, as* $F_{665} + F_{666} = F_{667}$ *and* $F_{666} + F_{667} = F_{668}$.

Exercise 65. *Determine, with explanation, whether the following infinite series converge absolutely, converge conditionally, or diverge.*

(a) $\displaystyle\sum_{k=1}^{\infty} \dfrac{2k+1}{k-1}$

(b) $\displaystyle\sum_{k=1}^{\infty} \dfrac{k^2}{\sqrt{k^6 + 1} - 1}$

(c) $\displaystyle\sum_{k=1}^{\infty} \dfrac{(-1)^{k-1}}{k}$

(d) $\displaystyle\sum_{k=1}^{\infty} \left(\dfrac{6k^3 + 2019}{5k^3 + 20\sqrt{19k^2 + 9}} \right)^k$

(e) $\displaystyle\sum_{k=1}^{\infty} \dfrac{\sin\left(\dfrac{k\pi}{2} \right)}{k}$

Solution 65. *(a) The series* $\boxed{\text{diverges}}$ *because the limit of the terms is not equal to 0; in fact,* $\displaystyle\lim_{k \to \infty} \dfrac{2k+1}{k-1} = 2$.

(b) By the Limit Comparison Test with $a_k = \dfrac{k^2}{\sqrt{k^6 + 1} - 1}$, $b_k = \dfrac{1}{k}$, $\lim_{k \to \infty} \dfrac{a_k}{b_k} = 1 > 0$, *the series* $\boxed{\text{diverges}}$ *(since* $\sum b_n$ *diverges as the harmonic series).*

(c) This is the alternating harmonic series, which $\boxed{\text{converges conditionally}}$ because the harmonic series diverges.

(d) Using the Root Test, we can immediately see that the series $\boxed{\text{diverges}}$ since the limit $\lim_{k\to\infty} \sqrt[k]{a_k} = \frac{6}{5} > 1$.

(e) Since $\sin\left(\frac{k\pi}{2}\right) = 1$ for odd k and -1 for even k, this is just the alternating series $\sum_{k=1}^{\infty} \frac{(-1)^k}{k}$. By the AST, this series $\boxed{\text{converges conditionally}}$ (c.f. part (c) of this exercise).

Exercise 66. *Determine the radius and interval of the convergence of the series*

$$\sum_{k=1}^{\infty} \frac{x^{2k+1} \cdot (2k+1)}{(x-2)!}.$$

(Make sure to account for the endpoints as well!)

Solution 66. *Using the Ratio Test, let* $a_k = \dfrac{x^{2k+1} \cdot (2k+1)}{(x-2)!}$, $a_{k+1} = \dfrac{x^{2k+3} \cdot (2k+3)}{(x-2)!}$. *Then*

$$\lim_{k\to\infty} \left| \frac{a_{k+1}}{a_k} \right| = \lim_{k\to\infty} \left| x^2 \cdot \frac{2k+3}{2k+1} \right| \implies |x^2| < 1 \implies -1 < x < 1$$

But we must also test $x = \pm 1$ *individually. At both endpoints, the series diverges (as the denominator is undefined). Hence the interval of convergence is* $\boxed{x \in (-1, 1)}$ *and the associated radius of convergence is* $\boxed{R = 1}$.

Exercise 67.

(a) *Find the partial fraction decomposition of* $\dfrac{x^3 + 1}{x^2 - 4}$.

(b) *Find the partial fraction decomposition of* $\dfrac{x^2 + 3x + 3}{x^3 + 3x^2 + 3x + 1}$.

(c) *Evaluate the sum*

$$\sum_{k=1}^{\infty} \frac{1}{2k^2 + k}.$$

(Hint: Use the fact that $\dfrac{1}{1} - \dfrac{1}{2} + \dfrac{1}{3} - \dfrac{1}{4} + \ldots = \ln 2$.)

Solution 67.

(a) *Since the degree of the numerator is larger than that of the numerator, we first rewrite the fraction as* $x + \dfrac{4x+1}{x^2-4}$. *Then observe that* $x^2 - 4 = (x+2)(x-2)$, *so write* $\dfrac{4x+1}{x^2-4} = \dfrac{A}{x+2} + \dfrac{B}{x-2}$ *and equate coefficients to get* $A(x+2) + B(x-2) = 4x+1 \implies A + B = 4, 2A - 2B = 1 \implies (A,B) = \left(\dfrac{9}{4}, \dfrac{7}{4}\right)$.

Thus, $\dfrac{x^3+1}{x^2-4} = \boxed{x + \dfrac{9}{4(x-2)} + \dfrac{7}{4(x+2)}}$.

(b) *This time, it is not necessary to rewrite first. Write the denominator as* $(x+1)^3$, *then set the fraction equal to* $\dfrac{A}{x+1} + \dfrac{B}{(x+1)^2} + \dfrac{C}{(x+1)^3}$. *This yields* $A(x+1)^2 + B(x+1) + C = x^2 + 3x + 3 \implies A = 1, B = 1, C = 1$, *so the partial fraction decomposition is* $\boxed{\dfrac{1}{x+1} + \dfrac{1}{(x+1)^2} + \dfrac{1}{(x+1)^3}}$.

(c) We have $\dfrac{1}{2k^2 + k} = \dfrac{1}{k(2k+1)} = \dfrac{A}{k} + \dfrac{B}{2k+1} \implies A(2k+1) + Bk = 1 \implies 2A + B = 0, A = 1 \implies$ $B = -2$. Hence, we end up with

$$\sum_{k=1}^{\infty} \frac{1}{k} - \frac{2}{2k+1}$$

$$= \left(\frac{1}{1} - \frac{2}{3}\right) + \left(\frac{1}{2} - \frac{2}{5}\right) + \left(\frac{1}{3} - \frac{2}{7}\right) + \dots$$

$$= 1 + (1 - 2\ln 2) = \boxed{2 - 2\ln 2}$$

Exercise 68. Let $\{S_n\}_{n=1}^{\infty}$ be the sequence of partial sums of the series

$$\sum_{n=1}^{\infty} \arctan(n+1) - \arctan(n).$$

(a) Determine $\lim\limits_{n \to \infty} S_n$.

(b) Find a general formula for S_n.

Solution 68. (a) As the series telescopes, with $S_1 = \arctan(2) - \arctan(1)$, $S_2 = \arctan(3) - \arctan(2)$, $\dots$, we have $\lim_{n\to\infty} S_n = \boxed{\infty}$.

(b) In general, $S_n = \arctan(n+1) - \arctan(1) = \arctan(n+1) - \dfrac{\pi}{4}$.

Exercise 69. Determine whether or not the following series converge. (If a series converges, you do not need to evaluate it.)

(a) $\displaystyle\sum_{k=1}^{\infty} \frac{2 + \sin k}{k^2}$

(b) $\displaystyle\sum_{k=1}^{\infty} \frac{k!}{k^k}$

(c) $\displaystyle\sum_{k=3}^{\infty} \frac{1}{\sqrt{k^2 - 5}}$

(d) $\displaystyle\sum_{k=2}^{\infty} \frac{1}{\sqrt{k^2 + 5}}$

Solution 69. (a) Observe that $\dfrac{2 + \sin k}{k^2} \le \dfrac{3}{k^2}$ for all k, and $\dfrac{3}{k^2}$ converges, since $\dfrac{1}{k^2}$ also converges. Hence, by the Direct Comparison Test, $\dfrac{2 + \sin k}{k^2}$ $\boxed{converges}$.

(b) Using the Ratio Test, let $a_k = \dfrac{k!}{k^k}$, $a_{k+1} = \dfrac{(k+1)!}{(k+1)^{k+1}}$. Then

$$\lim_{k\to\infty} \left| \frac{a_{k+1}}{a_k} \right| = \lim_{k\to\infty} (k+1) \cdot \frac{k^k}{(k+1)^{k+1}}$$

$$= \lim_{k\to\infty} \frac{k^k}{(k+1)^k}$$

$$= \lim_{k\to\infty} \left(\frac{k}{k+1} \right)^k = \frac{1}{e} < 1$$

using the fact that $\lim_{k\to\infty} \left(1 + \dfrac{1}{k}\right)^k = e$ and the fact that the expression $\left(\dfrac{k}{k+1}\right)^k$ is continuous for all $k \ge 1$. Hence, the series $\boxed{converges}$.

334

(c) Since $\dfrac{1}{\sqrt{k^2-5}} > \dfrac{1}{k}$ for all integers $k \geq 3$, $\displaystyle\sum_{k=3}^{\infty} \dfrac{1}{\sqrt{k^2-5}} > \sum_{k=3}^{\infty} \dfrac{1}{k}$. By the comparison test, since the harmonic series diverges, $\dfrac{1}{\sqrt{k^2-5}} > \dfrac{1}{k}$ $\boxed{diverges}$ as well.

(d) By the Limit Comparison Test with $a_k = \dfrac{1}{\sqrt{k^2+5}}$ and $b_k = \dfrac{1}{k}$, we have $\lim_{k\to\infty} \dfrac{k}{\sqrt{k^2+5}} = 1 > 0$. Since $\sum b_k$ diverges, so does $\sum a_k = \sum \dfrac{1}{\sqrt{k^2+5}}$.

Exercise 70. *(2016 AMC 12B) Let $ABCD$ be a unit square. Let Q_1 be the midpoint of $\overline{CD}$. For $i = 1, 2, \ldots,$ let P_i be the intersection of $\overline{AQ_i}$ and $\overline{BD}$, and let Q_{i+1} be the foot of the perpendicular from P_i to $\overline{CD}$. What is*

$$\sum_{i=1}^{\infty} \text{Area of } \triangle DQ_iP_i?$$

Solution 70. *Let $A = (0,1), B = (1,1), C = (1,0), D = (0,0)$. Then $Q_1 = \left(\dfrac{1}{2}, 0\right)$ and $P_1 = \left(\dfrac{1}{3}, \dfrac{1}{3}\right)$. We then have $[\triangle DQ_1P_1] = \dfrac{1}{12}$. Hence, $Q_2 = \left(\dfrac{1}{3}, 0\right)$, $P_2 = \left(\dfrac{1}{4}, \dfrac{1}{4}\right)$, and $[\triangle DQ_2P_2] = \dfrac{1}{24}$. Similarly, $[\triangle DQ_3P_3] = \dfrac{1}{40}$. We observe that, in general, $[\triangle DQ_iP_i] = \dfrac{1}{2(i+1)(i+2)}$, so we want to compute the sum*

$$\frac{1}{2}\sum_{i=1}^{\infty} \frac{1}{(i+1)(i+2)} = \frac{1}{2}\sum_{i=1}^{\infty} \frac{1}{i+1} - \frac{1}{i+2}$$

$$= \frac{1}{2}\left(\left(\frac{1}{2} - \frac{1}{3}\right) + \left(\frac{1}{3} - \frac{1}{4}\right) + \ldots\right) = \frac{1}{2}\cdot\frac{1}{2} = \boxed{\frac{1}{4}}$$

Extension Exercise 24. *(2018 Berkeley Math Tournament) Find the value of*

$$\frac{1}{\sqrt{2}} + \frac{4}{(\sqrt{2})^2} + \frac{9}{(\sqrt{2})^3} + \ldots$$

Extension Solution 24. *For brevity, let $x = \dfrac{1}{\sqrt{2}}$. Then the sum S becomes $S = x + 4x^2 + 9x^3 + 16x^4 + \ldots$ and with some manipulation, we find that $S(1-x) = S - xS = x + 3x^2 + 5x^3 + 7x^4 + \ldots$ Repeating the manipulation process yields $S(1-x)^2 = x + 2(x^2 + x^3 + x^4 + \ldots)$ Since $x^2 + x^3 + x^4 + \ldots = \dfrac{x^2}{1-x}$, we have $S(1-x)^2 = x + \dfrac{2x^2}{1-x}$. Back-substituting $x = \dfrac{1}{\sqrt{2}}$ reveals that $S = \boxed{24 + 17\sqrt{2}}$.*

Extension Exercise 25. *(2018 Berkeley Math Tournament) Let $F_1 = 0$, $F_2 = 1$, and $F_n = F_{n-1} + F_{n-2}$ [for $n \geq 3$]. Compute*

$$\sum_{n=1}^{\infty} \frac{\displaystyle\sum_{i=1}^{n} F_i}{3^n}.$$

Extension Solution 25. *We have*

$$\sum_{n=0}^{\infty} \sum_{i=1}^{n} \frac{F_i}{3^n} = \sum_{i=0}^{\infty} \sum_{j=0}^{\infty} \frac{F_i}{3^{i+j}}$$

$$= \sum_{i=0}^{\infty} \frac{F_i}{3^i} \sum_{j=0}^{\infty} \frac{1}{3^j}$$

$$= \frac{3}{5} \cdot \frac{3}{2} = \boxed{\frac{9}{10}}$$

where $\sum_{i=0}^{\infty} \frac{F_i}{3^i} = \frac{3}{5}$ comes from splitting the Fibonacci term $F_i = F_{i-1} + F_{i-2}$ into two separate sums.

Extension Exercise 26. *(2013 Putnam B1) For positive integers n, let the numbers $c(n)$ be determined by the rules $c(1) = 1$, $c(2n) = c(n)$, and $c(2n + 1) = (-1)^n \cdot c(n)$. Find the value of*

$$\sum_{n=1}^{2013} c(n)c(n + 1)c(n + 2)$$

Extension Solution 26. *Observe that $c(2n+1)c(2n+3) = (-1)^n c(n) \cdot (-1)^{n+1} c(n+1) = -c(n)c(n+1) = -c(2n)c(2n+2)$ for all n. Thus, $\sum_{n=2}^{2013} c(n)c(n+2) = \sum_{i=1}^{1006} c(2i)c(2i+2) + c(2i+1)c(2i+3) = 0$, so the requested sum simplifies to $c(1)c(3) = \boxed{-1}$.*

Section 5.7: Generalizing Vieta's Formulas to Higher-Degree Polynomials

Exercise 71.

(a) *Compute the sum and product of the roots of $(x^2 + 1)^3$.*

(b) *Compute the sum and product of the roots of $(x^2 + 1)^{2019}$.*

Solution 71. *Notice that the sum of the roots of the polynomial $x^2 + 1$ is 0 because the coefficient of the x^{2-1} or x term is 0. Raising $x^2 + 1$ to any positive integral power does not create any new roots; instead, the multiplicity of each root increases by a fixed number determined by the exponent. Thus, the sum of the roots of $(x^2 + 1)^k$ with $k \in \mathbb{Z}^+$ (for instance, k can equal 3 or 2019) is $\boxed{0}$. In addition, the product of the roots remains unaffected by the value of k, and it is equal to $\boxed{1}$, the constant term of the original polynomial (because it has an even degree).*

Exercise 72. *Let $f(x) = (x + 3)^3 + (x + 2)^2 + (x + 1)$.*

(a) *Compute the sum of the roots of $g(x) = f(x + 3)$.*

(b) *Compute the sum of the reciprocals of the roots of $g(x)$.*

(c) *Let a, b, and c be the roots of f. Find a polynomial whose roots are a^2, b^2, and c^2.*

Solution 72.

(a) *Substituting $x + 3$ in place of x in the definition of f yields $g(x) = x^3 + 19x^2 + 119x + 245$, so the sum of its roots is $\boxed{-19}$.*

(b) *This is the sum of the roots of the polynomial with coefficients in the reverse order as $g(x)$, i.e.*
$h(x) = 245x^3 + 119x + 19x + 1$, which has a sum of roots of $-\frac{119}{245} = \boxed{-\frac{17}{35}}$.

(c) *Note that $f(x) = x^3 + 10x^2 + 32x + 32$ has roots a, b, c, so $h(x) = x^3 - 10^2 + 32x - 32$ has roots $-a, -b, -c$ by Vieta's formulas. Then $f(x)h(x) = (x^2 - a^2)(x^2 - b^2)(x^2 - c^2)$, and substituting x^2 with x yields $\boxed{x^3 - 36x^2 + 384x - 1024}$.*

Exercise 73. *The positive real root of*

$$\left(\prod_{i=1}^{8} (x - i) \right) - 1001$$

can be written in the form $\dfrac{a + \sqrt{b}}{c}$, where a, b, and c are positive integers with $\gcd(a, c) = 1$ and b squarefree. Compute $a + b + c$.

Solution 73. *Notice that the expansion of the product can be written as* $(x - 1)(x - 8) \cdot (x - 2)(x - 7) \cdot (x - 3)(x - 6) \cdot (x - 4)(x - 5)$, *with each term re-writing as a difference of squares:* $(x - 4.5 + 3.5)(x - 4.5 - 3.5)(x - 4.5 + 2.5)(x - 4.5 - 2.5)(x - 4.5 + 1.5)(x - 4.5 - 1.5)(x - 4.5 + 0.5)(x - 4.5 - 0.5)$. *Then we have* $((x - 4.5)^2 - 3.5^2)((x - 4.5)^2 - 2.5^2)((x - 4.5)^2 - 1.5^2)((x - 4.5)^2 - 0.5^2) = 1001$, *or*

$$((x - 4.5)^2 - 2^2 - 1.5^2)((x - 4.5)^2 - 2^2 + 1.5^2)((x - 4.5)^2 - 2^2 - 0.5^2)((x - 4.5)^2 - 2^2 + 0.5^2) = 1001$$

$$((x - 4.5)^2 - 4 + 1.5)((x - 4.5)^2 - 4 - 1.5)((x - 4.5)^2 - 4 + 0.5)((x - 4.5)^2 - 4 - 0.5) = 1001$$

$$\implies x = \frac{9 + \sqrt{53}}{2}$$

Hence, $a + b + c = 9 + 53 + 2 = \boxed{64}$.

Exercise 74. *Let* a *and* b *be real numbers such that the equation* $x^3 + x^2 + ax + b = 0$ *has real solutions* $1 \pm \sqrt{2}$. *Compute the product of all solutions to the equation.*

Solution 74. *The sum of the solutions is* -1, *so the other root must be* -3. *Hence, the product of the solutions is* $\boxed{3}$.

Exercise 75. *(2019 AIME I) For distinct complex numbers* $z_1, z_2, \ldots, z_{673}$, *the polynomial*

$$(x - z_1)^3 (x - z_2)^3 \cdots (x - z_{673})^3$$

can be expressed as $x^{2019} + 20x^{2018} + 19x^{2017} + g(x)$, *where* $g(x)$ *is a polynomial with complex coefficients and with degree at most 2016. The value of*

$$\left| \sum_{1 \le j < k \le 673} z_j z_k \right|$$

can be expressed in the form $\frac{m}{n}$, *where* m *and* n *are relatively prime positive integers. Find* $m + n$.

Solution 75. *Let*

$$\sum_{1 \le i,j \le 673} z_j z_k = S$$

By Vieta's, $\sum_{i=1}^{673} z_i = -\frac{20}{3}$. *Henceforth, consider* $19x^{2017}$. *To produce the product of two roots, the roots roots can both be of the form* $z_o i$, *or of the form* z_j, z_k *for* $j < k$. *If they are both* z_i's, *there are 3 ways for this to occur; otherwise,* $3^2 = 9$ *ways. Thus,* $19 = 3\sum_{i=1}^{673} z_i^2 + 9S = 3\left(\left(-\frac{20}{3}\right)^2 - 2S\right) + 9S = \frac{400}{3} + 3S \implies$

$S = -\frac{343}{9} \implies \boxed{352}$.

Extension Exercise 27. *(2018 HMMT February Guts Round) Let* a, b, c *be positive integers. All the roots of each of the quadratics*

$$ax^2 + bx + c, ax^2 + bx - c, ax^2 - bx + c, ax^2 - bx - c$$

are integers. Over all triples (a, b, c), *find the triple with the third smallest value of* $a + b + c$.

Extension Solution 27. *By the quadratic formula, all roots are given by* $\frac{\pm b \pm \sqrt{b^2 \pm 4ac}}{2a}$. *Since all eight possible expressions are integers, we also have that* $\frac{b}{a}$ *and* $\frac{\sqrt{b^2 \pm 4ac}}{a}$ *are integers. Let* $\frac{b}{a} = b'$, $\frac{c}{a} = c'$. *We have that* $b', \sqrt{b'^2 \pm c'} \in \mathbb{N} \implies c' \in \mathbb{N}$. *Then let* $n^2 = b'^2 + c', m^2 = b'^2 - c'$.

We have $a + b + c = a(1 + b' + \frac{1}{4}c')$, *so we seek small ordered pairs* (b', c'). *This extends to computing ordered triples* (m, b', n) *such that* m^2, b'^2, n^2 *are in arithmetic progression. Using the fact that odd squares* *are congruent to 1 mod 4 and even squares to 0 mod 4, we can pare down our search to all of m, b', n odd.*

Plugging in $b' = 3, 5, 7, 9$ yields an interesting result: $(1, 5, 7)$ works, as does $(2, 10, 14)$, which yields the third-smallest solution, $\boxed{(1, 10, 24)}$ (from $(1, 10, 96)$ for (m, b', n)).

Extension Exercise 28. *(2007 HMMT February Algebra Test) The complex numbers α_1, α_2, α_3, and α_4 are the four distinct roots of the equation $x^4 + 2x^3 + 2 = 0$. Determine the unordered set*

$$\{\alpha_1\alpha_2 + \alpha_3\alpha_4, \alpha_1\alpha_3 + \alpha_2\alpha_4, \alpha_1\alpha_4 + \alpha_2\alpha_3\}$$

Extension Solution 28. *Using the symmetric sums of a quartic, we set up the polynomial*

$$P(x) = (x - (\alpha_1\alpha_2 + \alpha_3\alpha_4))(x - (\alpha_1\alpha_3 + \alpha_2\alpha_4))(x - (\alpha_1\alpha_4 + \alpha_2\alpha_3))$$

Observe that P is symmetric WRT each root, so $P(x) = x^3 - 8x + 8 = (x - 2)(x^2 - 2x - 4)$ after substituting in the definitions of the symmetric sums, from which it follows that $\boxed{x = -2, 1 \pm \sqrt{5}}$.

Section 5.8: Introduction to Logarithms

Exercise 76. *Given that $\log_{12} x = 20$ and $\log_3 y = 60$, compute $\log_{18} xy$.*

Solution 76. *We employ the change-of-base formula for each given equation: $\log_{12} x = \dfrac{\log_{18} x}{\log_{18} 12}$ and $\log_3 y = \dfrac{\log_{18} y}{\log_{18} 3}$. That is, $\log_{18} x = 20 \log_{18} 12$ and $\log_{18} y = 60 \log_{18} 3$. Applying the multiplication rule, $\log_{18} xy = \log_{18} x + \log_{18} y = 20 \log_{18} 12 + 60 \log_{18} 3$. Finishing off the logarithmic simplification, $20 \log_{18} 12 + 60 \log_{18} 3 = \log_{18} 12^{20} + \log_{18} 3^{60} = \log_{18}(12^{20} 3^{60}) = \log_{18}(2^{40} 3^{20+60}) = \log_{18} 18^{40} = \boxed{40}$.*

Exercise 77.

(a) *Solve the equation $\ln x^2 - \ln(x^2 - 1) = 1$ for x.*

(b) *Solve the equation $2 \log_{10}(x - 6) + \log_{10}(x^2 + 6x + 9) = 2$ for x.*

Solution 77. (a) *From properties of logarithms, we want to solve $\ln\left(\dfrac{x^2}{x^2 - 1}\right) = 1$, or $\dfrac{x^2}{x^2 - 1} = e$. That is, $x^2 = e(x^2 - 1)$, or $(e - 1)x^2 = e$. Solving for x, we find that $x = \boxed{\pm\sqrt{\dfrac{e}{e - 1}}}$.*

(b) *First, note that $x > 6$ because $2 \log_{10}(x - 6)$ must be defined. We can use properties of logarithms to rewrite the equation we wish to solve as $\log_{10}((x - 6)^2(x^2 + 6x + 9)) = 2$, or $(x - 6)^2(x + 3)^2 = 10^2$. Taking the square root of both sides, we want to find x when $(x - 6)(x + 3) = 10$ or $(x - 6)(x + 3) = -10$. The former equation has $x = -4$ and $x = 7$ as solutions, while the latter has $x = \dfrac{3 \pm \sqrt{41}}{2}$ as solutions. However, since $x > 6$, the only valid solution is $x = \boxed{7}$.*

Exercise 78. *(2015 ARML Local Individual Round #6) The line $x = a$ intersects the graphs of $f(x) = \log_{10}(x)$ and $g(x) = \log_{10}(x + 5)$ at points B and C, respectively. If $BC = 1$, compute a.*

Solution 78. *This effectively simplifies to $\log_{10}(x + 5) - \log_{10}(x) = 1$, which yields $\dfrac{x + 5}{x} = 10$ by the properties of logarithms. We then obtain $x = \boxed{\dfrac{5}{9}}$.*

Exercise 79. *(2018 AMC 12A) The solution to the equation $\log_{3x} 4 = \log_{2x} 8$, where x is a positive real number other than $\dfrac{1}{3}$ or $\dfrac{1}{2}$, can be written as $\dfrac{p}{q}$ where p and q are relatively prime positive integers. What is $p + q$?*

Solution 79. *We have* $\log_4(3x) = \log_8(2x)$ *by the change-of-base formula, then* $\log_4(3x) = \log_2(\sqrt{3x})$ *and* $\log_8(2x) = \log_2(\sqrt[3]{2x}) \implies \sqrt{3x} = \sqrt[3]{2x}$, *which has solution* $27x^3 = 4x^2 \implies x = \dfrac{4}{27} \implies p + q = 4 + 27 = \boxed{31}$.

Exercise 80.

(a) *Prove that* $x^{\log y} = y^{\log x}$ *for all* $x, y > 0$, *where* $\log x$ *denotes the natural logarithm of* x.

(b) *Prove that* $a^{\log_b c} = c^{\log_b a}$ *for all* $a, b, c > 0$.

Solution 80.

(a) *Taking the logarithm of both sides yields* $\log y \log x = \log x \log y$, *which is clearly true.*

(b) *We have* $a^{\log_b c} = e^{\ln a \cdot \log_b c} = e^{\frac{\ln a \cdot \ln c}{\ln b}} = c^{\frac{\ln a}{\ln b}} = c^{\log_b a}$ *by the change-of-base formula.*

Exercise 81. *(1984 AIME) Determine the value of* ab *if* $\log_8 a + \log_4 b^2 = 5$ *and* $\log_8 b + \log_4 a^2 = 7$.

Solution 81. *We have* $\log_8 a = \log_2 \sqrt[3]{a}$, $\log_4 b^2 = \log_2 b$, $\log_8 b = \log_2 \sqrt[3]{b}$, *and* $\log_4 a^2 = \log_2 a$, *so* $b\sqrt[3]{a} = 2^5 = 32$ *and* $a\sqrt[3]{b} = 2^7 = 128$. *Then multiplying the two yields* $(ab)^{\frac{4}{3}} = 2^{12} \implies ab = 2^9 = \boxed{512}$.

Exercise 82. *(2014 AIME II) Let* $f(x) = (x^2 + 3x + 2)^{\cos(\pi x)}$. *Find the sum of all positive integers* n *for which* $\left| \displaystyle\sum_{k=1}^{n} \log_{10} f(k) \right| = 1$.

Solution 82. *We seek all* n *such that* $\prod_{k=1}^{n} f(k) = 10, \dfrac{1}{10}$. *Note that* $\cos(\pi x) = -1$ *when* x *is odd, and* 1 *when* x *is even. Writing out the first few terms results in a telescoping series,* $\dfrac{1}{2 \cdot 3} \cdot (3 \cdot 4) \cdot \dfrac{1}{4 \cdot 5} \cdots$. *If* n *is odd, then the product telescopes to* $\dfrac{1}{2(n+2)}$, *where* $n = 3$. *If* n, *then it telescopes to* $\dfrac{n+2}{2}$, *in which case* $n = 18$ *and the sum of possible values is* $\boxed{21}$.

Extension Exercise 29. *Can we generalize the formula* $\lim_{n \to \infty} \left(1 + \dfrac{1}{n}\right)^n$ *to the limit* $\lim_{n \to \infty} \left(1 + \dfrac{k}{n}\right)^n$ *for any* $k \in \mathbb{R}$? *Or, for that matter, what happens if we replace the 1 with another real constant?*

Extension Solution 29. *We have*

$$\lim_{n \to \infty} \left(1 + \frac{k}{n}\right)^n = \lim_{n \to \infty} \left(1 + \frac{1}{\left(\frac{n}{k}\right)}\right)^n$$

$$= \lim_{n \to \infty} \left(\frac{1}{+}\frac{1}{m}\right)^{km} = e^k, m = \frac{n}{k}$$

If we change the constant term 1, we can simply scale upwards by that scalar factor.

Extension Exercise 30. *Prove that* $\dfrac{x}{1+x} \leq \ln(1+x) \leq x$ *for all* $x > -1$.

Extension Solution 30. *Let* $g(x) = \ln(1 + x) - x$; *then* $g'(x) = \dfrac{1}{1+x} - 1$, *and* $g'(x) = 0 \implies x = 0$, *which yields a maximum. Similarly, let* $h(x) = \dfrac{x}{1+x} - \ln(1+x)$, *with* $h'(x) = -\dfrac{x}{(1+x)^2}$. *Then* $h'(x) = 0 \implies x = 0$ *at which we again have a maximum. Hence,* $g(x) < g(0)$ *and* $h(x) < h(0)$, *proving the inequality for all* x.

Section 5.9: Complex Numbers, Part 1

Exercise 83.

(a) *Plot the points corresponding to* $2 + i$, $-2 + 3i$, *and* $6 + 5i$ *on an Argand diagram.*

(b) *What is the area of the triangle formed by connecting all three points?*

(c) (i) *Let* $z = 3 + 4i$. z *is then rotated* $90°$ *about the origin to obtain* z', *the image of* z. *Compute* z'.

    (ii) z' *is further rotated* $120°$ *about the origin to obtain* z''. *Compute* z''.

Solution 83.

(a)

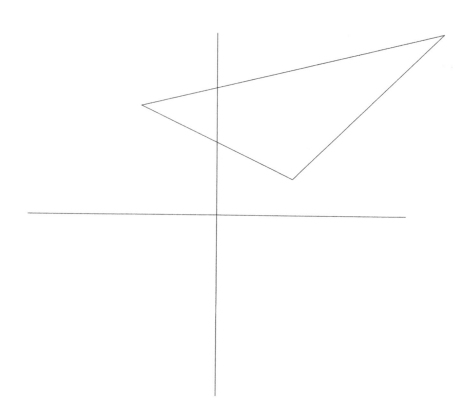

(b) *This is the area of the triangle in* $\mathbb{R}^2$ *with vertices* $(2, 1)$, $(-2, -3)$, $(6, 5)$, *which is* $\boxed{12}$ *by the Shoelace Theorem.*

(c) *Rotating the point* $z = (3, 4) \in \mathbb{R}^2$ *by* $90°$ *about* $(0, 0)$ *yields* $z' = \boxed{(-4, 3)}$.

Exercise 84. *Let* $S = \{z = a + bi \mid a, b \in \mathbb{R}, |z| \leq 1\}$. *If a complex number* z *is chosen randomly from* S, *what is the probability that* $a + b \leq 1$?

Solution 84. *The locus of points in* S *is the unit circle; furthermore,* $a + b \leq 1$ *is the line* $x + y = 1$ *in* $\mathbb{R}^2$. *The area of* S *is* π, *and the area of the region in* S *that is also below* $x + y = 1$ *is* $\dfrac{3\pi}{4} + \dfrac{1}{2} = \dfrac{3\pi + 2}{4}$. *The desired probability is the ratio of the areas, or* $\boxed{\dfrac{3\pi + 2}{4\pi}}$.

Exercise 85. *If* x *and* y *are complex numbers such that* $x^4 + y^4 = -1$ *and* $x^8 + y^8 = 1$, *how many ordered pairs* (x, y) *are possible?*

Solution 85. *Making the substitutions $a = x^4$, $b = y^4$, we get $a + b = -1$ and $a^2 + b^2 = 1$. Squaring the first expression yields $a^2 + 2ab + b^2 = 1$, or $2ab = 0 \implies a = 0$ or $b = 0$.*

Exercise 86. *(2017 AMC 12A) There are 24 different complex numbers z such that $z^{24} = 1$. For how many of these is z^6 a real number?*

Solution 86. *We have $z^{12} = \pm 1$. If $z^{12} = 1$, then $z^6 = \pm 1$; if $z^{12} = -1$, then $z^6 = \pm i$. The number of complex numbers z for which $z^{12} = 1$ is $\boxed{12}$ (these being the 12th roots of unity).*

Exercise 87. *Let $x, y \in \mathbb{C}$, $|x| = 2$, and $|y| = 7$. Determine the minimum value of $|x - y| \cdot |x + y|$.*

Solution 87. *Let $x = a + bi$, $y = c + di$. We have $a^2 + b^2 = 4, c^2 + d^2 = 49$, and want to minimize*

$$\sqrt{(a - c)^2 + (b - d)^2}\sqrt{(a + c)^2 + (b + d)^2} = \sqrt{((a^2 - 2ac + c^2) + (b^2 - 2bd + d^2))((a^2 + 2ac + c^2) + (b^2 + 2bd + d^2))}$$

$$= \sqrt{(53 - 2(ac + bd))(53 + 2(ac + bd))} = \sqrt{53^2 - 4(ac + bd)^2}$$

By Cauchy-Schwarz, $ac + bd \leq 2 \cdot 7 = 14$, so the minimum is $\boxed{45}$.

Exercise 88. *(1989 AIME) Given a positive integer n, it can be shown that every complex number of the form $r + si$, where r and s are integers, can be uniquely expressed in the base $-n + i$ using the integers $1, 2, \ldots, n^2$ as digits. That is, the equation*

$$r + si = a_m(-n + i)^m + a_{m-1}(-n + i)^{m-1} + \cdots + a_1(-n + i) + a_0$$

is true for a unique choice of non-negative integer m and digits $a_0, a_1, \ldots, a_m$ chosen from the set $\{0, 1, 2, \ldots, n^2\}$, with $a_m \neq 0$. We write

$$r + si = (a_m a_{m-1} \ldots a_1 a_0)_{-n+i}$$

to denote the base $-n + i$ expansion of $r + si$. There are only finitely many integers $k + 0i$ that have four-digit expansions

$$k = (a_3 a_2 a_1 a_0)_{-3+i}$$

$$a_3 \neq 0$$

Find the sum of all such k.

Solution 88. *We compute the first three powers of $-3 + i$, namely $(-3 + i)^1 = -3 + i$, $(-3 + i)^2 = 8 - 6i$, and $(-3 + i)^3 = -18 + 26i$. Hence, we must solve the equation $a_1 - 6a_2 - 26a_3 = 0$ over the integers. Since the a_i are digits less than or equal to 9, $a_1 - 6a_2 = 26a_3 \geq -54$. Casework on a_3 yields the four-digit numbers $(292a_0)_{-3+i}$ and $(154a_0)_{-3+i}$, plugging in the digits 292 and 154 into base 10 to get 345 and 145, respectively $\implies 345 + 145 = \boxed{490}$.*

Exercise 89. *(2018 HMMT November Guts Round) Let z be a complex number. In the complex plane, the distance from z to 1 is 2, and the distance from z^2 to 1 is 6. What is the real part of z?*

Solution 89. *We have $|z - 1| = 2$ and $|z^2 - 1| = 6$, from which it folows that $|z + 1| = 3$. The distance from z to -1 int he complex plane is 3. Hence, z and ± 1 are the vertices of a triangle with side lengths 2, 3, 3 (whose area is $\frac{3}{4}\sqrt{7}$). Then the distance between 1 and the foot from z to the real line is $\frac{1}{4}$ by Pythagorean Theorem, and the distance from 0 to the foot is $1 + \frac{1}{4} = \boxed{\frac{5}{4}}$.*

Exercise 90. *(2005 AIME II) For how many positive integers n less than or equal to 1000 is $(\sin t + i \cos t)^n = \sin nt + i \cos nt$ true for all real t?*

Solution 90. *We use the identity $e^{it} = \cos t + i \sin t$, with $\sin t + i \cos t = i(\cos t - i \sin t) = ie^{-it}$. Thus, $(\sin t + i \cos t)^n = \sin(nt) + i \cos(nt)$ becomes $(ie^{-it})^n = ie^{-itn}$. Thus, we need $i^n = i$, which occurs $\frac{1}{4}$ of the time (namely when $n \equiv 1 \pmod 4$). Hence the answer is $\boxed{250}$. (It is also possible to use de Moivre's Theorem here.)*

Extension Exercise 31. *(2004 AIME I) The polynomial $P(x) = (1 + x + x^2 + \cdots + x^{17})^2 - x^{17}$ has 34 complex roots of the form $z_k = r_k[\cos(2\pi a_k) + i\sin(2\pi a_k)]$, $k = 1, 2, 3, \ldots, 34$, with $0 < a_1 \le a_2 \le a_3 \le \cdots \le a_{34} < 1$ and $r_k > 0$. Given that $a_1 + a_2 + a_3 + a_4 + a_5 = m/n$, where m and n are relatively prime positive integers, find $m + n$.*

Extension Solution 31. *Write*

$$P = \left(\frac{x^{18} - 1}{x - 1}\right)^2 - x^{17}$$
$$= \frac{(x^{19} - 1)(x^{17} - 1)}{(x - 1)^2}$$

which has roots at the 17th and 19th roots of unity (excluding 1). (As $\gcd(17, 19) = 1$, there are no duplicate roots.) Hence, the a_i are the smallest fractions with denominators 17 and 19, namely $\frac{1}{19} + \frac{2}{19} + \frac{3}{19} + \frac{1}{17} + \frac{2}{17} = \frac{159}{323} \implies \boxed{482}$.

Extension Exercise 32. *(1994 AIME) The points $(0, 0)$, $(a, 11)$, and $(b, 37)$ are the vertices of an equilateral triangle. Find the value of ab.*

Extension Solution 32. *Consider these points on an Argand diagram. Then $b + 37i$ is a rotation of $a + 11i$ 60 degrees about the origin, so*

$$(\operatorname{cis} 60°)(a + 11i) = (a + 11i)\left(1 + \frac{i\sqrt{3}}{2}\right) = b + 37i$$

Equating coefficients yields $b = \frac{a}{2} - \frac{11\sqrt{3}}{2}$, $37 = \frac{11}{2} + \frac{a\sqrt{3}}{2}$, so $(a, b) = (21\sqrt{3}, 5\sqrt{3})$ and $ab = \boxed{315}$.

Section 5.10: Functional Equations, Part 1

Exercise 91. *(2014 Berkeley Math Tournament) Find $f(2)$ given that f is a real-valued function that satisfies the equation*

$$4f(x) + \left(\frac{2}{3}\right)(x^2 + 2)f\left(x - \frac{2}{x}\right) = x^3 + 1$$

Solution 91. *Substituting $x = 2$ into the equation yields $4f(2) + 4f(1) = 9$, and subsequently substituting $x = 1$ yields $4f(1) + 2f(-1) = 2$. Then substituting $x = -1$ yields $4f(-1) + 2f(1) = 0$; solving the system formed by the last two equations yields $f(1) = \frac{2}{3}$, $f(-1) = -\frac{1}{3}$. Then $f(2) = \boxed{\frac{19}{12}}$.*

Exercise 92. *(2015 Caltech-Harvey Mudd Math Competition) Let f be a function taking real numbers to real numbers such that for all reals $x \ne 0, 1$, we have*

$$f(x) + f\left(\frac{1}{1 - x}\right) = (2x - 1)^2 + f\left(1 - \frac{1}{x}\right)$$

Compute $f(3)$.

Solution 92. *Similarly as in the previous exercise, we substitute x, $\frac{1}{1 - x}$, and $1 - \frac{1}{x}$. Then we obtain two equations that we can sum together to obtain $2f(x) = (2x - 1)^2 + \left(2\left(1 - \frac{1}{x}\right) - 1\right)^2$ Therefore, $f(3) = \boxed{\frac{113}{9}}$.*

Exercise 93. *The function $\xi : \mathbb{R}\backslash\{0\} \mapsto \mathbb{R}$ satisfies $\xi(n) + 2\xi\left(\frac{2}{n}\right) = 4n$ for all $n \in \mathbb{R}\backslash\{0\}$. For what value of $n > 0$ does $\xi(n) = n$?*

Solution 93. *We substitute* $n, \dfrac{2}{n}$ *to obtain* $\xi\left(\dfrac{2}{n}\right) + 2\xi(n) = \dfrac{8}{n}$. *This implies* $\xi(n) + \xi\left(\dfrac{2}{n}\right) = \dfrac{4n + \frac{8}{n}}{3} = \dfrac{4n^2 + 8}{3n}$, *and thus* $\xi(n) = \dfrac{16 - 4n^2}{3n}$. *This equals* n *when* $n = \boxed{\dfrac{4}{\sqrt{7}}}$.

Exercise 94. *(2006 AMC 12A) The function f has the property that for each real number x in its domain, $\dfrac{1}{x}$ is also in its domain and $f(x) + f\left(\dfrac{1}{x}\right) = x$. What is the largest set of real numbers that can be in the domain of f?*

Solution 94. *By inspection, $x = \pm 1$ are in the domain of f. Plugging in $\dfrac{1}{x}$ in place of x yields $f\left(\dfrac{1}{x}\right) + f(x) = \dfrac{1}{x}$. But this implies $x = \dfrac{1}{x}$, which yields no additional solutions. Hence the largest set that can be in the domain is* $\boxed{\{-1, 1\}}$.

Exercise 95. *Find all functions $f : \mathbb{R} \mapsto \mathbb{R}$ such that $2f(x) + 3f\left(\dfrac{1}{x}\right) = x^2$.*

Solution 95. *Substituting $\dfrac{1}{x}$ for x gives $2f\left(\dfrac{1}{x}\right) + 3f(x) = \dfrac{1}{x^2}$. Then adding this with the given equation and performing algebraic manipulation yields $f(x) = $* $\boxed{\dfrac{3 - 2x^4}{5x^2}}$.

Exercise 96. *Solve the functional equation $f(x^2 - y^2) = f(x + y) \cdot f(x - y)$ over $\mathbb{R}$.*

Solution 96.

Exercise 97. *(2016 AIME I) Let $P(x)$ be a nonzero polynomial such that $(x - 1)P(x + 1) = (x + 2)P(x)$ for every real x, and $(P(2))^2 = P(3)$. Then $P\left(\dfrac{7}{2}\right) = \dfrac{m}{n}$, where m and n are relatively prime positive integers. Find $m + n$.*

Solution 97. *Plugging in $x = 1$ yields $P(1) = 0$, plugging in $x = 0$ yields $P(0) = 0$, and plugging in $x = -1$ yields $P(-1) = 0$ as well. Hence, for some polynomial $Q(x)$, we have $P(x) = Q(x) \cdot x(x - 1)(x + 1)$. From the given equation, we obtain*

$$(x - 1)P(x + 1) = (x + 2)P(x)$$
$$\implies (x - 1)((x + 1)(x)(x + 2)Q(x + 1))$$
$$= (x + 2)((x)(x - 1)(x + 1)Q(x))$$
$$\implies (x - 1)(x + 1)(x)(x + 2)Q(x + 1) = (x + 2)(x)(x - 1)(x + 1)Q(x) \implies Q(x + 1) = Q(x)$$

Then $Q(x) = c$ is a constant polynomial; therefore, $P(x) = cx(x + 1)(x - 1)$. As $P(2)^2 = P(3)$, we obtain $(6c)^2 = 24c$, or $c = 0, \dfrac{2}{3}$. It follows that $P(x) = \frac{2}{3}x(x - 1)(x + 1)$, leading to the final answer of $\boxed{109}$.

Exercise 98. *Prove that $f : \mathbb{C} \mapsto \mathbb{C}$ is a bijection if $f(f(x)) = x$ for all $x \in \mathbb{C}$.*

Solution 98. *Since $f(f(x))$ exists, $f(x)$ must lie in the domain of f. Hence, f must some set S to itself. By this observation alone, f is a bijection, as it is both an injection ($f(y) = y$, so $f(x) = f(y) \implies x = y$) and a surjection (for all $x \in S$, $f(x)$ is defined, and so $f(x) = x$ by injectivity).*

Extension Exercise 33. *(2016 USAJMO P6) Find all functions $f : \mathbb{R} \to \mathbb{R}$ such that for all real numbers x and y,*

$$(f(x) + xy) \cdot f(x - 3y) + (f(y) + xy) \cdot f(3x - y) = (f(x + y))^2.$$

Extension Solution 33. $x = y = 0$ yields $f(0) = 0$, and just $x = 0$ yields $f(y)f(-y) = f(y)^2$. In particular, if $f(y) \neq 0$, then $f(y) = f(-y)$, and so $f(y) = 0 \implies f(-y) = 0$. Setting $x = 3y$ then yields $f(8y) \cdot (f(y) + 3y^2) = f(4y)^2$. If we let $8y = t$, we get $f(t) = 0 \implies f\left(\dfrac{t}{2}\right) = 0$ for all $t \in \mathbb{R}$.

Next, set $y = -x$ to obtain $f(4x) \cdot (f(x) + f(-x) - 2x^2) = 0$. We can then derive $f(x) = 0, x^2$ for all x, as $2x62 = f(x) + f(-x) = 2f(x)$. Then $f(t) = 0 \implies f(2t) = 0$, since we can make the substitution $4y = t$.

Furthermore, observe that $f(a) = f(b) = 0 \implies f(b - a) = 0$, and if $f(t) \neq 0$, $f(t) = 0 \implies t = 0$.

These together implies that $f(y) = y^2, f(x - 3y) = (x - 3y)^2, f(3x - y) = (3x - y)^2$. It follows that $(x^2 + xy)(x - 3y)^2 + (y^2 + xy)(3x - y)^2 = 0 \implies (x + y)^4 = 0$, but this is a contradiction, since we have assumed $x + y \neq 0$. Hence, the only solutions are $f(x) = 0, x^2$ for all $x \in \mathbb{R}$. Indeed, both work, so we are done.

Extension Exercise 34. *(2012 USAMO P4)* Find all functions $f : \mathbb{Z}^+ \to \mathbb{Z}^+$ (where $\mathbb{Z}^+$ is the set of positive integers) such that $f(n!) = f(n)!$ for all positive integers n and such that $m - n$ divides $f(m) - f(n)$ for all distinct positive integers m, n.

Extension Solution 34. We have $f(1) = f(1!) = f(1)!, f(2) = f(2!) = f(2)! \implies f(1), f(2) = 1, 2$. We also have $n \cdot n! = (n + 1)! - n! \mid (f(n + 1)! - f(n)!)$ for all $n \in \mathbb{N}$. Assume that, for some $n \geq 2$, $f(n) = 1$. Indeed, we have $f(n + 1)! \equiv 1 \pmod{n \cdot n!}$, which follows by induction (this, in turn, implies $f(n + 1) = 1$).

In the case where $f(1) = f(2) = 1$, this yields $f(n) = 1$. If $f(1) = 2, f(2) = 1$, this implies $f(n) = 1$ for $n \geq 2$, but then $(3 - 1) \mid (f(3) - f(1))$, which is a contradiction.

If $f(1) = 1, f(2) = 2$, induction implies that $f(n) = n$. Finally, if $f(1) = f(2) = 2$, induction similarly implies $f(n) = 2$ for all n. Thus, there are three solutions: $f(n) = 1, 2, n$.

Section 5.11: Proving Function Theorems

Exercise 99. *Consider the function $f(x) = -20x^2 + 19x + 2$ on the interval $[0, 1]$.*

(a) *Compute $f(0)$, $f(1)$, $f'(x)$, $f'(0)$, and $f'(1)$.*

(b) *Show that f is continuous (and, by extension, differentiable) on $[0, 1]$.*

(c) *Find the value of $c \in [0, 1]$ that satisfies the conclusion of the MVT.*

(d) *Show that $f'(x) = 0$ for some $x \in [0, 1]$.*

Solution 99.

(a) *We have $f(0) = 2$, $f(1) = 1$, $f'(x) = -40x + 19$ (by the Power Rule), $f'(0) = 19$, and $f'(1) = -21$.*

(b) *This is an exercise in the δ-ϵ proof (and a special case of the result that polynomials are everywhere-continuous), as f is well-defined on $[0, 1]$ (no discontinuities).*

(c) *By the MVT, $f'(c) = \dfrac{f(b) - f(a)}{b - a} = \dfrac{f(1) - f(0)}{1 - 0} = -1 \implies c = \boxed{\dfrac{1}{2}}$.*

(d) *Of course, we could conclude that $x = \dfrac{19}{40} \in [0, 1]$ by inspection, but a more rigorous method would be to use the IVT to conclude that $f(b) = f(a)$. Then, we can say that the conditions of the IVT are satisfied (namely that f is continuous on $[0, 1]$), which allows us to arrive at the desired result.*

Exercise 100. *Use the MVT to show that, for all $a, b \in \mathbb{R}$, $|\cos a - \cos b| \leq |a - b|$.*

Solution 100. *Assume $a < b$. Note that $\cos$ is differentiable on $\mathbb{R}$, so it is continuous on $[a, b]$ (and, in particular, differentiable on (a, b)). By the MVT, $\exists c \in (a, b)$ such that $\cos'(c) = \dfrac{\cos x - \cos y}{x - y}$. Since $\cos'(c) = -\sin(c)$, and $|-\sin(c)| \leq 1$, so we have $\left| \dfrac{\cos a - \cos b}{a - b} \right| \leq 1$. Hence, $|\cos a - \cos b| \leq |a - b|$.*

Exercise 101. *Show that $x^5 + 20x + 19 = 0$ has exactly one real solution.*

Solution 101. *Assume for contradiction that the equation has two solutions $x = a, b$, with $a < b$. By Rolles Theorem, $\exists c \in (a, b)$ such that $f(c) = 0$. But $f(x) = 5x^4 + 20 = 0$ has no real solutions, so we know that f has 0 or 1 solution. Using the IVT, we have $f(-1) = -2$ and $f(0) = 19$, so $\exists c \in (-1, 0)$ such that $f(c) = 0$. Hence there is exactly one real solution.*

Exercise 102.

(a) *Consider the inverse function f^{-1} of f. Under what conditions are both f and f^{-1} differentiable?*

(b) *Let $f(x) = x^5 + 20x + 19$. Evaluate $(f^{-1})'(-2)$.*

Solution 102.

(a) *If f is differentiable, then f' must be continuous, and nonzero a point x, for $f^{-1}(x)$ to exist.*

(b) *The derivative of the inverse at x is equal to $\dfrac{1}{f'(f^{-1}(x))}$, so $f'(x) = \dfrac{1}{f^{-1}(f(x))}$. Then $f(x) = -2$ when $x = -1$, so $(f^{-1})'(-2) = \dfrac{1}{f'(-1)} = \dfrac{1}{25}$.*

Exercise 103. *Is there necessarily a real root of the polynomial $x^{2019} - 20x^{2018} + 19$ in the open interval $(0, 1)$?*

Solution 103. *No. By the same logic as in two exercises above, we evaluate at each endpoint, and get that, for $x \in (0, 1)$, $x^{2019} - 20x^{2018} + 19 > 0$, so no real root is guaranteed.*

Extension Exercise 35. *Prove that the converse of the IVT is false by constructing a counterexample.*

Extension Solution 35. *The standard counterexample is $f(x) = \sin x$ for $x \neq 0$ and 0 for $x = 0$. Note that f is a classic example of a function that is not continuous everywhere.*

Extension Exercise 36. *Prove that Rolle's Theorem is equivalent to the Mean Value Theorem. That is, do both of the following:*

(a) *Show that Rolle's Theorem is a special case of the MVT.*

(b) *Using Rolle's Theorem, prove the MVT.*

Extension Solution 36.

(a) *This is achieved by setting $(a) = f(b)$ as per the condition, with the expression in the MVT automatically becoming 0.*

(b) *Consider the function $g(x) = f(x) - (x - a) \cdot \dfrac{f(b) - f(a)}{b - a}$. g has the same differentiability properties as f, and $g(a) = g(b) = f(a)$. Rolle's Theorem states that there exists $c \in [a, b]$ with $g'(c) = 0$. We conclude that $g'(c) = 0 \implies f'(c) = \dfrac{f(b) - f(a)}{b - a}$, completing the proof.*

Extension Exercise 37. *(1955 Putnam B2) Let $\mathbb{R}$ be the reals. $f : \mathbb{R} \mapsto \mathbb{R}$ is twice differentiable, f'' is continuous and $f(0) = 0$. Define $g : \mathbb{R} \mapsto \mathbb{R}$ by $g(x) = \dfrac{f(x)}{x}$ for $x \neq 0$, $g(0) = f'(0)$. Show that g is differentiable and that g' is continuous.*

Extension Solution 37. *This is an application of MVT in which we must also consider the edge case* $x = 0$. *Namely, we have* $g'(x) = \dfrac{xf'(x) - f(x)}{x^2}$ *elsewhere. Given an* x, *we can find* δ *with* $f(x) = f(0) + xf'(0) + \dfrac{x^2}{2}f''(\delta)$. *Hence* $\lim_{x\to 0} \dfrac{g(x) - f'(0)}{x} = \lim_{x\to 0} \dfrac{1}{2}f''(\delta) = \dfrac{1}{2}f''(0)$, *so* g *is differentiable at 0.*

Furthermore, $\lim_{x\to 0} g'(x) = \lim_{x\to 0} \dfrac{xf'(x) - f(x)}{x^2}$. *We have* $f'(x) = f'(0) + xf''(\delta)$, *and* $f(x) = xf'(0) + \dfrac{1}{2}x^2 f''(\epsilon)$ *for* $\epsilon \in (0, x)$. *Thus,* $\lim_{x\to 0} g'(x) = \lim_{x\to 0} f''(\epsilon) - \dfrac{1}{2}f''(\delta) = \dfrac{1}{2}f''(0) = g'(0)$. *It follows that* g' *is continuous at 0, as desired.*

Extension Exercise 38. *(2004 Putnam A1) Basketball star Shanille O'Keal's team statistician keeps track of the number,* $S(N)$, *of successful free throws she has made in her first* N *attempts of the season. Early in the season,* $S(N)$ *was less than 80% of* N, *but by the end of the season,* $S(N)$ *was more than 80% of* N. *Was there necessarily a moment in between when* $S(N)$ *was exactly 80% of* N?

Extension Solution 38. *Yes. For contradiction, assume that there is an* N *with* $S(N) < \dfrac{4}{5}$ *but* $S(N+1) > \dfrac{4}{5}$. *Then if she made* n *of her first* N *free throws,* $\dfrac{n}{N} < \dfrac{4}{5}$ *but* $\dfrac{n+1}{N+1} > \dfrac{4}{5}$, *which implies* $5n < 4N$ *but* $5n + 1 > 4N$. *This is impossible; hence, contradiction.*

Section 5.12: Bézout's Identity

Exercise 104. *Show that, if* a, b, *and* c *are positive integers such that* a *and* c *are relatively prime, and* b *and* c *are relatively prime, then* ab *and* c *are relatively prime.*

Solution 104. *a and c have no prime factors in common, and b and c have no prime factors in common. Hence, ab has no prime factors in common with c, which forces* $\gcd(ab, c) = 1$.

Exercise 105. *Show that, if* m *and* n *are relatively prime positive integers, then there exists an integer* x *such that* $mx \equiv 1 \bmod n$.

Solution 105. *We have* $\gcd(m, n) = 1$, *so by Bézout's Identity, there exist* $x, y \in \mathbb{Z}$ *with* $mx + ny = 1$. *Hence,* $mx \equiv 1 \pmod{n}$ *as desired.*

Exercise 106. *(2013 AIME I) Ms. Math's kindergarten class has 16 registered students. The classroom has a very large number, N, of play blocks which satisfies the conditions:*

(a) *If 16, 15, or 14 students are present in the class, then in each case all the blocks can be distributed in equal numbers to each student, and*

(b) *There are three integers $0 < x < y < z < 14$ such that when x, y, or z students are present and the blocks are distributed in equal numbers to each student, there are exactly three blocks left over.*

Find the sum of the distinct prime divisors of the least possible value of N satisfying the above conditions.

Solution 106. *Note that N is a multiple of 14, 15, and 16, so it has 2, 3, 5, and 7 as prime factors. Hence, it is also a multiple of 1, 2, 3, 4, 5, 6, 7, 8, 10, and 12, meaning that $x = 9$, $y = 11$, $z = 13$ (as N is not guaranteed to be a multiple of 9; there is only 1 factor of 3 from the 15). Hence, this reduces to a Chinese Remainder Theorem problem, with $N \equiv 3 \pmod{1287}, 0 \pmod{1680}$. We find that the desired prime factors are 2 3, 5, 7, and additionally 131, leading to a final answer of* $\boxed{148}$.

Extension Exercise 39. *(1985 AIME) The numbers in the sequence 101, 104, 109, 116,... are of the form $a_n = 100 + n^2$, where $n = 1, 2, 3, \ldots$ For each n, let d_n be the greatest common divisor of a_n and a_{n+1}. Find the maximum value of d_n as n ranges through the positive integers.*

Extension Solution 39. *We have $a_{n+1} = 100 + (n+1)^2 = n^2 + 2n + 101$. This problem is essentially an exercise in using the Euclidean Algorithm, until we have reduced it to computing $\gcd(2n+1, 100 + n^2)$. Then we want $n^2 - 200n = 0$, or $n = 200$. Thus, our GCD is $200 \cdot 2 + 1 =$* $\boxed{401}$.

Extension Exercise 40. *(2018 HMMT February Guts Round) Find the number of unordered pairs (a, b), where $a, b \in \{0, 1, 2, \ldots, 108\}$ such that 109 divides $a^3 + b^3 - ab$.*

Extension Solution 40. *If either a or b is 0, then we get $a^3 \equiv 0$, i.e. $a, b = 0$. Hence assume that $a, b \neq 0$. Multiplying through by $a^{-1}b^{-2}$ yields $(ab^{-1})^{-2} + a^{-1}b \equiv ab \pmod{109}$. Let $a = xy^{-1}, b = y^{-1}$; then $y \equiv x^2 + x^{-1}$, and all solutions are of the form $((x + x^2)^{-1}, (x^2 + x^1)^{-1})$. it suffices to determine all pairs of the form $(x + x^{-2}, x^2 + x^{-1})$. Casework - and accounting for cases in which $y \equiv x^{-1}$ - yields $\boxed{54}$ cases (52 in the main case, $x \neq 1, 46, 64, 108$ (i.e. the values satisfying $x^3 + 1 \equiv 0$), 1 for $x = 1$, and 1 for $(a, b) = (0, 0)$).*

Extension Exercise 41. *(2005 IMO P4) Determine all positive integers relatively prime to all the terms of the infinite sequence*
$$a_n = 2^n + 3^n + 6^n - 1, \ n \geq 1.$$

Extension Solution 41. *If $p > 3$, then we have $2^{p-2} + 3^{p-2} + 6^{p-2} \equiv 1 \pmod{p}$ by Fermat's Little Theorem. Hence, $p \mid a_{n-2}$. We also have $2 \mid a_1, 3 \mid a_2$, so 1 is the only positive integer relatively prime to all the a_n.*

Section 5.13: Matrix Algebra, Part 1

Exercise 107. *In the equation $(AB)^T \neq A^T B^T$, one side might not exist even if the other does. Can you come up with some examples?*

Solution 107. *Some guidelines/hints for constructing examples: Though $(AB)^T$ exists, $A^T B^T$ does not, since the number of columns of A^T and the number of rows of B^T are unequal. Furthermore, is it possible for AB to not exist, but $A^T B^T$ to exist?*

Exercise 108. *Consider the system of equations*

$$2x_1 + 3x_2 - 4x_3 + x_4 = 0$$

$$-2x_1 - 2x_2 + 5x_3 - x_4 = 3$$

$$3x_1 - x_3 + 2x_4 = 7$$

$$x_2 + 6x_3 - 7x_4 = -8$$

(a) Represent the system as an augmented matrix.

(b) Solve the system.

(c) Verify that your solution in (b) is correct by reducing the augmented matrix you obtained in (a) to RREF form.

(d) If we change the coefficients of the variables, when will the system have no, one, or infinite solutions?

(e) Give an example of a system for each scenario in (d).

Solution 108.

(a)

$$\begin{bmatrix} 2 & 3 & -4 & 1 & 0 \\ -2 & -2 & 5 & -1 & 3 \\ 3 & 0 & -1 & 2 & 7 \\ 0 & 1 & 6 & -7 & -8 \end{bmatrix}$$

(b) By summing equations and using the traditional solution methods, we obtain the solution $\boxed{(1, 1, 2, 3)}$.

(c) Reducing the matrix to RREF form using elementary row operations yields

$$\begin{bmatrix} 1 & 0 & 0 & 0 & 1 \\ 0 & 1 & 0 & 0 & 1 \\ 0 & 0 & 1 & 0 & 2 \\ 0 & 0 & 0 & 1 & 3 \end{bmatrix}$$

(d) The system will be inconsistent (have no solutions)) when the system has equal coefficients and differing augments, i.e. two contradictory equations. It will have one solution in most cases, except for when two rows are identical (including augments), in which case it will have infinitely many solutions.

(e) An example of an inconsistent system (no solutions) is $a + 2b = 1, 2a + 4b = 3$. An example of a system with a single solution is $a + b = 1, 2a + 3b = 4$. An example of a system with infinite solutions is $a + b = 1, 2a + 2b = 2$.

Exercise 109. *Prove that row equivalence is an equivalence relation.*

Solution 109. *Let A, B, and C be matrices with the same dimensions. We prove each of the following:*

(a) Reflexive property (each matrix is row equivalent to itself).

(b) Symmetric property (if A is row equivalent to B, then B is row equivalent to A).

(c) Transitive property (if A is row equivalent to B, and B is row equivalent to C, then C is row equivalent to A).

(a) Reflexive property: this is trivial.

(b) Symmetric property: we may perform the same row operations that we used to transform A into B, except in reverse.

(c) Transitive property: If we perform a sequence of row operations R_1 to transform A to B, and a sequence R_2 to further transform B to C, then we can apply the reverse of R_2, followed by the reverse of R_1, to transform C back to A.

Exercise 110. *Consider the matrix*

$$M = \begin{bmatrix} 5 & 9 \\ 4 & 7 \end{bmatrix}.$$

(a) What is $\det M$?

(b) Is the matrix M singular or invertible? If the former, provide a proof. If the latter, provide a proof and compute the inverse of M, M^{-1}.

(c) What is M^2? Compute $\det M^2$.

(d) What can we conclude about $\det M^n$ for some positive integer n?

(e) Consider the matrix

$$A = \begin{bmatrix} 5 & 9 & 5 & 9 \\ 4 & 7 & 4 & 7 \\ 5 & 9 & 5 & 9 \\ 4 & 7 & 4 & 7 \end{bmatrix}$$

What is $\det A$? Can we derive any relationship between $\det M$ and $\det A$? In general, what effect does "copy-pasting" an $n \times n$ square matrix three times to form a $2n \times 2n$ square matrix have on the determinant of the original matrix?

Solution 110. *(a) $\det M = 5 \cdot 7 - 9 \cdot 4 = 35 - 36 = \boxed{-1}$.*

(b) *M is invertible because* $\det M \neq 0$. *Since* M *is a* 2×2 *matrix,* $M^{-1} = \dfrac{1}{\det M} \begin{bmatrix} 7 & -9 \\ -4 & 5 \end{bmatrix} = \boxed{\begin{bmatrix} -7 & 9 \\ 4 & -5 \end{bmatrix}}.$

To check that M^{-1} *is truly the inverse of* M, *we compute* MM^{-1} *and* $M^{-1}M$:

$MM^{-1} = \begin{bmatrix} 5 & 9 \\ 4 & 7 \end{bmatrix} \begin{bmatrix} -7 & 9 \\ 4 & -5 \end{bmatrix} = \begin{bmatrix} 5(-7) + 9 \cdot 4 & 5 \cdot 9 + 9(-5) \\ 4(-7) + 7 \cdot 4 & 4 \cdot 9 + 7(-5) \end{bmatrix} = \begin{bmatrix} 1 & 0 \\ 0 & 1 \end{bmatrix}.$

$M^{-1}M = \begin{bmatrix} -7 & 9 \\ 4 & -5 \end{bmatrix} \begin{bmatrix} 5 & 9 \\ 4 & 7 \end{bmatrix} = \begin{bmatrix} (-7)5 + 9 \cdot 4 & (-7)9 + 9 \cdot 7 \\ 4 \cdot 5 + (-5)4 & 4 \cdot 9 + (-5)7 \end{bmatrix} = \begin{bmatrix} 1 & 0 \\ 0 & 1 \end{bmatrix}.$

Since $MM^{-1} = M^{-1}M = \begin{bmatrix} 1 & 0 \\ 0 & 1 \end{bmatrix} = I$, M^{-1} *truly is the inverse of* M.

(c) $M^2 = MM = \begin{bmatrix} 5 & 9 \\ 4 & 7 \end{bmatrix} \begin{bmatrix} 5 & 9 \\ 4 & 7 \end{bmatrix} = \begin{bmatrix} 5 \cdot 5 + 9 \cdot 4 & 5 \cdot 9 + 9 \cdot 7 \\ 4 \cdot 5 + 7 \cdot 4 & 4 \cdot 9 + 7 \cdot 7 \end{bmatrix} = \boxed{\begin{bmatrix} 61 & 108 \\ 48 & 85 \end{bmatrix}}.$

$\det M^2 = 61 \cdot 85 - 48 \cdot 108 = \boxed{1}.$

(d) $\det M^n = (\det M)^n$. *We can extend the idea that* $\det(M^2) = \det(MM) = \det(M) \cdot \det(M) \det(M^2)$ *to any power* n.

(e) $\det A = \boxed{0}$. *Since copy-pasting a matrix in this manner effectively allows us to eliminate all rows except for the topmost (by setting them equal to zeros), and the matrix hence becomes upper triangular, its determinant is the product of its diagonal entries, which will be 0.*

Exercise 111.

(a) *Determine whether the vectors* $\mathbf{x} = (1, 2, 3)$, $\mathbf{y} = (4, 5, 6)$, *and* $\mathbf{z} = (7, 8, 9)$ *are linearly independent or linearly dependent.*

(b) *Do* $\mathbf{x}$, $\mathbf{y}$, *and* $\mathbf{z}$ *span* $\mathbb{R}^3$? *That is, does* $\mathbb{R}^3 = Span\{x, y, z\}$?

(c) *Show that, if a vector space* V *has a basis* $\{v_1, v_2, v_3, \ldots, v_n\}$, *then any set consisting of more than* n *vectors must be linearly dependent.*

Solution 111.

(a) *This is equivalent to determining whether the matrix*

$$\begin{bmatrix} 1 & 4 & 7 \\ 2 & 5 & 8 \\ 3 & 6 & 9 \end{bmatrix}$$

has a determinant of zero. Indeed, through computation, we find that $\det M = 0$, *so the vectors are linearly dependent.*

(b) *Since reducing the matrix* M *above to RREF form results in a free variable, the vectors do not span* $\mathbb{R}^3$.

(c) *Since* $\{v_1, v_2, v_3, \ldots, v_n\}$ *is a basis for* V, *and any set* $\{w_1, w_2, w_3, \ldots, w_m\}$ *has* $w_i \in V$ *for* $i = 1, 2, 3, \ldots, n$, *there exist constants* $a_{i,j}$ *with* $1 \leq i \leq n, 1 \leq j \leq m$ *such that* $w_i = \sum_{k=1}^{n} a_{i,j} v_k$. *Consider the linear combination of* w_is, $\sum_{i=1}^{m} c_i w_i = 0$. *Then this implies that*

$$\sum_{i=1}^{n} \left(\sum_{j=1}^{m} a_{i,j} c_j \right) v_i = 0$$

Since the v_is *are linearly independent by virtue of forming a basis for* V, *we have* $Mc = 0$ *where* M *is the* $n \times m$ *matrix which is equipped with the column vector* c *(with* $M_{i,j} = a_{i,j}$*).* M *has more columns than rows, so its null space is non-trivial, and we must have* $c_j \neq 0$ *with the linear combination summing to zero. Hence, the set of vectors* v_i *is linearly dependent.*

349

Exercise 112.

(a) *Find a basis for P_n, the space of polynomials with degree n.*

(b) *Find a basis for $M_{2\times 2}(\mathbb{R})$, the space of 2×2 matrices with elements in $\mathbb{R}$.*

(c) *Find a basis for $M_{k\times k}(\mathbb{R})$, the space of $k \times k$ matrices with elements in $\mathbb{R}$.*

(d) *Find a basis for $T_{k\times k}(\mathbb{R})$, the space of $k \times k$ matrices with elements in $\mathbb{R}$ and trace zero.*

Solution 112.

(a) *This is the basis $\{1, x, x^2, \ldots, x^n\}$.*

(b) *The basis consists of the matrices*

$$\begin{bmatrix} 1 & 0 \\ 0 & 1 \end{bmatrix}$$

$$\begin{bmatrix} 0 & 1 \\ 1 & 0 \end{bmatrix}$$

(c) *The basis is the set with size k^2 of the matrices with a 1 in each possible position.*

(d) *Since the diagonal elements must sum to zero, part of the basis will be matrices with a 1 in one entry, and 0s for all others off-diagonal entries. On the diagonal, we add the set of matrices for which each individual entry is 1 and another is -1, and all other entries are zero.*

Exercise 113. *Let $\mathbf{u} = \langle 1, -3 \rangle$, $\mathbf{v} = \langle 4, 2 \rangle$.*

(a) *Compute $\mathbf{u} \cdot \mathbf{v}$.*

(b) *Compute $\|\mathbf{u}\|$ and $\|\mathbf{v}\|$.*

Solution 113. (a) $\mathbf{u} \cdot \mathbf{v} = 1 \cdot 4 + (-3)2 = \boxed{-2}$.

(b) $\|\mathbf{u}\| = \sqrt{1^2 + (-3)^2} = \boxed{\sqrt{10}}$;

$\|\mathbf{v}\| = \sqrt{4^2 + 2^2} = \sqrt{20} = \boxed{2\sqrt{5}}$.

Exercise 114. *Prove that $(A^T)^{-1} = (A^{-1})^T$ for all invertible matrices $A \in M_{n\times n}(\mathbb{R})$.*

Solution 114. *We have $A^T(A^{-1})^T = (A^{-1}A)^T = I^T = I$, as well as $(A^{-1})^T A^T = (AA^{-1})^T = I^T = I$. Hence, $(A^T)^{-1} = (A^{-1})^T$.*

Extension Exercise 42. *Let V be a vector space, and let $M, N \subseteq V$ with $\dim V < \infty$. Show that $\dim(M + N) = \dim M + \dim N - \dim(M \cap N)$.*

Extension Solution 42. *Sketch: Let $\{m_1, m_2, m_3, \ldots, m_n\}$ be a basis for $M \cap N$. Note that this set is linearly independent as a basis, and so is linearly independent in M. Then M has an extended basis $\beta = \{m_1, m_2, m_3, \ldots, n_1, n_2, n_3, \ldots, n_j\}$. Thus, $\dim(M) = n + j$. Similarly, $\dim(N) = n + k$. Note that $\dim(M) + \dim(N) - \dim(M \cap N) = n + j + k$, so we must show that $\dim(M + N) = n + j + k$. This can be done by proving that β is indeed a basis of $M + N$.*

Specifically, we can show that β spans $M + N$ and is linearly independent on $M + N$. Let $v \in M + N$. Then $v = m + n$ for $m \in M$, $n \in N$. Hence, $v \in Span(m_1, m_2, m_3, \ldots, n_1, n_2, n_3, \ldots, n_j\})$. We then have that $c_1 n_1 + c_2 n_2 + \ldots$ is a linear combination of the vectors in β (which is a basis of M). By this logic, $\sum_i c_i n_i \in M \cap N$ as well, and $\{m_1, m_2, m_3, \ldots\}$ is a basis for $M \cap N$. By the definitions of linear (in)dependence, we can reduce the problem to the equation $\sum_i a_i m_i + b_i n_i = 0$. As β is linearly independent, all $a_i = b_i = 0$.

It follows that $\dim(M + N) = n + j + k$, as desired.

Extension Exercise 43. *Let A be a square $n \times n$ matrix, and let $b \in \mathbb{R}^n$ be a vector such that the equation $Ax = b$ has a unique solution for x. Prove that A is invertible.*

Extension Solution 43. *Let A be an invertible $n \times n$ matrix. By definition, there is a matrix A^{-1} with $A^{-1}A = AA^{-1} = I_n$. Considering the equation $Ax = b$, multiplying both sides by A^{-1} yields $x = A^{-1}b$, which yields a solution to the matrix equation $Ax = b$. (If we suppose the existence of a second solution, we arrive at a contradiction, however, since left-multiplication just produces the same solution as before.)*

Extension Exercise 44. *Let M be a 2×2 matrix with complex elements. For each of the following, either prove the statement or provide a counterexample:*

(a) *If $M^4 = I$, then $M = I, -I, i \cdot I, -i \cdot I$ (where $i = \sqrt{-1}$).*

(b) *If $M^4 = M^2$, then $M = 0, I, -I$.*

Extension Solution 44.

(a) *False; see the below for a matrix that generates a counterexample.*

(b) *False; a counterexample is*

$$\begin{bmatrix} 2 & 3 \\ -1 & -2 \end{bmatrix}$$

(or, for that matter, any matrix

$$\begin{bmatrix} a & b \\ c & d \end{bmatrix}$$

with $a^2 + bc = bc + d^2 = 1$, $ab + bd = ac + cd = 0$.

Chapter 5 Review Exercises

Exercise 115. *Suppose x, y, and z are real numbers such that $x^2 + y^2 + z^2 = 1$ with $x + 4y + 9z$ maximized. Compute $x + y + z$.*

Solution 115. *By Cauchy-Schwarz, $x + 4y + 9z \leq \sqrt{x^2 + y^2 + z^2}\sqrt{1^2 + 4^2 + 9^2} = \sqrt{98(x^2 + y^2 + z^2)} = \sqrt{98}$, with equality at $y = 4x$ and $z = 9x$. Thus, $x = \dfrac{\sqrt{2}}{14}$, $y = \dfrac{2\sqrt{2}}{7}$, and $z = \dfrac{9\sqrt{2}}{14}$, giving $x + y + z = \boxed{\sqrt{2}}$.*

Exercise 116. *Consider the 2×2 matrix*

$$M = \begin{bmatrix} 1 & 3 \\ 4 & 8 \end{bmatrix}$$

(a) (i) *Show that the RREF form of M is the identity matrix I.*

 (ii) *Show that any 2×2 matrix can be RREF simplified to the identity matrix I.*

(b) *Compute M^2 and M^3. Compute M^T, $\det M$, and $\det M^T$.*

(c) (i) *If M is invertible, compute M^{-1}. Otherwise, show that M is singular.*

 (ii) *Prove that $\det M^{-1} = (\det M)^{-1}$ for all invertible matrices M.*

(d) (i) *What are the rank and nullity of M?*

 (ii) *State the Rank-Nullity Theorem and use it to verify your answer.*

(e) *Find all row and column bases for M.*

(f) (i) *What is the characteristic polynomial of M?*

 (ii) *What are the eigenvalues and eigenvectors of M?*

Solution 116.

(a) (i) First we can divide the second row by 4, then subtract it from the first row, interchange the two rows, and finally subtract twice the (now)-top row from the now-bottom row to get the identity matrix.

   (ii) In general, the same procedure as above works: scalar multiplication can reduce a row to 1 and some other term, and we can subtract and scalar multiply again if necessary, and finally subtract once more.

(b) Exercise in computation.

$$M^2 = \begin{bmatrix} 13 & 27 \\ 36 & 76 \end{bmatrix}$$

$$M^3 = \begin{bmatrix} 121 & 255 \\ 340 & 716 \end{bmatrix}$$

$$M^T = \begin{bmatrix} 1 & 4 \\ 3 & 8 \end{bmatrix}$$

$\det M = 1 \cdot 8 - 4 \cdot 3 = -4$, $\det M^T = 1 \cdot 8 - 3 \cdot 4 = -4$. (These are easily checkable using an online calculator.)

(c) (i) We have

$$M^{-1} = -\frac{1}{4} \begin{bmatrix} 8 & -4 \\ -3 & 1 \end{bmatrix} = \begin{bmatrix} -2 & 1 \\ \frac{3}{4} & \frac{1}{4} \end{bmatrix}$$

   (ii) Use the fact that $\det(A \cdot B) = \det(A) \cdot \det(B)$, along with the fact that $AA^{-1} = I_n$. We know that $\det(A \cdot A^{-1}) = \det I = 1$. Rearranging yields the desired result.

(d) (i) $\text{rank}(M) = \boxed{2}$, nullity of $M = \boxed{0}$.

   (ii) Rank-Nullity Theorem: $\dim M = \text{rank} M + \text{nullity} M$. Indeed, this checks out with the answer to the above part.

(e) The row/column bases are just the standard basis vectors corresponding to the identity matrix, as the RREF form of M is I_2.

(f) (i) $P_M = (1 - \lambda)(8 - \lambda) - 12 = -4 - 9\lambda + \lambda^2$.

   (ii) The eigenvalues $\lambda_1, \lambda_2 = \dfrac{9 \pm \sqrt{97}}{2}$ are the roots of $P_M = 0$. Then the eigenvectors can be directly computed to be $\boxed{\left(\dfrac{7 - \sqrt{97}}{8}, 1 \right), \left(\dfrac{-7 - \sqrt{97}}{8}, 1 \right)}$.

Exercise 117. (1979 AHSME) For each positive number x, let

$$f(x) = \frac{\left(x + \frac{1}{x}\right)^6 - \left(x^6 + \frac{1}{x^6}\right) - 2}{\left(x + \frac{1}{x}\right)^3 + \left(x^3 + \frac{1}{x^3}\right)}.$$

What is the minimum value of $f(x)$?

Solution 117. Make the substitutions $a = \left(x + \dfrac{1}{x}\right)^3$, $b = x^3 + \dfrac{1}{x^3}$. Then we can write

$$f(x) = \frac{a^2 - b^2}{a + b} = a - b = \left(x + \frac{1}{x}\right)^3 - \left(x^3 + \frac{1}{x^3}\right) = 3\left(x + \frac{1}{x}\right)$$

By AM-GM, $x + \dfrac{1}{x} \geq 2$, so the minimum of f is $3 \cdot 2 = \boxed{6}$.

Exercise 118. *(1990 AIME) The sets $A = \{z : z^{18} = 1\}$ and $B = \{w : w^{48} = 1\}$ are both sets of complex roots of unity. The set $C = \{zw : z \in A \text{ and } w \in B\}$ is also a set of complex roots of unity. How many distinct elements are in C?*

Solution 118. *A quick route: note that the sets A and B effectively reduce to modulo 18 and 48, respectively. So we seek the union of these sets, which will contain elements with respect to modulo $\operatorname{lcm}(18, 48) = 144$. 144 will also be $|C|$.*

Exercise 119. *(2014 AMC 12A) The domain of the function*

$$f(x) = \log_{\frac{1}{2}}(\log_4(\log_{\frac{1}{4}}(\log_{16}(\log_{\frac{1}{16}} x))))$$

is an interval of length $\dfrac{m}{n}$, where m and n are relatively prime positive integers. What is $m + n$?

Solution 119. *We require $\log_4(\log_{\frac{1}{4}}(\log_{16}(\log_{\frac{1}{16}} x))) > 0$, i.e. $\log_{\frac{1}{4}}(\log_{16}(\log_{\frac{1}{16}} x)) > 1$, or $\log_{16}(\log_{\frac{1}{16}} x) > \frac{1}{4} \implies \log_{\frac{1}{16}} x > 2 \implies x > \frac{1}{256}$. At the same time, however, for $x \geq \frac{1}{16}$, $f(x)$ is also undefined. Hence the length of the domain interval is $\dfrac{15}{256} \implies \boxed{271}$.*

Exercise 120. *(2014 AMC 12B) For how many positive integers x is $\log_{10}(x - 40) + \log_{10}(60 - x) < 2$?*

Solution 120. *We require $(x-40)(60-x) < 10^2 = 100$, or $-x^2 + 100x - 2400 < 100 \implies x^2 - 100x + 2500 > 0 \implies (x - 50)^2 > 0 \implies x \neq 50$. At the same time, $40 < x < 60$ for both logarithms to be defined, so there are $\boxed{18}$ permissible integers (namely 41-49, 51-59).*

Exercise 121. *(2011 AIME II) Let M_n be the $n \times n$ matrix with entries as follows: for $1 \leq i \leq n$, $m_{i,i} = 10$; for $1 \leq i \leq n - 1$, $m_{i+1,i} = m_{i,i+1} = 3$; all other entries in M_n are zero. Let D_n be the determinant of matrix M_n. Then $\displaystyle\sum_{n=1}^{\infty} \dfrac{1}{8D_n + 1}$ can be represented as $\dfrac{p}{q}$, where p and q are relatively prime positive integers. Find $p + q$.*

Solution 121. *We have that $\det M_1 = 10$, $\det M_2 = 91$, and $\det M_3 = 820$. Indeed, we can prove that $\det M_{i+1} = 9 \det M_i + 1$: this involves forming a recurrence relation from expanding over the first column. At this point, we can confirm that $D_n = \dfrac{9^{n+1} - 1}{8}$ (sum of finite geometric series formula), leading to an answer in the final form of an infinite geometric series with sum $\dfrac{1}{72} \implies \boxed{73}$.*

Exercise 122. *(2013 AMC 12A) The sequence $\log_{12} 162$, $\log_{12} x$, $\log_{12} y$, $\log_{12} z$, $\log_{12} 1250$ is an arithmetic progression. What is x?*

Solution 122. *Observe that $y = \sqrt{162 \cdot 1250} = \sqrt{2^2 \cdot 3^4 \cdot 5^4} = 2 \cdot 3^2 \cdot 5^2 = 450$. Hence, $x = \sqrt{162 \cdot 450} = \boxed{270}$ by the properties of logs.*

Exercise 123. *(2002 AIME I) Consider the sequence defined by $a_k = \dfrac{1}{k^2 + k}$ for $k \geq 1$. Given that $a_m + a_{m+1} + \ldots + a_{n-1} = \dfrac{1}{29}$, for positive integers m and n with $m < n$, find $m + n$.*

Solution 123. *Note that $\sum a_k$ is the telescoping series $\dfrac{1}{k} - \dfrac{1}{k+1}$, so $a_m + a_{m+1} + \ldots + a_{n-1} = \dfrac{1}{m} - \dfrac{1}{n}$. Let $n = 29t$; then $t = 28$ after some manipulation, and $(m, n) = 28, n = 29 \cdot 28$, so $m + n = 30 \cdot 28 = \boxed{840}$.*

Exercise 124. *(2009 AIME II) The sequence (a_n) satisfies $a_0 = 0$ and $a_{n+1} = \dfrac{8}{5}a_n + \dfrac{6}{5}\sqrt{4^n - a_n^2}$ for $n \geq 0$. Find the greatest integer less than or equal to a_{10}.*

Solution 124. *First make the substitution $a_n = b_n \cdot 2^n$, then observe that b_n is periodic (with period 2). From there, it is not too hard to compute the answer of $\boxed{983}$.*

Exercise 125. *(2004 AIME I) Let S be the set of ordered pairs (x, y) such that $0 < x \leq 1, 0 < y \leq 1$, and $\left\lfloor \log_2\left(\frac{1}{x}\right) \right\rfloor$ and $\left\lfloor \log_5\left(\frac{1}{y}\right) \right\rfloor$ are both even. Given that the area of the graph of S is $\frac{m}{n}$, where m and n are relatively prime positive integers, find $m + n$. The notation $\lfloor z \rfloor$ denotes the greatest integer that is less than or equal to z.*

Solution 125. *For $\frac{1}{2} < x \leq 1$, $\frac{1}{5} < y \leq 1$, $(x, y) \in S$. In addition, points (x, y) such that $\frac{1}{8} < x \leq \frac{1}{4}$, $\frac{1}{125} < y \leq \frac{1}{25}$ are both points in S as well. We can actually express the areas of these regions as infinite geometric series, getting $\frac{5}{9}$ as the final area of the graph of $S \implies \boxed{14}$.*

Exercise 126.

(a) *Briefly explain how Bézout's Identity could be generalized to n variables, for $n \geq 3$. You do not need to prove this generalized form.*

(b) *Briefly explain how Bézout's Identity could be applied to single-variable polynomials. (Hint: Use the Fundamental Theorem of Algebra.)*

Solution 126.

(a) *If $\gcd(a_1, a_2, a_3, \ldots, a_n) = d$, then $\exists x_1, x_2, x_3, \ldots, x_)n \in \mathbb{Z}$ such that $d = \sum_{i=1}^{n} a_i x_i$ and every number of this form is a multiple of d, with d minimal.*

(b) *For polynomials f, g of a single variable with coefficients in some field $\mathbb{F}$, there exist polynomials a and b such that $af + bg = 1$ iff f and g have no common roots in $\mathbb{F}$.*

Extension Exercise 45. *(2018 HMMT November Guts Round) Over all real numbers x and y, find the minimum possible value of $(xy)^2 + (x + 7)^2 + (2y + 7)^2$.*

Extension Solution 45. *First expand the expression as $f(x, y) = x^2y^2 + x^2 + 14x + 49 + 4y^2 + 28y + 49$, then simplify this to $(x^2 + 4)(y^2 + 1) + 14(x + 2y) + 94$. From here, use the Cauchy-Schwarz Inequality on $(x^2 + 4)(1 + y^2) = (x^2 + 2^2)(1^2 + y^2)$ to conclude that $(x^2 + 4)(1 + y^2) \geq ((x \cdot 1 + 2 \cdot y))^2 = (x + 2y)^2$; hence, $f(x, y) \geq (x + 2y)^2 + 14(x + 2y) + 94 = (x + 2y + 7)^2 + 45$. Clearly, $f(x, y)$ is minimized when $x + 2y = -7$, and indeed, this yields real values for x and y. Thus, $f(x, y)$ has a minimum at $\boxed{45}$.*

Calculus solution. In order to find the minimum of a multi-variable function $f(x_1, x_2, x_3, \ldots, x_n)$ we must determine the critical points of f; i.e. where $\frac{\delta f}{\delta x_i} = 0$ (i.e. the partial derivative of f with respect to the variable x_i, which treats all other variables as constants) for all $i = 1, 2, 3, \ldots, n$. In this case, $\frac{\delta f}{\delta x} = 2xy^2 + 2x + 14$ and $\frac{\delta f}{\delta y} = 2x^2y + 8y + 28$. Thus, $2xy^2 + 2x = -14$ and $x^2y + 4y = -14$, or $2xy^2 - x^2y + 2x - 4y = 0 \implies xy(2y - x) - 2(2y - x) = 0 \implies (xy - 2)(2y - x) = 0$.

We then consider two cases: $xy = 2$, or $2y = x$. (These cases are mutually exclusive; assuming both gives the solution $(x, y) = (2, 1)$, or $f(x, y) = 166$. Clearly this is not optimal; we can observe this from evaluating $f(0, 0) = 98$!) If $xy = 2$, then given $xy^2 + x = -7$, we have $2y + x = -7$, and since $y = \frac{2}{x}$, this yields $\frac{4}{x} + x = -7$. We can re-write this as $x^2 + 7x + 4 = 0$, which has real solutions $x = \frac{-7 \pm \sqrt{33}}{2}$. Both yield $f(x, y) = 45$. If we check the second case, $2y = x$, as well, then we obtain $y = -\frac{7}{4}$ and $x = -\frac{7}{2}$, which yields a non-optimal value of $f(x, y) = \frac{3969}{64} > 45$.

Thus, the minimum value of $f(x, y) = (xy)^2 + (x + 7)^2 + (2y + 7)^2$ for $x, y \in \mathbb{R}$ is $\boxed{45}$.

Extension Exercise 46. *(2011 AIME II) Let $P(x) = x^2 - 3x - 9$. A real number x is chosen at random from the interval $5 \leq x \leq 15$. The probability that $\lfloor \sqrt{P(x)} \rfloor = \sqrt{P(\lfloor x \rfloor)}$ is equal to $\dfrac{\sqrt{a} + \sqrt{b} + \sqrt{c} - d}{e}$ where a, b, c, d and e are positive integers and none of a, b, or c is divisible by the square of a prime. Find $a + b + c + d + e$.*

Extension Solution 46. *Sketch: Construct a table of values for $P(x)$, then consider when $P(\lfloor x \rfloor)$ is a perfect square. Next, do casework on the permissible x-values, which will eventually lead to several intervals whose lengths can be summed divided by 10 to get a final answer of* $\boxed{850}$.

Extension Exercise 47. *(1981 IMO P1) P is a point inside a given triangle ABC. D, E, F are the feet of the perpendiculars from P to the lines BC, CA, AB, respectively. Find all P for which $\frac{BC}{PD} + \frac{CA}{PE} + \frac{AB}{PF}$ is least.*

Extension Solution 47. *Note that $BC \cdot PD + CA \cdot PE + AB \cdot PF$ is twice the triangle's area. By a direct application of Cauchy-Schwarz, it follows that $(BC \cdot PD + CA \cdot PE + AB \cdot PF)\left(\frac{BC}{PD} + \frac{CA}{PE} + \frac{AB}{PF}\right)$ $\geq (BC + CA + AB)^2$, with equality when $PD = PE = PF$, in which case P is the incenter.*

Extension Exercise 48. *(2012 AIME I) Complex numbers a, b, and c are zeros of a polynomial $P(z) = z^3 + qz + r$, and $|a|^2 + |b|^2 + |c|^2 = 250$. The points corresponding to a, b, and c in the complex plane are the vertices of a right triangle with hypotenuse h. Find h^2.*

Extension Solution 48. *By Vietas, $a + b + c = 0$, so $\dfrac{a + b + c}{3} = 0$ and the centroid of the triangle is at the origin as a result. Let the right triangle have leg lengths x, y, and WLOG let $\overline{ac}$ be the triangles hypotenuse. $|a|, |b|, |c|$ are all $\dfrac{2}{3}$ of the median lengths, as the centroid trisects each median. Then $|a|^2 = \dfrac{x^2 + 4y^2}{9}$, $|c|^2 = \dfrac{4x^2 + y^2}{9}$. The b-median has length $\dfrac{ac}{2}$, or $\dfrac{x^2 + y^2}{9}$. Therefore, $|a|^2 + |b|^2 + |c|^2 = \dfrac{2x^2 + 2y^2}{3} = 250 \implies x^2 + y^2 = h^2 = \boxed{375}$.*

Extension Exercise 49. *(2004 USAMO P5) Let a, b, and c be positive real numbers. Prove that*

$$(a^5 - a^2 + 3)(b^5 - b^2 + 3)(c^5 - c^2 + 3) \geq (a + b + c)^3$$

Extension Solution 49. *First note that $x^5 + 1 \geq x^3 + x^2$ by Rearrangement. It then suffices to show that $(a^3 + 2)(b^3 + 2)(c^3 + 2) \geq (a + b + c)^3$. By Hölder's Inequality,*

$$(a^3 + 1 + 1)(b^3 + 1 + 1)(c^3 + 1 + 1) \geq \left((a^3 \cdot 1 \cdot 1)^{\frac{1}{3}}(b^3 \cdot 1 \cdot 1)^{\frac{1}{3}}(c^3 \cdot 1 \cdot 1)^{\frac{1}{3}}\right)^3$$

with equality for $a = b = c = 1$.

Extension Exercise 50. *(1977 USAMO P3) If a and b are two of the roots of $x^4 + x^3 - 1 = 0$, prove that ab is a root of $x^6 + x^4 + x^3 - x^2 - 1 = 0$.*

Extension Solution 50. *Let $x^4 + x^3 - 1 = 0$ have roots a, b, c, d. Then $a + b + c + d = -1$, $ab + ac + ad + bc + bd + cd = 0$, $abcd = -1$, which implies $cd = -\dfrac{1}{ab}$ and $c + d = -1 - (a + b)$. We also have $ab + cd + (a + b)(c + d) = 0 \implies ab - \dfrac{1}{ab} + (a + b)(-1 - (a + b)) = 0$. Making substitutions for $ab, a + b$ and making algebraic manipulations using the fact that a, b are roots yields the desired conclusion.*

Chapter 6: Advanced Algebra

Section 6.1: Binomial Theorem, Part 3: The Multinomial Theorem

Exercise 127. *What connection can we make between the Multinomial Theorem and the stars-and-bars/balls-and-urns formula?*

Solution 127. *In the stars-and-bars formula, the multinomial coefficients give the number of solutions to the equation $x_1 + x_2 + x_3 + \ldots + x_n = d$, where $n, d \in \mathbb{N}$. Essentially, we are counting the number of terms in a multinomial sum when we apply the method of stars-and-bars.*

Exercise 128. *Show that*

$$\sum_{\sum_{i=1}^{m} k_i = n} \binom{n}{k_1, k_2, k_3, \ldots, k_m} = m^n$$

for all $m, n \in \mathbb{N}$.

Solution 128. *This immediately follows from the substitution of $x_i = 1$ for all i into the statement of the Multinomial Theorem.*

Exercise 129. *(1991 AIME) Expanding $(1+0.2)^{1000}$ by the binomial theorem and doing no further manipulation gives $\binom{1000}{0}(0.2)^0 + \binom{1000}{1}(0.2)^1 + \binom{1000}{2}(0.2)^2 + \ldots + \binom{1000}{1000}(0.2)^{1000} = A_0 + A_1 + A_2 + \ldots + A_{1000}$, where $A_k = \binom{1000}{k}(0.2)^k$ for $k = 0, 1, 2, \ldots, 1000$. For which k is A_k the largest?*

Solution 129. *In order to change A_k to A_{k+1}, we multiply by $0.2 = \dfrac{1}{5}$ and, at the same time, by $\dfrac{1000 - k + 1}{k}$. Hence, we require $\dfrac{1000 - k + 1}{5k} > 1 \implies k < 166.8$, so the answer is $\boxed{166}$.*

Exercise 130.

   (a) *A strict composition of the non-negative integer n is its decomposition into $n = k_1 + k_2 + k_3 + \ldots$ where the k_i are positive integers. A weak composition is similar, except the k_i can also be zero. Show that the exponents in a multinomial expansion form a weak composition (and not a strong one).*

   (b) *Provide explicit formulas for the number of strict and weak compositions of a positive integer n.*

Solution 130.

   (a) *Since there can be 0 duplicate elements in a string of n letters, we have weak compositions instead of strong compositions.*

   (b) *For n a positive integer, n has 2^{n-1} distinct compositions (as we can place a plus/minus sign in $n - 1$ spaces in between n 1's adding up to n), which is also the sum $\sum_{k=1}^{n} \binom{n-1}{k-1}$. For weak compositions, the summand becomes $\sum_{k=1}^{n} \binom{n+k-1}{k-1}$.*

Extension Exercise 51. *Try deriving Pascal's n-dimensional simplexes using multinomial coefficients. (Hint: Pascal's Triangle is the Pascal 2-simplex.)*

Extension Solution 51. *For an m-term polynomial raised to some power n, the Pascal m-simplex consists of components that are each of the form $|x|^n = \sum_{|k|=n} \binom{n}{k} x^k$ where $|x| = \sum_{i=1}^{m} x_i$.*

Section 6.2: Generating Functions

Exercise 131. *Find the generating functions, or sequences associated to the given generating functions, of each of the following:*

(a) $1, -2, 3, -4, 5, \ldots$

(b) $64, 32, 16, 8, 4, \ldots$

(c) $f(x) = x^2 \sin x$

(d) $1, 0, -\dfrac{1}{2}, 0, \dfrac{1}{24}, 0, -\dfrac{1}{720}, \ldots$

(e) $f(x) = \dfrac{x^2}{x^2 + 1}$

Solution 131.

(a) $1 - 2x + 3x^2 - 4x^3 + 5x^4 - 6x^5 + \ldots = (1 - x + x^2 - x^3 + x^4 - x^5 + \ldots) + (-x + x^2 - x^3 + x^4 - x^5 + \ldots) + (x^2 - x^3 + x^4 - x^5 + \ldots) + \ldots = \dfrac{1}{1+x} - \dfrac{x}{1+x} + \dfrac{x^2}{1+x} = \boxed{\dfrac{1}{(1+x)^2}}$.

(b) $64 + 32x + 16x^2 + 8x^3 + 4x^4 + \ldots = \dfrac{64}{1 - \frac{x}{2}} = \boxed{\dfrac{128}{2 - x}}$.

(c) $x^2 \sin x = x^2 \cdot \sin x$. *Since* $\sin x = x - \dfrac{x^3}{3!} + \dfrac{x^5}{5!} - \dfrac{x^7}{7!} + \ldots$ *the Taylor series for* $x^2 \sin x$ *is* $x^3 - \dfrac{x^5}{3!} + \dfrac{x^7}{5!} - \dfrac{x^9}{7!} + \ldots$ *which has a corresponding sequence of* $\left(0, 0, 0, 1, 0, -\dfrac{1}{3!}, 0, \dfrac{1}{5!}, 0, -\dfrac{1}{7!}, \ldots\right)$.

(d) $1 + 0x - \dfrac{1}{2}x^2 + 0x^3 + \dfrac{1}{24}x^4 + 0x^5 - \dfrac{1}{720}x^6 + \ldots = \boxed{\cos x}$.

(e) $f(x) = \dfrac{x^2}{x^2 + 1} = x^2 \cdot \dfrac{1}{x^2 + 1} = x^2 \cdot \dfrac{1}{1 - (-x^2)} = x^2 \cdot (1 - x^2 + x^4 - x^6 + \ldots) = x^2 - x^4 + x^6 - x^8 + \ldots$, *which corresponds to* $(0, 0, 1, 0, -1, 0, 1, 0, -1, \ldots)$

Exercise 132. *What is the 2019th term of the sequence whose generating function is $f(x) = -\dfrac{1}{(x-1)^3}$?*

Solution 132. *Let the sequence $(a_n) = (a_0, a_1, a_2, a_3, \ldots)$ have generating function $f(x) = a_0 + a_1 x + a_2 x^2 + a_3 x^3 + \ldots$ We have $f(x) = \dfrac{1}{(1-x)^3} = \left(\dfrac{1}{1-x}\right)^3$. If we multiply out the expansion of $(1 + x + x^2 + x^3 + \ldots)$ with itself repeatedly, we see that the coefficients become the sequence $1, 4, 10, 20, \ldots$, i.e. the sums of triangular numbers. Hence, we have $1 + 3 + 6 + 10 + \ldots + (2019 \cdot 1010)$ as the 2019th term.*

Exercise 133. *(2016 AIME II) For polynomial $P(x) = 1 - \dfrac{1}{3}x + \dfrac{1}{6}x^2$, define $Q(x) = P(x)P(x^3)P(x^5)P(x^7)P(x^9) = \sum_{i=0}^{50} a_i x^i$. Then $\sum_{i=0}^{50} |a_i| = \dfrac{m}{n}$, where m and n are relatively prime positive integers. Find $m + n$.*

Solution 133. *Observe that all coefficients of odd-exponent terms result from multiplying together an odd number of odd degree terms, and all coefficients of even-exponent terms result from multiplying together an even number of odd degree terms. Every odd-degree term is negative, and every even-degree term is positive, so the sum is $Q(-1) = P(-1)^5 = \dfrac{243}{32} \implies \boxed{275}$.*

Exercise 134.

(a) *What is the Maclaurin series of the function* $f(x) = \sin x \cos x$?

(b) *What is the sequence whose generating function is given by* f?

(c) *What is the kth term of this sequence, in terms of k?*

(d) *Determine the generating function of the sequence of only non-zero coefficients of the Maclaurin series of f.*

Solution 134.

(a) *The Maclaurin series for* $\sin x \cos x = \dfrac{1}{2} \sin(2x)$ *results from Maclaurin series for* $\sin x$, *i.e.*

$$\sum_{n=0}^{\infty} \frac{(-1)^n \cdot x^{2n+1}}{(2n+1)!}$$

Substituting $2x$ for x yields

$$\frac{1}{2} \sum_{n=0}^{\infty} \frac{(-1)^n \cdot (2x)^{2n+1}}{(2n+1)!}$$

(b) *Extracting coefficients yields* $\left(0, 1, 0, -\dfrac{2}{3}, 0, \dfrac{2}{15}, 0, -\dfrac{4}{315}, \ldots \right)$.

(c) *The kth term a_k (where $a_0 = 0$) is given by* $\boxed{\dfrac{(-1)^{\frac{k-1}{2}} \cdot 2^{k-1}}{k!}}$ *if k is odd and 0 if k is even.*

(d) *The sequence of non-zero coefficients is* $\left(1, -\dfrac{2}{3}, \dfrac{2}{15}, -\dfrac{4}{316} \right)$, *which leads to the generating function*

$$1 - \frac{2}{3}x + \frac{2}{15}x^2 - \frac{4}{315}x^3 + \ldots = \sum_{n=0}^{\infty} \frac{(-4)^n \cdot x^n}{(2n+1)!}$$

Exercise 135. *(2007 HMMT February Combinatorics Test) Let S denote the set of all triples (i, j, k) of positive integers where $i + j + k = 17$. Compute*

$$\sum_{(i,j,k) \in S} ijk$$

Solution 135. *Let* $s_n = \sum_{i+j+k=n} ijk$. *Then*

$$\sum s_n x^n = \left(\sum n x_n \right)^3 = \left(\frac{x}{(1-x)^2} \right)^3 = \frac{x^3}{(1-x)^6}$$

Hence, $s_n = \dbinom{n+2}{5} \implies s_{17} = \boxed{\dbinom{19}{5}}$.

Exercise 136. *(2010 AIME I) Jackie and Phil have two fair coins and a third coin that comes up heads with probability $\dfrac{4}{7}$. Jackie flips the three coins, and then Phil flips the three coins. Let $\dfrac{m}{n}$ be the probability that Jackie gets the same number of heads as Phil, where m and n are relatively prime positive integers. Find $m + n$.*

Solution 136. *If we represent by x^n the outcome of flipping n tails, then the generating functions for the coins are $1 + x$, $1 + x$, $4 + 3x$, whose product is $4 + 11x + 10x^2 + 3x^3$. The square of the sum of coefficients (total outcomes) is 784, and the sum of the squares of the coefficients (desirable outcomes) is 246. Hence, the desired probability is* $\dfrac{123}{392} \implies \boxed{515}$.

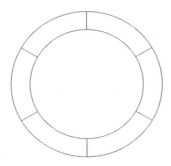

Figure 9.3: Diagram for this Exercise.

Exercise 137. *(2016 AIME II) The figure below shows a ring made of six small sections which you are to paint on a wall. You have four paint colors available and you will paint each of the six sections a solid color. Find the number of ways you can choose to paint the sections if no two adjacent sections can be painted with the same color.*

Solution 137. *Call the colors 0, 1, 2, 3. Then, between any two adjacent sections s_i and s_{i+1}, there exists $d_i = 1, 2, 3$ such that $s_{i+1} \equiv s_i + d_i \pmod{4}$. Each border then has GF $x + x^2 + x^3$, with each term representing increasing the color by 1, 2, or 3 mod 4. The GF for all six borders is then $(x + x^2 + x^3)^6$, where each coefficient of x^n indicates the number of colorings in which we increase the color number by n in total. We then want to compute the sum of the coefficients of x^n with $n \equiv 0 \pmod{4}$.*

If $P(x) = x^n$, then $P(x) + P(ix) + P(-x) + P(-ix) = 4x^n$ iff $n \equiv 0 \pmod{4}$. Hence, the answer is 4 times $\dfrac{3^6 + 3(-1)^6}{4}$, or just $\boxed{732}$.

Exercise 138. *Show that the generating function of the Fibonacci series F_n is equal to $F(z) = \dfrac{1}{1 - z - z^2}$.*

Solution 138. *Write*

$$1 + z + 2z^2 + 3z^3 + 5z^4 + 8z^5 + \ldots = \sum_{n=0}^{\infty} F_n z^n$$

Note that

$$(1 - z - z^2) \sum_{n=0}^{\infty} F_n z^n$$

$$= \sum_{n=0}^{\infty} F_n z_n - \sum_{n=0}^{\infty} F_n z^{n+1} - \sum_{n=0}^{\infty} F_n z^{n+2}$$

$$= \sum_{n=0}^{\infty} F_n z_n - \sum_{n=1}^{\infty} F_{n-1} z^n - \sum_{n=2}^{\infty} F_{n-2} z^n$$

$$= F_0 + z(F_1 - F_0) + \sum_{n=2}^{\infty} (F_n - F_{n-1} - F_{n-2}) z^n$$

Since $F_1 = F_0$ and $F_n = F_{n-1} + F_{n-2}$, we have

$$(1 - z - z^2) \sum_{n=0}^{\infty} F_n z^n = 1 \implies \sum_{n=0}^{\infty} F_n z^n = F_0 = \dfrac{1}{1 - z - z^2}$$

as claimed.

Extension Exercise 52. *Show that*

$$\sum_{k=0}^{n} \binom{2k}{k} \binom{2(n-k)}{n-k} = 4^n$$

359

Extension Solution 52. *Consider the set of length-$2n$ strings consisting of As and Bs; there are $2^{2n} = 4^n$ such sequences. We count these strings in another way, by defining the equality and inequality properties.*

Say such a string has the equality property if it consists of n As and n Bs. There are clearly $\binom{2n}{n}$ such strings. Furthermore, say a string has the inequality property if, from left to right, the string up to a certain point does not have the equality property. Let $I(2k)$ be the number of strings of length $2k$ with the inequality property (and define $E(2k)$ similarly).

For any length-$2n$ string, we can split it at a point such that the left side has the equality property, while the right side has the inequality property. For $k \in [0, n]$, there are $E(2k)I(2n - 2k)$ such strings (with left side of length $2k$). Hence,

$$\sum_{k=0}^{n} E(2k)I(2n - 2k) = \sum_{k=0}^{n} \binom{2k}{k} I(2n - 2k)$$

It remains to show that $I(2k) = \binom{2k}{k}$.

To prove this, assume $k > 0$ and consider all length-$2k$ strings beginning with A. Divide these strings into 4 groups, namely strings with the inequality property, strings beginning with substrings that have the equality property and have more As than Bs, strings beginning with substrings with the equality property and have more Bs than As, and strings with the equality property. (These groups are mutually exclusive.) Let S_1, S_2, S_3, S_4 be the numbers of strings in each of these groups, respectively.

We claim $S_2 = S_3$. This can be proven by supposing a string satisfies the second condition, in which case we let the length of the longest string with the equality property be $L \geq 2$. There are an equal number of As and Bs up to position L, so there are more As than Bs after L. We replace the As with Bs (and vice versa) after position L, so that the whole string now satisfies the third condition. We can also reverse this, hence $S_2 = S_3$.

Furthermore, the B count is greater than the A count iff the string satisfies condition 3. In order for the B count to exceed the A count in a string starting with A, at same point, the counts must be equal (this is an application of the Intermediate Value Theorem, in essence). Hence it begins with a substring having the equality property.

We then can compute S_3 to be the number of strings consisting of $k + 1$ Bs, $k + 2$ Bs, and so forth. This allows us to write

$$S_3 = \binom{2k - 1}{k + 1} + \binom{2k - 1}{k + 2} + \ldots + \binom{2k - 1}{2k - 1}$$

$$= \binom{2k - 1}{0} + \binom{2k - 1}{1} + \ldots + \binom{2k - 1}{k - 2} = S_2$$

Furthermore, we have $S_4 = \binom{2k - 1}{k}$, and so $S_1 + S_2 + S_3 + S_4 = \binom{2k - 1}{0} + \binom{2k - 1}{1} + \ldots + \binom{2k - 1}{2k - 1} = 2^{2k-1}$. This yields $S_1 = \binom{2k - 1}{k - 1}$, and so $I(2k) = 2S_1 = \binom{2k}{k}$ as desired. This completes the proof.

Extension Exercise 53. *(1986 USAMO P5) By a partition π of an integer $n \geq 1$, we mean here a representation of n as a sum of one or more positive integers where the summands must be put in nondecreasing order. (E.g., if $n = 4$, then the partitions π are $1 + 1 + 1 + 1$, $1 + 1 + 2$, $1 + 3$, $2 + 2$, and 4).*

For any partition π, define $A(\pi)$ to be the number of 1's which appear in π, and define $B(\pi)$ to be the number of distinct integers which appear in π. (E.g., if $n = 13$ and π is the partition $1 + 1 + 2 + 2 + 2 + 5$, then $A(\pi) = 2$ and $B(\pi) = 3$).

Prove that, for any fixed n, the sum of $A(\pi)$ over all partitions of π of n is equal to the sum of $B(\pi)$ over all partitions of π of n.

Extension Solution 53. *Define $S(n) = \sum A(\pi), T(n) = \sum B(\pi)$. We show that the OGFs of $S(n)$ and $T(n)$ are, in fact, equal.*

Each coefficient in the OGF of $S(n)$ is given by the total number of 1s in the partitions of n. This is the same as the number of partitions of $n - x$ without 1s, where x is the number of 1s in the partition. Hence,

$$S(n) = \sum_{k=1}^{n} k \cdot N$$

where N is the number of partitions of $n - k$ with no 1s.

The number of partitions of m with no 1s, then, is the coefficient of x^m in the expansion of

$$f(x) = (1 + x^2 + x^4 + \ldots)(1 + x^3 + x^6 + \ldots)(1 + x^4 + x^8 + \ldots)\cdots$$

From here, let c_m be this coefficient; then $f(x) = c_0 + c_1 x + c_2 x^2 + \ldots$ We want to determine the value of $S(n) = c_{n-1} + 2c_{n-2} + 3c_{n-3} + \ldots + n \cdot c_0$. If we define

$$g(x) = (x + 2x^2 + 3x^3 + 4x^4 + \ldots)f(x) = \frac{x}{1-x} \prod_{i=1}^{\infty} \frac{1}{1 - x^i}$$

then we find that the coefficient of x^n in $g(x)$ is equal to $c_{n-1} + 2c_{n-2} + \ldots + nc_0 = S(n)$. Hence, $g(x)$ is the OGF of $S(n)$.

Similarly, we determine the OGF of $T(n)$. This works similarly, with the modus operandi of defining another polynomial $P(x)$ that we expand so as to compute $T(n) = d_0 + d_1 + d_2 + \ldots + d_{n-1}$. If we again consider a power series expansion

$$h(x) = \frac{x}{1-x} \prod_{i=1}^{\infty} \frac{1}{1 - x^i}$$

we get that the coefficient of x^n in $h(x)$ is once again $T(n)$, so $h(x)$ is the OGF of $T(n)$, and we are done!

Extension Exercise 54. *In the decimal expansion of $\dfrac{1}{9,999,899,999}$, "blocks" of five digits after the decimal point contain the Fibonacci numbers:*

$$0.00000\ 00001\ 00001\ 00002\ 00003\ 00005\ 00008\ldots$$

For how many digits does this pattern continue?

Extension Solution 54. *Fakesolve: Note that the pattern can only continue as long as the decimal representation itself can continue to accommodate it in terms of space limitations! That is, as soon as we encounter a Fibonacci number longer than 5 digits, the decimal expansion can no longer support the pattern, as digits will begin to roll over. Hence, we seek the largest Fibonacci number less than 100,000. This is not too hard to compute by hand, and turns out to be $F_{25} = 75,025$. Hence, the answer is $5(25 + 1) = \boxed{130}$.*

Section 6.3: Functional Equations, Part 2

Exercise 139. *Find all functions $f : \mathbb{R} \mapsto \mathbb{R}$ such that $f(x^2 + y^2) = f(x)^2 + 2xy + f(y)^2$.*

Solution 139. *Plug in $(x, y) = (0, 0)$ to obtain $f(0) = 2f(0)^2 \implies f(0) = 0, \frac{1}{2}$. Then try $x = 1, y = 0$ to obtain $f(1) = f(1)^2 \implies f(1) = 0, 1$. Also try $x = 1, y = 1$ to get $f(2) = 2f(1)^2 + 2$, or $f(2) = 2$ if $f(1) = 0$, $f(2) = 4$ if $f(1) = 1$. These initial cases yield a contradiction in the case of an input such as $f(25)$, since $25 = 4^2 + 3^2 = 0^2 + 5^2$. Hence the equation has no solutions.*

Exercise 140. *(2012 IMO Shortlist A5) Find all functions $f : \mathbb{R} \mapsto \mathbb{R}$ that satisfy the conditions $f(1 + xy)f(x + y) = f(x)f(y)$ for all $x, y \in \mathbb{R}$, $f(-1) \neq 0$.*

Solution 140. *We claim that the sole solution is $f(x) = x - 1$.*

Define $g(x) = f(x) + 1$, and show that $g(x) = x$ for all real x. Define $C = g(-1) - 1$. Then letting $y = -1$ yields $g(1 - x) - g(x - 1) = C(g(x) - 1)$. Setting $x = 1$ yields $C(g(1) - 1) = 0$, and so $g(1) = 1$. $x = 0$ further reveals that $g(0) = 0$ and $x = 2$ yields $g(2) = 2$.

To prove the resulting equations $g(x) + g(2 - x) = g(x + 2) - g(x) = 2$, we substitute $1 - x$ in place of x in $g(1 - x) - g(x - 1) = C(g(x) - 1)$, since g is cyclic with period 2. We end up with the equations $g(x) - g(-x) = C(g(1 - x) - 1)$ and $g(-x) - g(x) = C(g(1 + x) - 1)$. Adding the two equations results in $C(g(1 - x) + g(1 + x) - 2) = 0$. Hence, we have proven that $g(1 - x) - g(x - 1) = C(g(x) - 1)$.

Henceforth, let there exist m, n with $m + n = 1$. If we consider $g(1 + xy) - g(x + y) = ((g(x) - 1)(g(y) - 1)$ for all $x, y \in \mathbb{R}$ with $g(-1) \neq 1$, which results from our initial conditions, along with the pairs $(m, n), (2 - m, 2 - n)$, we find that $g(1 + mn) - g(1) = (g(m) - 1)(g(n) - 1)$, $g(3 + mn) - g(3) = (g(2 - m) - 1)(g(2 - n) - 1)$. We have that $(g(m) - 1)(g(n) - 1) = (g(2 - m) - 1)(g(2 - n) - 1)$, so in conjunction with $m + n = 1$, we end up with $g(mn + 3) - g(mn + 1) = g(3) - g(1)$.

We can write $x = mn + 1$, where $m + n = 1$, whenever $x \leq \dfrac{5}{4}$ (using the Quadratic Formula). Thus, $g(x + 2) - g(x) = g(3) - g(1)$ on that interval. For $x = 0, 1, 2$, we have $g(x) = x$, so setting $x = 0$ gives $g(3) = 3$. If $x > \dfrac{5}{4}$, on the other hand, $-x < \dfrac{5}{4}$, which implies $g(2 - x) - g(-x) = 2$. We also have $g(x) = 2 - g(2 - x)$, alongside $g(x + 2) = 2 - g(-x)$, so that $g(x + 2) - g(x) = g(2 - x) - g(-x) = 2$.

Substituting $-x$ for x yields $g(-x) + g(2 + x) = 2$. This implies $g(x) + g(-x) = 0$, which further implies that $g(1 - xy) - g(y - x) = (g(x) + 1)(1 - g(y))$, $g(1 - xy) - g(x - y) = (1 - g(x))(g(y) + 1)$. Summing yields $g(1 - xy) = 1 - g(x)g(y) \implies g(1 + xy) = 1 + g(x)g(y)$, and so $g(x + y) = g(x) + g(y)$ for all $x, y \in \mathbb{R}$.

Using the property of additivity of g, we can write $g(1 + xy) = g(1) + g(xy)$. We already know that $g(1 + xy) = 1 + g(x)g(y)$, so $g(xy) = g(x)g(y)$. Setting $y = x$ yields $g(x^2) = g(x)^2 \geq 0$, or $g(x) \geq 0$. It follows that g is a linear function, with $g(x) = x$. Hence, $f(x) = g(x) - 1 = x - 1, x \in \mathbb{R}$.

Exercise 141. *(2010 IMO P1) Find all functions $f : \mathbb{R} \to \mathbb{R}$ such that for all $x, y \in \mathbb{R}$ the following equality holds*

$$f(\lfloor x \rfloor y) = f(x) \lfloor f(y) \rfloor$$

where $\lfloor a \rfloor$ is the greatest integer not greater than a.

Solution 141. *Begin by setting $x = y = 0$; then $f(0) = 0$ or $\lfloor f(0) \rfloor = 1$. If $\lfloor f(0) \rfloor = 1$, then setting $y = 0$ yields $f(x) = f(0)$, i.e. f is a constant function. Then $f(x) = 0$ for all real x, or $f(x) = a$, $a \in [1, 2)$. If $f(0) = 0$, setting $x = y = 1$ yields $f(1) = 0$ or $\lfloor f(1) \rfloor = 1$. In the former case, $x = 1$ yields $f(y) = 0$ for all y, which is a solution. In the latter, $y = 1$ implies $f(\lfloor x \rfloor) = f(x)$.*

In the original equation, setting $(x, y) = \left(2, \dfrac{1}{2}\right)$ gives $f(1) = f(2) \cdot \lfloor f\left(\dfrac{1}{2}\right) \rfloor$. From $f(\lfloor x \rfloor) = f(x)$, we have $f\left(\dfrac{1}{2}\right) = f(0) = 0$, implying $f(1) = 0$ - but this contradicts $\lfloor f(1) \rfloor = 1$!

Hence, $f(x) = 0$ for all x, or $f(x) = a$, for $a \in [1, 2)$. QED.

Extension Exercise 55. *(2012 IMO P4) Find all functions $f : \mathbb{Z} \to \mathbb{Z}$ such that, for all integers a, b, and c that satisfy $a + b + c = 0$, the following equality holds:*

$$f(a)^2 + f(b)^2 + f(c)^2 = 2f(a)f(b) + 2f(b)f(c) + 2f(c)f(a).$$

Extension Solution 55. *Set $a = b = c = 0$; then we get $f(0) = 0$. Then consider $c = 0$, $a = -b$. It then follows that $f(a) = f(-a)$. We can write $f(c)^2 - 2f(c)(f(a) + f(b)) + (f(a) - f(b))^2 = 0$, which suggests that*

$f(c) = f(-c) = f(a + b) = f(a) + f(b) \pm 2\sqrt{f(a)f(b)}$ by the Quadratic Formula. If $f(b) = 0$, then we have $f(a + b) = f(a)$, so f is periodic modulo b. We now split the solution of this equation into several cases.

Case 1: $f(1) = 0$. In this case, $f(x)$ is the zero function.

Case 2: $f(1) \neq 0$. If $f(1) \neq 0$, but $f(2) = 0$, then f is periodic with period 2 and implies that $f(x) = f(1)$ for odd x, and $f(x) = f(0)$ for even x. If instead $f(2) = 4f(1)$, then we solve for $f(3)$ to get that it is equal to either $f(1)$ or $9f(1)$. Repeating this process by induction reveals that $f(x) = x^2 f(1)$ for all x.

Extension Exercise 56. *(2010 Putnam B5) Is there a strictly increasing function $f : \mathbb{R} \mapsto \mathbb{R}$ such that $f'(x) = f(f(x))$ for all x?*

Extension Solution 56. *(Adapted from the official Putnam solution manual.) No; assume otherwise. f would need to be differentiable, as well as strictly increasing, so $\forall x, f'(x) > 0$. Indeed, $f'(x)$ is itself strictly increasing, with $f''(x) > 0$ for all x. Thus, if $y > x$, then $f(y) = f(f(y)) > f(f(x)) = f(x)$. $f(x) > 0$ for all x as well.*

For any x_0, if $f(x_0) = a, f(x_0) = b > 0$, then $f(x) > a$ whenever $x > x_0$. Thus, $f(x) > a(x - x_0) + b$ for $x > x_0$, from which we can say that either $b < x_0$ or $a = f(x_0) = f(f(x_0)) = f(b) \geq a(b - x_0) + b$. If $b \geq x_0$, then $b \leq \dfrac{a(x_0 + 1)}{a + 1} \leq x_0 + 1$. It follows that $f(x_0) \leq x_0 + 1$ for $x_0 \geq 0$.

Then $\forall x, f(f(x)) \leq 1$, as we cannot have $f(x) > x + 1$. Thence if $f(0) = b_0$ and $f(0) = a_0 > 0$, $f(x) > a_0 x + b_0$ for positive x. For $x > \max 0, -\dfrac{b_0}{a_0}$, $f(x) > 0$ and $f(f(x)) > a_0 x + b_0$. But it would then follow that $f(f(x)) > 1$ for some x, which is a contradiction.

Section 6.4: Proving Multi-Variable Inequalities

Exercise 142.

(a) *Provide an example of a non-constant homogeneous function of two variables with degree 5.*

(b) *Provide an example of a non-constant non-homogeneous function of two variables with degree 3.*

(c) *Show that, if f and g are two homogeneous functions of degree m and n respectively, then $h = \dfrac{f}{g}$ is also homogeneous, and of degree $m - n$, where $g \neq 0$.*

Solution 142.

(a) $x^4 y + x^3 y^2$

(b) $x^2 y + xy + x^3 + y^2$

(c) Let $t \in \mathbb{R}$. Then if we define $Q(x, y) = \dfrac{M(x, y)}{N(x, y)}$,

$$Q(tx, ty) = \frac{M(x, y)}{N(x, y)} = \frac{M(tx, ty)}{N(tx, ty)}$$

$$= \frac{t^n M(x, y)}{t^n N(x, y)} = Q(x, y)$$

This argument applies to M, N homogeneous polynomials of any degree.

Exercise 143. *(1991 AIME) For positive integer n, define S_n to be the minimum value of the sum*

$$\sum_{k=1}^{n} \sqrt{(2k - 1)^2 + a_k^2}$$

where $a_1, a_2, \ldots, a_n$ are positive real numbers whose sum is 17. There is a unique positive integer n for which S_n is also an integer. Find this n.

Solution 143. *Applying Minkowski's Inequality leads to the inequality*

$$S_n \geq \sqrt{\left(\sum_{k=1}^{n}(2k-1)\right)^2 + \left(\sum_{k=1}^{n} a_k\right)^2}$$

Using the sum formulae for the integers, and using the fact that $\sum_{i=1}^{17} a_i = 17$, we have $S_n \geq \sqrt{n^4 + 17^2}$. If this is an integer, then setting $n^4 + 17^2 = m^4$ yields $m = 145$, or $n = \boxed{12}$.

Exercise 144. *(2011 USAMO P1) Let a, b, c be positive real numbers such that $a^2 + b^2 + c^2 + (a+b+c)^2 \leq 4$. Prove that*

$$\frac{ab+1}{(a+b)^2} + \frac{bc+1}{(b+c)^2} + \frac{ca+1}{(c+a)^2} \geq 3.$$

Solution 144. *Rearranging yields $a^2 + b^2 + c^2 + ab + bc + ca \leq 2$. Note that*

$$\frac{2ab+2}{(a+b)^2} \geq \frac{2ab + a^2 + b^2 + c^2 + ab + bc + ca}{(a+b)^2} = \frac{(a+b)^2 + (a+c)(b+c)}{(a+b)^2}$$

Summing over all pairs in $\{a, b, c\}$ yields

$$\sum_{cyc} \frac{2ab+2}{(a+b)^2} = \sum_{cya} 3 + \frac{(a+c)(b+c)}{(a+b)^2} \geq 6$$

where the final step follows from the AM-GM Inequality. Dividing by 2 yields the desired result.

Exercise 145. *(2001 USAMO P3) Let $a, b, c \geq 0$ and satisfy $a^2 + b^2 + c^2 + abc = 4$. Show that $0 \leq ab + bc + ca - abc \leq 2$.*

Solution 145. *We first show that $ab + bc + ca - abc \geq 0$. WLOG assume $a \leq 1$, and also assume that either $b, c > 1$ or $b, c < 1$. Then $ab + bc + ca - abc = a(b+c) + bc(1-a) \geq 0$. From our WLOG assumption, we have $(b-1)(c-1) \geq 0$, so $a = \dfrac{-bc + \sqrt{(4-b^2)(4-c^2)}}{2}$ by Quadratic Formula. Then we can apply Cauchy-Schwarz to prove the upper bound and finish off the proof.*

Exercise 146. *(2018 USAMO P1) Let a, b, c be positive real numbers such that $a + b + c = 4\sqrt[3]{abc}$. Prove that*

$$2(ab + bc + ca) + 4\min(a^2, b^2, c^2) \geq a^2 + b^2 + c^2.$$

Solution 146. *WLOG assume $a \leq b \leq c$. The inequality is homogeneous, so we can assume $abc = 1$ and $a + b + c = 4$. We then must show $2ab + 2bc + 2ca \geq b^2 + c^2 - 3a^2$, which simplifies into $(a+b+c)^2 \geq 2b^2 + 2c^2 - 2a^2 \implies b^2 + c^2 - a^2 \leq 8$. This implies that $b^2 + c^2 - (4-b-c)^2 \leq 8$, or $8b + 8c - 2bc \leq 24 \implies 4(4-a) - \dfrac{1}{a} \leq 12 \implies (2a-1)^2 \geq 0$ which is true.*

Exercise 147. *In an extremely large group of people whose heights obey a standard normal distribution, approximately 95 percent of the group have heights lying between 64 and 76 inches, inclusive, with a mean height of 70 inches.*

   (a) *At least what fraction of people in the group must be at least 5 feet tall? (91/100)*

   (b) *At most what fraction of people in the group can be at least 6 feet tall? (troll; can't conclude anything for $k \leq 1$)*

   (c) *At least what fraction of people in the group must have heights between 58 and 79 inches, inclusive? (8/9, min of 8/9 and 15/16)*

If no statistically valid conclusion can be drawn, explain why.

Solution 147.

(a) *By the 68-95-99.7 rule of the standard normal distribution, the standard deviation is 3 inches. Hence, by Chebyshev's Inequality, at least* $1 - \dfrac{3}{10}^2 = \boxed{\dfrac{91}{100}}$ *of the population must lie within* $\dfrac{10}{3}$ *SDs of the mean.*

(b) *We cannot make any statistically valid conclusion, since it is entirely possible that the entire population is clustered so close to the mean that no one lies beyond 6 feet, or perhaps clustered between 70 and 76 inches.*

(c) *Chebyshev's Inequality says that the portion of the population lying within 4 SDs is at least* $\dfrac{15}{16}$, *and also that the minimum portion lying within 3 SDs is* $\dfrac{8}{9}$. *We do not know how many of the population lie between* -4 *and* -3 *SDs, so the answer is* $\boxed{\dfrac{8}{9}}$.

Exercise 148. *(2010 USA TST) If* a, b, c *are real numbers such that* $abc = 1$, *prove that*

$$\sum_{cyc} \frac{1}{a^5(b+2c)^2} \geq \frac{1}{3}$$

Solution 148. *Making the substitutions* $a = \dfrac{1}{x}, b = \dfrac{1}{y}, c = \dfrac{1}{z}$, *we have*

$$\sum_{cyc} \frac{x^3}{(2y+z)^2} \geq \frac{(x+y+z)^{3\,2}}{\sum(2y+z)}$$

$$= \frac{(x+y+z)^3}{9(x+y+z)^2} = \frac{x+y+z}{9} \geq \frac{1}{3}$$

Exercise 149. *(1976 USAMO P4) If the sum of the lengths of the six edges of a tri-rectangular tetrahedron* $PABC$ *(i.e.,* $\angle APB = \angle BPC = \angle CPA = 90°$*) is* S, *determine its maximum volume.*

Solution 149. *Let the side lengths* AP, BP, CP *be* a, b, c *respectively. Then* $S = a + b + c + \sqrt{a^2 + b^2} + \sqrt{b^2 + c^2} + \sqrt{c^2 + a^2}$, *and the volume* V *of the tetrahedron is* $\dfrac{abc}{6}$.

By AM-GM, $S \geq 3\sqrt[3]{abc} + \sqrt{2ab} + \sqrt{2bc} + \sqrt{2ca} \geq 3\sqrt[3]{abc} + 3\sqrt{2}\sqrt[3]{abc}$, *so* $\sqrt[3]{abc} \leq \dfrac{S}{3 + 3\sqrt{2}} = \dfrac{S(\sqrt{2}-1)}{3}$.

Then $\dfrac{abc}{6} \leq \boxed{\dfrac{S^3(\sqrt{2}-1)^3}{162}}$.

Extension Exercise 57. *(2000 USAMO P1) Call a real-valued function* f *very convex if*

$$\frac{f(x) + f(y)}{2} \geq f\left(\frac{x+y}{2}\right) + |x - y|$$

holds for all real numbers x *and* y. *Prove that no very convex function exists.*

Extension Solution 57. *For contradiction, assume the existence of a very convex function* f. *Observe that* f *convex* $\iff f(x) + c$ *convex for* $c \in \mathbb{R}$. *This implies* $f(0) = 0$.

Let $a = f(1), b = f(-1)$. *Then* $\dfrac{f(0) + f(2^{-n})}{2} \geq f(2^{-n-1}) + \dfrac{1}{2^n}$ *An induction argument shows that*

$f(2^{-n}) \leq \dfrac{a - 2n}{2^n}$ for $n \geq 0$. Along similar lines, $f(-2^{-n}) \leq \dfrac{b - 2n}{2^n}$. Hence, it follows that $f(2^{-n}) + f(-2^{-n}) \leq \dfrac{a + b - 4n}{2^n}$.

Take $n > \dfrac{a + b - 4}{4}$, which establishes the (strict) upper bound of $\dfrac{1}{2^{n-2}}$. But the very convex condition for f stipulates a lower bound of $\dfrac{1}{2^{n-2}}$, a contradiction. QED.

Extension Exercise 58. *(1999 IMO P2) Let $n \geq 2$ be a fixed integer.*

(a) *Find the least constant C such that for all non-negative real numbers $x_1, \ldots, x_n$,*

$$\sum_{1 \leq i < j \leq n} x_i x_j (x_i^2 + x_j^2) \leq C \left(\sum_{i=1}^{n} x_i \right)^4.$$

(b) *Determine when equality occurs for this value of C.*

Extension Solution 58. *Define $P_2 = \sum_{i=1}^{n} x_i^2$, $S_2 = \sum_{1 \leq i, j \leq n} x_i x_j$. Then*

$$\sum_{1 \leq i, j \leq n} x_i x_j (x_i^2 + x_j^2) \leq \sum_{1 \leq i, j \leq n} x_i x_j P_2 = \frac{(2\sqrt{2 P_2 S_2})^2}{8} = \leq \frac{(P_2 + 2S_2)^2}{8} = \frac{1}{8} \left(\sum_{i=1}^{n} x_i \right)^4$$

so $C = \boxed{1}8$. Equality holds for all but two of the x_i equal to 0, which reduces to showing that $(x_1 - x_2)^4 \geq 0$ (but this is, of course, trivial).

Extension Exercise 59. *If a, b, and c are real numbers with $\displaystyle\sum_{cyc} \frac{1}{1 + a} = 2$, show that*

$$\sqrt{3 + a + b + c} \geq \sqrt{a} + \sqrt{b} + \sqrt{c}.$$

Extension Solution 59. *This is equivalent to*

$$\sum_{cyc} \left(\frac{1}{1 + a} - 1 \right) = -1$$

$$\implies \sum_{cyc} \frac{a}{1 + a} = 1$$

We want to show that

$$\left(\sum_{cyc} \frac{a}{1 + a} \right) ((1 + a) + (1 + b) + (1 + c)) \geq (\sqrt{a} + \sqrt{b} + \sqrt{c})^2$$

but this follows from an application of Cauchy-Schwarz. Equality holds for $a = b = c = \dfrac{1}{2}$.

Section 6.5: Newton's Sums

Exercise 150. *If a and b are the roots of $x^2 + 2x + 19$, what is the value of $a^4 + b^4$?*

Solution 150. *We have $P_1 = a + b$, $P_2 = a^2 + b^2$, and so forth. Newton's Sums tell us that $P_1 + 2 = 0 \implies P_1 = -2$, $P_2 + 2P_1 + 38 = 0 \implies P_2 = -34$, $P_3 + 2P_2 + 19P_1 = 0 \implies P_3 = 106$, $P_4 + 2P_3 + 19P_2 = 0 \implies P_4 = \boxed{434}$.*

Exercise 151. *Let $P(x) = x^3 - 3x^2 + 6x - 9$ have roots r, s, t. Compute $r^5 + s^5 + t^5$.*

Solution 151. *(Nitty-gritty computational details left as an exercise to the reader; including them here is not especially instructive.) We can make good use of the symmetric sums for cubics, namely $S_1 = r + s + t$, $S_2 = rs + st + tr$, $S_3 = rst$. Then recursively apply Newton's Sums to obtain $r^5 + s^5 + t^5 = \boxed{108}$.*

Exercise 152. *Let $a + b + c = 1$, $a^2 + b^2 + c^2 = 2$, and $a^3 + b^3 + c^3 = 3$.*

(a) What is the value of abc?

(b) Find the smallest positive integer $n > 3$ such that $a^n + b^n + c^n$ is a positive integer.

Solution 152. *(a) For $k = 3$, we know from Newton's sums that $P_0 = 3$, $P_1 = e_1$, $P_2 = e_1 P_1 - 2e_2$, $P_3 = e_1 P_2 - e_2 P_1 + 3e_3$. Then $e_1 = 1, e_2 = -\dfrac{1}{2}$, and $e_3 = abc = \boxed{\dfrac{1}{6}}$.*

(b) We have $P_n = x^n + y^n + z^n$, $S_1 = x + y + z$, $S_2 = xy + yz + zx$, $S_3 = xyz$. Using Newtons sums, $P_4 = S_1 P_3 - S_2 P_2 - S_3 P_1 = \dfrac{25}{6}$, $P_5 = S_1 P_4 - S_2 P_3 - S_3 P_2 - S_4 P_1 = 6$. Hence the answer is $\boxed{5}$.

Exercise 153. *(2003 AIME II) Consider the polynomials $P(x) = x^6 - x^5 - x^3 - x^2 - x$ and $Q(x) = x^4 - x^3 - x^2 - 1$. Given that z_1, z_2, z_3, and z_4 are the roots of $Q(x) = 0$, find $P(z_1) + P(z_2) + P(z_3) + P(z_4)$.*

Solution 153. *Note that $\dfrac{P(x)}{Q(x)} = x^2 - x + 1$, so $P(z_1) = z_1^2 - z_1 + 1$. (This also goes for the roots of $Q(x)$.) Hence, $P(z_1) + P(z_2) + P(z_3) + P(z_4) = \sum_{cyc} z_1^2 - z_1 + 1$.*

Vietas formulas tell us that $z_1 + z_2 + z_3 + z_4 = 1$, and so Newtons Sums allow us to compute the sum of the roots of $Q(x)$ (doing so yields $s_2 = 3$, and so $P(z_1) + P(z_2) + P(z_3) + P(z_4) = \boxed{6}$).

Exercise 154. *Show that the only solutions over the complex numbers to the system*

$$x + y + z = 4$$
$$x^2 + y^2 + z^2 = 14$$
$$x^3 + y^3 + z^3 = 34$$

are $(-1, 2, 3)$ and its permutations.

Solution 154. *Let x, y, z be the roots of some cubic polynomial $P(x) = ax^3 + bx^2 + cx + d$. We have $P_1 = 4$, $P_2 = 14$, $P_3 = 34$, so $4a + b = 0$, $14a + 4b + 2c = 0$, $34a + 14b + 4c + 3d = 0$. The first two equations imply that $a = c$, $b = -4a$, so $d = 6a$. This is indeed the only satisfactory solution set. (Furthermore, assume WLOG that $a = c = 1$, so that $b = -4$ and $d = 6$; then $P(x) = x^3 - 4x^2 + x + 6$.)*

Exercise 155. *(1973 USAMO P4) Determine all the roots, real or complex, of the system of simultaneous equations*

$$x + y + z = x^2 + y^2 + z^2 = x^3 + y^3 + z^3 = 3$$

Solution 155. *Let $P(t) = t^3 - at^2 + bt - c$ have roots x, y, z. Then $P(x) + P(y) + P(z) = 3(1 - a + b - c) = 0$, so $P(1) = 0$. Thus, $x = 1$, $y = 1$, or $z = 1$; WLOG assume $x = 1$. We can then find that $y = z = 1$, so $(1, 1, 1)$ is the only solution.*

Extension Exercise 60. *(1988 USAMO P5) Let $p(x)$ be the polynomial $(1 - x)^a (1 - x^2)^b (1 - x^3)^c \cdots (1 - x^{32})^k$, where $a, b, \cdots, k$ are integers. When expanded in powers of x, the coefficient of x^1 is -2 and the coefficients of x^2, x^3, ..., x^{32} are all zero. Find k.*

Extension Solution 60. *The key observation in this problem is that reversing the coefficients of $P(x)$ turns it into a polynomial with reciprocal roots. Thus, let $q(x) = (x - 1)^a (x^2 - 1)^b (x^3 - 1)^c \cdots (x^{32} - 1)^k$, with a coefficient of -2 in front of x^{n-1}, and coefficients of 0 in front of x^{n-k} for $k \in [2, 32]$.*

By Newtons Sums, let P_n be the sum of the nth powers of the roots of $q(x)$. By Vietas, $P_1 = 2$ and

$$P_2 - 2P_1 = 0, P_3 - 2P_2 = 0, \ldots, P_{32} - 2P_{31} = 0 \implies P_n = 2^n, n \in [1, 32].$$

To compute a_{32}, note that the sum of the 32nd powers of the roots of q is the same as the sum of the 32nd powers of the roots of $(x-1)^{a_1}(x^2-1)^{a_2}(x^4-1)^{a_4}\cdots(x^{32}-1)^{a_{32}}$. The 32nd power of each of these is 1, so $P_{32} = 2^{32} = a_1 + 2a_2 + 4a_4 + 8a_8 + 16a_{16} + 32a_{32}$. Then $P_{16} = 2^{16} = a_1 + 2a_2 + 4a_4 + 8a_8 + 16a_{16}$. Thus, $a_{32} = \boxed{2^{27} - 2^{11}}$.

Section 6.6: Vieta Jumping

Exercise 156. Let x and y be positive integers such that $xy \mid (x^2 + y^2 + 1)$. Prove that $x^2 + y^2 + 1 = 3xy$.

Solution 156. Immediately, we observe that $(1,1)$, $(1,2)$, and $(2,5)$ are all solutions. If $x = y$, we have $x^2 \mid (2x^2 + 1)$, so $x = y = 1$. If $x \neq y$, suppose that (x, y) is a solution, with $1 \leq x < y$; let $x^2 + y^2 + 1 = k \cdot xy$.

Consider the quadratic equation $a^2 - kax + x^2 + 1 = 0$. Then $a = y$ is a solution, and the other root is $kx - y = \dfrac{x^2 + 1}{y} \in \mathbb{Z}$. Since this root is smaller than x, we have a smaller solution $\left(\dfrac{x^2+1}{y}, x\right)$.

As a result, we can descend from any solution until we reach $x = 1$, yielding $y \mid (y^2 + 2)$, or $y = 1, 2$. k remains constant, at $k = 3$ (as desired).

Exercise 157. (1981 IMO P3) Determine the maximum value of $m^2 + n^2$, where m and n are integers satisfying $m, n \in \{1, 2, \ldots, 1981\}$ and $(n^2 - mn - m^2)^2 = 1$.

Solution 157. First note that m and n are relatively prime, with $n \geq m$. Furthermore, (m, n) is a solution $\iff$ $(b, a+b)$ is a solution (since $(m+k)^2 - (m^2 + km) - m^2 = -(m^2 - km - k^2)$). If $n \in \mathbb{N}$, we require $n = 1, 2$, so we can backtrack for any given solution using the Euclidean Algorithm. By induction, all solutions have both components as Fibonacci numbers. Then the maximal value is $\boxed{987^2 + 1597^2}$.

Exercise 158. Find, with proof, the largest integer $n \leq 1000$ such that there exist integers $a, b \geq 0$ with $\dfrac{a^2 + b^2}{ab - 1} = n$.

Solution 158. Suppose (a, b) is a solution. Then we claim $\left(b, \dfrac{n+b^2}{a}\right)$ is also a solution. To prove this, consider the quadratic equation $x^2 - nbx + (b^2 - n) = 0$. If a is a root, then by Vietas, so is $\dfrac{n+b^2}{a} = \dfrac{a^2 + ab^2}{a^2b - a}$. The numerator is a multiple of $ab - 1$, as well as of a, and since a and $ab - 1$ are relatively prime, $a(ab-1) \mid (a^2 + ab^2)$. Thus, the root is an integer, as desired.

In addition, we claim that if $a \geq b \geq 3$, then $ab > b^2 + n$. Since $b \geq 3$, $2b < b^2 - 1$, and so

$$\frac{b^3 + b}{b^2 - 1} = b + \frac{2b}{b^2 - 1} < b + 1 \leq a$$

$$\implies b^3 - b < a^2 - a \implies a + b^3 < ab^2 - b \implies b > \frac{a + b^3}{ab - 1} \implies b > \frac{b^2 + n}{a}$$

Let (a, b) be the desired solution; we necessarily have $0 \leq b \leq 2$. If $b = 0$, then $n \leq 0$. If $b = 1$, $a = 0, 2, 3$ and $n = -1, 5, 5$ respectively. Finally, if $b = 2$, then $17 \mid (2a - 1)$ and $a = 1, 9 \implies n = 5$ in both cases. Thus, $n \leq \boxed{5}$ (note that the 1000 restriction is, in fact, a red herring).

Exercise 159. (Brilliant.org) Find the number of pairs of non-negative integers (n, m) such that $1 \leq n < m \leq 100$, $n \mid (m^2 - 1)$, and $m \mid (n^2 - 1)$.

Solution 159. We first show that the ordered pair $(k, n) = \left(\dfrac{n^2 - 1}{m}, n\right)$ satisfies $1 \leq k \leq n$, $k \mid (n^2 - 1)$, and $n \mid (k^2 - 1)$.

Note that $m^2 \equiv 1 \pmod{n}$ and $mk \equiv -1 \pmod{n}$, so $k^2 \equiv 1 \pmod{n}$. Applying this multiple times produces smaller pairs that eventually lead to $(1, m)$ for some $c > 1$. (We can also perform this construction in reverse, by setting $m = \dfrac{n^2 - 1}{k}$ iteratively.) By inspection, there are $\boxed{208}$ pairs: 99 with $n = 1$, 98 with $m = n + 1$, $n \geq 2$, and 11 more.

Extension Exercise 61. *(2007 IMO P5) Let a and b be positive integers. Show that if $4ab - 1$ divides $(4a^2 - 1)^2$, then $a = b$.*

Extension Solution 61. *Assume that (a, b) is a solution. Then note that $(4ab - 1) \mid b^2(4a^2 - 1)^2 - (4ab - 1)(4a^3b - 2ab + a^2) = (a - b)^2$.*

Fix some positive constant $c = \dfrac{(a - b)^2}{4ab - 1}$, and consider the solutions (x, y) to $c = \dfrac{(a - b)^2}{4xy - 1}$. Let (m, n) be the ordered pair with $m + n$ minimal. WLOG assume $m > n$, and consider $x^2 - (2n + 4kn)x + n^2 + k = 0$. This equation has 2 solutions, namely m and $2n + 4kn - m = \dfrac{n^2 + k}{m}$. We have $n \geq m$, which contradicts $m > n$.

Hence, there are no solutions with $a \neq b$, and so $a = b$.

Extension Exercise 62. *(1988 IMO P6)[82] Let a and b be positive integers such that $ab + 1$ divides $a^2 + b^2$. Show that $\dfrac{a^2 + b^2}{ab + 1}$ is the square of an integer.*

Extension Solution 62. *Classic example of Vieta jumping! Here, we employ the tactic of choosing integers a, b, k with $a^2 + b^2 = k(ab + 1)$. Fix k, and among all pairs (a, b), pick the one with the smallest $\min(a, b)$. Let $a' = \max(a, b)$, $b' = \min(a, b)$. Then $a'^2 - ka'b' + b'^2 - k = 0$ is quadratic in a'. Assuming the existence of another root c', we would have $b'c' \leq a'c' = b^2 - k < b'^2$, which implies $c' < b'$. Thus, $c' \notin \mathbb{N}$. But $c' = kb' - a'$, so we do have $c' \in \mathbb{Z}$. Hence, $c' \leq 0$.*

We also have $(a' + 1)(c' + 1) = a'c' + a' + c' + 1 = b'^2 - k + b'k + 1 = b'^2 + k(b' - 1) + 1 \geq 1$, so $c' > -1$. It follows that $c' = 0$ exactly, and subsequently that $b'^2 = k$. Thus, k is always a perfect square.

Section 6.7: Complex Numbers, Part 2

Exercise 160. *(2018 AIME I) Let N be the number of complex numbers z with the properties that $|z| = 1$ and $z^{6!} - z^{5!}$ is a real number. Find the remainder when N is divided by 1000.*

Solution 160. *Let $a = z^{120}$. Then $a^6 - a \in \mathbb{R} \implies Im(a^6) = Im(a)$. Let $\theta = \arg(a)$ in the complex plane. Observe that, for each $0 \leq \theta < 2\pi$, there is exactly one a associated to θ. This holds for 12 values of a. Thus, there are $120 \cdot 12 = 1440$ solutions for $z \implies \boxed{440}$.*

Exercise 161. *(2014 AIME I) Let w and z be complex numbers such that $|w| = 1$ and $|z| = 10$. Let $\theta = \arg\left(\dfrac{w - z}{z}\right)$. The maximum possible value of $\tan^2 \theta$ can be written as $\dfrac{p}{q}$, where p and q are relatively prime positive integers. Find $p + q$. (Note that $\arg(w)$, for $w \neq 0$, denotes the measure of the angle that the ray from 0 to w makes with the positive real axis in the complex plane.)*

Solution 161. *Observe that $\left|\dfrac{w}{z}\right| = \dfrac{1}{10}$, so $\dfrac{w - z}{z} = \dfrac{w}{z} - 1$ lies on the circle γ with radius $\dfrac{1}{10}$ and center at $(-1, 0)$ in the complex plane.*

Let P, Q be the points $\dfrac{w}{z} - 1$, -1 in $\mathbb{C}^2$, and let O be the origin. To maximize $\tan^2(\theta)$, we must maximize $\angle POQ$, which occurs when $\overline{PO}$ is tangent to γ. By the Pythagorean Theorem, $PO^2 = \dfrac{99}{100}$, or $\tan^2(\theta) = \dfrac{1}{99} \implies \boxed{100}$.

---

[82]Yes, the notorious, ultra-difficult one that was once considered the hardest IMO problem ever. That one!

Exercise 162. *(2017 AIME I)* Let $z_1 = 18 + 83i$, $z_2 = 18 + 39i$, and $z_3 = 78 + 99i$, where $i = \sqrt{-1}$. Let z be the unique complex number with the properties that $\dfrac{z_3 - z_1}{z_2 - z_1} \cdot \dfrac{z - z_2}{z - z_3}$ is a real number and the imaginary part of z is the greatest possible. Find the real part of z.

Solution 162. *This can be bashed algebraically, by first computing* $\dfrac{z_3 - z_1}{z_2 - z_1} = \dfrac{15i - 4}{11}$. *Let* $z = a + bi$; *then* $\dfrac{z - z_2}{z - z_3} = \dfrac{(a - 18) + i(b - 39)}{(a - 78) + i(b - 99)}$. *Proceed to algebra bash and set the product equal to a real number (i.e. the imaginary part equal to zero). Eventually, we arrive at* $a^2 - 112a + b^2 - 122b + \dfrac{1989}{5} = 0$, *which upon completing the square yields* $a = \boxed{56}$ *upon maximizing* $(b - 61)^2$.

Exercise 163. *Prove that for any* $x, y, z \in \mathbb{C}$, $|x - y| = |y - z| = |z - x| \implies (x-y)^2 + (y-z)^2 + (z-x)^2 = 0$.

Solution 163. *Let* $x = a + bi$, $y = c + di$, $z = e + fi$. *Then* $|x - y| = |y - z| = |z - x| \implies (a - c)^2 + (d - b)^2 = (e - c)^2 + (f - d)^2 = (a - e)^2 + (b - f)^2$. *It follows through algebraic manipulation that* $(x - y)^2 + (y - z)^2 + (z - x)^2 = 0$.

Exercise 164. *(2001 AIME II)* There are $2n$ complex numbers that satisfy both $z^{28} - z^8 - 1 = 0$ and $|z| = 1$. These numbers have the form $z_m = \cos\theta_m + i\sin\theta_m$, where $0 \leq \theta_1 < \theta_2 < \ldots < \theta_{2n} < 360$ and angles are measured in degrees. Find the value of $\theta_2 + \theta_4 + \ldots + \theta_{2n}$.

Solution 164. *Let* $z = \operatorname{cis}\theta$, *and rearrange to form* $\operatorname{cis}(28\theta) = \operatorname{cis}(8\theta) + 1$. *As* $Re(\operatorname{cis}(28\theta)) = Re(\operatorname{cis}(8\theta)) + 1$, *but their imaginary parts are equal, we either have* $\operatorname{cis}(28\theta) = \dfrac{1 + i\sqrt{3}}{2}$, $\operatorname{cis}(8\theta) = \dfrac{-1 + i\sqrt{3}}{2}$, *or* $\operatorname{cis}(28\theta) = \dfrac{1 - i\sqrt{3}}{2}$, $\operatorname{cis}(8\theta) = \dfrac{-1 - i\sqrt{3}}{2}$. *In the first case,* $z^{28} = \operatorname{cis}60°$ *and* $z^8 = \operatorname{cis}120°$. *This yields solutions of* $15°, 105°, 195°, 285°$. *In the second case,* $z^{28} = \operatorname{cis}300°$ *and* $z^8 = \operatorname{cis}240°$, *which yields solutions of* $75°, 165°, 255°, 345°$. *Hence,* $\theta_2 + \theta_4 + \ldots + \theta_{2n} = 75° + 165° + 255° + 345° = \boxed{840°}$.

Exercise 165. *(2016 AIME I)* For integers a and b consider the complex number

$$\frac{\sqrt{ab + 2016}}{ab + 100} - \left(\frac{\sqrt{|a + b|}}{ab + 100}\right) i$$

Find the number of ordered pairs of integers (a, b) such that this complex number is a real number.

Solution 165. *We either require* $ab \geq -2016$, $\dfrac{\sqrt{|a + b|}}{ab + 100} = 0$ *or* $ab < -2016$, $\dfrac{\sqrt{-ab - 2016} - \sqrt{|a + b|}}{ab + 100} = 0$.

In the former, $-\dfrac{\sqrt{|a + b|}}{ab + 100} = 0 \implies a = -b, ab = -a^2 \implies a^2 < 2016$. *Thus,* $a \in [-44, 44]$, *with the exception of* $a = \pm 10$. *This yields 87 values in this case.*

In the latter, we have $2016 - ab = |a + b|$. *If* $a > 0$ *but* $b < 0$, *let* $-b = c$ *and if* $c > a$, $ac - 2016 = c - a$ *yields 4 cases by SFFT. Furthermore, if* $c < a$, *SFFT again yields 4 additional cases. Finally, if* $a < 0$, $b > 0$, *there are another 8 solutions, for 16 in this case, and a total of* $\boxed{103}$ *ordered pairs.*

Exercise 166. *(1999 AIME)* A function f is defined on the complex numbers by $f(z) = (a + bi)z$, where a and b are positive numbers. This function has the property that the image of each point in the complex plane is equidistant from that point and the origin. Given that $|a + bi| = 8$ and that $b^2 = m/n$, where m and n are relatively prime positive integers. Find $m + n$.

Solution 166. *For* $z = 1$, $f(1) = a + bi$. *This means that* $a = \dfrac{1}{2}$, *by equidistance. Since* $|a + bi| = 8$, *we have* $b^2 = \dfrac{255}{4} \implies \boxed{259}$.

Exercise 167. *(2018 HMMT February Algebra/Number Theory Round) Let ω_1, ω_2, ..., ω_{100} be the roots of $\dfrac{x^{101} - 1}{x - 1}$ (in some order). Consider the set*

$$S = \{\omega_1^1, \omega_2^2, \omega_3^3, \ldots, \omega_{100}^{100}\}$$

Let M be the maximum possible number of unique values in S, and let N be the minimum possible number of unique values in S. Find MN.

Solution 167. *First, let $N = 1$, and ω a 101st (primitive) root of unity. Further, let $\omega_n = \sqrt[n]{\omega}$. Then S really only has one element, namely ω.*

$M = 100$ is not possible (briefly, if we fix a 101st root of unity, we discover a contradiction modulo 100). However, $M = 99$ works, since we can actually attain $\dfrac{n}{n+1} = \dfrac{m}{m+1} \iff m = n$. Hence, our final answer is $99 - 1 = \boxed{98}$.

Exercise 168. *(2017 HMMT November General Test) Given that a, b, c are integers with $abc = 60$, and that complex number $\omega \neq 1$ satisfies $\omega^3 = 1$, find the minimum possible value of $|a + b\omega + c\omega^2|$.*

Solution 168. *ω is a third root of unity, so let $z = a + b\omega + c\omega^2$, with $\bar{z} = a + c\omega + b\omega^2$ (and $z\bar{z} = |z|^2$). Through some computation, it follows that $|z|^2 = \dfrac{(a-b)^2 + (b-c)^2 + (c-a)^2}{2}$, using the fact that $\omega + \omega^2 = 1$. This is minimal when a, b, c are as close together as possible, namely $(a, b, c) = (3, 4, 5)$. Hence, $|z|^2 = 3$ and $|z| = \boxed{\sqrt{3}}$.*

Extension Exercise 63. *(2012 USAMO P5) Let P be a point in the plane of triangle ABC, and γ a line passing through P. Let A', B', C' be the points where the reflections of lines PA, PB, PC with respect to γ intersect lines BC, AC, AB, respectively. Prove that A', B', C' are collinear.*

Extension Solution 63. *(Note: The most enriching part of this exercise is arguably scanning through the below and thinking about how to relate it to what we have learned about complex coordinates and "bashing!" We don't want to give it away or make it too crystal-clear to the point where it's no longer fun or meaningful.)*

Via Law of Sines on $\triangle ABP$, we obtain $AB = AP \cdot \dfrac{\sin \angle APB}{\sin \angle ABP}$. Similarly, we obtain alike expressions for BC, CA, AB, BC, CA. It follows that $\dfrac{AB}{BC} \cdot \dfrac{CA}{AB} \cdot \dfrac{BC}{CA}$ is a product of sine ratios simplifying to 1. Menelaus Theorem finishes the problem, with A, B, C collinear.

Extension Exercise 64. *Let $G = \{z \in \mathbb{C} \mid \exists n \in \mathbb{N} \text{ s.t. } z^n = 1\}$. Prove that G is a group under multiplication.*

Extension Solution 64. *We have the identity 1, an inverse, closure under multiplication, and associativity. The identity is self-explanatory; the inverse results from the fact that if x is an nth root of unity, then so is $\dfrac{1}{x}$, because $\left(\dfrac{1}{x}\right)^n = \dfrac{1}{x^n} = 1$. The closure results from having two roots of unity, namely x, y, and having xy also a root of unity. Finally, associativity is inherited from $\mathbb{C}$ as a whole.*

Extension Exercise 65. *Let $z_1, z_2, z_3 \in \mathbb{C}$ be distinct. When are z_1, z_2, z_3 the vertices of an equilateral triangle in the complex plane?*

Extension Solution 65. *We claim that this occurs iff $z_1^2 + z_2^2 + z_3^2 = z_1 z_2 + z_2 z_3 + z_3 z_1$. A sufficient condition for this is that $z_2 - z_1$ is at a $60°$ angle to $z_3 - z_1$, and likewise for $z_1 - z_3$ to $z_2 - z_3$. After converting to polar form using Euler's formula $e^{i\pi} + 1 = 0$, we thus have $\dfrac{z_2 - z_1}{z_1 - z_3} = \dfrac{z_3 - z_1}{z_2 - z_3}$, which implies the desired result.*

The necessary condition. Let $z_1^2 + z_2^2 + z_3^2 = z_1 z_2 + z_2 z_3 + z_3 z_1$. Then rearranging/manipulating yields $\dfrac{z_2 - z_1}{z_1 - z_3} = \dfrac{z_3 - z_1}{z_2 - z_3}$, which implies that $z_2 - z_1, z_3 - z_1$ are at the same angle to each other as are $z_1 - z_3$ and $z_2 - z_1$. It follows that all three angles of the triangle are equal, which makes it equilateral.

Extension Exercise 66. *(2014 IMO Shortlist G6) Let $\triangle ABC$ be a fixed acute-angled triangle. Consider some points E and F lying on the sides $\overline{AC}$ and $\overline{AB}$, respectively, and let M be the midpoint of $\overline{EF}$. Let the perpendicular bisector of $\overline{EF}$ intersect the line $\overline{BC}$ at K, and let the perpendicular bisector of $\overline{MK}$ intersect the lines $\overline{AC}$ and $\overline{AB}$ at S and T, respectively. We call the pair (E, F) interesting, if the quadrilateral $KSAT$ is cyclic. Suppose that the pairs (E_1, F_1) and (E_2, F_2) are interesting. Prove that $\dfrac{E_1 E_2}{AB} = \dfrac{F_1 F_2}{AC}$.*

Extension Solution 66. *Consider the diagram in the complex plane with A at the origin. Let $\triangle EFK$ be an interesting triangle, with $\angle KEF = \angle KFE = \angle A$ and $\dfrac{K-E}{F-E}$ invariant over all interesting triangles. Hence, $K = \alpha \cdot E + \beta \cdot F$, where $\alpha + \beta = 1$ are scalar weights summing to 1.*

Choose points X, Y on $\overline{AB}, \overline{AC}$ respectively such that $\angle CXA = \angle AYB = \angle KEF = \angle A$. Then $\triangle AXC, \triangle YAB$ are both similar to any interesting triangle. Thus, it follows that $C = \alpha \cdot A + \beta \cdot X = \beta \cdot X$, $B = \alpha \cdot Y + \beta \cdot A = \alpha \cdot Y$. (Also $\dfrac{X}{Y} = \dfrac{\overline{C}}{\overline{B}}$.)

One also derives $E = \gamma \cdot Y, F = \delta \cdot X, K = \lambda \cdot B + (1-\lambda) \cdot C$ for some reals γ, δ, λ. We now have $B(\lambda - gamma) = C(\delta + \lambda - 1)$. The coefficients vanish (as $|B| \neq |C|$), so $\lambda = \gamma$, $\delta = 1 - \lambda$. This means that $\dfrac{E}{Y} + \dfrac{F}{X} = 1$.

If (E_1, F_1) and (E_2, F_2) are different interesting pairs, then we get $\dfrac{E_1 - E_2}{Y} = \dfrac{F_1 - F_2}{X}$, so $\dfrac{E_1 - E_2}{F_1 - F_2} = \dfrac{Y}{X} = \dfrac{\overline{B}}{\overline{C}}$. This leads to the desired conclusion.

Section 6.8: Matrix Algebra, Part 2

Exercise 169. *Show that A and A^T have the same eigenvalues.*

Solution 169. *We use the fact that, for any matrix A, $\det A = \det A^T$. Note that the eigenvalues λ come from solving the equation $\det(A - \lambda I) = 0$, and $(A - \lambda I)^T = A^T - \lambda \cdot I^T = A^T - \lambda \cdot I$. Hence, $\det(A - \lambda I) = \det(A^T - \lambda \cdot I)$. It follows that A and A^T have the same characteristic polynomial, so they must have the same eigenvalues.*

Exercise 170. *What are all eigenvalues and eigenvectors of*

$$\begin{bmatrix} 0 & 6 & 0 \\ 2 & 1 & 0 \\ 0 & 2 & 1 \end{bmatrix}$$

Solution 170. $\lambda = 4, -3, 1$, $v_1 = (9, 6, 4)$, $v_2 = (4, -2, 1)$, $v_3 = (0, 0, 1)$.

Exercise 171. *Let $x \in [0, 1]$, and define $f(x) = x^2$, $g(x) = 1 - cx$, $c \in \mathbb{R}$. For what value of c is $f \perp g$ in the inner product space $C([0, 1])$ of continuous, real-valued functions on $[0, 1]$?*

Solution 171. *Recall that, over $C([a, b])$, the inner product is defined as*

$$\langle f, g \rangle = \int_a^b f(x)\overline{g(x)}\, dx$$

Hence, we want to find $c \in \mathbb{R}$ such that

$$\int_0^1 x^2 - cx^3\, dx = 0$$

$$\implies \frac{x^3}{3} - \frac{cx^4}{4} = 0 \implies c = \boxed{\frac{4}{3}}$$

Exercise 172. *Obtain an orthonormal basis for the set of vectors $S = \{v_1 = \langle 2, 4 \rangle, v_2 = \langle 5, 3 \rangle\}$.*

Solution 172. *This is a computational example involving the Gram-Schmidt algorithm. We have* $u_1 = v_1 = \langle 2, 4 \rangle$, $u_2 = v_2 - \dfrac{\langle u_1, v_2 \rangle}{\langle u_1, u_1 \rangle} u_1 = \langle 3, 5 \rangle - \dfrac{26}{20} \cdot \langle 2, 4 \rangle = \langle \dfrac{2}{5}, -\dfrac{1}{5} \rangle$. *Finally, we have* $e_1 = \dfrac{u_1}{||u_1||} = \langle \dfrac{\sqrt{5}}{5}, \dfrac{2\sqrt{5}}{5} \rangle$, $e_2 = \langle \dfrac{2\sqrt{5}}{5}, -\dfrac{\sqrt{5}}{5} \rangle$.

Exercise 173. *Let V be a finite-dimensional inner product space. Prove that V has an orthonormal basis.*

Solution 173. *Suppose V is finite-dimensional, and choose a basis of V. Apply the GramSchmidt Algorithm to it; this will produce an orthonormal list with $\dim V$ elements. This orthonormal list is then an orthonormal basis of V.*

Exercise 174. *Show that $S = \{(x, 4x) \mid x \in \mathbb{R}\}$ is a vector space.*

Solution 174. *The elements of S are associative and commutative; these are not hard to verify. There additionally exists the additive identity element 0, as well as the inverse $(-x, -4x)$. In addition, there is the multiplicative identity 1, in tandem with compatibility of scalar multiplication. Finally, S is equipped with the distributive property, so it is a vector space.*

Extension Exercise 67. *(2013 USAMO P2) For a positive integer $n \geq 3$ plot n equally spaced points around a circle. Label one of them A, and place a marker at A. One may move the marker forward in a clockwise direction to either the next point or the point after that. Hence there are a total of $2n$ distinct moves available; two from each point. Let a_n count the number of ways to advance around the circle exactly twice, beginning and ending at A, without repeating a move. Prove that $a_{n-1} + a_n = 2^n$ for all $n \geq 4$.*

Extension Solution 67. *Choose the tilings for the first/second passes around the circle simultaneously. We consider whether we are in a continuation of a length-2 block on both passes. Using the idea of Markov transition matrices, we can set up the matrix*

$$\begin{bmatrix} 0 & 1 & 1 \\ 1 & 0 & 1 \\ 1 & 1 & 0 \end{bmatrix}$$

representing the three states (non, non), (non, cont), (cont, cont). Then $a_n = T_{1,1}^n + T_{2,3}^n$; it suffices at this point to compute T^n. We have

$$T_n = (\mathbf{1} - I)^n = \sum_{i=0}^{n} \mathbf{1}^i (-1)^{n-i} \binom{n}{i} =$$

$$\mathbf{1} \sum_{i=1}^{n} 3^{i-1} (-1)^{n-i} = \frac{1}{3} \mathbf{1} \cdot (2^n - (-1)^n) + I \cdot (-1)^n$$

which eventually leads to the desired conclusion, $a_n + a_{n-1} = 2^n$. ($\mathbf{1}$ is the matrix of all ones.)

Extension Exercise 68. *Let A and B be linear transformations on a vector space V with $\dim V < \infty$. Show that $\dim \ker AB \leq \dim \ker A + \dim \ker B$.*

Extension Solution 68. *Let $\{e_1, e_2, e_3, \ldots, e_n\}$ be a basis of $\ker B$. As $\ker B$ is a subspace of $\ker AB$, there exist vectors $v_1, v_2, v_3, \ldots, v_n$ with $\{e_1, e_2, e_3, \ldots, v_1, v_2, v_3, \ldots, v_n\}$ forming a basis for $\ker AB$. We then have that $\dim \ker AB = m + n$.*

Henceforth, note that the set of Bv_i is linearly independent. If there exist scalars λ_i with $\sum_i \lambda_i B v_i = 0$, then $\sum_i \lambda_i v_i \in \ker B \iff \lambda_i = 0$ for all i. As $\{Bv_1, Bv_2, \ldots, Bv_n\} \in \ker A$, we have $\dim \ker A \geq m$. Therefore, $\dim \ker A \geq \dim \ker AB - \dim \ker B$, as desired.

Extension Exercise 69. *Is there a matrix $M \in \mathcal{M}_{n \times n}(\mathbb{R})$ with trace 0 and $M^2 + M^T = I$?*

Extension Solution 69. *We claim no such matrix exists. For contradiction, assume its existence; then it would follow that $M^T = I - M^2$, or $MM^T = M - M^3 = M^T M$. Thus, M is normal, and there exists a unitary matrix U and a diagonal matrix D such that $M = UDU^*$. Hence, $I = A^2 - A^T = A^2 - A^* = UD^2U^* - UDU^*$, or equivalently, $I = D^2 - D$. As D is a diagonal matrix, all of its diagonal entries $d_1, d_2, d_3, \ldots, d_n$ are such that $d_i^2 - d_i = 1$ for $i \in [1, n]$. Thus, $d_i = \dfrac{1 \pm \sqrt{5}}{2}$, but no combination of these can satisfy $\sum_i d_i = 0$.*

Extension Exercise 70. *(1997 IMO P4) Call an $n \times n$ matrix with elements from the set $S = \{1, 2, \ldots, 2n-1\}$ a silver matrix if, for each $i = 1, 2, \ldots, n$, the ith row and the ith column together contain all elements of S. Show that no silver matrix exists for $n = 1997$, and furthermore, that silver matrices exist for infinitely many values of n.*

Extension Solution 70. *In general, let M be an $n \times n$ silver matrix. For $i \in [1, n]$, let S_n be the set consisting of elements in the nth row and in the nth column, not including the diagonal element $M_{n,n}$. Let $e = M_{i,j}$ be an off-diagonal element, which is subsequently in both S_i and S_j. Hence, e partitions the S_i into pairs of S_i, S_j, so n would need to be even, which 1997 is not - contradiction.*

To show the second statement, as before, let M be a silver matrix with dimensions $n \times n$. We can construct a $2n \times 2n$ silver matrix by positioning two "copies of M on the diagonal, and two other matrices A and B with A having entries from $2n$ to $3n - 1$, and with B having entries from $3n$ to $4n - 1$. This new matrix is indeed a $2n \times 2n$ silver matrix, since it contains all elements from 1 to $2n$ in each of its rows and columns. We can repeat this process infinitely; QED.

Chapter 6 Review Exercises

Exercise 175. *(2002 AMC 12A) Consider the sequence of numbers: $4, 7, 1, 8, 9, 7, 6, \ldots$ For $n > 2$, the nth term of the sequence is the units digit of the sum of the two previous terms. Let S_n denote the sum of the first n terms of this sequence. Compute the smallest value of n for which $S_n > 10,000$.*

Solution 175. *The sequence repeats as follows: $4, 7, 1, 8, 9, 7, 6, 3, 9, 2, 1, 3, 4, 7, 1, 8, \ldots$ The sequence repeats every 12 terms, and in each block of 12, the sum is 60. Hence, $S_{1992} = 9960$. $4+7+1+8+9+7+6 = 42 > 40$ for the first time, so the answer is $1992 + 7 = \boxed{1999}$.*

Exercise 176. *Let s_n be the positive integer consisting of the first n triangular numbers concatenated (strung together), so that $s_1 = 1$, $s_2 = 13$, $s_3 = 136$, and $s_4 = 13610$. Let r_n be the remainder when s_n is divided by 9. Find the remainder when $\displaystyle\sum_{k=1}^{2018} r_k$ is divided by 1000.*

Solution 176. *Use recursion, as well as the fact that the remainder when dividing by 9 is the same as the remainder when dividing the sum of digits by 9. Eventually, we obtain a cycle, which leads us to form blocks and sum them (as in the previous exercise). The answer is $\boxed{72}$.*

Exercise 177. *(2007 AMC 12A) Call a set of integers spacy if it contains no more than one out of any three consecutive integers. How many subsets of $\{1, 2, 3, \ldots, 12\}$, including the empty set, are spacy?*

Solution 177. *This is an exercise in recursion/dynamic programming, as the spacy subsets of S_{n+1} are either in a group G_1 not containing $n + 1$, or in the group G_2 containing $n + 1$. Clearly, $|G_1| = S_n$. Then $|G_2| = S_{n-2}$, as if we remove $n + 1$, we end up with another spacy set, whose elements are at most $n - 2$. Thus, $S_{n+1} = S_n + S_{n-2}$, and this recursion leads to the answer of $\boxed{129}$.*

Exercise 178. *What is the interval of positive k such that the quartic $x^4 + 10x^3 + kx^2 + 10x + 1$ has all real roots?*

Solution 178. *Dividing through by x^2, we get $x^2 + 10x + k + \dfrac{10}{x} + \dfrac{1}{x^2} = 0$. Make the substitution $t = \dfrac{x}{1} + \dfrac{1}{x}$ to obtain $t^2 + 10t + (k - 2) = 0$, and then use Quadratic Formula to get $t = -5 \pm \sqrt{27 - k}$. Clearly, $k \leq \boxed{27}$. (We must also verify that x is real, which it is for all $k \leq 27$.)*

Exercise 179. *(2015 AMC 12A) For each positive integer n, let $S(n)$ be the number of sequences of length n consisting solely of the letters A and B, with no more than three As in a row and no more than three Bs in a row. What is the remainder when $S(2015)$ is divided by 12?*

Solution 179. *Another exercise in recursion/DP, only we must also consider the remainders. Let $A(n)$ be the number of length-n sequences ending with an A, and define $B(n)$ analogously. Observe that $A(n) = B(n)$ for all n, and $S(n) = A(n) + B(n)$, so $S(n) = 2A(n)$. Any sequence that is counted as part of $A(n)$ must consist of a series of As after a string ending with a B, which is where the recursion comes in. Indeed, since there can be no more than three consecutive Bs, we have the recurrence $A(n) = B(n-1)+B(n-2)+B(n-3) = A(n-1) + A(n-2) + A(n-3)$. This is the tough part; from here, we can just make a table of values, spot the pattern, and use some elementary modular arithmetic, to arrive at the answer of $\boxed{8}$.*

Exercise 180. *Let T_n be the nth triangular number. Define $Q(n)$ to be the numerator of the quotient of T_n and $n!$ in simplest form. Given that the sum of all 306 prime numbers from 1 to 2019, inclusive, is 283,081, find the remainder when*

$$\sum_{k=1}^{2019} Q(k)$$

is divided by 1000.

Solution 180. *Note that $T_n = \dfrac{n(n+1)}{2}$, so $Q(n) = \dfrac{n+1}{2(n-1)!}$. Note that, whenever $n+1$ is prime, it will not divide $(n-1)!$, so $Q(n+1) = n+1$. For $n = 1$, clearly $Q(n) = 1$, and for all other n, $Q(n) = 1$ as well. Thus, we sum the primes from 2 to 2019, but subtract 2, since $Q(1) = 1$ and not 2. Then we add the $2019 - 306 + 1 = 1714$ instances of 1 to obtain a remainder of $\boxed{793}$ (mod 1000).*

Exercise 181. *(1994 AIME) The equation*

$$x^{10} + (13x - 1)^{10} = 0$$

has 10 complex roots $r_1, \overline{r_1}, r_2, \overline{r_2}, r_3, \overline{r_3}, r_4, \overline{r_4}, r_5, \overline{r_5}$, where the bar denotes complex conjugation. Find the value of

$$\frac{1}{r_1\overline{r_1}} + \frac{1}{r_2\overline{r_2}} + \frac{1}{r_3\overline{r_3}} + \frac{1}{r_4\overline{r_4}} + \frac{1}{r_5\overline{r_5}}$$

Solution 181. *Dividing both sides by x^{10} yields $\left(13 - \dfrac{1}{x}\right)^{10} = -1$, so $13 - \dfrac{1}{x} = \omega = e^{i\left(\frac{\pi n}{5} + \frac{\pi}{10}\right)}$. Since $\dfrac{1}{x} = 13 - \omega$, we have $\dfrac{1}{x\overline{x}} = (13-\omega)(13-\overline{\omega}) = 169 - 13(\omega + \overline{\omega}) + \omega\overline{\omega} = 170 - 13(\omega + \overline{\omega})$. Taking the sum over all roots yields $5 \cdot 170 - 13\left(e^{\frac{i\pi}{10}} + \ldots + e^{i\frac{9\pi}{5} + \frac{\pi}{10}}\right)$. The sum of complex terms is just 0, so we are left with $5 \cdot 170 = \boxed{850}$.*

Exercise 182. *(2015 AIME II) There are $2^{10} = 1024$ possible 10-letter strings in which each letter is either an A or a B. Find the number of such strings that do not have more than 3 adjacent letters that are identical.*

Solution 182. *Rehash of 2015 AMC 12A #22! Note that $a_n = a_{n-1} + a_{n-2} + a_{n-3}$ for all $n \geq 4$, with $a_1 = 1, a_2 = 2, a_3 = 4$. Hence, $a_{10} = 274$, and $a_{10} + b_{10} = 2 \cdot 274 = \boxed{548}$.*

Exercise 183. *We can generalize the concept of Fibonacci numbers to the sequences of n-acci numbers, with $n > 1$ a positive integer. The n-acci sequence starts with n 1's, and has each successive term equal to the sum of the n terms before it. For example, the 3-acci sequence begins $1, 1, 1, 3, 5, 9, 17, \ldots$ For how many values of $m \leq 2019$ does an m-acci sequence contain a multiple of 13 less than or equal to $13m$?*

Solution 183. *We split the problem into two cases: one where $m \leq 13$, and one where $14 \leq m \leq 2019$.*

Case 1: $m \leq 13$. This case serves to establish a potential pattern for m, thereby simplifying computation. For $m = 2$, we obtain the Fibonacci sequence, which contains 13, and is less than $2 \cdot 13 = 26$. For

$m = 3$, however, the resulting sequence does not have this property. When $m \geq 4$, however, we can "shortcut the process temporarily by multiplying the preceding term by 2 and subtracting 1 to obtain the next term. Doing so, we obtain $m \equiv 0, 4, 7, 9 \pmod{13}$ as the satisfactory m-values. In addition, $m = 5$ works as well, giving 6 values for $m \leq 13$.

Case 2: Applying the result of Case 1 above, and setting an upper bound of $2015 = 13 \cdot 155$, we have 154 groups of 4 numbers out of every 13, which satisfy the condition. Thus, we add an extra 617 numbers, for a total of $\boxed{623}$ values of m.

Exercise 184. *(1990 AIME) Find $ax^5 + by^5$ if the real numbers a, b, x, and y satisfy the equations*

$$ax + by = 3$$
$$ax^2 + by^2 = 7$$
$$ax^3 + by^3 = 16$$
$$ax^4 + by^4 = 42$$

Solution 184. *A recurrence relation of the form $t_n = at_{n-1} + bt_{n-2}$ has closed form $ax^n + by^n$, with $t_1 = 3, t_2 = 7$, so $t_3 = ax^3 + by^3 = 16 = 7a + 3b$, and similarly, $t_4 = 42 = 16a + 7b$. Then $(a,b) = (-14, 38)$, so $t_5 = -14 \cdot 42 + 38 \cdot 16 = \boxed{20}$.*

Exercise 185. *Using generating functions, show that the closed form of F_n (the nth Fibonacci number, where $F_1 = F_2 = 1$) is*

$$F_n = \frac{1}{\sqrt{5}}\left(\left(\frac{1 + \sqrt{5}}{2}\right)^n - \left(\frac{1 - \sqrt{5}}{2}\right)^n\right)$$

Solution 185. *We have previously shown that the OGF of the Fibonacci sequence is $\dfrac{1}{1 - z - z^2}$. The roots of $1 - x - x^2$ are $A = \dfrac{-1 - \sqrt{5}}{2}$ and $B = \dfrac{-1 + \sqrt{5}}{2}$, so we can obtain a partial fraction decomposition of the OGF as $\dfrac{1}{\sqrt{5}}\left(\dfrac{B}{x + B} - \dfrac{A}{x + A}\right)$. Expressing each of the two terms as a geometric series yields the desired closed form, $F_n = \dfrac{1}{\sqrt{5}}(A^n - B^n)$.*

Exercise 186. *(2011 Stanford Math Tournament) How many polynomials P of degree 4 satisfy $P(x^2) = P(x)P(-x)$?*

Solution 186. *Observe that, if r is a root of P, then r^2 is also a root of P. Hence, $r, r^2, r^4, r^8, \ldots$ are all roots. P has a finite number of roots, so two should be equal. Hence, either $r = 0$ or $r^n = 1, n > 0$. If all roots are 0 or 1, then we can write $P = ax^b(x - 1)^{4-b}$, where $b \in [0, 4]$.*

Assume that this does not hold. Then let q be the largest integer with $r = e^{\frac{2p\pi i}{q}}$ a root with $\gcd(p, q) = 1$. The only satisfactory q are $q = 3, 5$.

This is because if r is a root, then so is either $\sqrt{r}$ or $-\sqrt{r}$. If q is even, then one of $e^{\frac{p\pi i}{q}}$ or $e^{\frac{2\pi i \cdot (p+q)}{2q}}$ is a root of p, but both $\gcd(p, q) = 1$ and $\gcd(p + q, 2q) = 1$ by the Euclidean Algorithm. Hence, q is not maximal. Thus, q must be odd, but if $q \geq 7$, then $\dfrac{1}{r^2}, \dfrac{1}{r}, r, r^2, r^4$ should be distinct. It follows that $q = 3$ or $q = 5$.

If we assume $q = 3$ first, then $(x^2 + x + 1) \mid P(x)$, and so $P(x) = c(x^2 + x + 1)Q(x)$ for $c \in \mathbb{R}$, $Q(x)$ a polynomial. Inductively, we can show that $Q(x) = x^2 + x + 1$ or $Q(x) = x^b(x - 1)^{2-b}$. If $q = 5$, however, then the roots of P are the 5th roots of unity, so let $P = c(x^4 + x^3 + x^2 + x + 1)$.

It remains to show that all 10 resulting polynomial classes actually satisfy the criteria; this is straightforward algebra. Indeed, there are exactly $\boxed{10}$ satisfactory quartic polynomials.

Exercise 187.

(a) Prove that, in any triangle $\triangle ABC$, $\sin A + \sin B + \sin C \le \dfrac{3\sqrt{3}}{2}$.

(b) What, if anything, can we conclude about the sums $\cos A + \cos B + \cos C$ and $\tan A + \tan B + \tan C$?

Solution 187.

(a) On the interval $x \in [0, \pi]$, $f(x) = \sin x$ is concave. By Jensen's Inequality, $\dfrac{f(A) + f(B) + f(C)}{3} \le$ $f\left(\dfrac{A+B+C}{3}\right) = \sin\left(\dfrac{\pi}{3}\right) = \dfrac{\sqrt{3}}{2}$. (Equality holds for $A = B = C = \dfrac{\pi}{3}$.)

(b) Note that $\cos x$ and $\tan x$ are also concave functions on $[0, \pi]$. Hence, Jensen's Inequality also guarantees upper bounds of $\dfrac{3}{2}$ for cos and $3\sqrt{3}$ for tan.

Exercise 188. *(2009 USAMO P4)* For $n \ge 2$ let $a_1, a_2, \ldots, a_n$ be positive real numbers such that $(a_1 + a_2 + \ldots + a_n)\left(\dfrac{1}{a_1} + \dfrac{1}{a_2} + \ldots + \dfrac{1}{a_n}\right) \le \left(n + \dfrac{1}{2}\right)^2$ Prove that $\max(a_1, a_2, \ldots, a_n) \le 4\min(a_1, a_2, \ldots, a_n)$.

Solution 188. WLOG assume that the a_i are in decreasing order. We show that $a_1 \le 4a_n$.

By Cauchy-Schwarz,

$$(a_n + a_2 + a_3 + \ldots + a_{n-1} + a_1)\left(\dfrac{1}{a_1} + \dfrac{1}{a_2} + \ldots + \dfrac{1}{a_n}\right) \ge \left(\sqrt{\dfrac{a_n}{a_1}} + (n-2) + \sqrt{\dfrac{a_1}{a_n}}\right)^2$$

$$\implies \left(n + \dfrac{1}{2}\right)^2 \ge \left(\sqrt{\dfrac{a_n}{a_1}} + (n-2) + \sqrt{\dfrac{a_1}{a_n}}\right)^2$$

$$\implies n + \dfrac{1}{2} \ge (n-2) + \sqrt{\dfrac{a_n}{a_1}} + \sqrt{\dfrac{a_1}{a_n}}$$

$$\implies \dfrac{5}{2} \ge \sqrt{\dfrac{a_n}{a_1}} + \sqrt{\dfrac{a_1}{a_n}}$$

$$\implies \dfrac{17}{4} \ge \dfrac{a_1}{a_n} + \dfrac{a_n}{a_1}$$

$$\implies 0 \ge (a_1 - 4a_n)\left(a_1 - \dfrac{a_n}{4}\right)$$

As $a_1 \ge a_n$, $a_1 - \dfrac{a_n}{4} > 0$, so $0 \ge a_1 - 4a_n \implies 4a_n \ge a_1$, as desired.

Exercise 189. *(1994 USAMO P4)* Let $a_1, a_2, a_3, \ldots$ be a sequence of positive real numbers satisfying $\sum_{j=1}^{n} a_j \ge \sqrt{n}$ for all $n \ge 1$. Prove that, for all $n \ge 1$,

$$\sum_{j=1}^{n} a_j^2 > \dfrac{1}{4}\left(1 + \dfrac{1}{2} + \cdots + \dfrac{1}{n}\right).$$

Solution 189. By minimizing a_j^2, we are effectively clustering the a_js as close together as possible. But (a_j) is increasing, so we should actually have $a_j = \sqrt{j} - \sqrt{j-1} = \dfrac{1}{\sqrt{j} + \sqrt{j+1}}$ (by radical conjugation). Note that $a_j > \dfrac{1}{2\sqrt{j}} > \left(\dfrac{1}{\sqrt{4j}}\right)^2$ so the sum we want to minimize is greater than $\sum_n \dfrac{1}{4n}$.

Exercise 190. *(2015 IMO P5)* Let $\mathbb{R}$ be the set of real numbers. Determine all functions $f : \mathbb{R} \mapsto \mathbb{R}$ satisfying the equation $f(x + f(x + y)) + f(xy) = x + f(x + y) + yf(x)$ for all real numbers x and y.

377

Solution 190. *Setting $x = y = 0$ yields $f(0 + f(0)) + f(0) = f(0) \implies f(f(0)) = 0$. If $f(0) = c$, then $f(c) = 0$.*

Setting $x = 0, y = c$. Then similarly, $f(0 + f(c)) + f(0) = 0 + f(c) + cf(0)$. We have $f(c) = 0$ and $f(0) = c$, so $f(0) + f(0) = f(0)^2 \implies f(0) = 0, 2$.

Should $f(0) = 0$, let $x = 0, y = x$. Then $f(f(x)) = f(x)$. Let $f(x) = z$; then $f(z) = z$, so $f(x) = x$ works.

If $f(0) = 2$, then we can use the same construction as before and get that $f(f(x)) + 2 = f(x) + 2x$. Then $\deg f = 1$, so it is linear and $f(x) = 2 - x$. This, along with $f(x) = x$, are the only solutions, so we are done.

Exercise 191. *(2001 IMO P2) Let $a, b, c \in \mathbb{R}^+$. Prove that $\dfrac{a}{\sqrt{a^2 + 8bc}} + \dfrac{b}{\sqrt{b^2 + 8ca}} + \dfrac{c}{\sqrt{c^2 + 8ab}} \geq 1$.*

Solution 191. *There are actually tons of solutions and different directions we can take! Here is one of many using Cauchy: Since the inequality is homogeneous, we may assume WLOG that $a + b + c = 3$. By Cauchy-Schwarz,*

$$\left(\sum_{cyc} \frac{a}{\sqrt{a^2 + 8bc}} \right) \left(\sum_{cyc} a\sqrt{a^2 + 8bc} \right) \geq (a + b + c)^2 = 9$$

We then want to show that

$$\sum_{cyc} a\sqrt{a^2 + 8bc} \leq 9$$

after some manipulation. This implies

$$\left(\sum_{cyc} a\sqrt{a^2 + 8bc} \right)^2 \leq 81$$

and another application of Cauchy shows that $\sum_{cyc} a^3 + 8bc \leq 27$. Then we have

$$(a + b + c)^3 - 3 \left(\sum_{sum} a^2 b \right) + 18abc \leq 27$$

and $a + b + c = 3$ implies $\left(\sum_{sym} a^2 b \right) \geq 6abc$ which is true by AM-GM. QED.

Extension Exercise 71. *(2010 Putnam B1) Is there an infinite sequence of real numbers $a_1, a_2, a_3, \ldots$ such that $a_1^m + a_2^m + a_3^m + \ldots = m$ for every positive integer m?*

Extension Solution 71. *The answer is $\boxed{no}$. For contradiction, if we assume existence, then Cauchy-Schwarz would immediately lead to*

$$8 = (a_1^2 + a_2^2 + \ldots)(a_1^4 + a_2^4 + \ldots) \geq (a_1^3 + a_2^3 + \ldots) \geq 9,$$

which is a contradiction.

Extension Exercise 72. *(2017 USAMO P6) Find the minimum possible value of*

$$\frac{a}{b^3 + 4} + \frac{b}{c^3 + 4} + \frac{c}{d^3 + 4} + \frac{d}{a^3 + 4}$$

given that a, b, c, d are nonnegative real numbers such that $a + b + c + d = 4$.

Extension Solution 72. *Subtract* $\dfrac{a+b+c+d}{4} = 1$ *from both sides to obtain*

$$\sum_{cyc} \frac{ab^3}{b^3 + 4} \le \frac{4}{3}.$$

As $\dfrac{b^3}{b^3 + 4} \le \dfrac{b}{3}$ *(provable by induction), we have* $b(b-1)(b-2)^2 \ge 0$ *for* $b \ge 0$. *Then*

$$\sum_{cyc} \frac{ab}{3} \le \frac{4}{3} \implies (a+b+c+d)^2 > 4(ab+bc+cd+da) \implies (a-b+c-d)^2 \ge 0$$

which is clearly true. The minimum is $\boxed{\dfrac{2}{3}}$. *(Equality holds when two of the variables are 2, and the other two are 0; interestingly, not when all four are 1!)*

Extension Exercise 73. *(1973 USAMO P2) Let* $\{X_n\}$ *and* $\{Y_n\}$ *denote two sequences of integers defined as follows:*

$$X_0 = 1, X_1 = 1, X_{n+1} = X_n + 2X_{n-1}(n = 1, 2, 3, \dots)$$
$$Y_0 = 1, Y_1 = 7, Y_{n+1} = 2Y_n + 3Y_{n-1}(n = 1, 2, 3, \dots)$$

Thus, the first few terms of the sequences are:

$$X : 1, 1, 3, 5, 11, 21, \dots$$
$$Y : 1, 7, 17, 55, 161, 487, \dots$$

Prove that, except for the "1", there is no term which occurs in both sequences.

Extension Solution 73. *Consider both sequences* (mod 8). *Note that* $X \equiv 1, 1, 3, 5, 3, 5 \dots$ *while* $Y \equiv 1, 7, 1, 7, 1, 7, \dots$ *We observe that* $X_i \not\equiv Y_j$ *for* $i \ge 3, j \ge 2$. *Hence the result follows.*

Extension Exercise 74. *(2017 Putnam A2) Let* $Q_0(x) = 1$, $Q_1(x) = x$, *and*

$$Q_n(x) = \frac{(Q_{n-1}(x))^2 - 1}{Q_{n-2}(x)}$$

for all $n \ge 2$. *Show that, whenever* n *is a positive integer,* $Q_n(x)$ *is equal to a polynomial with integer coefficients.*

Extension Solution 74. *Define the sequence of polynomials* $P_n(x)$ *recursively such that* $P_0(x) = 1$, $P_1(x) = x$, $P_n(x) = xP_{n-1}(x) + P_{n-2}(x)$ *for all* $n \ge 2$. *We claim that* $P_n(x) = Q_n(x)$ *for all* n.

It suffices to verify that Q_n *obeys the same recursion as does* P_n. *For* $n \ge 1$, *let the matrix*

$$M_n = \begin{bmatrix} P_{n-1}(x) & P_n(x) \\ P_{n-2}(x) & P_{n-1}(x) \end{bmatrix}$$

so that $\det M_n = P_{n-1}(x)^2 - P_{n-2}(x)P(x)$. *Furthermore, define*

$$N = \begin{bmatrix} x & -1 \\ 1 & 0 \end{bmatrix}$$

so $\det N = 1$ *and* $NM_n = M_{n+1}$. *By induction,* $\det(M_n) = \det(M_1) = 1$, *which leads to the desired conclusion.*

Extension Exercise 75. *(2008 USAMO P6) At a certain mathematical conference, every pair of mathematicians are either friends or strangers. At mealtime, every participant eats in one of two large dining rooms. Each mathematician insists upon eating in a room which contains an even number of his or her friends. Prove that the number of ways that the mathematicians may be split between the two rooms is a power of two (i.e. is of the form* 2^k *for some positive integer* k*).*

Extension Solution 75. *Reintrepret the problem in terms of graph theory. That is, for each vertex $v \in G$, with G a graph, let $f(v) = 0$ if v is in the first partition/room and $f(v) = 1$ if v is in the second partition. Then consider f in the context of the field F_2 with $|F_2| = 2$. Let V be the vector space of all such functions f, and define $T : V \mapsto V$ such that $(Tf)(v) = \sum_{v \sim w} f(v) - f(w)$ where $v \sim w$ denotes that v and w are adjacent. Then f is a partition iff $(Tf)(v) = \deg v$.*

Observe that, if $Af = \deg v$, and $Ag = \deg v$, then $f - g \in N(A)$ (the null space of A). Then the number of solutions is the size of $N(A)$, or $2^{\dim(N(A))}$. Let i be such that $\forall v, i(v) = 1 \implies Ai = 0 \implies \dim(N(A)) > 0$. This shows that the number of ways to do this is 2^k.

Extension Exercise 76. *$\triangle ABC$ has side lengths a, b, and c. Let $n \geq 1$ be an integer. Prove that*

$$\sum_{cyc} \frac{a^n}{b+c} \geq \frac{(a+b+c)^{n-1}}{2 \cdot 3^{n-2}}$$

Extension Solution 76. *First apply Chebyshevs Inequality to get*

$$\sum_{cyc} \frac{a^n}{b+c} \geq \frac{a^n + b^n + c^n}{3} \cdot \left(\sum_{cyc} \frac{1}{a+b} \right).$$

Then, subsequently using Cauchy-Schwarz yields

$$2(a+b+c)\left(\frac{1}{a+b} + \frac{1}{b+c} + \frac{1}{c+a} \right) \geq 9$$

and thus,

$$\frac{a^n + b^n + c^n}{3} \geq \left(\frac{a+b+c}{3} \right)^n.$$

Hence, it follows that

$$\sum_{cyc} \frac{a^n}{b+c} \geq \left(\frac{a+b+c}{3} \right)^n \left(\sum_{cyc} \frac{1}{a+b} \right)$$

$$\geq \frac{1}{3} \cdot \frac{1}{2} \cdot \left(\frac{2s}{3} \right)^{n-1} \cdot 9 = \frac{2s^{n-1} \cdot 2^{n-2}}{3^{n-2}}$$

as desired (after rearranging).

Part 3: Geometry

Chapter 7: Beginning Geometry

Section 7.1: An Introduction to Geometric Terminology

Exercise 192. *What is the area of a regular polygon with side length 1, each of whose interior angles measure 120°? (Stuck? See §7.3.2 for the angle measure formulas of a regular polygon.)*

Solution 192. *The regular polygon whose angles each measure 120° is the hexagon (with 6 sides), since the formula for the interior angle measure of a regular n-gon is $\left(\dfrac{180(n-2)}{n}\right)^{\circ}$. Furthermore, the area of a regular hexagon with side length s is $\dfrac{3s^2\sqrt{3}}{2}$ (try splitting it into six equilateral triangles with side length s). Hence, its area for $s = 1$ is $\boxed{\dfrac{3\sqrt{3}}{2}}$.*

Exercise 193. *(2016 AMC 10A/12A) Find the area of the shaded region.*

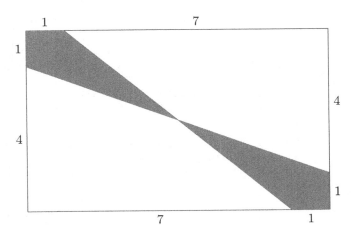

Solution 193. *Label the vertices of the gray regions such that the top-left corner has coordinates $(0,5)$ and the bottom-right corner has coordinates $(8,0)$. Then by setting up equations of lines through $(1,5)$ and $(7,0)$, and through $(0,4)$ and $(8,1)$, we know that the point of intersection is $\left(4,\dfrac{5}{2}\right)$. We can draw dashed lines from this point of intersection in all four directions to create rectangles with extraneous triangles (in white); subtracting off the areas as appropriate yields the area of $\boxed{\dfrac{13}{2}}$.*

Exercise 194. *A rectangle has perimeter P and area A. What is the length of its diagonal, in terms of P and A?*

Solution 194. *Let the rectangle's side lengths be x and y. Then $2(x+y) = P \implies x+y = \dfrac{P}{2}$ and $xy = A$. Squaring the first equation (a common tactic in problems such as this!) yields $x^2 + 2xy + y^2 = \dfrac{P^2}{4}$, and subtracting twice the second equation from this yields $x^2 + y^2 = \dfrac{P^2 - 8A}{4}$. By the Pythagorean Theorem, the length of the diagonal is $\sqrt{x^2 + y^2}$, or $\boxed{\dfrac{\sqrt{P^2 - 8A}}{2}}$.*

Exercise 195. *An equiangular hexagon has side lengths 1, 2, 2, 1, 2, 2, in clockwise order. What is its area?*

Solution 195. *The hexagon has interior angle measures of $120°$ (see solution 306), so circumscribe a rectangle around the hexagon and cut out four 30-60-90 right triangles from it. Each of these triangles has diagonal 2, and hence leg lengths of 1 and $\sqrt{3}$ (i.e. area $\dfrac{\sqrt{3}}{2}$ apiece, for a total of $2\sqrt{3}$ extraneous area). Subtracting this from the area of the rectangle (namely $(1 + 2 \cdot 1)(2\sqrt{3}) = 6\sqrt{3}$) yields the final answer of $\boxed{4\sqrt{3}}$.*

Exercise 196. *Find the area bounded by the graphs of $y = |x + 2|$ and $y = |2x + 6|$.*

Solution 196. *We perform casework based on where the graphs intersect. For $x + 2, 2x + 6 \geq 0$, we have $x + 2 = 2x + 6 \implies x = -4$, which is a contradiction. For $x + 2 \geq 0$, but $2x + 6 \leq 0$, we have $x + 2 = -2x - 6$, or $x = -\dfrac{8}{3}$, which is again a contradiction. For $x + 2 < 0, 2x + 6 \geq 0$, we have $-x - 2 = 2x + 6 \implies x = -\dfrac{8}{3}$ again, but since $y = x + 2 = -\dfrac{2}{3} < 0$, this is indeed a valid point of intersection. Finally, if $x + 2, 2x + 6 < 0$, then $-x - 2 = -2x - 6$, which yields $x = -4$ (valid). The vertices of the triangle representing the desired area are thus $(-4, 2), \left(-\dfrac{8}{3}, \dfrac{2}{3}\right), (-3, 0)$. Using the method of circumscribing a rectangle and subtracting off extraneous area, we obtain an area of $\boxed{\dfrac{2}{3}}$.*

Exercise 197. *(2002 AMC 12B) Let $f(x) = x^2 + 6x + 1$, and let R denote the set of points (x, y) in the coordinate plane such that*

$$f(x) + f(y) \leq 0 \qquad and \qquad f(x) - f(y) \leq 0$$

The area of R is closest to [what positive integer?]

Solution 197. *Given the first inequality $f(x) + f(y) \leq 0$, we can write $(x + 3)^2 + (y + 3)^2 \leq 16$, the closed disk with radius 4 centered at $(-3, -3)$. The second condition, similarly, yields $(x - y)(x + y + 6) \leq 0$, which leads to either $x - y \geq 0, x + y + 6 \leq 0$ or $x - y \leq 0, x + y + 6 \geq 0$. Graphing these inequalities yields a region with have the circles area, or $8\pi \approx \boxed{25}$.*

Extension Exercise 77. *Circle Γ_1 has equation $(x - 2)(x - 4) + (y - 1)(y - 3) = 167$. Circles Γ_2 and Γ_3 have center $(4, 3)$ and radii $\sqrt{157}$ and $\sqrt{185}$ respectively. Compute the area of the quadrilateral with vertices at the points of intersection of the circles.*

Extension Solution 77. *The equation of circle Γ_1 can be rewritten as*

$$(x - 3)^2 + (y - 2)^2 = 169$$

from which we know that it has center $(3, 2)$ and radius 13. Circles Γ_2 and Γ_3 have respective equations

$$\Gamma_2 := (x - 4)^2 + (y - 3)^2 = 157$$

$$\Gamma_3 := (x - 4)^2 + (y - 3)^2 = 185$$

Clearly, Γ_2 and Γ_3 are concentric, so they do not intersect. Thus, we consider the intersection points of Γ_1 and Γ_2, and of Γ_1 and Γ_3.

Intersections between Γ_1 *and* Γ_2: *We seek all ordered pairs* (x, y) *such that* $(x-3)^2 + (y-2)^2 = 169$ *and* $(x-4)^2 + (y-3)^2 = 157$. *Expanding both equations yields*

$$(x^2 - 6x + 9) + (y^2 - 4y + 4) = 169 \implies x^2 - 6x + y^2 - 4y = 156$$

$$(x^2 - 8x + 16) + (y^2 - 6y + 9) = 157 \implies x^2 - 8x + y^2 - 6y = 132$$

from which we obtain $2x + 2y = 24 \implies x + y = 12$. *Repeatedly adding this equation to either of the two aboev yields* $x^2 - 2x + y^2 = 204$; *completing the square gives* $(x-1)^2 + y^2 = 205$. *At this point, we can decompose each of the numbers on the RHS into sums of two perfect squares:* $169 = 12^2 + 5^2$, $157 = 11^2 + 6^2$, $205 = 14^2 + 3^2$. *If we set* $(x, y) = (15, -3)$, *we obtain a point of intersection. Likewise,* $(x, y) = (-2, 14)$ *satisfies the system of equations.* Γ_1 *and* Γ_2 *intersect at only two points, since we have only two choices for each of* x *and* y.

Intersections between Γ_1 *and* Γ_3: *We seek all* (x, y) *such that* $(x-3)^2 + (y-2)^2 = 169$ *and* $(x-4)^2 + (y-3)^2 = 185$. *Since* $185 = 13^2 + 4^2$, *we may set either* $x - 3 = -12$, $y - 2 = -5$ *or* $x - 3 = 5$, $y - 2 = -12$ *to get* $(x, y) = (-9, 7)$ *and* $(8, -10)$, *respectively.*

We now know all four of the quadrilateral's vertices, namely $(15, -3)$, $(-2, 14)$, $(-9, 7)$, *and* $(8, -10)$. *If we plot these points on the plane, we observe that the quadrilateral is in fact a rectangle with side lengths* $17\sqrt{2}$ *and* $7\sqrt{2}$. *Therefore, its area is* $\boxed{238}$.

Extension Exercise 78. *(2013 AIME I) Let* $\triangle PQR$ *be a triangle with* $\angle P = 75^o$ *and* $\angle Q = 60^o$. *A regular hexagon* $ABCDEF$ *with side length 1 is drawn inside* $\triangle PQR$ *so that side* $\overline{AB}$ *lies on* $\overline{PQ}$, *side* $\overline{CD}$ *lies on* $\overline{QR}$, *and one of the remaining vertices lies on* $\overline{RP}$. *There are positive integers* $a, b, c,$ *and* d *such that the area of* $\triangle PQR$ *can be expressed in the form* $\frac{a + b\sqrt{c}}{d}$, *where* a *and* d *are relatively prime, and* c *is not divisible by the square of any prime. Find* $a + b + c + d$.

Extension Solution 78. *Compute* $\angle R$ *to be* $45°$, *then draw hexagon* $ABCDEF$. *Then observe that* Q *is adjacent to* C, D, *so the height of the hexagon is* $\sqrt{3}$ *and* $QR = 2 + \sqrt{3}$. *If we assign coordinates such that* $RP := y = x$, $PQ := y = (2 - x)\sqrt{3} + 3$. *Hence, the resulting system yields* $y = x = \dfrac{3 + \sqrt{3}}{2}$. *Therefore,*

$$|\triangle PQR| = \frac{5\sqrt{3} + 9}{4} \implies \boxed{21}.$$

Section 7.2: Translating Algebra to Geometry: The Coordinate Plane

Exercise 198.

(a) *At what point do the graphs of the lines* $y = 2x + 9$ *and* $y = -3x - 1$ *intersect in the Cartesian plane?*

(b) *What is the equation of the line perpendicular to the line* $y = 10x + 5$ *passing through the point in (a)?*

(c) *The line in (b) passes through all quadrants except for one. What is this quadrant? Show algebraically that the line never passes through this quadrant.*

(d) (i) *Give an example of a line that passes through two or fewer quadrants, or show that no such line exists.*

(ii) *Give an example of a line that passes through all four quadrants, or show that no such line exists.*

Solution 198.

(a) *We set* $2x + 9 = -3x - 1 \implies x = -2, y = 5$. *Hence, the point of intersection is* $\boxed{(-2, 5)}$.

(b) *A line perpendicular to a line with slope 10 has slope* $-\dfrac{1}{10}$; *if it passes through* $(-2, 5)$, *its equation is*

$$\boxed{y = -\frac{1}{10}x + \frac{24}{5}}.$$

(c) *The line never passes through quadrant III, as if $x < 0$, then $y > 0$.*

(d) (i) *Any constant function passes through two quadrants ($y = 0$ excluded).*

(ii) *No such line exists. If it were to exist, it would have to change slope.*

Exercise 199. *Show that, if two lines intersect in the xy-plane, they must intersect at either zero, one, or infinitely many distinct points.*

Solution 199. *If the lines are parallel, and have the same y-intercept, then they will clearly intersect at infinitely many points, as they are the exact same line. If they are parallel but not coincident, namely $y = mx + b$ and $y = mx + c$, then we end up with $b = c$, but this is a contradiction; hence, no solutions. Finally, in all other cases, there will be one solution, since setting the equations equal yields a linear equation (which, by the Fundamental Theorem of Algebra, has 1 solution, since a linear polynomial is a degree-1 polynomial).*

Exercise 200. *Suppose that points $A = (x, y)$ and $B = (y, x)$ are equidistant from the origin, with AB also equal to this common distance. Given that $y > x$, what is $\frac{y}{x}$?*

Solution 200. *Let D_A, D_B be the distances from A and B to the origin; then $D_A = D_B = \sqrt{x^2 + y^2} = \sqrt{2(x-y)^2}$. This simplifies to $x^2 + y^2 = 2(x^2 - 2xy + y^2)$, or $x^2 + y^2 - 4xy = 0$. Subtracting x^2 from both sides, we obtain the equation $y^2 - 4xy = -x^2$; completing the square yields $y^2 - 4xy + 4x^2 = 3x^2 \implies (y - 2x)^2 = 3x^2 \implies y - 2x = \pm\sqrt{3}x \implies y = (\pm\sqrt{3} + 2))x$. Because $2 - \sqrt{3} < 1$, $\frac{y}{x} = \boxed{2 + \sqrt{3}}$.*

Exercise 201. *(2009 AIME I) In parallelogram $ABCD$, point M is on $\overline{AB}$ so that $\frac{AM}{AB} = \frac{17}{1000}$ and point N is on $\overline{AD}$ so that $\frac{AN}{AD} = \frac{17}{2009}$. Let P be the point of intersection of $\overline{AC}$ and $\overline{MN}$. Find $\frac{AC}{AP}$.*

Solution 201. *Fakesolve: WLOG assume the parallelogram is just a straight line. Then the problem becomes trivialized; the answer is $\dfrac{3009}{17} = \boxed{177}$.*

Exercise 202. *The x-coordinates of the points of intersection of the circles with radii 6 and centers $(-2, 3)$ and $(2, -3)$ can be written in the form $\pm\sqrt{\dfrac{a}{b}}$, where a and b are positive integers such that $\gcd(a, b) = 1$. Find $a + b$.*

Solution 202. *Let the circles with centers $(-2, 3)$ and $(2, -3)$ be C_1 and C_2 respectively. We have*

$$C_1 : (x + 2)^2 + (y - 3)^2 = 36$$

$$C_2 : (x - 2)^2 + (y + 3)^2 = 36$$

Expanding and setting the equations equal to each other, we obtain

$$(x^2 + 4x + 4) + (y^2 - 6y + 9) = (x^2 - 4x + 4) + (y^2 + 6y + 9) = 36$$

$$8x - 12y = 0 \implies y = \frac{2}{3}x$$

$$(x + 2)^2 + \left(\frac{2}{3}x - 3\right)^2 = 36 \implies x = \pm\sqrt{\frac{207}{13}} \implies \boxed{220}$$

Extension Exercise 79. *Prove that any three non-collinear points uniquely determine a circle in the plane.*

Extension Solution 79. *In a circle, if we draw a line between any two points in the circle, and then bisect it, the bisector will pass through the center of the circle. Hence, for any three points, we can draw two bisectors that will intersect at a unique point, which will then be the center of the circle.*

Extension Exercise 80. *(2011 AIME I) Let L be the line with slope $\frac{5}{12}$ that contains the point $A = (24, -1)$, and let M be the line perpendicular to line L that contains the point $B = (5, 6)$. The original coordinate axes are erased, and line L is made the x-axis and line M the y-axis. In the new coordinate system, point A is on the positive x-axis, and point B is on the positive y-axis. The point P with coordinates $(-14, 27)$ in the original system has coordinates (α, β) in the new coordinate system. Find $\alpha + \beta$. (Hint: Use the point-to-line distance formula.)*

Extension Solution 80. *The equation for line L is $y + 1 = \frac{5}{12}(x - 24)$ in point-slope form. $M \perp L$ and contains $(5, 6)$, so its equation in point-slope form is $y - 6 = -\frac{12}{5}(x - 5)$. Then we can also write $L := 5x - 12y - 132 = 0$, $M := 12x + 5y - 90 = 0$. Using the point-to-line distance formula on P to lines L, M yields distances to L, M of $\frac{526}{13}$ and $\frac{123}{13}$, respectively, the difference of which is $\boxed{31}$.*

Extension Exercise 81. *Try to generalize the point-to-line distance formula to dimensions ≥ 3! (Hint: interpret lines as vectors.)*

Extension Solution 81. *We'll begin with 3 dimensions. In $\mathbb{R}^3$, we specify a line by two points $P_1 = (x_1, y_1, z_1)$ and $P_2 = (x_2, y_2, z_2)$; hence, a vector along the line assumes the form*

$$v = \begin{bmatrix} x_1 + t(x_2 - x_1) \\ y_1 + t(y_2 - y_1) \\ z_1 + t(z_2 - z_1) \end{bmatrix}$$

The point-line distance is subsequently

$$\sqrt{((x_1 - x_0) + t(x_2 - x_1))^2 + ((y_1 - y_0) + t(y_2 - y_1))^2 + ((z_1 - z_0) + t(z_2 - z_1))^2}$$

In 4 and higher dimensions, we can instead consider the point-plane distance. The process itself is similar, though a plane must pass through three distinct points in order to be uniquely determined.

Section 7.3: Proving Formulas in Cartesian Coordinate Geometry

Exercise 203. *Classify each of the following sets of triangle side lengths as corresponding to an equilateral, isosceles, scalene, or degenerate triangle. If non-degenerate, further classify them as corresponding to an acute, right, or obtuse triangle.*

(a) $\{3, 4, 6\}$

(b) $\{1, 2, 2\}$

(c) $\{7, 24, 25\}$

(d) $\{99, 100, 101\}$

(e) $\{20, 19, 50\}$

(f) $\{x, 1.5x, 2x\}$ *for some* $x \in \mathbb{N}$

Solution 203.

(a) Scalene, obtuse; $3^2 + 4^2 < 6^2$.

(b) Isosceles, acute; $1^2 + 2^2 > 2^2$.

(c) Scalene, right; $7^2 + 24^2 = 25^2$.

(d) Scalene, acute; $99^2 + 100^2 > 101^2$.

(e) Degenerate; $20 + 19 \leq 50$.

(f) Scalene, obtuse; $x^2 + (1.5x)^2 = 3.25x^2 < (2x)^2 = 4x^2$ for all $x \in \mathbb{N}$.

Exercise 204. *(2014 Berkeley Math Tournament) Call two regular polygons supplementary if the sum of an internal angle from each polygon adds up to $180°$. For instance, two squares are supplementary because the sum of the internal angles is $90° + 90° = 180°$. Find the other pair of supplementary polygons. Write your answer in the form (m, n) where m and n are the number of sides of the polygons and $m < n$.*

Solution 204. *One of the interior angles must be less than 90°, and the other greater (since 90°, 90° has already been given in the problem). But the only n for which a regular n-gon has interior angles less than 90° is a triangle, which has interior angle measure 60° if it is equilateral. Hence the other angle measures is $120^{P\circ}$, which corresponds to a regular hexagon (i.e. n = 6. Hence, the desired ordered pair of supplementary polygons is* $\boxed{(3,6)}$.

Exercise 205. *What is the area of the quadrilateral with vertices at $(0,2)$, $(3,5)$, $(9,8)$, and $(8,6)$ (in that order)?*

Solution 205. *Shoelace Theorem immediately gives* $\frac{1}{2}|(0 \cdot 5 + 3 \cdot 8 + 9 \cdot 6 + 2 \cdot 8) - (2 \cdot 3 + 5 \cdot 9 + 8 \cdot 8 + 6 \cdot 0)| = \boxed{\frac{21}{2}}$.

Exercise 206.

(a) *A regular n-gon has 2019 times as many diagonals as edges. What is the value of n?*

(b) *For the value of n in (a), set up (but do not evaluate) an expression for the area of the polygon. Fully justify your answer with supporting work.*

Solution 206.

(a) *The number of diagonals in a regular n-gon is* $\frac{n(n-3)}{2}$, *so* $\frac{n-3}{2} = 2019 \implies n = \boxed{4041}$.

(b) *In general, the area of a regular n-gon is given by n times the area of an isosceles triangle with vertex angle measure* $\left(\frac{360}{n}\right)^\circ$, *i.e. n times the sine of* $\left(\frac{180}{n}\right)^\circ$ *times the cosine of* $\left(\frac{180}{n}\right)^\circ$, *as the diagonal (the hypotenuse of each right triangle formed) has length 1. So we plug in n = 4041, and we are done (we do not evaluate).*

Exercise 207. *(2017 ARML Individual Round #2) Trapezoid ARML has AR ∥ ML. Given that AR = 4, $RM = \sqrt{26}$, ML = 12, and $LA = \sqrt{42}$, compute AM.*

Solution 207. *Let the height of trapezoid ARML be h, let the foot of the perpendicular from R be P (and the foot from A be Q), and let PM = x (so that LQ = 8 − x). Then $x^2 + h^2 = 26, (8-x)^2 + h^2 = 42 \implies 64 - 16x = 16 \implies x = 3, h^2 = 17$. Then $AM = \sqrt{AQ^2 + QM^2} = \sqrt{h^2 + (12 - LQ)^2} = \sqrt{17 + 7^2} = \boxed{\sqrt{66}}$.*

Exercise 208. *A rectangle is made out of unit squares as shown below. Find the total area in square units of the magenta shaded region.*

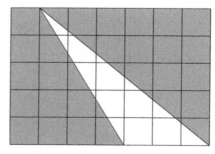

Solution 208. *The area of the largest rectangle is 5·7 = 35, and subtracting off the unshaded area ($\frac{3 \cdot 5}{2} = \frac{15}{2}$ yields $\boxed{\frac{55}{2}}$. (Alternatively, we can add up the two triangles and the 1 × 5 rectangle in magenta to get the same answer.)*

Exercise 209. *(2016 Berkeley Math Tournament) Let S be the set of all non-degenerate triangles with integer side lengths, such that two of the sides are 20 and 16. Suppose we pick a triangle, at random, from this set. What is the probability that it is acute?*

Solution 209. *By the Triangle Inequality, the third side length can be anywhere from $20 - 16 + 1 = 5$ to $20 + 16 - 1 = 35$, inclusive. Let the side length be x. Of $x \in [5, 35]$, we require either $20^2 + 16^2 > x^2 \implies x < \sqrt{656} \in [25, 26) \implies x \le 25$ or $16^2 + x^2 > 20^2 \implies x \ge 13$. Hence, $x \in [13, 25]$ is the permissible interval, for a total of 13 good values out of 31, yielding a probability of $\boxed{\dfrac{13}{31}}$.*

Exercise 210.

(a) *Is there an isosceles triangle with integer side lengths and area 12?*

(b) *Is there an isosceles triangle with integer side lengths and area 18?*

(c) *For what positive integers k does there exist an isosceles triangle with integer side lengths and area k?*

Solution 210.

(a) *Yes; consider the 5-5-6 triangle (which can be split into two 3-4-5 triangles).*

(b) *No; we can test this by trying Pythagorean triples on the altitude and half the base length.*

(c) *Consider the altitude length h and half the base length $\dfrac{b}{2}$. If $\sqrt{\dfrac{b^2 + 4}{+} h^2}$ is an integer, and $k = bh$, with $b \in \mathbb{Z}$, then such a triangle is constructible for that k.*

Extension Exercise 82. *Prove the Shoelace Theorem by induction.*

Extension Solution 82. *We show that if the shoelace theorem holds for some polygon $A_1 A_2 A_3 \cdots A_n$, then it will also holds for $A_1 A_2 A_3 \cdots A_{n+1}$. This can be done by sub-dividing $A_1 A_2 A_3 \cdots A_{n+1}$ into two polygons, namely $A_1 A_2 A_3 \cdots A_n$ and $A_1 A_n A_{n+1}$. Let $A_i = (x_i, y_i)$; then applying Shoelace on the two polygons yields*

$$[A_1 A_2 A_3 \cdots A_n] = \frac{1}{2} \sum_{i=1}^{n} (x_i y_{i+1} - x_{i+1} y_i)$$

$$[A_1 A_n A_{n+1}] = \frac{1}{2} (x_1 y_2 + x_2 y_3 + x_3 y_1 - x_1 y_3 - x_2 y_1 - x_3 y_2)$$

Therefore, we get that

$$[A_1 A_2 A_3 \cdots A_n A_{n+1}] = \frac{1}{2} \sum_{i=1}^{n} (x_i y_{i+1} - x_{i+1} y_i)$$

as desired.

Extension Exercise 83. *(2015 ARML Individual Round #9) An (a, r, m, l)-trapezoid is a trapezoid with bases of length a and r, and other sides of length m and l. Compute the number of positive integer values of l such that there exists a $(20, 5, 15, l)$-trapezoid.*

Extension Solution 83. *This is ultimately an extension of the Triangle Inequality to quadrilaterals, which says that $l \ge 20 - 5 - 15 + 1 = 1$ and $l \le 20 + 15 - 5 - 1 = 29$. Hence, there are $\boxed{29}$ possible values for l. (Sketch of the proof of the quadrilateral inequality: consider rearranging the formula to yield a parallel to the familiar result for triangles. Also consider dropping auxiliary lines to form triangles within the interior of the quadrilateral.)*

Extension Exercise 84. *If a, b, and c are the side lengths of a triangle with $a \ge b \ge c$, prove that $\min \left(\dfrac{a}{b}, \dfrac{b}{c} \right) \in [1, \phi)$ where $\phi = \dfrac{1 + \sqrt{5}}{2}$.*

Extension Solution 84. *The lower bound is obvious. For the upper bound r, WLOG let the side lengths be $r^2, r, 1$. Then $1 + r > r^2 \implies 1 + r - r^2 > 0 \implies r < \dfrac{1 + \sqrt{5}}{2} = \phi$.*

Extension Exercise 85. *Propose a generalization of Pick's Theorem for a polygon with n holes (where a hole of a polygon is defined to be a polygon with a positive area $\mathcal{A}$ whose union of interior and boundary points is contained strictly within the polygon).*

Extension Solution 85. *For n holes in the interior of a polygon, the area becomes $I + \dfrac{B}{2} + n - 1$. We apply Picks Theorem to each hole separately. Alternatively, we split a holed polygon into two disjoint hole-free polygons by way of a path that connects two arbitrary boundary points and incorporates one or more boundary points lying along each hole.*

Section 7.4: Working With Similarity

Exercise 211. *(2017 AMC 8) In the figure below, choose point D on $\overline{BC}$ so that $\triangle ACD$ and $\triangle ABD$ have equal perimeters. What is the area of $\triangle ABD$?*

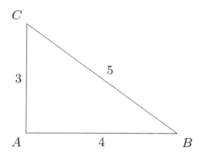

Figure 9.4: Diagram for this Exercise.

Solution 211. *Note that $\overline{AD}$ does not affect the difference between the perimeters, so we require $CD = BD + 1$, i.e. $BD = 2$. Hence, by triangle similarity, $[\triangle ABD] = \boxed{\dfrac{12}{5}}$.*

Exercise 212. *(2018 AMC 10A) Two circles of radius 5 are externally tangent to each other and are internally tangent to a circle of radius 13 at points A and B, as shown in the diagram. The distance AB can be written in the form $\dfrac{m}{n}$, where m and n are relatively prime positive integers. What is $m + n$?*

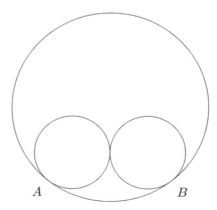

Figure 9.5: Diagram for this Exercise.

Solution 212. *Let O be the center of the largest circle. Then the circle tangent at A will have center O_1, and the remaining circle (tangent at B) will have center O_2. Then $\triangle OO_1O_2$ is similar to $\triangle OAB$. Hence,*
$$\frac{OO_1}{OA} = \frac{OO_2}{OB} \implies \frac{13-5}{13} = \frac{5+5}{AB} \implies AB = \frac{65}{4} \implies \boxed{69}.$$

Exercise 213. *(2016 AMC 8) A semicircle is inscribed in an isosceles triangle with base 16 and height 15 so that the diameter of the semicircle is contained in the base of the triangle as shown. What is the radius of the semicircle?*

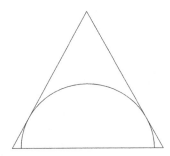

Figure 9.6: Diagram for Exercise 101.

Solution 213. *Draw the triangle's altitude; this will split it into two 8-15-17 triangles. Notice that the semicircles radius is $\dfrac{8 \cdot 15}{17} = \boxed{\dfrac{120}{17}}$, since the altitude of the 8-15-17 triangle creates two similar triangles.*

Exercise 214. *(2008 AIME I) Square $AIME$ has sides of length 10 units. Isosceles triangle GEM has base EM, and the area common to triangle GEM and square $AIME$ is 80 square units. Find the length of the altitude to EM in $\triangle GEM$.*

Solution 214. *Let $\overline{GE}$ and $\overline{GM}$ intersect $\overline{AI}$ at P, Q respectively. The area inside $AIME$ but outside $\triangle GEM$ is 20 square units, which implies that $\dfrac{10 - PQ}{2} = 2 \implies PQ = 6$. Using similarity ratios, we get that the altitude of $\triangle GEM$ is $\boxed{25}$.*

Exercise 215. *(2001 AIME I) In triangle $\triangle ABC$, angles A and B measure 60 degrees and 45 degrees, respectively. The bisector of angle A intersects $\overline{BC}$ at T, and $AT = 24$. Find the area of triangle ABC.*

Solution 215. *We note that $\angle C = 75°$, so $\angle CAT = 30°$ and $\angle CTA = 75°$. Thus, $\triangle CAT$ is isosceles, and $AC = 24$. After dropping the altitude from C to a point D, we can apply similarity of 30-60-90 triangles to get that $AD = 12, CD = 12\sqrt{3}$. $\triangle CDB$ is isosceles right, and CD is congruent to BD. Hence, $BD = 12\sqrt{3}$ as well, which makes $AB = 12 + 12\sqrt{3}$. Then it follows that $[\triangle ABC] = 216 + 72\sqrt{3} \implies \boxed{291}$.*

Exercise 216. *In triangle $\triangle ABC$, let $AB = 13$, $AC = 14$, $BC = 15$. Point P_0 lies on $\overline{AC}$ such that $\overline{BP_0} \perp \overline{AC}$. Point P_1 lies on $\overline{BC}$ such that $\overline{P_0P_1} \perp \overline{BC}$, point P_2 lies on $\overline{AC}$ with $\overline{P_1P_2} \perp \overline{AC}$, and so forth. What is the value of the sum*
$$\sum_{i=0}^{\infty} [P_iP_{i+1}P_{i+2}]$$

(where $[A]$ denotes the area of A)?

Solution 216. *By either computing the first few areas, or using similarity to conclude that each side is multiplied by $\dfrac{3}{5}$ per iterate, we obtain an infinite geometric series with sum $\dfrac{\frac{7776}{625}}{1 - \frac{9}{25}} = \boxed{\dfrac{486}{25}}$.*

Exercise 217. *(2011 AMC 10B) Let T_1 be a triangle with sides $2011, 2012$, and 2013. For $n \geq 1$, if $T_n = \triangle ABC$ and D, E, and F are the points of tangency of the incircle of $\triangle ABC$ to the sides AB, BC and AC, respectively, then T_{n+1} is a triangle with side lengths AD, BE, and CF, if it exists. What is the perimeter of the last triangle in the sequence (T_n)?*

Solution 217. *Draw the angle bisectors and the inradii to see that $AD = AF, BD = BE, CE = CF$. Let $AD = x, BD = y, CE = z$. Then $x + y = a - 1, x + z = a, y + z = a + 1 \implies (x, y, z) = \left(\frac{a}{2} - 1, \frac{a}{2}, \frac{a}{2} + 1\right)$. Hence, T_2 has side lengths $1005, 1006, 1007$. By extension, T_3 has sides $502, 503, 504$, and so forth. This continues until T_{11}, which is a degenerate triangle by Triangle Inequality. The perimeter of T_{10} is $\boxed{\dfrac{1509}{128}}$.*

Exercise 218. *(2009 AIME I) In right $\triangle ABC$ with hypotenuse $\overline{AB}$, $AC = 12$, $BC = 35$, and $\overline{CD}$ is the altitude to $\overline{AB}$. Let ω be the circle having $\overline{CD}$ as a diameter. Let I be a point outside $\triangle ABC$ such that $\overline{AI}$ and $\overline{BI}$ are both tangent to circle ω. The ratio of the perimeter of $\triangle ABI$ to the length AB can be expressed in the form $\dfrac{m}{n}$, where m and n are relatively prime positive integers. Find $m + n$.*

Solution 218. *Let P, Q be the intersections of ω with $\overline{BI}, \overline{AI}$ respectively. We have $AB = 37$, $CD = \dfrac{420}{37}$, and the inradius of $\triangle ABI$ equal to $\dfrac{210}{37}$.*

As triangles $\triangle CDB$ and $\triangle ACB$ are similar, we have $\dfrac{BC}{BD} = \dfrac{AB}{BC}$, so $BD = \dfrac{BC^2}{AB} = \dfrac{35^2}{37}$. Similarly, $AD = \dfrac{12^2}{37}$. Also note that $IP = IQ = x$, $BP = BD$, and $AQ = AD$. Then $AI = \dfrac{144}{37} + x, BI = \dfrac{1225}{37} + x$. Computing the area of $\triangle ABI$ using both Herons formula and the formula $A = rs$ (where r is the inradius and s is the semi-perimeter), we get $3x^2 + 74x - 37^2 = 0 \implies x = \dfrac{37}{3}$. Hence, the ratio of the perimeter to AB is $\dfrac{8}{3} \implies \boxed{11}$.

Extension Exercise 86. *(2002 AIME II) The perimeter of triangle APM is 152, and the angle $\angle PAM$ is a right angle. A circle of radius 19 with center O on $\overline{AP}$ is drawn so that it is tangent to $\overline{AM}$ and $\overline{PM}$. Given that $OP = \dfrac{m}{n}$ where m and n are relatively prime positive integers, find $m + n$.*

Extension Solution 86. *Let B be the intersection point of $\overline{PM}$ with the circle. Observe that $\triangle MPA$ and $\triangle OPB$ are similar triangles. Then by Power of a Point, $AM = BM$, so $\dfrac{19}{AM} = \dfrac{152 - 2AM}{152} \implies AM = 38$, and so the side lengths are in the ratio $2 : 1$. Thus, $2OP = PB + 38$, and $2PB = OP + 19$. It follows that $OP = \dfrac{95}{3} \implies \boxed{98}$.*

Extension Exercise 87. *(2004 AIME II) Let $ABCDE$ be a convex pentagon with $AB \parallel CE, BC \parallel AD, AC \parallel DE, \angle ABC = 120°, AB = 3, BC = 5$, and $DE = 15$. Given that the ratio between the area of triangle ABC and the area of triangle EBD is m/n, where m and n are relatively prime positive integers, find $m + n$.*

Extension Solution 87. *Suppose that $\overline{AD}$ and $\overline{CE}$ intersect at point F. Since $AB \parallel CE, BC \parallel AD$, $ABCF$ is a parallelogram, so $\triangle ABC \cong \triangle CFA$. It also follows that $\triangle ABC$ is similar to $\triangle EFD$. Using Law of Cosines, $AC = 7$, and so $\dfrac{AC}{ED} = \dfrac{7}{15}$.*

If we let h_1, h_2 be the altitude lengths to $\overline{AC}$ and $\overline{DE}$ respectively, then $\dfrac{[ABC]}{[BDE]} = \dfrac{7h_1}{15h_2}$. Hence, the ratio of the altitude of $\triangle ABC$ to the altitude of $\triangle BDE$ is $\dfrac{7}{2 \cdot 7 + 15} = \dfrac{7}{29}$, and the desired area ratio is $\dfrac{7}{15} \cdot \dfrac{7}{29} = \dfrac{49}{435} \implies \boxed{484}$.

Extension Exercise 88. *(2013 AIME I) Triangle AB_0C_0 has side lengths $AB_0 = 12$, $B_0C_0 = 17$, and $C_0A = 25$. For each positive integer n, points B_n and C_n are located on $\overline{AB_{n-1}}$ and $\overline{AC_{n-1}}$, respectively, creating three similar triangles $\triangle AB_nC_n \sim \triangle B_{n-1}C_nC_{n-1} \sim \triangle AB_{n-1}C_{n-1}$. The area of the union of all triangles $B_{n-1}C_nB_n$ for $n \geq 1$ can be expressed as $\dfrac{p}{q}$, where p and q are relatively prime positive integers. Find q.*

Extension Solution 88. *First compute* $[\triangle ABC] = 90$. *For any* $n \geq 0$, *all of the lines* $\overline{B_nC_n}$ *are parallel, so we want the ratio* $r = \dfrac{[\triangle B_0B_1C_1]}{[\triangle B_0B_1C_1] + [\triangle C_1C_0B_0]}$. *Observe that* $r = 1 - [\triangle B_1C_1A] - [\triangle B_0C_0C_1]$, *and* $1 - r = \left(\dfrac{17}{25}\right)^2$. *Then the ratio of the successive areas from* $\triangle B_1C_1A$ *is* $\left(\dfrac{336}{625}\right)^2$. *All that remains is to compute* $90r$, *which will have denominator* $\boxed{961}$ *after some computation.*

Section 7.5: Proving Angle Theorems

Exercise 219. *(2016 AMC 12A) In* $\triangle ABC$, $AB = 6$, $BC = 7$, *and* $CA = 8$. *Point* D *lies on* $\overline{BC}$, *and* $\overline{AD}$ *bisects* $\angle BAC$. *Point* E *lies on* $\overline{AC}$, *and* $\overline{BE}$ *bisects* $\angle ABC$. *The bisectors intersect at* F. *What is the ratio* $AF : FD$?

Solution 219. *By the angle bisector theorem, we have* $\dfrac{CD}{AC} = \dfrac{BD}{AB}$, *so* $3CD = 4BD$. *Since* $BD + CD = 7$, $BD = 3$ *and* $CD = 4$. *Apply angle bisector again to* $\triangle ABD$, *with* $\dfrac{AB}{AF} = \dfrac{BD}{FD} \implies AF = 2FD$, *or a ratio of* $\boxed{2:1}$.

Exercise 220. *(2018 AMC 12B) Side* $\overline{AB}$ *of* $\triangle ABC$ *has length* 10. *The bisector of angle* A *meets* $\overline{BC}$ *at* D, *and* $CD = 3$. *The set of all possible values of* AC *is an open interval* (m, n). *What is* $m + n$?

Solution 220. *Let* $BD = x$; *then* $AC = \dfrac{30}{x}$ *by the Angle Bisector Theorem. We then apply Triangle Inequality to set up three inequalities that lead to* $x \in (3, 15) \implies m + n = \boxed{18}$.

Exercise 221. *(2010 AMC 12A) Nondegenerate* $\triangle ABC$ *has integer side lengths,* $\overline{BD}$ *is an angle bisector,* $AD = 3$, *and* $DC = 8$. *What is the smallest possible value of the perimeter?*

Solution 221. *By Angle Bisector Theorem,* $8AB = 3BC$. *But* $AB = 3, BC = 8$ *leads to a degenerate triangle. However,* $AB = 6, B = 16$ *works, leading to the answer of* $\boxed{33}$.

Exercise 222. *(2017 AIME II) A triangle has vertices* $A(0,0)$, $B(12,0)$, *and* $C(8,10)$. *The probability that a randomly chosen point inside the triangle is closer to vertex* B *than to either vertex* A *or vertex* C *can be written as* $\frac{p}{q}$, *where* p *and* q *are relatively prime positive integers. Find* $p + q$.

Solution 222. *The desired region is the area to the right of the perpendicular bisector of* $\overline{AB}$ *and below the perpendicular bisector of* $\overline{BC}$. *We can compute the coordinates of the vertices of the resulting quadrilateral, then use Shoelace Theorem to obtain* $\dfrac{109}{300} \implies \boxed{409}$.

Exercise 223. *Let* $\triangle ABC$ *be a triangle with* $AB = 3$, $BC = 4$, $CA = 5$. *Point* D *is on* $\overline{BC}$ *such that the circumcircles of* $\triangle ABD$ *and* $\triangle ADC$ *have areas that differ by* $\dfrac{37\pi}{16}$. *The angle bisector of* $\triangle ABD$ *hits* $\overline{AC}$ *at* E. *Compute the area of* $\triangle ADE$.

Solution 223. *Use circumcenter/circumradius formulas (namely,* $A = \dfrac{abc}{4R}$*) to get a generalized distance of* $\dfrac{\pi}{4}(k^2 + 9)$ *for* $D = (k, 0)$. *Setting this equal to* $\dfrac{37\pi}{16}$ *yields* $D = \left(\dfrac{1}{2}, 0\right)$, *and an angle bisector* $y = x$ *which hits* $\overline{AC}$ *at* $E = \left(\dfrac{3}{7}, \dfrac{3}{7}\right)$. *Thus,* $[\triangle ADE] = \boxed{\dfrac{9}{4}}$.

Exercise 224. *(1990 AIME) A triangle has vertices* $P = (-8, 5)$, $Q = (-15, -19)$, *and* $R = (1, -7)$. *The equation of the bisector of* $\angle P$ *can be written in the form* $ax + 2y + c = 0$. *Find* $a + c$.

Solution 224. *Using the Angle Bisector Theorem, we compute that the bisector of* $\angle P$ *splits* $\overline{QR}$ *into segments of length* $\dfrac{25}{x} = \dfrac{15}{20 - x} \implies x = \dfrac{15}{2}, \dfrac{25}{2}$. *Hence, if we designate the intersection point by* S, *then* $\dfrac{SQ}{SR} = \dfrac{5}{3}$, *and* $S = \left(-5, -\dfrac{23}{2}\right)$. *The equation of line* PS *is* $11x + 2y + 78 = 0 \implies \boxed{89}$.

Exercise 225. *(2011 AIME II) In triangle ABC, $AB = \frac{20}{11}AC$. The angle bisector of $\angle A$ intersects BC at point D, and point M is the midpoint of AD. Let P be the point of the intersection of AC and BM. The ratio of CP to PA can be expressed in the form $\frac{m}{n}$, where m and n are relatively prime positive integers. Find $m + n$.*

Solution 225. *Let there be a point D on $\overline{AC}$ with BP $\parallel$ DD. Then $\triangle BPC$ is similar to $\triangle DDC$, so*
$$\frac{CP}{CD} = 1 + \frac{BD}{BC} = 1 + \frac{AB}{AC} = \frac{31}{11}. \text{ Furthermore, } AP = PD, \text{ so } \frac{PC}{PA} = \frac{PC}{PD} = \frac{1}{1 - \frac{CD}{CP}} = \frac{31}{20} \implies \boxed{51}.$$
(Alternatively, we can directly apply Menelaus Theorem with transversal line $\overline{PB}$ on $\triangle ACD$.)

Exercise 226. *(2009 AIME I) Triangle ABC has $AC = 450$ and $BC = 300$. Points K and L are located on $\overline{AC}$ and $\overline{AB}$ respectively so that $AK = CK$, and $\overline{CL}$ is the angle bisector of angle C. Let P be the point of intersection of $\overline{BK}$ and $\overline{CL}$, and let M be the point on line BK for which K is the midpoint of $\overline{PM}$. If $AM = 180$, find LP.*

Solution 226. *Since K is the midpoint of both $\overline{AC}$ and $\overline{PM}$, it follows that AMCP is a parallelogram, and so AM $\parallel$ LP, with $\triangle AMB \sim \triangle LPB$. Hence $\frac{AM}{LP} = \frac{AL}{LB} + 1$. By Angle Bisector, $\frac{AL}{LB} = \frac{AC}{BC} = \frac{3}{2}$, so $\frac{180}{LP} = \frac{5}{2}$ and $LP = \boxed{72}$.*

Exercise 227. *(2005 AIME II) In triangle ABC, $AB = 13, BC = 15$, and $CA = 14$. Point D is on $\overline{BC}$ with $CD = 6$. Point E is on $\overline{BC}$ such that $\angle BAE \cong \angle CAD$. Given that $BE = \frac{p}{q}$ where p and q are relatively prime positive integers, find q.*

Solution 227. *Let the A-altitude intersect $\overline{BC}$ at P. Then drop altitudes from D and E to hit their respective side lengths at Q and R respectively. Using Herons formula, we have $DS = \frac{24}{5}$ from the similarity formed by the altitudes, as well as $SC = \frac{18}{5}, AS = 14 - SC = \frac{52}{5}$. On $\triangle AQB$, $BQ = \frac{33}{5}$.*

Moreover, $RB = 13 - AR$, which is again conducive to similar triangles. $\triangle ARE \sim \triangle ASD$, so $\frac{RE}{RA} = \frac{24}{53}$. Also, $\triangle BRE \sim \triangle BQA$, yielding $\frac{RE}{13 - AR} = \frac{56}{33}$. Solving for x and y yields $(x, y) = \left(\frac{2184}{463}, \frac{4732}{463}\right)$. Only the denominator is important, so the answer is $\boxed{463}$.

Extension Exercise 89. *Triangle $\triangle ABC$ has circumcircle Ω. The internal angle bisectors of $\angle A$, $\angle B$, and $\angle C$ are all extended to intersect Ω at points D, E, F respectively. What is the area of $\triangle DEF$?*

Extension Solution 89. *Note that, by the subtended arc theorem, we have $\angle ABE = \angle ADE = \frac{1}{2}\angle ABC$, as well as $\angle ACF = \angle ADF = \frac{1}{2}\angle ACB$, as they are subtended by the same arcs. Hence, Law of Sines says that $EF = 2R\sin\angle EDF = 2R\sin\left(\frac{1}{2}\angle ABC + \frac{1}{2}\angle ACB\right)$. From here, use the fact $[\triangle DEF] = \frac{DE \cdot EF \cdot FD}{4R}$ where R is the circumradius to finish this problem off. The actual answer is*

$$\boxed{2R^2 \sin\left(\frac{A+B}{2}\right)\sin\left(\frac{B+C}{2}\right)\sin\left(\frac{C+A}{2}\right)}$$

Extension Exercise 90. *Let $\triangle ABC$ have side lengths a, b, and c opposite vertices A, B, and C, respectively. Compute the length of each internal angle bisector of $\triangle ABC$ in terms of a, b, and c.*

Extension Solution 90. *This result is in fact a dual consequence of the Angle Bisector Theorem and Stewarts Theorem (see §9). We begin with $\frac{BD}{CD} = \frac{b}{c}$ by ABT, then derive $BD = \frac{ac}{b+c}, CD = \frac{ac}{b+c}$. Applying Stewarts Theorem (with $b^2 \cdot BD + c^2 \cdot BC = ad^2 + a \cdot BD \cdot DC$) yields $d = \frac{\sqrt{bc}}{b+c}\sqrt{(b+c)^2 - a^2}$ after substituting in BD, CD above. Repeat for the other two bisector lengths.*

Extension Exercise 91. *(1992 USAMO P4) Chords AA', BB', and CC' of a sphere meet at an interior point P but are not contained in the same plane. The sphere through A, B, C, and P is tangent to the sphere through A', B', C', and P. Prove that $AA' = BB' = CC'$.*

Extension Solution 91. *Consider the plane passing through A, A, B, B. This plane also passes through point P. This allows us to determine that $\triangle APB$ is isosceles, as the base angles are congruent. Hence $AP = BP$, and so $AP = BP$ by similar logic. It follows that $AA = BB$ as well. Thus, by symmetry, $BB = CC$ (and, by the transitive property, $CC = AA$). Hence we have equality, as desired.*

Extension Exercise 92. *(1959 IMO P5b) An arbitrary point M is selected in the interior of the segment AB. The squares $AMCD$ and $MBEF$ are constructed on the same side of AB, with the segments AM and MB as their respective bases. The circles about these squares, with respective centers P and Q, intersect at M and also at another point N. Let N' denote the point of intersection of the straight lines AF and BC. Prove that the straight lines MN pass through a fixed point S independent of the choice of M.*

Extension Solution 92. *Observe that $\dfrac{AM}{BM} = \dfrac{CM}{BM} = \dfrac{AN}{NB}$, since $\triangle ABN \sim \triangle BCM$. Then $\overline{MN}$ bisects $\triangle ANB$.*

Consider the circle ω with diameter $\overline{AB}$. $\angle ANB$ is right, so N lies on ω. Since $\overline{MN}$ bisects $\triangle ANB$, its intercepting arcs are then congruent. That is, it passes through the constant point of the bisector of arc AB.

Section 7.6: Introduction to 3D Geometry

Exercise 228. *A cube has a surface area of 96 square feet. A sphere is inscribed within that cube; then, another cube is inscribed within the sphere. Find the surface area, in square feet, of the smaller cube.*

Solution 228. *Since the surface area of a cube is equal to 6 times the square of its side length, the outermost cube has a side length of 4 feet. Then the inscribed sphere has a radius of $\dfrac{4}{2} = 2$ feet. The innermost cube would have a space diagonal of length 4 feet, which is equal to $\sqrt{3}$ times its side length. Therefore, the innermost cube would have a surface area of $6\left(\dfrac{4}{\sqrt{3}}\right)^2 = \boxed{32}$.*

Exercise 229. *Determine the sum of the edge lengths of a cube whose volume is numerically equal to six times its surface area (ignoring units).*

Solution 229. *We have $s^3 = 6(6s^2) = 36s^2 \implies s = 36$, so the sum of edge lengths is $12s = \boxed{432}$.*

Exercise 230. *(2000 AMC 12) The point $P = (1, 2, 3)$ is reflected in the xy-plane, then its image Q is rotated by $180°$ about the x-axis to produce R, and finally, R is translated by 5 units in the positive-y direction to produce S. What are the coordinates of S?*

Solution 230. *Reflecting a point in $\mathbb{R}^3$ about the xy-plane changes the sign of the z-coordinate, so at this point we have $(1, 2, -3)$. The second step, rotation $180°$ about the x-axis, changes the signs of both y and z, so that we have $(1, -2, 3)$. Finally, we add 5 to the y-coordinate, ending up with $\boxed{(1, 3, 3)}$.*

Exercise 231. *(2014 AMC 10A) Four cubes with edge lengths 1, 2, 3, and 4 are stacked [at the lower-left corner]. What is the length of the portion of $\overline{XY}$ contained in the cube with edge length 3 [where X is the lower-left, uppermost point of the smallest cube and Y is the upper-right, lowermost point of the largest cube]?*

Solution 231. *This length will be $\dfrac{3}{10}$ of the diagonal's length, which is $\sqrt{10^2 + 4^2 + 4^2} = \sqrt{132} = 2\sqrt{33}$ by the Distance Formula. Hence, our answer is $\boxed{\dfrac{3\sqrt{33}}{5}}$.*

Exercise 232. *Three points, P_1, P_2, and P_3, lie on the surface of a cube with side length 1. Find the maximal area of $\triangle P_1 P_2 P_3$ and prove that your construction is optimal.*

Solution 232. *Let the vertices be (x_i, y_i, z_i), $i = 1, 2, 3$. Then one of the squares of a side length is $S_3 = (x_1 - x_2)^2 + (y_1 - y_2)^2 + (z_1 - z_2)^2$, and the others are similar. By AM-GM, we have $\sqrt[3]{S_1 S_2 S_3} \leq \dfrac{S_1 + S_2 + S_3}{3}$. Note that the RHS has three terms, one of which is $(x_1 - x_2)^2 + (x_2 - x_3)^2 + (x_3 - x_1)^2$, so we expand and aim to minimize $2(x_1^2 + x_2^2 + x_3^2 - x_1 x_2 - x_2 x_3 - x_3 x_1)$ given $0 \leq x_i \leq 1$. This maximum turns out to be $\boxed{2}$.*

Exercise 233. *(2008 AMC 12B) A pyramid has a square base $ABCD$ and vertex E. The area of square $ABCD$ is 196, and the areas of $\triangle ABE$ and $\triangle CDE$ are 105 and 91, respectively. What is the volume of the pyramid?*

Solution 233. *Let h be the height of the pyramid, and let d be the distance from the altitude to $\overline{CD}$. By the Pythagorean Theorem, $13^2 - (14 - a)^2 = h^2$ and $15^2 - a^2 = h^2$, so $a = 9$ and $h = 12$. Hence, the pyramids volume is $\boxed{784}$.*

Exercise 234. *(2009 ARML Individual Round #9) A cylinder with radius r and height h has volume 1 and total surface area 12. Compute $\dfrac{1}{r} + \dfrac{1}{h}$.*

Solution 234. *We have $\pi r^2 h = 1 \implies r^2 h = \dfrac{1}{\pi}$, and $2\pi r(h + r) = 12 \implies r(h + r) = \dfrac{6}{\pi}$. Then $\dfrac{1}{r} + \dfrac{1}{h} = \dfrac{r + h}{rh} = \dfrac{r(h + r)}{r^2 h} = \boxed{6}$.*

Extension Exercise 93. *(2013 AMC 12A) Six spheres of radius 1 are positioned so that their centers are at the vertices of a regular hexagon of side length 2. The six spheres are internally tangent to a larger sphere whose center is the center of the hexagon. An eighth sphere is externally tangent to the six smaller spheres and internally tangent to the larger sphere. What is the radius of this eighth sphere?*

Extension Solution 93. *Form an isosceles triangle from the center of the last sphere and the two diametrically opposed ends of the hexagon. Form another triangle from the points of tangency of the seventh and eighth spheres, and the points of tangency between the seventh and two of its adjacent spheres. Let r be the radius of the eighth sphere, and let h be the height of the first isosceles triangle. Then we can set up the equations $(1 + r)^2 = 2^2 + h^2$, $18 = 3^2 + (h + r)^2$, which together imply $r = \boxed{\dfrac{3}{2}}$.*

Extension Exercise 94. *(2018 Berkeley Math Tournament) A plane cuts a sphere of radius 1 into two pieces, one of which has three times the surface area of the other. What is the area of the disk that the sphere cuts out of the plane?*

Extension Solution 94. *Observe that the plane also cuts the perpendicular diameter of the sphere into the ratio $3 : 1$. Hence the distance from the plane to the spheres center is $\dfrac{1}{2}$, and so the disks radius is $\dfrac{\sqrt{3}}{2} \implies \dfrac{3\pi}{4}$ as the area.*

Extension Exercise 95. *(2018 AMC 12B) Ajay is standing at point A near Pontianak, Indonesia, $0°$ latitude and $110°$ E longitude. Billy is standing at point B near Big Baldy Mountain, Idaho, USA, $45°$ N latitude and $115°$ W longitude. Assume that Earth is a perfect sphere with center C. What is the degree measure of $\angle ACB$?*

Extension Solution 95. *Assume that Earth is a unit sphere with center at the origin of $\mathbb{R}^3$. We can then let $A = (1, 0, 0)$, $B = \left(-\dfrac{1}{2}, \dfrac{1}{2}, \dfrac{\sqrt{2}}{2}\right)$ Note that the angle θ between A and B is such that $\cos\theta = A \cdot B = -\dfrac{1}{2}$ (dot product), or $\theta = \boxed{120°}$.*

Extension Exercise 96. *Compute the number of lattice points lying within the interior of the hypersphere in $\mathbb{R}^4$ with radius $2\sqrt{2}$.*

Extension Solution 96. *Sketch: Casework on the integer quadruples (x, y, z, w) satisfying $x^2 + y^2 + z^2 + w^2 < 8$. This is not too hard to do going by the fact that each variable is constrained between -2 and 2, inclusive.*

Extension Exercise 97. *Let H_n denote the n-dimensional unit hypercube, and denote by $H_n(m)$ the number of m-dimensional faces contained within H_n. Show that*

$$\sum_{m=0}^{n} H_n(m) = 3^n$$

for all positive integers n.

Extension Solution 97. *Consider the k-cubes that emanate from each vertex of an n-cube. In the n-cube, there are n edges with an endpoint at each vertex of the cube; we can choose any k od them to form the edges with a common endpoint of the k-cube. Thus, there are $\binom{n}{k}$ k-cubes emanating from each vertex. Furthermore, we are counting each k-cube 2^k times, once for each of its vertices. Thus, the total number of k-cubes is $2^{n-k}\binom{n}{k}$. Summing this expression from 0 to n over the variable k yields 3^n, as desired. (Hint: Use the binomial expansion of $(1 + 2)^n$ to finish the proof rigorously.)*

Chapter 7 Review Exercises

Exercise 235. *The perimeter of a sector of a circle is equal to 2.018 times its radius. In radians, what is the angle measure of the sector's arc?*

Solution 235. *The perimeter consists of 2 radii, so the actual arc length is $\dfrac{9}{500}$ of the circumference length, or just $\boxed{\dfrac{9}{500}}$ radians by the definition of a radian.*

Exercise 236. *(2009 AMC 12A) One dimension of a cube is increased by 1, another is decreased by 1, and the third is left unchanged. The volume of the new rectangular solid is 5 less than that of the cube. What was the volume of the cube?*

Solution 236. *Let the original side length be s; then $s^3 - 5 = s(s+1)(s-1) = s(s^2-1) = s^3 - s \implies s = 5 \implies s^3 = \boxed{125}$ cubic units.*

Exercise 237. *In rectangle $ABCD$ with $AB = 4$ and $BC = 3$, M is the midpoint of $\overline{CD}$. Lines $\overline{AC}$ and $\overline{BM}$ form four distinct regions in the interior of the rectangle: R_1, R_2, R_3, and R_4, whose areas are in decreasing order from R_1 to R_4. What is $[R_1] - [R_2] + [R_3] - [R_4]$? ($[A]$ denotes the area of A.)*

Solution 237. *Since $[R_1 + R_2 + R_3 + R_4] = 12$, it suffices to compute $[R_2]$ and $[R_4]$. We have $[R_2] = 4$ by coordinates, and $[R_4] = 1$ by similarity. Hence, the desired answer is $\boxed{2}$.*

Exercise 238. *Rectangle $MATH$ has $MA = 20$ and $AT = 18$. Point P is on diagonal $\overline{MT}$ such that $TP = 3PM$, point R is on diagonal $\overline{AH}$ such that $AR = 7RH$, and points O and F are the centroids of triangles $\triangle MAP$ and $\triangle TOR$. What is the area of quadrilateral $PROF$?*

Solution 238. *Let $M = (0, 18), A = (20, 18), T = (20, 0), H = (0, 0)$. Then it follows that $P = \left(5, \dfrac{9}{2}\right)$, $R = \left(\dfrac{5}{2}, \dfrac{9}{4}\right)$, $O = \left(\dfrac{25}{3}, \dfrac{27}{2}\right)$, $F = \left(\dfrac{185}{18}, \dfrac{21}{4}\right)$. Shoelace then yields $\boxed{30}$.*

Exercise 239. *In $\triangle ABC$, $AB = 10$, $BC = 17$, $CA = 21$. Let M be the midpoint of $\overline{AB}$. The perpendicular bisector of $\overline{AB}$ passing through M cuts $\overline{CA}$ at D. Compute the ratio of the area of $\triangle AMD$ to the area of $\triangle ABC$.*

Solution 239. *WLOG assign* $A = (0,0), B = (6,8), C = (21,0)$. *Then the equation of the perpendicular bisector of* $\overline{AB}$ *is* $y = -\frac{3}{4}x + \frac{25}{4}$, *which intersects* $\overline{BC}$ *(i.e. the x-axis) at* $x = \frac{25}{3}$. *The area of* $\triangle AMD$ *is thus* $\frac{1}{2} \cdot \frac{25}{3} \cdot 4 = \frac{50}{3}$, *while the area of* $\triangle ABC$ *is 84 by Herons. Hence the desired answer is* $\boxed{\dfrac{25}{126}}$.

Exercise 240. *What is the tangent of the acute angle formed by* $y = 20x + 18$ *and* $y = 4x + 2$?

Solution 240. *The lines intersect at the point* $(x, y) = (-1, -2)$, *and* $20x + 18$ *hits the x-axis at* $\left(-\frac{9}{10}, 0\right)$; $4x + 2$ *hits the x-axis at* $\left(-\frac{1}{2}, 0\right)$. *Hence, we can draw a right triangle with vertices* $(-1, -2), (-1, 0), (0, 0)$ *and another right triangle with vertices* $(-1, -2), (-1, 0), \left(-\frac{9}{10}\right)$. *Computing the tangents of the vertex angles of each triangle, then applying the tangent subtraction formula, yields* $\boxed{\dfrac{16}{81}}$.

Exercise 241. *(2018 AMC 12B) Square ABCD has side length 30. Point P lies inside the square so that* $AP = 12$ *and* $BP = 26$. *The centroids of* $\triangle ABP$, $\triangle BCP$, $\triangle CDP$, *and* $\triangle DAP$ *are the vertices of a convex quadrilateral. What is the area of that quadrilateral? (Note: The centroid of a polygon is the point whose coordinates are the averages of the respective coordinates of the vertices.)*

Solution 241. *The centroid of a triangle has coordinates that are the means of the coordinates of the triangles vertices, so the lengths of the diagonals are* $\frac{2}{3}$ *of the triangles altitudes (the sum of which is 30, the squares side length). Then the area is* $\frac{2}{3} \cdot \frac{2}{3} \cdot \frac{1}{2} = \frac{2}{9}$ *that of the square, or just* $\boxed{200}$. *(Note that* $AP = 12$ *and* $BP = 26$ *are actually irrelevant here. The only thing you need is the fact that the centroid is the mean of the coordinates of the vertices! Furthermore, while it is possible to coord-bash, it isnt recommended here - the numbers are simply too messy to work with under actual test conditions.)*

Exercise 242. *(2002 AMC 12A) Triangle ABC is a right triangle with* $\angle ACB$ *as its right angle,* $m\angle ABC = 60°$, *and* $AB = 10$. *Let P be randomly chosen inside ABC, and extend* $\overline{BP}$ *to meet* $\overline{AC}$ *at D. What is the probability that* $BD > 5\sqrt{2}$?

Solution 242. *We have* $BC = 5$ *and* $AC = 5\sqrt{3}$. *Take P and obtain D such that* $BD = 5\sqrt{2}$ *and* $CD = 5$. *For* $BD > 5\sqrt{2}$, $CD > 5$, *so we end up with an isosceles right triangle with hypotenuse length* $5\sqrt{2}$. *It follows that P lies in the interior of* $\triangle ABD$. *The probability that this will occur is in fact the ratio* $\dfrac{[\triangle ABD]}{[\triangle ABC]} = \dfrac{AD}{AC}$ *(since both triangles have equal altitude lengths). This ratio is* $\dfrac{AC - CD}{AC} = 1 - \dfrac{CD}{AC} = 1 - \dfrac{5}{5\sqrt{3}} = \boxed{\dfrac{3 - \sqrt{3}}{3}}$.

Exercise 243. *A right triangle has integer side lengths, and its longer leg length is 1 less than its hypotenuse length. Let r be its inradius, and let R be its circumradius. If* $1 < |R - r| < 100$, *find the sum of the possible values for the triangle's hypotenuse.*

Solution 243. *Let the hypotenuse length be x, so the longer leg has length* $x - 1$ *(and the shorter leg has length* $\sqrt{2x + 1}$). *Therefore,* $2x + 1$ *is a perfect square. We know that* $r = \dfrac{\sqrt{2x+1} - 1}{2}$, $R = \dfrac{x}{2}$, *and that* $r = \dfrac{A}{s}$ *where A is the triangles area and s is half the perimeter. Testing small values of x reveals that* $x = 5$ *doesnt quite work (as the lower bound of 1 is excluded), but* $x = 13$ *works, as in the 5-12-13 triangle. In addition,* $x = 25$ *works, as do* $x = 41$, $x = 61$, $x = 85$, *and so forth. This continues up until the 19-180-181 triangle, so* $x = 5, 13, 25, 41, 61, 85, 113, 145, 181 \implies \boxed{669}$.

Exercise 244. *(2002 AMC 12A) In triangle* $\triangle ABC$, *side* $\overline{AC}$ *and the perpendicular bisector of* $\overline{BC}$ *meet in point D, and* $\overline{BD}$ *bisects* $\angle ABC$. *If* $AD = 9$ *and* $DC = 7$, *what is the area of triangle* $\triangle ABD$?

Solution 244. *The perpendicular bisector of $\triangle BCD$ touches vertex B, implying that the triangle is isosceles. Let $\angle C = x$, so that $\angle B = 2x$. Then $\angle ABD = x$ and $\angle ADB = 2x$. Thence $\triangle ABD$ is similar to $\triangle ACB$, so $\dfrac{16}{AB} = \dfrac{AB}{9} \implies AB = 12$. Finally, we have $[\triangle ABD] = \boxed{14\sqrt{5}}$.*

Extension Exercise 98. *(2013 AIME I) A rectangular box has width 12 inches, length 16 inches, and height $\frac{m}{n}$ inches, where m and n are relatively prime positive integers. Three faces of the box meet at a corner of the box. The center points of those three faces are the vertices of a triangle with an area of 30 square inches. Find $m + n$.*

Extension Solution 98. *Let the boxs height be h. Using the Pythagorean Theorem repeatedly, we can obtain a triangle with sides $10, \sqrt{\dfrac{h^2}{4} + 64}, \sqrt{\dfrac{x^2}{4} + 36}$. Since the triangles area is 30, the altitude from the length-10 side must be 6. The altitude creates two more triangles, so we again apply Pythagoras to get $\sqrt{28 + \dfrac{x^2}{4}} + \dfrac{x}{2} = 10$ which can be solved to obtain $x = \dfrac{36}{5} \implies \boxed{41}$.*

Extension Exercise 99.

(a) *Does a tetrahedron exist with edge lengths 1, 2, 2, 3, 4, 5? If so, construct an example of such a tetrahedron. If not, prove that no such tetrahedron exists.*

(b) *(2014 Caltech-Harvey Mudd Math Competition) What's the greatest pyramid volume one can form using edges of length 2, 3, 3, 4, 5, 5, respectively?*

Extension Solution 99.

(a) *No such tetrahedron exists. Because of the 1, the only possible side lengths neighboring it would be 2 and 2, and in turn, the only possibility for the other adjacent side length is 3 by the Triangle Inequality. For the 2-3 triangle, the only possibility is 4 by the same lines, but then filling the remaining edge with a 5 would lead to a degenerate 1-3-5 triangle. Hence this configuration is impossible.*

(b) *By the Triangle Inequality, given a triangle side length of 2, the other two side lengths may only be $(3,3)$, $(3,4)$, $(4,5)$, or $(5,5)$. If two sides share the length-2 edge, then only the three cases $(3,3)$ and $(5,5)$ (case 1), $(3,3),(4,5)$ (case 2), or $(3,4),(5,5)$ (case 3) are acceptable.*

In case 1, the volume is $\dfrac{8\sqrt{2}}{3}$, since the pyramid has 2 right angles. In case 2, there are 2 possibilities for the pyramid, but their volumes are equal, and less than $\dfrac{8\sqrt{2}}{3}$. In the final case, the outcome is the same as in case 2. Therefore the answer is $\boxed{\dfrac{8\sqrt{2}}{3}}$.

Extension Exercise 100. *Recall that, in a triangle, the intersection of the angle bisectors determines the incenter. Is it necessarily true that the intersection of the angle bisector planes in a tetrahedron determines the center of the inscribed sphere? What about for higher-dimensional analogues of the tetrahedron (known as n-polytopes, where $n \geq 4$)?*

Extension Solution 100. *Yes; the first statement is true. More specifically, we are dealing with dihedral angles in dimensions higher than 2. If A, B, C, D are the tetrahedrons vertices, to obtain the bisector plane of two planes, construct lines AC, AD perpendicular to $\overline{AB}$. The resulting plane from the bisector of $\angle CAD$ with $\overline{AB}$ is the bisector plane, as desired. Repeat this construction for ABC, ACD and for ABC, DBC.*

Extension Exercise 101. *(1972 USAMO P2) A given tetrahedron $ABCD$ is isosceles, that is, $AB = CD$, $AC = BD$, $AD = BC$. Show that the faces of the tetrahedron are acute-angled triangles.*

Extension Solution 101. *For contradiction assume $\angle BAC \geq 90°$; then, as all three sides of $\triangle BAC, \triangle CDB$ are congruent, $\triangle BAC \cong \triangle CDB$, and so $\angle BDC = \angle BAC > 90°$. If we were to construct a sphere with $\overline{BC}$ as its diameter, then since $\angle BAC, \angle BDC < 90°$, A, D both lie within the closed ball. (Ideally, $AD = BC$. This occurs when $\overline{AD}$ is also a diameter of the sphere. For this to hold, all four points should be coplanar, but this is a contradiction, since $ABCD$ would then be degenerate. The desired conclusion follows.*

Extension Exercise 102. *(2010 ARML Individual Round #8) In square $ABCD$ with diagonal 1, E is on AB and F is on BC with $\angle BCE = \angle BAF = 30°$. If CE and AF intersect at G, compute the distance between the incenters of triangles AGE and CGF.*

Extension Solution 102. *Observe that $\triangle AGE$, $\triangle CGF$ are congruent and isosceles. Let M be the midpoint of $\overline{AC}$, such that $AM = \dfrac{1}{2}$ and $AG = \dfrac{AM}{\cos 15°} = \dfrac{\sqrt{6} - \sqrt{2}}{2}$. Then let I_1, I_2 be the incenters of $\triangle AGE, \triangle CGF$ respectively. We have $\cos 15° = \dfrac{\sqrt{6} - \sqrt{2}}{4 I_1 G} \implies I_1 G = 2 - \sqrt{3}$. Finally, $I_1 I_2 = I_1 G + G I_2 = 2 I_1 G = \boxed{4 - 2\sqrt{3}}$.*

Extension Exercise 103. *Right triangle $\triangle ABC$ with $\angle B = 90°$ and hypotenuse $\overline{AC}$ has square $ACDE$ with $\overline{AC}$ as an edge, and not overlapping with the interior of $\triangle ABC$. If $(AD + BE)(AD - BE) = 24$, and $\triangle ABC$ has integer side lengths, compute the area of pentagon $ABCDE$.*

Extension Solution 103. *Let $B = (0, 0), A = (0, y), C = (x, 0)$. Then $D = (x + y, x), E = (y, x + y)$. Note that $(AD + BE)(AD - BE) = AD^2 - BE^2 = 24$, with $AD^2 = (x + y)^2 + (x - y)^2$ and $BE^2 = y^2 + (x + y)^2$. This implies that $(x - y)^2 - y^2 = 24 \implies x^2 - 2xy = 24 \implies x(x - 2y) = 24$. Thus, x is a factor of 24. In addition, $x^2 + y^2$ is a perfect square, since the hypotenuse must also be an integer. Indeed, $x = 12, y = 5$ works, so the answer is $30 + 13^2 = \boxed{199}$.*

Chapter 8: Intermediate Geometry

Section 8.1: Law of Sines

Exercise 245. *If $\triangle ABC$ is a 13-14-15 triangle, compute $\sin A + \sin B + \sin C$.*

Solution 245. *By the Extended Law of Sines, we have $\dfrac{13}{\sin A} = \dfrac{14}{\sin B} = \dfrac{15}{\sin C} = 2R$, with $R = \dfrac{13 \cdot 14 \cdot 15}{336} = \dfrac{65}{8}$. Thus, $\sin A + \sin B + \sin C = \dfrac{4}{5} + \dfrac{56}{65} + \dfrac{12}{13} = \boxed{\dfrac{168}{65}}$.*

Exercise 246. *Show that the area of $\triangle ABC$ is $\dfrac{abc}{4R}$, where a, b, and c are the side lengths and R is the circumradius.*

Solution 246. *Let $\triangle ABC$ be inscribed in circle ω with center O, and let D be the intersection point of $\overline{BO}$ with ω. Then $\angle BAD$ is right, as $\overline{BD}$ is a diameter of ω. Also, $\angle ADB = \angle BCA$ by virtue of subtending arc AB. Hence, $\triangle BAD$ and $\triangle BEC$ are AA similar, so it follows that $\dfrac{BD}{BA} = \dfrac{BC}{BE}$, which implies $\dfrac{a}{h} = \dfrac{2R}{c}$. Some substitutions with area $= \dfrac{bh}{2}$ yield $R = \dfrac{abc}{4A}$ as desired.*

Exercise 247. *(2014 AMC 10A) In rectangle $ABCD$, $\overline{AB} = 20$ and $\overline{BC} = 10$. Let E be a point on $\overline{CD}$ such that $\angle CBE = 15°$. What is $\overline{AE}$?*

Solution 247. *By a direct application of Law of Sines, $CE = 10 \cdot \dfrac{\sin 15°}{\sin 75°} = 20 - 10\sqrt{3}$, so $DE = 10\sqrt{3}$. Since $AD = 10$, $AE = \boxed{20}$.*

Exercise 248. *Prove that the area of $\triangle ABC$ is $\dfrac{ab \sin C}{2}$.*

Solution 248. *This is apparent when we observe that the height is the hypotenuse of one of the right triangles it forms, multiplied with the sine of its opposite angle a priori.*

Exercise 249. *Show that, in $\triangle ABC$, $\sin A$, $\sin B$, and $\sin C$ must obey the Triangle Inequality.*

Solution 249. *By Law of Sines, the side lengths are proportional to the sine values. Hence, if the sine values do not follow the Triangle Inequality, neither can the side lengths.*

Extension Exercise 104. *(2003 AIME I) Triangle ABC is isosceles with $AC = BC$ and $\angle ACB = 106°$. Point M is in the interior of the triangle so that $\angle MAC = 7°$ and $\angle MCA = 23°$. Find the number of degrees in $\angle CMB$.*

Extension Solution 104. *WLOG assume $CA = CB = 1$; then using LoS on $\triangle AMC$ to obtain $\dfrac{1}{\sin 150°} = \dfrac{MC}{\sin 7°} \implies MC = 2\sin 7°$. Thence using LoC on $\triangle MCB$, we get that $MB^2 = 4\sin^2 7° - 4\sin 7° \cdot \cos 83° + 1 = 1$, as $\sin 7° = \cos 83°$. Then $\triangle MCB$ is isosceles, with $\angle CMB = \boxed{83°}$.*

Extension Exercise 105. *For $\triangle ABC$, prove the following:*

(a) $a = b\cos C + c\cos B$

(b) $a(\sin B - \sin C) + b(\sin C - \sin A) + c(\sin A - \sin B) = 0$

Extension Solution 105.

(a) If $\triangle ABC$ is acute, then $a = BC = BD + CD$ and $\cos B = \dfrac{DB}{AB} \implies DB = AB\cos B \implies DB = c\cos B$. $\cos C = \dfrac{DC}{AC}$, so $CD = AC\cos C = b\cos C$. Substitution directly yields the desired result. (The process for when $\triangle ABC$ is obtuse is similar, except with $a = BC = CD - BD$.) If $\triangle ABC$ is right, it is worth mentioning that $a = BC$ and $\cos B = \dfrac{CB}{AB}$, with $a = c\cos B = c\cos B + b\cos C$ (since $\cos C = \cos 90° = 0$, assuming WLOG that $\angle C$ is the right angle).

(b) Note that the above actually holds for all three sides, not just a. Hence, $b = c\cos A + a\cos C$ and $c = a\cos B + b\cos A$ as well. The given equation simplifies to $a\sin B - a\sin C + b\sin C - b\sin A + c\sin A - c\sin B = 0$, which holds after substitution of all three formulae.

Extension Exercise 106. *(2001 HMMT February Geometry Test) Equilateral triangle $\triangle ABC$ with side length 1 is drawn. A square is drawn such that its vertex at A is opposite to its vertex at the midpoint of $\overline{BC}$. Find the area enclosed within the intersection of the insides of the triangle and square.*

Extension Solution 106. *Let D be the midpoint of $\overline{BC}$, let F be the point of intersection of the square and triangle that lies on $\overline{AC}$. Furthermore, let l be the length of $\overline{FC}$, s be the triangles side length, and x the length of $\overline{AD}$. Applying law of sines on $\triangle CDF$ yields $\dfrac{2\sin 75^\circ}{s} = \dfrac{\sin 45^\circ}{l} \implies l = \dfrac{s\sqrt{2}}{4\sin 75^\circ}$.*

Furthermore, the desired area evaluates to $\dfrac{x(s-l)}{2} = \dfrac{1}{2}\cdot(sx)\cdot\left(1 - \dfrac{\sqrt{2}}{4\sin 75^\circ}\right)$. Then by the Pythagorean Theorem on $\triangle ABD$, we have $y = \dfrac{\sqrt{3}}{2}x$, and the area is $\boxed{\dfrac{3\sqrt{3}-3}{8}}$.

Section 8.2: Law of Cosines

Exercise 250. *(2017 AMC 12B) Let ABC be an equilateral triangle. Extend side $\overline{AB}$ beyond B to a point B' so that $BB' = 3AB$. Similarly, extend side $\overline{BC}$ beyond C to a point C' so that $CC' = 3BC$, and extend side $\overline{CA}$ beyond A to a point A' so that $AA' = 3CA$. What is the ratio of the area of $\triangle A'B'C'$ to the area of $\triangle ABC$?*

Solution 250. *Let $x = AB = BC = CA$. By LoC, setting $AB = BC = CA = y$, we have $y^2 = (3x)^2 + (4x)^2 - 2(3x)(4x)(\cos 120^\circ) = 25x^2 + 12x^2 = 37x^2$. By AA similarity, the answer is $\boxed{37:1}$.*

Exercise 251. *A triangle has side lengths 4 and 8, and it has an area of $3\sqrt{15}$. Find the possible lengths of the third side.*

Solution 251. *Using the area formula $\dfrac{1}{2}ab\sin C$, with $a = 4$ and $b = 8$, we have $3\sqrt{15} = 16\sin C \implies \sin C = \dfrac{3\sqrt{15}}{16} \implies \cos C = \dfrac{11}{16}$ as $\sin^2 C + \cos^2 C = 1$. Hence, LoC implies that $c^2 = 4^2 + 8^2 - 2\cdot8\cdot4\cdot\cos C = 80 - 64\cos C = 36 \implies c = \boxed{6}$.*

Exercise 252. *(2006 AMC 12B) Isosceles $\triangle ABC$ has a right angle at C. Point P is inside $\triangle ABC$, such that $PA = 11$, $PB = 7$, and $PC = 6$. Legs $\overline{AC}$ and $\overline{BC}$ have length $s = \sqrt{a + b\sqrt{2}}$, where a and b are positive integers. What is $a + b$?*

Solution 252. *Let $BC = CA = s$. Using LoC on $\triangle PBC$, we get $PB^2 = BC^2 + PC^2 - 2BC\cdot PC\cdot\cos(\theta) = 49$, where $\theta = \angle BCP$. Then $36 + s^2 - 12s\cos(\theta) = 49 \implies \cos(\theta) = \dfrac{s^2 - 13s}{12s}$. Applying LoC again on $\triangle PAC$, we get $PA^2 = AC^2 + PC^2 - 2AC\cdot PC\cdot\cos(90^\circ - \theta) = 121 \implies 36 + s^2 - 12s\sin(\theta) = 121 \implies \sin(\theta) = \dfrac{s^2 - 85}{12s}$. Use the fact that $\sin^2(\theta) + \cos^2(\theta) = 1$ to obtain a quartic in terms of s, and solve in terms of s^2. Then discard the extraneous solution and simplify to obtain $s = \sqrt{85 + 42\sqrt{2}} \implies \boxed{127}$.*

Exercise 253. *Derive the Triangle Inequality from the Law of Cosines. (Hint: Use the fact that $\cos$ is bounded.)*

Solution 253. *Note that the side lengths of the triangle can be written as $a = \|u\|, b = \|v\|, c = \|u - v\|$. The cosine of the angle θ between u, v is $\dfrac{u\cdot v}{\|u\|\,\|v\|} = \dfrac{ab\cos(\theta)}{ab} = \cos(\theta)$.*

Since $\cos(\theta) \geq -1 \implies -\cos(\theta) \leq 1$, we have $c^2 \leq a^2 + b^2 + 2ab = (a + b)^2$ by LoC. Both $c > 0$ and $a + b > 0$, taking square roots preserves the inequality. Hence, $c \leq a + b$, or $\|u - v\| \leq \|u\| + \|v\|$. Equality holds when $\cos(\theta) = -1 \implies \theta = \pi$.

Exercise 254. $\triangle ABC$ *has perimeter 19. We know that* $\angle A = 60°$, *and* $BC = 9$. *What is the area of* $\triangle ABC$?

Solution 254. *Let the other two side lengths be* b *and* c. *By Law of Cosines, we know that* $9^2 = b^2 + c^2 - 2bc \cdot \cos 60° \implies 81 = b^2 + c^2 - bc$. *We also know that* $b + c = 10$; *squaring this yields* $b^2 + 2bc + c^2 = 100$, *so* $bc = \dfrac{19}{3}$, *and solving the resulting system yields* $b = \dfrac{15 + 2\sqrt{42}}{3}$, $c = \dfrac{15 - 2\sqrt{42}}{3}$ *(in either order). Hence, Heron's formula yields the desired area of* $\boxed{\dfrac{19\sqrt{3}}{12}}$.

Extension Exercise 107. *(2013 AIME I) A paper equilateral triangle ABC has side length 12. The paper triangle is folded so that vertex A touches a point on side $\overline{BC}$ a distance 9 from point B. The length of the line segment along which the triangle is folded can be written as $\frac{m\sqrt{p}}{n}$, where m, n, and p are positive integers, m and n are relatively prime, and p is not divisible by the square of any prime. Find $m + n + p$.*

Extension Solution 107. *Let P, Q be the points on $\overline{AB}$ and $\overline{AC}$ onto which the paper is folded. Let D be on $\overline{BC}$ be the point where the fold from A intersects $\overline{BC}$. Let $AP = a, AQ = b, PQ = x$. Then $PD = a, QD = b, BP = 12 - a, CQ = 12 - b, BD = 9, CD = 3$. LoC on $\triangle BPD$ says that $a^2 = (12 - a)^2 + 9^2 - 2(12 - a) \cdot 9 \cdot \cos 60° \implies a = \dfrac{39}{5}$. Similarly, LoC on $\triangle CQD$ says that $b = \dfrac{39}{7}$. A third application of LoC on $\triangle DPQ$ reveals that $x = \dfrac{39\sqrt{39}}{35} \implies \boxed{113}$.*

Extension Exercise 108. *(2013 HMMT November Team Round) Consider triangle $\triangle ABC$ with side lengths $AB = 4$, $BC = 7$, and $AC = 8$. Let M be the midpoint of segment $\overline{AB}$, and let N be the point on the interior of segment $\overline{AC}$ that also lies on the circumcircle of triangle $\triangle MBC$. Compute BN.*

Extension Solution 108. *Denote $\theta = \angle BAC$. Then $\cos\theta = \dfrac{4^2 + 8^2 - 7^2}{2 \cdot 4 \cdot 8}$ by Law of Cosines. $AM = 2$, and power of a point yields $AB \cdot AM = AC \cdot AN$, i.e. $AN = 1 \implies NC = 7$. LoC on $\triangle BAN$ yields $BN^2 = \dfrac{105}{8}$, so $BN = \boxed{\dfrac{\sqrt{210}}{4}}$.*

Extension Exercise 109. *(2018 AIME I) In $\triangle ABC$, $AB = AC = 10$ and $BC = 12$. Point D lies strictly between A and B on $\overline{AB}$ and point E lies strictly between A and C on $\overline{AC}$ so that $AD = DE = EC$. Then AD can be expressed in the form $\dfrac{p}{q}$, where p and q are relatively prime positive integers. Find $p + q$.*

Extension Solution 109. *Let $x = AD$, and let F be the foot of the A-altitude. Write $AE = 10 - x$, so that we can see that $\triangle ABF$ is a 6-8-10 right triangle. Hence $\sin \angle BAF = \dfrac{3}{5} \implies \cos \angle BAF = \dfrac{4}{5}$. It remains to compute $\cos \angle BAC$, which is just $\cos 2\angle BAF = \dfrac{7}{25}$ (using the double-angle identity for $\cos$). Using Law of Cosines allows us to set up the following:*

$$DE^2 = AD^2 + AE^2 - 2 \cdot AD \cdot AE \cdot \cos \angle BAC$$

$$x^2 = x^2 + (10 - x)^2 - 2x(10 - x) \cdot \frac{7}{25} \implies x = \frac{250}{39} \implies \boxed{289}$$

Extension Exercise 110.

(a) *(2014 HMMT February Algebra Test) Given that a, b, and c are complex numbers satisfying*

$$a^2 + ab + b^2 = 1 + i$$

$$b^2 + bc + c^2 = -2$$

$$c^2 + ca + a^2 = 1$$

compute $(ab + bc + ca)^2$.

(b) Can you generalize this to any system of the form

$$a^2 + ab + b^2 = z$$

$$b^2 + bc + c^2 = x$$

$$c^2 + ca + a^2 = y$$

Extension Solution 110. *We claim that the answer in general is* $\dfrac{1}{16}\left((x+y+z)^2 - 2(x^2+y^2+z^2)\right)$ *(which yields the answer of* $\boxed{-\dfrac{11+4i}{3}}$ *for part (a)).*

Plug in the substitutions we have made for x, y, z to obtain a polynomial f in terms of a, b, c. We show that $g(a, b, c) = 0$ for all choices of $a, b, c \in \mathbb{R}^+$. By Law of Cosines, let point $P \in \triangle ABC$ with $PA = a, PB = b, PC = c$, and $\angle PAB = \angle PBC = \angle PCA = 120°$, and with $x = BC^2, y = CA^2, z = AB^2$. Herons formula yields $[\triangle ABC]^2 = \dfrac{1}{16}\left((x+y+z)^2 - 2(x^2+y^2+z^2)\right)$, as desired.

Section 8.3: Strategies for Angle Chasing

Exercise 255. *(2017 CMIMC Geometry Test) Cyclic quadrilateral $ABCD$ satisfies $\angle ABD = 70°$, $\angle ADB = 50°$, and $BC = CD$. Suppose $\overline{AB}$ intersects $\overline{CD}$ at point P, while $\overline{AD}$ intersects $\overline{BC}$ at point Q. Compute $\angle APQ - \angle AQP$.*

Solution 255. *Observe that $\angle BAD = 180 - \angle ABD - \angle ADB = 60°$, so $\angle PCQ = 120°$. Thus, as $BC = CD$, $\overline{AC}$ is an angle bisector of $\angle BAD$. If I is the incenter of $\triangle APQ$, then $\angle PIQ = 90° + \dfrac{\angle PAQ}{2} = 120°$, where $PCIQ$ is a cyclic quadrilateral. A, I, C are collinear, so C actually coincides with I. Furthermore, $\angle APD = \angle ABD - \angle BDC = 40°$, and $\angle AQB = \angle ADB - \angle DBC = 20°$, we have $\angle APQ - \angle AQP = 2(\angle APC - \angle AQC) = \boxed{40°}$.*

Exercise 256. *(2008 AIME I) Let $ABCD$ be an isosceles trapezoid with $\overline{AD} \parallel \overline{BC}$ whose angle at the longer base $\overline{AD}$ is $\dfrac{\pi}{3}$. The diagonals have length $10\sqrt{21}$, and point E is at distances $10\sqrt{7}$ and $30\sqrt{7}$ from vertices A and D, respectively. Let F be the foot of the altitude from C to $\overline{AD}$. The distance EF can be expressed in the form $m\sqrt{n}$, where m and n are positive integers and n is not divisible by the square of any prime. Find $m + n$.*

Solution 256. *Since points A, D, E are collinear (which is an argument by contradiction assuming $\triangle ADE$ is non-degenerate involving the circle with radius A), $AD = 20\sqrt{7}$ and $\triangle ADC \cong \triangle ACF$ (both 30-60-90 triangles). Hence, $AF = 15\sqrt{7}$, and $EF = EA + AF = 25\sqrt{7} \implies \boxed{32}$.*

Exercise 257. *Let $\triangle ABC$ be an isosceles right triangle with $\angle B = 90°$. Points D and E lie in the plane of $\triangle ABC$ such that $ACDE$ is a square that does not overlap with $\triangle ABC$. Let O be the center of $ACDE$. Compute $\angle OBC$ in degrees.*

Solution 257. *Note that O is directly below E and above C; hence $\triangle OBC$ is itself isosceles right, and so $\angle C = 90°$. Thus, $\angle OBC = \boxed{45°}$.*

Exercise 258. *(2018 AMC 12A) In $\triangle PAT$, $\angle P = 36°$, $\angle A = 56°$, and $PA = 10$. Points U and G lie on sides $\overline{TP}$ and $\overline{TA}$, respectively, so that $PU = AG = 1$. Let M and N be the midpoints of segments $\overline{PA}$ and $\overline{UG}$, respectively. What is the degree measure of the acute angle formed by lines MN and PA?*

Solution 258. *Extend $\overline{PN}$ to a point Q with $QN = PN$. Then $\overline{MN}$ is an angle bisector of $\triangle PAQ$, such that $\angle QAP = \angle NMP$. Subsequently observe that $\triangle PDN \cong QGN$, and so $QG = AG = DP = 1$. It also follows that $\angle AQG = \angle QAG$, and $\angle GQN = \angle NPD$. Along with $\angle GQN + \angle QPA = \angle QPD + \angle QPA = \angle DPA = 36°$, this implies that $2\angle QAG = 180° - 56° - 36° = 88°$. Finally, $\angle QAP = \angle QAG + \angle TAP = 56° + 44° = 100°$, with the acute angle being $\boxed{80°}$.*

Extension Exercise 111. *(2013 HMMT February Guts Round) Consider triangle $\triangle ABC$ with $\angle A = 2\angle B$. The angle bisectors from A and C intersect at D, and the angle bisector from C intersects $\overline{AB}$ at E. If $\frac{DE}{DC} = \frac{1}{3}$, compute $\frac{AB}{AC}$.*

Extension Solution 111. *For brevity, denote $x = AE, y = BE$. Angle bisector theorem on $\triangle ACE$ yields $\frac{x}{DE} = \frac{AC}{DC} \implies AC = 3x$. Angle chasing gives that $\angle ADE = \angle AED$, so $\triangle AED$ is isosceles with $AD = x = AE$. Further note that $\triangle CDA$ and $\triangle CEB$ are similar, which yields $\frac{y}{DC + DE} = \frac{x}{DC} \implies \frac{y}{x} = \frac{4}{3}$, so $AB = x + \frac{4}{3}x = \frac{7}{3}x$. Thus, $\frac{AB}{AC} = \boxed{\frac{7}{9}}$.*

Extension Exercise 112. *(2011 USAJMO P5) Points A, B, C, D, E lie on a circle ω and point P lies outside the circle. The given points are such that*

(i) lines PB and PD are tangent to ω,

(ii) P, A, C are collinear, and

(iii) $\overline{DE} \parallel \overline{AC}$.

Prove that $\overline{BE}$ bisects $\overline{AC}$.

Extension Solution 112. *If O is the circles center and $\overline{AC}$ and $\overline{BE}$ intersect at X, and $\angle OPA = x, \angle OPD = y$, then $\angle OPB = \angle OPD = y, 2\angle BED = \angle DOB = 180 - 2y$, and $\angle ODE = \angle PDE - 90° = (90 - x - y)°$. Thus, $\angle OBE = (90 - x)° = \angle OPA$. It follows that $PXBO$ is cyclic, and hence, $\angle OXP = \angle OBP = 90°$, with X the midpoint of $\overline{AC}$, as desired.*

Section 8.4: Proving Triangle Theorems: Part 1

Exercise 259. *Prove that every triangle has an orthocenter (which need not lie within the interior of the triangle).*

Solution 259. *Let the A-altitude hit $\overline{BC}$ at D, and define E, F on $\overline{AB}, \overline{AC}$ respectively. We show that, given that $\overline{AD}, \overline{CE}$ intersect at O, $\overline{BF}$ also passes through O.*

The equation of $\overline{AD}$ is just $x = d$, where d is the x-coordinate of A. The equation of $\overline{CE}$ is $y = -\frac{dx}{a} + \frac{dc}{a}$, where a is the y-coordinate of A and c is the x-coordinate of c.

The x-coordinate of O must be $x = d$, and substitution yields $y = \frac{dc - d^2}{a}$. We check to say that this point lies on $\overline{BF}$, and indeed it does! Hence the orthocenter must exist.

Exercise 260. *Suppose $\triangle ABC$ has area 2019. Let D, E, and F be points on $\overline{AB}$, $\overline{BC}$, and $\overline{CA}$, respectively, such that $AD = 3BD$, $BE = 3CE$, and $CF = 3AF$. Compute the area of $\triangle DEF$.*

Solution 260. *This is a direct application of Rouths Theorem, which states that the area is $2019 \cdot \frac{3^3 - 1}{(3^2 + 3 + 1)^3} = 2019 \cdot \frac{26}{13^3} = \boxed{\frac{4038}{169}}$.*

Exercise 261. *Show that Routh's Theorem implies that the medians of a triangle are concurrent.*

Solution 261. *Essentially, we must prove the existence of the centroid using Rouths Theorem. This is achieved by setting $x = y = z = 1$, since the numerator simplifies to 0 (i.e. the degenerate case).*

Exercise 262. *Triangles $\triangle ABC$ and $\triangle DEF$ are similar (with AB corresponding to $\overline{DE}$, $\overline{BC}$ corresponding to $\overline{EF}$, and $\overline{CA}$ corresponding to $\overline{FA}$). Show that the ratio of correspondent medians is the same as the ratio of correspondent sides.*

Solution 262. *By similarity, $\frac{AB}{BC} = \frac{DE}{EF}$. Since the midpoint M of $\overline{BC}$ corresponds to M of $\overline{EF}$, we also have $\frac{AB}{BM} = \frac{DE}{EM}$. Since $\angle B = \angle B$, it follows that $\triangle ABM$ and $\triangle DEM$ are similar. Hence, $\frac{AB}{DE} = \frac{AM}{DM}$.*

Exercise 263. *(2011 Stanford Math Tournament) Let $\triangle ABC$ be any triangle, and D, E, F be points on $\overline{BC}$, $\overline{CA}$, $\overline{AB}$ such that $CD = 2BD$, $AE = 2CE$ and $BF = 2AF$. $\overline{AD}$ and $\overline{BE}$ intersect at X, $\overline{BE}$ and $\overline{CF}$ intersect at Y, and $\overline{CF}$ and $\overline{AD}$ intersect at Z. Find $\frac{Area(\triangle ABC)}{Area(\triangle XYZ)}$.*

Solution 263. *Apply Menelaus Theorem on $\triangle ABD$ with C, F, X collinear; this yields $3DX = 4XA$. Applying Menelaus again on $\triangle ADC$ with B, Y, E collinear yields $AY = 6YD$. Thus, $AX : XY : YD = 3 : 3 : 1$. By symmetry, we can extend this to $CZ : ZX : XF$ and $BY : YZ : ZE$. We find that $[\triangle ABC] = \frac{3}{2}[\triangle ADC] = \frac{3}{2} \cdot \frac{7}{3}[\triangle XYC] = \boxed{7}[\triangle XYZ]$ (since $\triangle XYC$ and $\triangle XYZ$ share the same base, $\overline{XY}$).*

Exercise 264. *Prove that the internal angle bisectors of two angle of a triangle, and the external angle bisector of the last angle, intersect their opposite sides at three collinear points.*

Solution 264. *Since $\overline{AE}, \overline{BD}$ are internal bisectors, we have $\frac{AD}{DC} = \frac{AB}{BC}$ as well as $\frac{CE}{EB} = \frac{CA}{AB}$. Since $\overline{CF}$ is the external bisector, we have $\frac{BF}{FA} = -\frac{BC}{CA}$ by Ceva. Thus, $\frac{AD}{DC} \cdot \frac{CE}{EB} \cdot \frac{BF}{FA} = -\frac{AB}{BC} \cdot \frac{CA}{AB} \cdot \frac{BC}{CA} = -1$, which implies the collinearity of D, E, F by the converse of Menelaus.*

Extension Exercise 113. *(1996 USAMO P5) Let ABC be a triangle, and M an interior point such that $\angle MAB = 10°$, $\angle MBA = 20°$, $\angle MAC = 40°$ and $\angle MCA = 30°$. Prove that the triangle is isosceles.*

Extension Solution 113. *We have $\angle AMB = 150°, \angle AMC = 110°$ to begin with. By Law of Sines on $\triangle ABM$ and $\triangle ACM$, we have $2AB = \frac{AM}{\sin 20°}$ and $\frac{AC}{\sin 110°} = 2AM$. Then $\frac{AB}{AC} = \frac{\frac{1}{4}}{\sin 20° \sin 110°}$.*

WLOG let $AB = \frac{1}{4}$, $AC = \sin 20° \sin 110°$. Then Law of Cosines states that $BC^2 = AB^2 + AC^2 - 2 \cdot AB \cdot BC \cdot \cos \angle BAC$, which leads to $BC = \frac{1}{4}$, or $AB = BC$, as desired.

Section 8.5: Power of a Point

Exercise 265. *(1971 CMO P1) Arc DEB is a chord of a circle such that $DE = 3$ and $EB = 5$. Let O be the center of the circle. Join $\overline{OE}$ and extend $\overline{OE}$ to cut the circle at C. Given $EC = 1$, find the radius of the circle.*

Solution 265. *Extend $\overline{CO}$ to intersect the circle at P. Let the circles radius be r; then applying Power of a Point yields $CE \cdot EP = BE \cdot ED \implies 2r - 1 = 15 \implies r = \boxed{8}$.*

Exercise 266. *(1995 AHSME) In the figure, $\overline{AB}$ and $\overline{CD}$ are diameters of the circle with center O, $\overline{AB} \perp \overline{CD}$, and chord $\overline{DF}$ intersects $\overline{AB}$ at E. If $DE = 6$ and $EF = 2$, then the area of the circle is*

Solution 266. *Let the radius of the circle be r, and let $OE = x$. By Pythagorean Theorem, $OD^2 + OE^2 = DE^2 \implies r^2 + x^2 = 36$. Power of a point says that $AE \cdot EB = DE \cdot EF$, or $r^2 - x^2 = 12$. Then $r^2 = 24$ and the area is $\pi r^2 = \boxed{24\pi}$.*

Exercise 267. *(2017 AMC 12A) Quadrilateral $ABCD$ is inscribed in circle O and has side lengths $AB = 3, BC = 2, CD = 6$, and $DA = 8$. Let X and Y be points on $\overline{BD}$ such that $\frac{DX}{BD} = \frac{1}{4}$ and $\frac{BY}{BD} = \frac{11}{36}$. Let E be the intersection of line AX and the line through Y parallel to $\overline{AD}$. Let F be the intersection of line CX and the line through E parallel to $\overline{AC}$. Let G be the point on circle O other than C that lies on line CX. What is $XF \cdot XG$?*

Solution 267. *First compute $\frac{XY}{BD} = \frac{4}{9}$, then observe the similarity of $\triangle AXD, \triangle EXY$ to get $\frac{DX}{XY} = \frac{9}{16}$, as well as the similarity of $\triangle ACX, \triangle EFX$ to get $\frac{AX}{XE} = \frac{9}{16}$; thus it follows that $\frac{XF}{CX} = \frac{16}{9}$.*

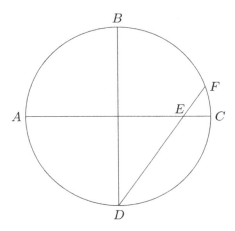

Figure 9.7: Diagram for this Exercise.

Then since the quadrilateral is cyclic, LoC applies in finding BD: $BD^2 = 40 + 24\cos\angle BAD \implies \cos\angle BAD = \frac{11}{24}, BD = \sqrt{51}$. From power of a point, $CX \cdot XG = DX \cdot XB = \frac{153}{16}$. Therefore $XF \cdot XG = \frac{51}{3} = \boxed{17}$.

Exercise 268. *(1992 AHSME) A circle of radius r has chords $\overline{AB}$ of length 10 and $\overline{CD}$ of length 7. When $\overline{AB}$ and $\overline{CD}$ are extended through B and C, respectively, they intersect at P, which is outside of the circle. If $\angle APD = 60°$ and $BP = 8$, then compute r^2.*

Solution 268. *Power of a Point on P yields $PC = 9, PD = 16$. We have that $PD = 2BP$ and $\angle BPD = 60°$; hence, $\triangle BPD$ is a 30-60-90 triangle (with $\angle B = 90°$). Thus, $BD = 8\sqrt{3}$, with $\triangle ABD$ a right triangle. Finally, $AD = 2r = 2\sqrt{73}$ and $r^2 = \boxed{73}$.*

Extension Exercise 114. *(2016 AIME I) Circles ω_1 and ω_2 intersect at points X and Y. Line ℓ is tangent to ω_1 and ω_2 at A and B, respectively, with line AB closer to point X than to Y. Circle ω passes through A and B intersecting ω_1 again at $D \neq A$ and intersecting ω_2 again at $C \neq B$. The three points C, Y, D are collinear, $XC = 67$, $XY = 47$, and $XD = 37$. Find AB^2.*

Extension Solution 114. *(Note: this solution makes use of the Radical Axis Theorem.)*

Let $Z = XY \cap AB$. By the Radical Axis Theorem, AD, XY, and BC all intersect at a point P. Angle chasing leads to the conclusion that $\triangle DXE \sim \triangle EXC$, with $EX = y$ and $XZ = z$. Then $y^2 = 37 \cdot 67$, and from there, Power of a Point leads to $x(y-x) = \frac{AB^2}{4}$, and $x(x+47) = \frac{AB^2}{4}$ as well. Hence, $y - x = x + 47$, and $y = x + 94$. Thus, $AB = \boxed{270}$ after some more straightforward algebra.

Extension Exercise 115. *(1998 USAMO P2) Let $\mathcal{C}_1$ and $\mathcal{C}_2$ be concentric circles, with $\mathcal{C}_2$ in the interior of $\mathcal{C}_1$. From a point A on $\mathcal{C}_1$ one draws the tangent AB to $\mathcal{C}_2$ ($B \in \mathcal{C}_2$). Let C be the second point of intersection of AB and $\mathcal{C}_1$, and let D be the midpoint of AB. A line passing through A intersects $\mathcal{C}_2$ at E and F in such a way that the perpendicular bisectors of DE and CF intersect at a point M on AB. Find, with proof, the ratio AM/MC.*

Extension Solution 115. *Note that $4AD = 2AB = AC$. Since B, E, F are concyclic, $AE \cdot AF = AB^2 = AC \cdot AD$ by Power of a Point. Thus, $\triangle ACF \sim \triangle AED$, so $\angle ACF = \angle AED$. Hence, $CFED$ is a cyclic quadrilateral with circumcenter M. This shows that $2MC = CD$, which implies that $\frac{AM}{MC} = \frac{AC - MC}{MC} = \frac{AC - \frac{CD}{2}}{\frac{CD}{2}} = \frac{\frac{5AC}{8}}{\frac{3AC}{8}} = \boxed{\frac{5}{3}}$.*

Section 8.6: Cyclic Quadrilaterals

Exercise 269. *(2004 AMC 10B) In triangle $\triangle ABC$ we have $AB = 7$, $AC = 8$, $BC = 9$. Point D is on the circumscribed circle of the triangle so that $\overline{AD}$ bisects angle $\angle BAC$. What is the value of $\dfrac{AD}{CD}$?*

Solution 269. *Let $BD = x$; then $CD = x$ as well, as $\angle BAD, \angle DAC$ intercept equal-length arcs. Ptolemys Theorem implies $7x + 8x = 9AD \implies$ a ratio of $\boxed{\dfrac{5}{3}}$.*

Exercise 270. *It's time for reverse cyclic quads!*

(2000 AIME II) A circle is inscribed in quadrilateral $ABCD$, tangent to $\overline{AB}$ at P and to $\overline{CD}$ at Q. Given that $AP = 19$, $PB = 26$, $CQ = 37$, and $QD = 23$, find the square of the radius of the circle.

Solution 270. *Let O be the center of the circle. Draw the lines from O tangent to the sides, and to the vertices of $ABCD$. This forms four pairs of congruent right triangles. Hence, $\angle AOP + \angle POB + \angle COQ + \angle QOD = 180°$, i.e. $\arctan \dfrac{19}{r} + \arctan 26r + \arctan 37r + \arctan 23r = 180°$. Take the tangent of both sides and use the tangent addition identity to get $\dfrac{45}{r\left(1 - \frac{494}{r^2}\right)} + \dfrac{60}{r\left(1 - \frac{851}{r^2}\right)} = 0 \implies r^2 = \boxed{647}$.*

Exercise 271. *(2018 AMC 12A) Triangle ABC is an isosceles right triangle with $AB = AC = 3$. Let M be the midpoint of hypotenuse $\overline{BC}$. Points I and E lie on sides $\overline{AC}$ and $\overline{AB}$, respectively, so that $AI > AE$ and $AIME$ is a cyclic quadrilateral. Given that triangle EMI has area 2, the length CI can be written as $\dfrac{a-\sqrt{b}}{c}$, where a, b, and c are positive integers and b is not divisible by the square of any prime. What is the value of $a + b + c$?*

Solution 271. *$\triangle EMI$ is an isosceles right triangle, with M the midpoint of arc $\overline{EI}$, so $MI = 2$ and $MC = \dfrac{3\sqrt{2}}{2}$. $\angle MCI = 45°$, so Law of Cosines gives $CI = \dfrac{3 \pm \sqrt{7}}{2}$. BE is similar; since $CI < BE$, $CI = \dfrac{3 - \sqrt{7}}{2} \implies \boxed{12}$.*

Exercise 272. *(2013 AIME II) A hexagon that is inscribed in a circle has side lengths 22, 22, 20, 22, 22, and 20 in that order. The radius of the circle can be written as $p + \sqrt{q}$, where p and q are positive integers. Find $p + q$.*

Solution 272. *Draw diameter $AD = d$. By Ptolemys Theorem on $ADEF$, $AD \cdot EF + AF \cdot DE = AE \cdot DF$. Both diagonals of $ADEF$ are equal, say to x. Then $20d + 22^2 = x^2$. $\angle AED$ is subtended by $\overline{AD}$; hence $\angle AED = 90°$. By Pythagorean Theorem, $x^2 + 22^2 = d^2$ and solving yields $d = 2\sqrt{267} + 10 \implies r = \sqrt{267} + 5 \implies \boxed{272}$.*

Exercise 273. *(2014 AMC 12B) Let $ABCDE$ be a pentagon inscribed in a circle such that $AB = CD = 3$, $BC = DE = 10$, and $AE = 14$. The sum of the lengths of all diagonals of $ABCDE$ is equal to $\frac{m}{n}$, where m and n are relatively prime positive integers. What is $m + n$?*

Solution 273. *Let a be the side opposite the 14 and 3, b the side opposite the 14 and 10, and c the side opposite 3/10. By Ptolemy, we obtain $c^2 = 3a + 100 = 10b + 9$, $ab = 30 + 14c$, $ac = 3c + 140$, $bc = 10c + 42$. It follows that $a = \dfrac{c^2 - 100}{3}$, $b = \dfrac{c^2 - 9}{10}$. Substitution and algebraic manipulation yields a factorization of $(c - 12)(c + 7)(c + 5) = 0$. Anything other than $c = 12$ is absurd. (Indeed, it is not too difficult to verify that this actually works.) Hence $a = \dfrac{44}{3}$ and $b = \dfrac{27}{2}$. Finally, $3c + a + b$, the desired quantity, is $\dfrac{385}{6} \implies \boxed{391}$.*

Extension Exercise 116. *(APMO 2007) Let $\triangle ABC$ be an acute-angled triangle with $\angle BAC = 60°$ and $AB > AC$. Let I be the incenter and H the orthocenter of the triangle $\triangle ABC$. Prove that $2\angle AHI = 3\angle ABC$.*

Extension Solution 116. *Let P, Q, R be the intersection points of BH, CH, AH with AC, AB, BC in that order. Then it follows that $\angle IBH = \angle ICH$. Then $\angle IBH = \angle ABP - \angle ABI = 30° - \dfrac{\angle ABC}{2}$. Similarly, $\angle ICH = \angle ACI - \angle ACH = 30 - \dfrac{\angle ABC}{2}$, since $\angle ABH = \angle ACH = 30°$ and $\angle ABC + \angle ACB = 120°$. $BIHC$ is cyclic, and so $2\angle BHI = \angle ACB$. But we also have $\angle BHR = 90° - \angle HBR = 120° - \angle ABC$. Thus, $\angle AHI = 180 - \angle BHI - \angle BHR = \dfrac{3}{2}\angle ABC$.*

Extension Exercise 117. *(2018 AIME I) David found four sticks of different lengths that can be used to form three non-congruent convex cyclic quadrilaterals, A, B, C, which can each be inscribed in a circle with radius 1. Let φ_A denote the measure of the acute angle made by the diagonals of quadrilateral A, and define φ_B and φ_C similarly. Suppose that $\sin\varphi_A = \frac{2}{3}$, $\sin\varphi_B = \frac{3}{5}$, and $\sin\varphi_C = \frac{6}{7}$. All three quadrilaterals have the same area K, which can be written in the form $\dfrac{m}{n}$, where m and n are relatively prime positive integers. Find $m + n$.*

Extension Solution 117. *(Adapted from Art of Problem Solving.)*

Let the four side lengths correspond to arc lengths of $2a, 2b, 2c, 2d \implies a + b + c + d = 180°$. Then we want to find which arc is opposed to the $2a°$ arc. We have $\phi_A = a + c, \phi_B = a + b, \phi_C = a + d$. We have $ABCD$ with $AB = 2a°, BC = 2b°, CD = 2c°, DA = 2d°$.

By an application of Law of Sines, $AC = 2\sin\left(\dfrac{ABC}{2}\right) = 2\sin(a + b)$ and similarly, $BD = 2\sin(a + d)$. Thus $K = \dfrac{1}{2} \cdot AC \cdot BD \cdot \sin(\phi_A) = 2\sin\phi_A \sin\phi_B \sin\phi_C = \dfrac{24}{35} \implies \boxed{59}$.

Extension Exercise 118. *Suppose that $ABCD$ is not a cyclic quadrilateral.*

(a) *Provide a construction of such a quadrilateral (that is, specify its side and diagonal lengths).*

(b) *Prove an inequality pertaining to this quadrilateral, and show that Ptolemy's Theorem does not hold for it.*

Extension Solution 118.

(a) *Any non-square rhombus is automatically non-cyclic; take a rhombus with sides 20, 19, 20, 19, for instance.*

(b) *Ptolemys inequality holds for non-cyclic quadrilaterals (as well as cyclic quadrilaterals in the equality case). Namely, we have $AB \cdot CD + BC \cdot DA \geq AC \cdot BD$. For any four points in the plane, we can apply the triangle inequality through an inversion centered at one of the points. (Also, we can place the points in the complex plane and apply the identity $(A - B)(C - D) + (B - C)(A - D) = (A - C)(B - D)$ so as to form a triangle with side lengths that are the products of the side lengths of the original quadrilateral, then using triangle inequality once more.)*

Extension Exercise 119.

(a) *From Ptolemy's Theorem, derive the addition and subtraction formulas for $\sin$ and $\cos$.*

(b) *Can we also accomplish the converse? How so?*

Extension Solution 119.

(a) *Consider quadrilateral $ABDC$ inscribed in the unit circle, with $\overline{BC}$ as a diameter. Define $\alpha = \angle ABC$, $\beta = \angle CBD$. Then the $\sin$ addition/subtraction formulae follow directly from Ptolemys Theorem, with $BC = 1, AB = \cos\alpha, AC = \sin\alpha, BD = \cos\beta, DC = \sin\beta, AD = \sin(\alpha + \beta)$: $\sin(\alpha + \beta) = \sin(\alpha)\cos(\beta) + \cos(\alpha)\sin(\beta)$. Then $\sin\left(\dfrac{\pi}{2} - \alpha\right) = \cos\alpha$, and $\cos\left(\dfrac{\pi}{2} - \alpha\right) = \sin\alpha$. This proves the corresponding $\cos$ identities as well.*

(b) *Yes; indeed, the above is quite easily reversible if we consider $\sin\alpha, \cos\alpha, \cos\beta, \sin\beta$ as side lengths in accordance with Ptolemys Theorem.*

Section 8.7: Mass Points

Exercise 274. *(2004 AMC 10B) In $\triangle ABC$ points D and E lie on BC and AC, respectively. If AD and BE intersect at T so that $\frac{AT}{DT} = 3$ and $\frac{BT}{ET} = 4$, what is $\frac{CD}{BD}$?*

Solution 274. *Let A have a mass of 1, D a mass of 3, and subsequently, T a mass of 4. Then B has a mass of 1 and E a mass of 4, so that T has a mass of 5. Then to "balance A and C on E, we would assign C a mass of $4 - \frac{5}{4} = \frac{11}{4}$. The desired ratio is thus $\boxed{\dfrac{4}{11}}$.*

Exercise 275. *In triangle $\triangle ABC$, $AB = 7$, $BC = 8$, $CA = 9$. Point D is the midpoint of $\overline{BC}$. Point E trisects $\overline{CA}$, and is closer to C. The angle bisector of $\angle A$ hits $\overline{BC}$ at point F. What is the area of $\triangle DEF$?*

Solution 275. *Set $B = (0,0)$, and then assign it a mass of 1. Then E will have a mass of 3, with A having a mass of 2 and C a mass of 1. D has mass 2, which allows us to compute the mass of F as $\frac{16}{7}$ from $AC = 9$ and $AB = 7$. Finally, use Shoelace to compute the area of $\triangle DEF$ from these masses by converting them to Cartesian points; the final area should be $\boxed{\dfrac{\sqrt{5}}{4}}$.*

Exercise 276. *$\triangle ABC$ has $AB = 13$, $BC = 15$, $CA = 14$. Let $\overline{BD}$ be an altitude of $\triangle ABC$. The bisector of $\angle C$ hits $\overline{AB}$ at E and $\overline{BD}$ at F. Compute $\frac{CE}{EF}$.*

Solution 276. *Using mass points, $15AE = 14EB$, and if we assume WLOG that $A = (0,0)$, then the x-coordinate of E is $\frac{70}{29}$. Thus, $\frac{CE}{EF} = \frac{14 - \frac{70}{29}}{5 - \frac{70}{29}} = \boxed{\dfrac{112}{25}}$.*

Exercise 277. *(1975 AHSME) In $\triangle ABC$, M is the midpoint of side $\overline{BC}$, $AB = 12$ and $AC = 16$. Points E and F are taken on $\overline{AC}$ and $\overline{AB}$, respectively, and lines $\overline{EF}$ and $\overline{AM}$ intersect at G. If $AE = 2AF$ then compute $\frac{EG}{GF}$.*

Solution 277. *Let $\overline{CP} \parallel \overline{EF}$ and intersect $\overline{AB}$ at P. Then we have $\frac{AP}{AC} = \frac{AF}{AE}$, so $AP = 8$. By the Law of Sines, $\frac{BM}{\sin \angle MAB} = \frac{12}{\sin \angle MAB}$ and $\frac{BM}{\sin \angle CAM} = \frac{16}{\sin \angle MAB}$, so $\frac{\sin \angle CAM}{\sin \angle MAB} = \frac{3}{4}$. Furthermore, let Q be the intersection of $\overline{AM}$ and $\overline{CP}$. Then $\frac{PQ}{\sin \angle MAB} = \frac{8}{\sin AQP}$ and $\frac{QC}{\sin \angle CAM} = \frac{16}{\sin \angle AQP}$. Hence, $\frac{EG}{GF} = \boxed{\dfrac{3}{2}}$.*

Exercise 278. *In $\triangle ABC$ with $AB = 7$, $BC = 3$, and $CA = 5$, angle bisectors $\overline{AD}$ and $\overline{BE}$ intersect at point P. Compute $\frac{BP}{PE}$.*

Solution 278. *The answer is $\boxed{2}$; either use coordinates or assign masses as normal according to the distances.*

Exercise 279. *Three similar problems, one from MathCounts and two from the 2016 AMC 10 tests:*

(a) *(2012 MathCounts State Sprint) In rectangle $ABCD$, point M is the midpoint of side $\overline{BC}$, and point N lies on $\overline{CD}$ such that $DN : NC = 1 : 4$. Segment $\overline{BN}$ intersects $\overline{AM}$ and $\overline{AC}$ at points R and S, respectively. If $NS : SR : RB = x : y : z$, where x, y and z are positive integers, what is the minimum possible value of $x + y + z$?*

(b) *(2016 AMC 10A) In rectangle $ABCD$, $AB = 6$ and $BC = 3$. Point E between B and C, and point F between E and C are such that $BE = EF = FC$. Segments $\overline{AE}$ and $\overline{AF}$ intersect $\overline{BD}$ at P and Q, respectively. The ratio $BP : PQ : QD$ can be written as $r : s : t$ where the greatest common factor of r, s, and t is 1. What is $r + s + t$?*

(c) (2016 AMC 10B) Rectangle $ABCD$ has $AB = 5$ and $BC = 4$. Point E lies on $\overline{AB}$ so that $EB = 1$, point G lies on $\overline{BC}$ so that $CG = 1$. and point F lies on $\overline{CD}$ so that $DF = 2$. Segments $\overline{AG}$ and $\overline{AC}$ intersect $\overline{EF}$ at Q and P, respectively. What is the value of $\frac{PQ}{EF}$?

Note: These, like many other exercises in this part, can be coordinate bashed, but we recommend using synthetic methods in favor of bashing.

Solution 279.

(a) Let $AB = 5n$; then $NC = 4n$ (equivalent to assigning masses of 4 and 5). Triangles $\triangle ABS$ and $\triangle CNS$ are similar, so $4BS = 5NS$. Extending $\overline{AM}$ and $\overline{DC}$ to intersect at point P, we find that P has a mass of 14, so it follows that $SR = BS - BR = \frac{25}{126}BN$. In addition, $BR = \frac{5}{14}BN$ and $RN = \frac{9}{14}BN$, so the ratio is $56 : 25 : 45 \implies \boxed{126}$.

(b) Let $\overline{AC}, \overline{BD}$ intersect at R. In $\triangle ABC$, we have $2EC = BE$, and $AK = KC$. Using mass points, we also have $BP = PK$, and $2FC = BF$. Again using mass points, we compute $BQ = 4QK$.

Let $BP = x = PR, BQ = 4y, QR = y$. Then $y = \frac{2}{5}x$, and so $QR = \frac{2}{5}x, PQ = \frac{3}{5}x, BP = x$. The desired ratio becomes $5 : 3 : 12 \implies \boxed{20}$.

(c) Using similar triangles at first, extend $\overline{AG}$ to hit $\overline{CD}$ at point H. If $x = HC$, then $\triangle HCG \sim \triangle HDA \implies x = \frac{5}{3}$. In addition, $\triangle AEQ \sim \triangle HFQ, \triangle AEP \sim \triangle CFP$. Then $\frac{AE}{HF} = \frac{EQ}{QF} \implies \frac{EQ}{EF} = \frac{6}{13}$ and $\frac{AE}{CF} = \frac{PE}{PF} \implies \frac{PF}{PE} = \frac{3}{7}$. Hence, $\frac{PQ}{EF} = 1 - \frac{6}{13} - \frac{3}{7} = \boxed{\frac{10}{91}}$. (What does this have to do with mass points? Similar triangles are, in fact, the very manifestation of mass points! Consider the fact that the side lengths are in direct proportion with each other.)

Exercise 280. (2001 AIME I) Triangle ABC has $AB = 21$, $AC = 22$ and $BC = 20$. Points D and E are located on $\overline{AB}$ and $\overline{AC}$, respectively, such that $\overline{DE}$ is parallel to $\overline{BC}$ and contains the center of the inscribed circle of triangle ABC. Then $DE = m/n$, where m and n are relatively prime positive integers. Find $m + n$.

Solution 280. Let I be the incenter, and draw bisector $\overline{AP}$ to its intersection with $\overline{BC}$ (and call the intersection point F). By the angle bisector theorem, $\frac{BF}{CF} = \frac{21}{22}$, so B has a mass of 22 and C has a mass of 21; hence, F's mass is 43. Similarly, A's mass is 20, and P then has a mass of 63. Thus, $DE = \frac{43}{63}BC$, and $DE = 20 \cdot \frac{43}{63} = \frac{860}{63} \implies \boxed{923}$.

Extension Exercise 120. (1989 AIME) Point P is inside $\triangle ABC$. Line segments APD, BPE, and CPF are drawn with D on $\overline{BC}$, E on $\overline{AC}$, and F on $\overline{AB}$. Given that $AP = 6$, $BP = 9$, $PD = 6$, $PE = 3$, and $CF = 20$, find the area of $\triangle ABC$.

Extension Solution 120. Observe that E's mass is 3, B's is 1, and both A and D have mass 2. Then C has mass 1, and F has mass 3. Thus, $CP = 15, PF = 5$.

Hence, $DC = DB$, and $\overline{DP}$ is a median to $\overline{BC}$ in $\triangle BCP$. By Stewart's Theorem, $BC^2 + 144 = 2(15^2 + 9^2) \implies BC = 6\sqrt{13}$. Then $2[\triangle BCP] = [\triangle ABC]$. Since $[\triangle BCP] = 54$, $[\triangle ABC] = \boxed{108}$.

Section 8.8: 3D Geometry, Part 2

Exercise 281. (1996 AHSME) On a $4 \times 4 \times 3$ rectangular parallelepiped, vertices A, B, and C are adjacent to vertex D. What is the perpendicular distance from D to the plane containing A, B, and C?

Solution 281. *Let d be the required distance. Consider pyramid $ABCD$. If ABC is the base, then the volume of $ABCD$ is $\frac{x}{3} \cdot [\triangle ABC]$. However, if BCD is the base instead, then the volume is $\frac{1}{3} \cdot [\triangle BCD] \cdot AD = 8$. Hence, $x \cdot [\triangle ABC] = 24$.*

To compute $[\triangle ABC]$ itself, note that $AB = 5$, $AC = 5$, $BC = 4\sqrt{2}$. So the altitude to $\overline{BC}$ has length $\sqrt{17}$ by the Pythagorean Theorem, so $[\triangle ABC] = 2\sqrt{34}$. Then $x = \boxed{\dfrac{6\sqrt{34}}{17}}$. (Alternatively, we can use the point-to-plane distance formula.)

Exercise 282. *(2012 AIME I) Cube $ABCDEFGH$ [with E above A, F above B, and likewise for G and H] has edge length 1 and is cut by a plane passing through vertex D and the midpoints M and N of $\overline{AB}$ and $\overline{CG}$ respectively. The plane divides the cube into two solids. The volume of the larger of the two solids can be written in the form $\frac{p}{q}$, where p and q are relatively prime positive integers. Find $p + q$.*

Solution 282. *In $\mathbb{R}^3$, let D be at the origin, so that $M = (.5, 1, 0), N = (1, 0, .5)$. Then the plane passing through D, M, N has equation $2x - y - 4z = 0$, so that the intersection point of the plane with BF is $(1, 1, .25)$. Let $P = (1, 2, 0)$, so that P lies on the plane, as well as on the lines $\overline{DM}, \overline{NQ}, \overline{BC}$. Hence $DPCN$ is its own pyramid, and is similar to $MPBQ$. Thus, the volume of $DMBCQN$ equals $[DPCN] - [MPBQ] = [DPCN] - \frac{1}{8}[DPCN] = \frac{7}{8}[DPCN]$. The volume of $DPCN$ is $\frac{1}{6}$, so $[DMBCQN] = \frac{7}{48}$. The larger solid then has volume $\frac{41}{48} \implies \boxed{89}$.*

Exercise 283. *(2000 AIME II) The points A, B and C lie on the surface of a sphere with center O and radius 20. It is given that $AB = 13$, $BC = 14$, $CA = 15$, and that the distance from O to $\triangle ABC$ is $\frac{m\sqrt{n}}{k}$, where m, n, and k are positive integers, m and k are relatively prime, and n is not divisible by the square of any prime. Find $m + n + k$.*

Solution 283. *This is essentially the distance that, when squared and summed with the square of the circumradius, yields $400 = 20^2$. Since $R = \frac{65}{8}$ by the formula $[\triangle ABC] = \frac{abc}{4R}$, we have $OD = \frac{15\sqrt{95}}{8} \implies \boxed{118}$.*

Exercise 284. *(2016 HMMT November Guts Round) Create a cube C_1 with edge length 1. Take the centers of the faces and connect them to form an octahedron O_1. Take the centers of the octahedron's faces and connect them to form a new cube C_2. Continue this process infinitely. Find the sum of all the surface areas of the cubes and octahedrons.*

Solution 284. *The first cube and first octahedron have surface areas of 6 and $\sqrt{3}$ respectively. Note that the surface area is proportional to the square of the side length, so it is multiplied by $\frac{1}{3^2} = \frac{1}{9}$ each iteration. Hence the sum of surface areas is an infinite geometric series with first term $6 + \sqrt{3}$ and common ratio $\frac{1}{9}$, which has sum $\boxed{\dfrac{54 + 9\sqrt{3}}{8}}$.*

Exercise 285. *Let P_1 be the plane passing through the points $(4, 5, 3)$, $(2, 0, 1)$, $(1, 9, 4)$. Let P_2 pass through the points $(20, 19)$ and $(-4, -5)$ with arbitrary z-coordinate (i.e. the line $y = x - 1$ in $\mathbb{R}^2$ with $z \in \mathbb{R}$). Either compute the minimum distance between P_1 and P_2, or show that P_1 and P_2 overlap (and determine the locus of overlap).*

Solution 285. *Note that P_1 has equation $3x + 8y - 23z + 17 = 0$. If a distance exists between the planes at all, then they must be parallel; this would require that $y \neq x - 1$ for any point $(x, y, z) \in P_1$. But a counterexample to this is $(x, y, z) = \left(1, 0, \frac{20}{23}\right)$, so P_1 and P_2 intersect. The region of overlap is all (x, y, z) with $y = x - 1$.*

Exercise 286. *(2009 AMC 12A) A regular octahedron has side length 1. A plane parallel to two of its opposite faces cuts the octahedron into the two congruent solids. The polygon formed by the intersection of the plane and the octahedron has area $\frac{a\sqrt{b}}{c}$, where a, b, and c are positive integers, a and c are relatively prime, and b is not divisible by the square of any prime. What is $a + b + c$?*

Solution 286. *The planes intersection with the octahedron is a hexagon, so consider the net of the octahedron. The hexagon is transformed into a line, which is parallel to $\frac{1}{3}$ of the nets ages. Also, by symmetry, two of the hexagons vertices lie on the midpoint of the octahedrons sides. It follows that we have a regular hexagon with side length $\frac{1}{2}$, which has area $\frac{3\sqrt{3}}{8} \implies \boxed{14}$.*

Exercise 287. *(2011 AIME I) In triangle ABC, $BC = 23$, $CA = 27$, and $AB = 30$. Points V and W are on $\overline{AC}$ with V on $\overline{AW}$, points X and Y are on $\overline{BC}$ with X on $\overline{CY}$, and points Z and U are on $\overline{AB}$ with Z on $\overline{BU}$. In addition, the points are positioned so that $\overline{UV} \parallel \overline{BC}$, $\overline{WX} \parallel \overline{AB}$, and $\overline{YZ} \parallel \overline{CA}$. Right angle folds are then made along $\overline{UV}$, $\overline{WX}$, and $\overline{YZ}$. The resulting figure is placed on a level floor to make a table with triangular legs. Let h be the maximum possible height of a table constructed from triangle ABC whose top is parallel to the floor. Then h can be written in the form $\frac{k\sqrt{m}}{n}$, where k and n are relatively prime positive integers and m is a positive integer that is not divisible by the square of any prime. Find $k + m + n$.*

Solution 287. *By Heron's formula, the area of $\triangle ABC$ is $20\sqrt{221}$. Let h_i be the height from vertex i. Hence, $h_a = \frac{40}{23}\sqrt{221}$, $h_b = \frac{40}{27}\sqrt{221}$, $h_c = \frac{4}{3}\sqrt{221}$. Similar triangles yields that $h \leq \frac{h_a h_c}{h_a + h_c}$. This holds for any combination of a, b, c, and so we find that the minimum upper bound yields $h = \frac{40}{57}\sqrt{221} \implies \boxed{318}$.*

Extension Exercise 121. *(1960 IMO P5) Consider the cube $ABCDA'B'C'D'$ (with face $ABCD$ directly above face $A'B'C'D'$).*

(a) Find the locus of the midpoints of the segments XY, where X is any point of AC and Y is any point of $B'D'$;

(b) Find the locus of points Z which lie on the segment XY of part a) with $ZY = 2XZ$.

Extension Solution 121.

(a) WLOG let the cube have side length 2, with $A = (0,0,0), B = (0,2,0), C = (2,2,0), D = (2,0,0), A = (0,0,2), B = (0,2,2), C = (2,2,2), D = (2,0,2)$. Then $X = (t,t,2)$ for any $0 \leq t \leq 2$, and $Y = (t, 2-t, 0)$ for $0 \leq t \leq 2$. Then the locus of midpoints of $\overline{XY}$ is all points of the form $(t,1,1)$ for a parameter $t \in [0,2]$.

(b) For $ZY = 2XZ$, we have $Z = \left(t, \dfrac{t+2}{3}, \dfrac{4}{3}\right)$, $t \in [0,2]$.

Extension Exercise 122. *(1966 IMO P3) Prove that the sum of the distances of the vertices of a regular tetrahedron from the center of its circumscribed sphere is less than the sum of the distances of these vertices from any other point in space.*

Extension Solution 122. *Let the tetrahedron be $ABCD$, and let a point $P \in ABCD$. Let X be the midpoint of $\overline{CD}$, let Y be the midpoint of $\overline{AB}$, and let P' be the foot of the perpendicular from P to ABX. We claim that, if $P \neq P$, $PA + PB + PC + PD > P'A + P'B + P'C + P'D$.*

We initially have $PA > P'A$, as $\angle PP'A$ is right, and similarly, $PB > P'B$. Note that $\triangle P'CD$ is isosceles, but not $\triangle PCD$. However, P and P' are equidistant from $\overline{CD}$. Hence, $PC + PD > P'C + P'D$.

Finally, if P minimizes the sum of distances, then it lies in the plane ABX, and so also in the plane CDY. Thus, P lies on $\overline{XY}$. It then lies on the line joining the midpoints of another pair of opposite sites, and so must be the center of the tetrahedron $ABCD$.

411

Extension Exercise 123. *(2002 Putnam A2) Given any five points on a sphere, show that some four of them must lie on a closed hemisphere.*

Extension Solution 123. *Through two of the points, draw their great circle. Two closed hemispheres have the great circle as a boundary, and each of the other three points lies in one of those hemispheres. At least two of the three points lie in the same hemisphere, so one of the hemispheres has at least four of the five points.*

Chapter 8 Review Exercises

Exercise 288. *A right triangle has hypotenuse 15, and leg lengths 9 and 12. What is the product of the radii of its incircle and circumcircle?*

Solution 288. *The inradius r is given by $\dfrac{A}{s}$, where A is the triangles area and s is half its perimeter. We have $A = 54$ and $s = 18$, so $r = 3$. In addition, the circumdiameter is the hypotenuse, so the circumradius is half the hypotenuse, or $\dfrac{15}{2}$. Hence, the desired product is $\boxed{\dfrac{45}{2}}$.*

Exercise 289. *(2015 Berkeley Math Tournament) Let $\mathcal{C}$ be the sphere $x^2 + y^2 + (z-1)^2 = 1$. Point P on $\mathcal{C}$ is $(0,0,2)$. Let $Q = (14, 5, 0)$. If PQ intersects $\mathcal{C}$ again at Q_0, then find the length PQ_0.*

Solution 289. *In $\overline{PQ}$, for any $x \in \mathbb{R}$, we have $y = \dfrac{5x}{14}$ and $z = 2 - \dfrac{x}{7}$. Hence, $x^2 + y^2 + (z-1)^2 = 1 \implies x = 0, \dfrac{56}{225}$. Since $x = 0$ already has an intersection point (namely, P), we have $Q_0 = \left(\dfrac{56}{225}, \dfrac{4}{45}, \dfrac{442}{225} \right)$ and so $PQ_0 = \boxed{\dfrac{4}{15}}$.*

Exercise 290. *(2018 HMMT February Geometry Round) How many noncongruent triangles are there with one side of length 20, one side of length 17, and one $60°$ angle?*

Solution 290. *We can test three cases: one case each for the $60°$ angle opposite the 20, the 17, or the third side with unknown length. Using the Law of Cosines each time, we get that $c^2 = a^2 + b^2 - ab$, so either $c = \sqrt{349}, \dfrac{17 + \sqrt{733}}{2}$, and a non-real root - $\boxed{2}$ possibilities.*

Exercise 291. *Two line segments with a common endpoint have lengths a and b. If they span an angle of $\theta = \cos^{-1}\left(\dfrac{a}{b}\right)$, show that the triangle formed by drawing the line segment from the endpoints of the line segments cannot be a right triangle.*

Solution 291. *This follows from the Law of Cosines, with $\cos\theta = \dfrac{a}{b}$. We have $c^2 = a^2 + b^2 - 2a^2 = b^2 - a^2$, but then $c^2 \neq a^2 + b^2$ since $a \neq 0$; QED.*

Exercise 292. *$\triangle ABC$ has $AB = 13$, $BC = 14$, $CA = 15$. $\overline{AD}$ is an altitude of the triangle. What is the maximum x-distance between some two points on the circumcircles of $\triangle ABD$ and $\triangle ADC$?*

Solution 292. *WLOG assume $A = (5, 12), B = (0,0), C = (14, 0)$. Then $D = (5, 0)$. Since both $\triangle ABD$ and $\triangle ADC$ are right, the circumcircle of $\triangle ABD$ is $\left(x - \dfrac{5}{2} \right)^2 + (y-6)^2 = \dfrac{169}{4}$, and the circumcircle of $\triangle ADC$ is $\left(x - \dfrac{19}{2} \right)^2 + (y-6)^2 = \dfrac{225}{4}$.*

Then we want to maximize the x-distance, which occurs for $y = 6$. Solving yields a distance of $\boxed{21}$.

Exercise 293. *(2010 AIME II) Triangle ABC with right angle at C, $\angle BAC < 45°$ and $AB = 4$. Point P on $\overline{AB}$ is chosen such that $\angle APC = 2\angle ACP$ and $CP = 1$. The ratio $\dfrac{AP}{BP}$ can be represented in the form $p + q\sqrt{r}$, where p, q, r are positive integers and r is not divisible by the square of any prime. Find $p + q + r$.*

Solution 293. *Let $x = \angle ACP$. By Law of Sines, $BP = -\dfrac{\cos x}{\cos 3x}$, $AP = \dfrac{\sin x}{\sin 3x}$. Applying the triple-angle identities yields $\sin^2 x = \dfrac{4 \pm \sqrt{2}}{8}$ after some algebraic gymnastics; and substitution yields $\dfrac{AP}{PB} = 3 + 2\sqrt{2} \implies$ $\boxed{7}$.*

Exercise 294. *(2018 AMC 12B) Let $ABCDEF$ be a regular hexagon with side length 1. Denote by X, Y, and Z the midpoints of sides $\overline{AB}$, $\overline{CD}$, and $\overline{EF}$, respectively. What is the area of the convex hexagon whose interior is the intersection of the interiors of $\triangle ACE$ and $\triangle XYZ$?*

Solution 294. *The area consists of equilateral $\triangle MNO$ and three right triangles (namely, $\triangle MPN, \triangle NQO, \triangle ORM$). $\overline{AD}$ and $\overline{BC}$ are parallel, so $\overline{XY}$ divides $\overline{AB}$ and $\overline{CD}$ into equal segments. Hence, it does the same for $\overline{AC}, \overline{CO}$. By symmetry, this also holds for $\overline{CE}, \overline{EA}$, and $\overline{EM}, \overline{AN}$.*

In $\triangle ACE$, $\triangle MNO$ has $\dfrac{1}{4}$ the area of $\triangle ACE$ by similarity, and $\triangle MPN$ has $\dfrac{1}{8}$ the area. Then the desired area is $\left(\dfrac{1}{4} + \dfrac{3}{8}\right) \cdot \dfrac{3\sqrt{3}}{4} = \boxed{\dfrac{15\sqrt{3}}{32}}$.

Exercise 295. *(2018 AMC 12B) In $\triangle ABC$ with side lengths $AB = 13$, $AC = 12$, and $BC = 5$, let O and I denote the circumcenter and incenter, respectively. A circle with center M is tangent to the legs AC and BC and to the circumcircle of $\triangle ABC$. What is the area of $\triangle MOI$?*

Solution 295. *Assign coordinates with B at the origin. Then the incenter has coordinates $(2, 2)$ and the circumcenter has coordinates $\left(6, \dfrac{5}{2}\right)$. Let $M = (x, x)$; then $\sqrt{\left(\dfrac{5}{2} - x\right)^2 + (6 - x)^2} + x = \dfrac{13}{2}$. Solving yields $x = 4$, and from here, we can finish with Shoelace. The answer is $\boxed{\dfrac{7}{2}}$. (Fakesolving this is also possible by arbitrarily setting $M = C$, in which case we can directly apply Shoelace to get the answer.)*

Exercise 296. *(2014 Caltech-Harvey Mudd Math Competition) Find the area of the cyclic quadrilateral with side lengths given by the solutions to $x^4 - 10x^3 + 34x^2 - 45x + 19 = 0$.*

Solution 296. *By Vieta's, the semi-perimeter of the quadrilateral is 5; substituting this into the polynomial yields 19. Brahmaguptas formula says that it suffices to take the square root of this to get $\boxed{\sqrt{19}}$.*

Exercise 297. *(1999 USAMO P2) Let $ABCD$ be a cyclic quadrilateral. Prove that*

$$|AB - CD| + |AD - BC| \geq 2|AC - BD|.$$

Solution 297. *Arc AB of the circumcircle has degree measure $2x$, with arc BC having $2y$, CD having $2z$, and DA having $2w$. Then, for all $x + y + z + w = 180°$, $|\sin x - \sin z| + |\sin y - \sin w| \geq 2 |\sin(x + y) - \sin(y + z)|$. Simplifying this with trig identities results in $0 \leq \dfrac{x - z}{2} \leq \dfrac{x + z}{2} \leq 90° \implies \left|\sin \dfrac{x + z}{2}\right| > \left|\sin \dfrac{x - z}{2}\right|$ (and, similarly, $\left|\sin \dfrac{w + y}{2}\right| > \left|\sin \dfrac{y - w}{2}\right|$). After we multiply the first inequality through by $\left|\sin \dfrac{x - z}{2}\right|$, and the second by $\left|\sin \dfrac{y - w}{2}\right|$, the result follows.*

Exercise 298. *(2014 Stanford Math Tournament) In cyclic quadrilateral $ABCD$, $AB \cong AD$. If $AC = 6$ and $\dfrac{AB}{BD} = \dfrac{3}{5}$, find the maximum possible area of $ABCD$.*

Solution 298. *By Ptolemy's Theorem, $AB \cdot CD + BC \cdot AD = AC \cdot BD \implies BC + CD = AC \cdot \dfrac{BD}{AB} = 10$. Extend $\overline{CB}$ past B to E such that $\overline{BE} \cong \overline{CD}$. $ABCD$ is a cyclic quadrilateral, so $\angle ABC + \angle ADC = 180°$. Hence, $\angle ADC \cong \angle ABE$. Thus, $[ABCD] = [\triangle ACE]$. $\triangle ACE$ is an isosceles triangle with base 10 and leg 6, so the altitude is $\sqrt{11}$, and the area is $\boxed{5\sqrt{11}}$.*

Exercise 299. *(2017 HMMT February Geometry Test) Let $\triangle ABC$ be a triangle with circumradius $R = 17$ and inradius $r = 7$. Find the maximum possible value of $\sin \dfrac{A}{2}$.*

Solution 299. *Let I be the incenter, and let O be the circumcenter. By the triangle inequality, $AO \leq AI + IO \implies R \leq \dfrac{r}{\sin \dfrac{A}{2}} + \sqrt{R(R - 2r)}$. Substitution yields $\sin \dfrac{A}{2} \leq \boxed{\dfrac{17 + \sqrt{51}}{34}}$.*

Exercise 300. *(2015 AMC 10A/12A) Let S be a square of side length 1. Two points are chosen independently at random on the sides of S. The probability that the straight-line distance between the points is at least $\dfrac{1}{2}$ is $\dfrac{a - b\pi}{c}$, where $a, b,$ and c are positive integers and $\gcd(a, b, c) = 1$. What is $a + b + c$?*

Solution 300. *Pick one point on a fixed side; the probability of the second point lying on the same side is $\dfrac{1}{4}$, on an adjacent side is $\dfrac{1}{2}$, and on the opposite side is $\dfrac{1}{4}$. This lends itself to three disjoint cases.*

Case 1: same side. The probability that the distance is $\leq \dfrac{1}{2}$ is clearly $\dfrac{1}{4}$.

Case 2: adjacent side. This is essentially geometric probability; this corresponds to the region $(x, y) \in \mathbb{R}^2 \mid \sqrt{x^2 + y^2} \leq \dfrac{1}{2}$ in the xy-plane. The probability is $1 - \dfrac{\pi}{16}$.

Case 3: opposite side. This is trivial.

In total, the probability is $\dfrac{26 - \pi}{32} \implies \boxed{59}$.

Extension Exercise 124. *(2008 USAMO P2) Let ABC be an acute, scalene triangle, and let $M, N,$ and P be the midpoints of $\overline{BC}, \overline{CA},$ and $\overline{AB}$, respectively. Let the perpendicular bisectors of $\overline{AB}$ and $\overline{AC}$ intersect ray AM in points D and E respectively, and let lines BD and CE intersect in point F, inside of triangle ABC. Prove that points $A, N, F,$ and P all lie on one circle.*

Extension Solution 124. *WLOG assume that $AB > AC$. It suffices to show that $\angle OFA$ is right, since this is synonymous with A, P, O, F, N being concyclic. By Menelaus on $\triangle BFC$ with transversal line through E, M, D, we have $\dfrac{FE}{EC} \cdot \dfrac{CM}{MB} \cdot \dfrac{BD}{DF} = 1 \implies \dfrac{FE}{EC} = \dfrac{FD}{BD}$. (Here we use the fact that M is the midpoint of $\overline{BC}$.) Since E is also the midpoint of $\overline{AC}$, and D is the midpoint of $\overline{AB}$, it follows that $\overline{AF}$ bisects $\angle EFD$. $\overline{OE}$ is also a bisector of $\angle FED$, as is $\overline{OD}$ with respect to $\angle FED$; therefore, O is the F-excenter of $\triangle FED$, and $\overline{OF}$ is in turn a bisector of $\triangle EFD$. This means that $AF \perp OF$, as desired.*

Extension Exercise 125. *(2012 Putnam A1) Let $d_1, d_2, \ldots, d_{12}$ be real numbers in the open interval $(1, 12)$. Show that there exist distinct indices i, j, k such that d_i, d_j, d_k are the side lengths of an acute triangle.*

Extension Solution 125. *WLOG assume the d_i are non-decreasing. If $d_{i+2}^2 < d_i^2 + d_{i+1}^2$, this forces them to be the sides of an acute triangle; hence, the contrary holds. By induction on i, we can show that the d_i follow the Fibonacci recurrence, which leads to a contradiction with $d_{12}^2 \geq 144 d_1^2$.*

Extension Exercise 126. *(2015 AIME I) A block of wood has the shape of a right circular cylinder with radius 6 and height 8, and its entire surface has been painted blue. Points A and B are chosen on the edge of one of the circular faces of the cylinder so that arc AB on that face measures $120°$. The block is then sliced in half along the plane that passes through point A, point B, and the center of the cylinder, revealing a flat, unpainted face on each half. The area of one of these unpainted faces is $a\pi + b\sqrt{c}$, where $a, b,$ and c are integers and c is not divisible by the square of any prime. Find $a + b + c$.*

Extension Solution 126. *Let C, D be the intersection points of the plane with the top face of the cylinder. Let O be the center of the cylinder, with C, O, A collinear. Let M be the midpoint of $\overline{AB}$, and P the center of the bottom face. Then $OP = 4$ and $PM = 3$, so $OM = 5$.*

Let X and Y be the projections of C, D respectively onto the bottom face of the cylinder. Then $ABCD$ is a dilation of $ABXY$, with homothety ratio $\frac{5}{3}$. The area of $ABXY$ is $18\sqrt{3}+12\pi$, so $[ABCD] = 20\pi+30\sqrt{3} \implies$ $\boxed{53}$.

Extension Exercise 127. *(1998 Putnam A1) A cone has circular base radius 1, and vertex at height 3 directly above the center of the circle. A cube has four vertices in the base and four on the sloping sides. What is the cube's side length?*

Extension Solution 127. *Consider the plane that cuts vertically through a base diagonal. This produces a rectangle with length s and width $s\sqrt{2}$. By similar triangles, $3 = \dfrac{s}{1 - \frac{s}{\sqrt{2}}} \implies s = \boxed{\dfrac{9\sqrt{2}-6}{7}}$.*

Extension Exercise 128. *(1999 USAMO P6) Let $ABCD$ be an isosceles trapezoid with $AB \parallel CD$. The inscribed circle ω of triangle BCD meets CD at E. Let F be a point on the (internal) angle bisector of $\angle DAC$ such that $EF \perp CD$. Let the circumscribed circle of triangle ACF meet line CD at C and G. Prove that the triangle AFG is isosceles.*

Extension Solution 128. *Additionally, the excenter lies on the angle bisector of $\triangle DAC$, as well as along the perpendicular from $\overline{DC}$ to E. Then F is indeed the excenter. As a result, we can write $\angle GFA = \angle GCA = \angle DCA$. Henceforth, $\angle ACF = \dfrac{1}{2}\angle DCA + 90°$, so $\angle AGF = 90 - \dfrac{1}{2}\angle ACD$. Finally $\angle AGF = 90 - \dfrac{1}{2}\angle ACD$. It follows that $\triangle AFG$ is isosceles.*

Extension Exercise 129. *(1995 IMO Shortlist G3) $\triangle ABC$ is a triangle. The incircle touches $\overline{BC}$, $\overline{CA}$, $\overline{AB}$ at D, E, F respectively. X is a point inside the triangle such that the incircle of $\triangle XBC$ touches $\overline{BC}$ at D. It touches $\overline{CX}$ at Y and $\overline{XB}$ at Z. Show that $EFZY$ is cyclic.*

Extension Solution 129. *If $\overline{BC} \parallel \overline{EF}$, then $AB = AC$, D is the midpoint of $\overline{BC}$, and $AD \perp BC$. Then X is on $\overline{AD}$, and $\overline{BC} \parallel \overline{YZ}$.*

Hence, extend $\overline{EF}$ to hit $\overline{BC}$ at a point P. By Menelaus, $\dfrac{AF}{FB} \cdot \dfrac{BP}{CP} \cdot \dfrac{CE}{EA} = 1$, and $AE = AF$ (as both are tangent to the incircle), and $XY = XZ$ by the same logic (and, in addition, $BD = BZ, CE = CD = CY$). Thus, $\dfrac{XZ}{BZ} \cdot \dfrac{BP}{CP} \cdot \dfrac{CY}{XY} = 1$. Thus, Z, P, Y are collinear.

We conclude that $PE \cdot PF = PD^2 = PY \cdot PZ$, so $PE \cdot PF = PY \cdot PZ$ and $EFZY$ is cyclic.

Chapter 9: Advanced Geometry

Section 9.1: Proving Triangle Theorems, Part 2

Exercise 301. *(2012 Online Math Open) Let $\triangle ABC$ be a triangle with circumcircle ω. Let the bisector of $\angle ABC$ meet segment $\overline{AC}$ at D and circle ω at $M \neq B$. The circumcircle of $\triangle BDC$ meets line $\overline{AB}$ at $E \neq B$, and $\overline{CE}$ meets ω at $P \neq C$. The bisector of $\angle PMC$ meets segment $\overline{AC}$ at $Q \neq C$. Given that $PQ = MC$, determine the degree measure of $\angle ABC$.*

Solution 301. *Angle chasing yields $\angle ABM = \angle ACP = \angle DBE = \angle DCE = \angle MBC$. Then arcs AP, AM, AC are of equal length, so $AP = AM = MC \implies AP = PQ$. As $\overline{MQ}$ bisects angle $\angle PMC$, Q is the incenter of $\triangle PMC$. Further angle chasing leads to the observations that $PAMC = 3B$, so $\angle PMC = \frac{1}{2}PBC = 180 - \frac{3}{2}B$. Then $\angle PQC = 90 + \frac{1}{2}\angle PMC = 180 - \frac{3}{4}B$. Hence, $\angle AQP = 180 - \angle PQC = \frac{3}{4}B$. Moreover, $\angle PAQ = \angle PAC = \angle PMC = 180 - \frac{3}{2}B$. Since $AP = PQ$, $B = \boxed{80°}$.*

Exercise 302. *(2014 Online Math Open) Let $\triangle ABC$ be a triangle with $AB = 26$, $AC = 28$, $BC = 30$. Let X, Y, and Z be the midpoints of arcs BC, CA, AB (not containing the opposite vertices) respectively on the circumcircle of $\triangle ABC$. Let P be the midpoint of arc BC containing point A. Suppose lines $\overline{BP}$ and $\overline{XZ}$ meet at M, while lines $\overline{CP}$ and $\overline{XY}$ meet at N. Find the square of the distance from X to $\overline{MN}$.*

Solution 302. *Let I be the incenter; Pascals Theorem (applied to the hexagon $PBYXZC$) guarantees that I lies on $\overline{MN}$. In isosceles $\triangle BIZ$, we have $\angle MIB = \angle MBI = \angle PBY = \angle ICB$. Then $\overline{MI}$ (and $\overline{IN}$) are tangent to the circumcircle of $\triangle BIC$. Thus, $\overline{MN}$ is the tangent. Let X be the circumcenter; then I is the foot of the X-altitude down to $\overline{MN}$.*

All that remains is to compute $IX = BX = BC \cdot \dfrac{\sin\frac{A}{2}}{\sin A} = \sqrt{325}$, so $BX^2 = \boxed{325}$.

Exercise 303. *Let O be the circumcenter of $\triangle ABC$. O is tangent to the sides of $\triangle ABC$ at points X, Y, and Z. Let the orthic triangle of $\triangle ABC$ be $\triangle DEF$. Prove that the isogonal conjugate of O with respect to $\triangle DEF$ is the orthocenter of $\triangle XYZ$.*

Solution 303. *Let F be the intersection of the A-altitude with the circumcircle of $\triangle ABC$, and E the intersection with the bisector. Then let M_a be the midpoint of $\overline{BC}$. Observe that O, M_a, E are collinear, and $\angle AEO = \angle EAO$, so $\angle FAE = \angle AEO \implies \angle FAE = \angle EAO$. Apply this to the vertices of $\triangle ABC$ to obtain the desired result.*

Extension Exercise 130.

(a) *What are the trilinear coordinates of the Miquel point of a triangle?*

(b) *Determine when the Miquel point coincides with the circumcenter of the triangle.*

(If you're stuck, it may help to read through §9.3 first.)

Extension Solution 130.

(a) *The trilinear coordinates of the Miquel point are given by $(\alpha : \beta : \gamma)$, where $\alpha = a(-a^2 d_a d_a + b^2 d_a d_b + c^2 d_a d_c)$ and β, γ are defined analogously. This shares an inextricable relationship with the trilinear coordinates of the circumcenter, namely $(\cos\alpha : \cos\beta : \cos\gamma)$.*

(b) *When $d_a = d_b = d_c = \dfrac{1}{2}$ (i.e. when all the distances along the edges of the triangle to A, B, C are $\dfrac{1}{2}$), the Miquel point coincides with the circumcenter.*

Extension Exercise 131. *(2006 IMO P1) Let $\triangle ABC$ be a triangle with incenter I. A point P in the interior of the triangle satisfies $\angle PBA + \angle PCA = \angle PBC + \angle PCB$. Show that $AP \geq AI$, and that equality holds if and only if $P = I$.*

Extension Solution 131. *We have* $\angle IBP = \angle IBC - \angle PBC = \frac{1}{2}(\angle ABC - \angle PBC) = \frac{1}{2}(\angle PCB - \angle PCA)$, *and also* $\angle ICP = \angle PCB - \angle ICB = \frac{1}{2}(\angle PBA - \angle PBC)$. *Since* $\angle PBA + \angle PCA = \angle PBC + \angle PCB$, *it follows that* $\angle PBA - \angle PBC = \angle PCB - \angle PCA$.

Combining these, we get that $\angle IBP = \angle ICP$, *which is sufficient to show that* $BIPC$ *is cyclic. Let* $\overline{AI}$ *hit the circumcircle (which we have just now shown exists) of* $\triangle ABC$ *at* J; *then by Fact 5,* $JB = JC = JI = JP$. *Finally, we have* $AP + JP \geq AJ = AI + IJ$ *by Triangle Inequality, but* $JI = JP$. *Thus* $AP \geq AI$, *completing the proof (with equality iff the triangle is degenerate).*

Extension Exercise 132. *(2012 IMO P5) Let* $\triangle ABC$ *be a triangle with* $\angle BCA = 90°$, *and let* D *be the foot of the altitude from* C. *Let* X *be a point in the interior of the segment* CD. *Let* K *be the point on the segment* $\overline{AX}$ *such that* $BK = BC$. *Similarly, let* L *be the point on the segment* $\overline{BX}$ *such that* $AL = AC$. *Let* M *be the point of intersection of* $\overline{AL}$ *and* $\overline{BK}$. *Show that* $MK = ML$.

Extension Solution 132. *As* $AC^2 = AL^2 = AB \cdot AD$, *we have* $\triangle ABL \sim \triangle ALD$, *and so* $\angle ALD \cong \angle XBA$. *Extend* $\overline{DC}$ *past* C *to* R *such that* $DR \cdot DX = AD \cdot BD$. *As both* $\angle BDX, \angle RDA$ *are right, it follows that* $\triangle RAD$ *and* $\triangle BXD$ *are similar, and also that* $\angle XBD = \angle ARD$, *with* $\angle ALD = \angle ARD$. *Most importantly,* R, A, D, L *are concyclic and* $RADL$ *is cyclic.*

This implies $\angle RLA$ *is right, and so* $RL^2 = AR^2 - AL^2 = AR^2 - AC^2$. *Similarly,* $RK^2 = BR^2 - BC^2$. *Since* $RC \perp AB$, $AR^2 - AC^2 = BR^2 - BC^2$, *and so* $RL^2 = RK^2 \implies RL = RK$. *We then conclude that triangles* $\triangle RLM, \triangle LKM$ *are not only similar, but altogether congruent, and thusly that* $MK = ML$.

Extension Exercise 133. *(2006 USAMO P6) Let* $ABCD$ *be a quadrilateral, and let* E *and* F *be points on sides* AD *and* BC, *respectively, such that* $AE/ED = BF/FC$. *Ray* FE *meets rays* BA *and* CD *at* S *and* T *respectively. Prove that the circumcircles of triangles* SAE, SBF, TCF, *and* TDE *pass through a common point.*

Extension Solution 133. *Suppose that the circumcircles of* $\triangle SAE, \triangle SBF$ *intersect at* X, *and the circumcircles of* $\triangle TCF, \triangle TDE$ *meet at* Y. *We have* $\triangle XAE \sim \triangle XBF$, *with* X *the center of spiral similarity for* A, E, B, F. *Then, similarly,* $\triangle YED \sim \triangle YFC$, *with* Y *analogously the center of a spiral similarity with* E, D, F, C.

It then follows that $\frac{XE}{XF} = \frac{AE}{BF}$. *Given that* $\frac{ED}{EA} = \frac{CF}{BF}$, *we immediately obtain* $\frac{ED}{FC} = \frac{XE}{XF}$. *Then* $\triangle XED \sim \triangle XFC$, *with* X *again a center of spiral similarity.* $X = Y$, *which proves the statement.*

Section 9.2: (Some) Advanced Synthetic Geometry

Exercise 304. *Prove that any* $\triangle ABC \sim \triangle DEF$ *with the same orientation has a center* O *such that some similitude about* O *sends* $\triangle ABC \rightarrow \triangle DEF$ *given that none of the 15 lines formed by the 6 points* A, B, C, D, E, F *are parallel.*

Solution 304. *We use directed angles because we care about orientation.*

Let the center of similitude from $A, B \rightarrow D, E$ *be* O.

Then by the definition of similitude, $\triangle ABI \sim \triangle DEO$, *implying* $\angle(OA, AB) = \angle(OD, DE)$ *and* $\angle(OB, BA) = \angle(OE, ED)$ *as well as* $\frac{AB}{DE} = \frac{OA}{OD} = \frac{OB}{OE} = r$. *Then by angle addition, we have* $\angle(OB, BC) = \angle(OE, EF)$. *Using SAS similarity gives us* $\triangle BCO \sim \triangle EFO$, *and* $\frac{OB}{OE} = \frac{OC}{OF} = r$. *This gives us our constant ratio of similarity* r *and common rotational angle of* $\angle(AO, OD) = \angle(BO, OE) = \angle(CO, OF)$, *as desired.*

Exercise 305. *(2015 AIME I) Point* B *lies on line segment* $\overline{AC}$ *with* $AB = 16$ *and* $BC = 4$. *Points* D *and* E *lie on the same side of line* AC *forming equilateral triangles* $\triangle ABD$ *and* $\triangle BCE$. *Let* M *be the midpoint of* $\overline{AE}$, *and* N *be the midpoint of* $\overline{CD}$. *The area of* $\triangle BMN$ *is* x. *Find* x^2.

Solution 305. *This is not hard to coord-bash, but we want to tackle this synthetically. Observe that* $\triangle ABE$ *and* $\triangle DBE$ *are SAS congruent, and so a* $60°$ *spiral similarity rotation about* B *maps* $\triangle ABE$ *to* $\triangle DBC$, as well as M to N. Hence $BM = BN, \angle MBN = 60°$, making $\triangle BMN$ equilateral and allowing for the convenient use of Law of Cosines. Doing so yields $AE = 4\sqrt{21}$, or $AM = ME = 2\sqrt{21}$. Now applying Stewarts Theorem yields $BM = 2\sqrt{13}$, i.e. $[\triangle BMN] = 13\sqrt{3}$, or $x^2 = \boxed{507}$.

Exercise 306. *(2017 AIME II) Circle C_0 has radius 1, and the point A_0 is a point on the circle. Circle C_1 has radius $r < 1$ and is internally tangent to C_0 at point A_0. Point A_1 lies on circle C_1 so that A_1 is located $90°$ counterclockwise from A_0 on C_1. Circle C_2 has radius r^2 and is internally tangent to C_1 at point A_1. In this way a sequence of circles $C_1, C_2, C_3, \ldots$ and a sequence of points on the circles $A_1, A_2, A_3, \ldots$ are constructed, where circle C_n has radius r^n and is internally tangent to circle C_{n-1} at point A_{n-1}, and point A_n lies on C_n $90°$ counterclockwise from point A_{n-1}, as shown in the figure below. There is one point B inside all of these circles. When $r = \frac{11}{60}$, the distance from the center C_0 to B is $\frac{m}{n}$, where m and n are relatively prime positive integers. Find $m + n$.*

Solution 306. *Consider the figure as a whole. Note that performing a $90°$ rotation clockwise, then dilating by a scale factor r yields an invariance in the figure - hence spiral similarity! Equating coefficients from*
$$B = (x, y) = (1 - r(1 + y), rx) \text{ yields } x = \frac{1 - r}{r^2 + 1}, y = \frac{r - r^2}{r^2 + 1}, \text{ so the desired distance is } \frac{49}{61} \implies \boxed{110}.$$

Exercise 307. *Prove that two directly similar triangles are related by a spiral similarity, and that two triangles related by a spiral similarity are directly similar.*

Solution 307. *In the solution to the first exercise, we already have proved that two directly similar triangles are related by a spiral similarity. So all that remains is to prove two triangles related by a spiral similarity are directly similar.*

Without loss of generality, let our two triangles be $\triangle ABC \sim \triangle DEF$ such that $A \to B \to C, D \to E \to F$. Then by SAS, we have $ABO \sim BCO \sim DEO \sim EFO$. Since $\angle DEO = \angle ABO$ and $\angle FEO = \angle CBO$, we have $\angle DEF = \angle ABC$. By SAS, $\triangle ABC \sim \triangle DEF$.

Exercise 308. *Consider $\triangle ABO$ and $\triangle CDO$ such that $\triangle ABO \sim \triangle CDO$ with the same orientation. Then let AC intersect BD at X. Prove that $\angle AXB = \angle COD$.*

Solution 308. *Notice that O is the center of spiral similarity from $A \to C, B \to D$. By the Fundamental Theorem of Spiral Similarity, A, B, X, O and C, D, X, O are concyclic. By Inscribed Angle, $\angle AXB = \angle CXD = \angle COD$, as desired.*

Exercise 309. *(Fundamental Theorem of Similarity) Given $A_1 A_2 \ldots A_n \sim B_1 B_2 \ldots B_n$ with the same orientation, prove that for all $M_1, M_2 \ldots M_n$ such that $\frac{A_i M_i}{B_i M_i} = r$ where r is constant for all $1 \leq i \leq n$, $M_1 M_2 \ldots M_n \sim A_1 A_2 \ldots A_n \sim B_1 B_2 \ldots B_n$.*

Solution 309. *Let O be the center of spiral similarity such that $A_1, A_2 \to B_1, B_2$. Then O also is the center of spiral similarity that sends $A_1, B_1 \to A_2, B_2$. As $\triangle OA_1 M_1 \sim \triangle OA_2 M_2$, we notice that $\angle M_1 O M_2 = \angle M_2 O A_1 - \angle M_1 O A_1 = \angle M_2 O A_1 - \angle M_2 O A_2 = \angle A_1 O A_2$. Combining this with our ratio condition produces $\triangle OA_1 A_2 \sim \triangle OM_1 M_2$. Repeating this for every triangle, we are done.*

Exercise 310. *(2014 IMO P4) Points P and Q lie on side BC of acute-angled $\triangle ABC$ so that $\angle PAB = \angle BCA$ and $\angle CAQ = \angle ABC$. Points M and N lie on lines AP and AQ, respectively, such that P is the midpoint of AM, and Q is the midpoint of AN. Prove that lines BM and CN intersect on the circumcircle of $\triangle ABC$.*

Solution 310. *Let L be the midpoint of $\overline{BC}$. Angle chasing yields $\angle AQP = \angle APQ = \angle BAC$. Then $\cot \angle MBC - \cot \angle ABC = 2 \cot \angle BAC$. Hence $\cot \angle BAL = 2 \cot \angle BAC + \cot \angle ABC = \cot \angle MBC$.*

Note that $\cot$ has a period of π, but all angles are less than π, so it must be that $\angle BAL = angle MBC$. Similarly, $\angle LAC = \angle NCB$. If we let R be the intersection point of $\overline{BM}, \overline{CN}$, then $\angle BRC = 180 - \angle BAC$. It follows that $BACR$ is cyclic, which implies the desired result.

Exercise 311. *Consider directly similar $\triangle ABC \sim \triangle DEF$. Let X, Y, Z be the intersection of AD and BE, BE and CF, and CF and AD, respectively. Prove that $\triangle XYZ$ is isosceles.*

Solution 311. *Without loss of generality, let there exist a spiral similarity such that $A \to B \to C, D \to E \to F$. By the Fundamental Theorem of Spiral Similarity, X lies on the circumcircles of $\triangle ABO$ and $\triangle DEO$, and Y lies on the circumcircles of $\triangle DEO$ and $\triangle CFO$. By the Inscribed Angle Theorem, $\angle AOB = \angle AXB$ and $\angle DOE = \angle DYF$. Then notice that $\angle AXB = \angle DXE = \angle ZXY$ and $\angle EYF = \angle CYB = \angle ZYX$, so $\triangle XYZ$ is isosceles.*

Extension Exercise 134. *(1978 IMO P4) In the triangle $\triangle ABC$, $AB = AC$. A circle is tangent internally to the circumcircle of the triangle and also to $\overline{AB}$, $\overline{AC}$ at P, Q respectively. Prove that the midpoint of $\overline{PQ}$ is the center of the incircle of the triangle.*

Extension Solution 134. *By symmetry, M is on the bisector of $\angle A$, so it suffices to show it also lies on the bisector of $\angle B$. Let the bisector of $\angle A$ meet the circumcircle at $R \neq A$. Note that $\overline{AP}$ is tangent to the circle touching $\overline{AB}$ at P; hence, $\angle PRQ = \angle APQ = \angle ABC$. Then since $\angle PBR = \angle PMR = 90°$, $PBRM$ must be cyclic. Finally, $\angle PBM = \angle PRM = \frac{1}{2}\angle PRQ$, so $\overline{BM}$ is indeed a bisector of $\angle B$.*

Extension Exercise 135. *(1995 IMO P1) Let A, B, C, D be four distinct points on a line, in that order. The circles with diameters AC and BD intersect at X and Y. The line XY meets BC at Z. Let P be a point on the line XY other than Z. The line CP intersects the circle with diameter AC at C and M, and the line BP intersects the circle with diameter BD at B and N. Prove that the lines AM, DN, XY are concurrent.*

Extension Solution 135. *We have $\angle AMC$ right, so $\angle MCA = 90 - \angle A$. Then $\angle BND = 90°$ as well. Since $\overline{XY}$ is the radical axis of the circles with diameters $\overline{AC}, \overline{BD}$, we have $PB \cdot PN = PC \cdot PM$, which implies that $MNBC$ is cyclic (by the converse of Power of a Point). Hence $90 - \angle A = \angle MCA = \angle BNM$. It also follows that $\angle MND = 180 - \angle A$, with $AMND$ cyclic. Denote O as the circumcircle of $AMND$; then the radical axis of O is $\overline{AM}$, which is also the radical axis of the circle with diameter $\overline{BC}$. Furthermore, the radical axis of the circle with diameter $\overline{BD}$ and O is $\overline{DN}$. The radical axes of all three circles are pairwise concurrent, and so $\overline{AM}, \overline{DN}, \overline{XY}$ concur.*

Extension Exercise 136. *(2003 USAMO P4) Let ABC be a triangle. A circle passing through A and B intersects segments AC and BC at D and E, respectively. Lines AB and DE intersect at F, while lines BD and CF intersect at M. Prove that $MF = MC$ if and only if $MB \cdot MD = MC^2$.*

Extension Solution 136. *Sketch: Note that $MB \cdot MD = MC^2 \iff \frac{MB}{MC} = \frac{MC}{MD}$. This, in turn, is equivalent to $\triangle MCD \sim \triangle MBC$, which finally suffices to prove that $\angle CBA = \angle FCA$ bi-directionally. (Note that the similarity here is SAS similarity.)*

Extension Exercise 137. *(1981 IMO P5) Three congruent circles have a common point O and lie inside a given triangle. Each circle touches a pair of sides of the triangle. Prove that the incenter and the circumcenter of the triangle and the point O are collinear.*

Extension Solution 137. *Denote the triangle by $\triangle ABC$, and label its side lengths a, b, c opposite their respective angles as usual. Let the centers of the inscribed circles of $\angle A, \angle B, \angle C$ be O_A, O_B, O_C respectively. Then $\triangle O_A O_B O_C, \triangle ABC$ are homotheties of each other, since their corresponding sides are parallel. As O_A is on the bisector of $\angle A$ (and the same holds for O_B, O_C), the center of the homothety is the common incenter of both triangles. O is the circumcenter of $\triangle O_A O_B O_C$, so O is collinear with the incenter and circumcenter of $\triangle ABC$, as desired.*

Section 9.3: Polar, Cylindrical, and Barycentric Coordinates

Exercise 312. *(2013 AIME II) In $\triangle ABC$, $AC = BC$, and point D is on $\overline{BC}$ so that $CD = 3 \cdot BD$. Let E be the midpoint of $\overline{AD}$. Given that $CE = \sqrt{7}$ and $BE = 3$, the area of $\triangle ABC$ can be expressed in the form $m\sqrt{n}$, where m and n are positive integers and n is not divisible by the square of any prime. Find $m + n$.*

Solution 312. *Using Stewarts: set $AE = ED = m$, $BD = k$ with $CD = 3k$, $AC = 4k$. Then, applying Stewart's Theorem on $\triangle ACD$ with cevian $\overline{CE}$, we have $2m^2 + 14 = 25k^2$. Re-apply Stewart on $\triangle CEB$ with* *cevian* $\overline{DE}$ *to get* $2m^2 = 17 - 6k^2$; *i.e.* $k = 1, m = \dfrac{\sqrt{22}}{2}$. *Then by base ratios,* $[\triangle CED] = \dfrac{3}{8}[\triangle ABC]$, *so* $[\triangle ABC] = 3\sqrt{7} \implies \boxed{10}$.

We can also coordinate bash to get a much shorter solution. If we set $A = (-a, 0), B = (a, 0), c = (0, h)$, *then* $D = \left(\dfrac{3a}{4}, \dfrac{h}{4}\right), E = \left(-\dfrac{a}{8}, \dfrac{h}{8}\right)$. *In conjunction with* $EC^2 = 7$, *we end up with* $a^2 + 49h^2 = 448$, *and with* $EB = 3$, *this implies* $81a^2 + h^2 = 576$. *Then* $a = \sqrt{7}, h = 3$, *so* $ah = 3\sqrt{7} \implies \boxed{10}$.

Exercise 313. *(1992 AIME) In triangle* ABC, A', B', *and* C' *are on the sides* BC, AC, *and* AB, *respectively. Given that* AA', BB', *and* CC' *are concurrent at the point* O, *and that* $\dfrac{AO}{OA'} + \dfrac{BO}{OB'} + \dfrac{CO}{OC'} = 92$, *find* $\dfrac{AO}{OA'} \cdot \dfrac{BO}{OB'} \cdot \dfrac{CO}{OC'}$.

Solution 313. *This is made somewhat easier with mass points. In general, let the weights of* A, B, C *be* a, b, c. *Then* A, B, C *have weights of* $b + c, a + c, a + b$ *respectively. Then* $\dfrac{AO}{OA} = \dfrac{b+c}{a}, \dfrac{BO}{OB} = \dfrac{a+c}{b}, \dfrac{CO}{OC} = \dfrac{a+b}{c}$. *Thus, their product is* $2 + \sum_{cyc} \dfrac{a+b}{c} = 2 + \dfrac{AO}{OA} + \dfrac{BO}{OB} + \dfrac{CO}{OC} = \boxed{94}$.

Exercise 314. *(1999 AIME) Let* $\mathcal{T}$ *be the set of ordered triples* (x, y, z) *of nonnegative real numbers that lie in the plane* $x + y + z = 1$. *Let us say that* (x, y, z) *supports* (a, b, c) *when exactly two of the following are true:* $x \geq a, y \geq b, z \geq c$. *Let* $\mathcal{S}$ *consist of those triples in* $\mathcal{T}$ *that support* $\left(\frac{1}{2}, \frac{1}{3}, \frac{1}{6}\right)$. *The area of* $\mathcal{S}$ *divided by the area of* $\mathcal{T}$ *is* m/n, *where* m *and* n *are relatively prime positive integers. Find* $m + n$.

Solution 314. *The area in* $\mathbb{R}^3$ *for which* $x \geq \dfrac{1}{2}, y \geq \dfrac{1}{3}$ *is a triangle, as is the region with* $y \geq \dfrac{1}{3}, z \geq \dfrac{1}{6}$, *and the region with* $x \geq \dfrac{1}{2}, z \geq \dfrac{1}{6}$. *Then we can visualize the areas of intersection of the triangles with the plane as three equilateral triangles in two dimensions. The side length of the largest one is* $\dfrac{\sqrt{2}}{2}$, *the side length of the middle one is* $\dfrac{\sqrt{2}}{3}$, *and the side length of the smallest is* $\dfrac{\sqrt{2}}{6}$. *Hence the requested ratio is* $\dfrac{7}{18} \implies \boxed{25}$.

Exercise 315. *(1986 AIME) In* $\triangle ABC$, $AB = 425$, $BC = 450$, *and* $AC = 510$. *An interior point* P *is then drawn, and segments are drawn through* P *parallel to the sides of the triangle. If these three segments are of an equal length* d, *find* d.

Solution 315. *Let* D, D, E, E, F, F *lie on* $\overline{AB}, \overline{BC}, \overline{CA}$ *(with the Ds on* $\overline{AB}$, *the Es on* $\overline{BC}$, *and the Fs on* $\overline{CA}$) *such that the segments hit the triangle at those points. Each of these line segments is parallel to a side of the triangle, so all three of the small triangles are similar to the largest one; the other three regions are parallelograms. Then* $BE = \dfrac{15}{17}d, EC = \dfrac{18}{17}d$. $FD = BC - EE$, *so* $900 - \dfrac{33}{17}d = d \implies d = \boxed{306}$.

Exercise 316. *(2012 ARML Team Round #7) Given noncollinear points* A, B, C, *segment* $\overline{AB}$ *is trisected by points* D *and* E, *and* F *is the midpoint of segment* $\overline{AC}$. $\overline{DF}$ *and* $\overline{BF}$ *intersect* $\overline{CE}$ *at* G *and* H, *respectively. If* $[DEG] = 18$, *compute* $[FGH]$. *[Author's note:* $[A]$ *denotes the area of* A.*]*

Solution 316. *Let* B *have mass 1, and let* A, C *have mass 2. Then* $\triangle ABC$ *balances at* H, *so* F *has mass 4 and* $\dfrac{BH}{HF} = 4$. *Consider* $\triangle BEF$; *if* E *has mass 1 as well, then the triangle balances at* G, *and* $\dfrac{DG}{GF} = 2, \dfrac{EG}{GH} = 5$. *Thus,* $\dfrac{[\triangle DEG]}{[\triangle FHG]} = \dfrac{DG}{FG} \cdot \dfrac{EG}{HG} = 10$. *Hence,* $[\triangle FGH] = \dfrac{[\triangle DEG]}{10} = \boxed{\dfrac{9}{5}}$.

Extension Exercise 138. *(2006 Putnam A1) Find the volume of the region of points* (x, y, z) *such that*

$$(x^2 + y^2 + z^2 + 8)^2 \leq 36(x^2 + y^2)$$

420

Extension Solution 138. *Converting to cylindrical coordinates with $r = \sqrt{x^2 + y^2}$, we obtain $r^2 + z^2 + 8 \leq 6r$, i.e. $r^2 - 6r + 9 + z^2 \leq 1 \implies (r-3)^2 + z^2 \leq 1$. By Pappus Theorem, the volume of revolution about the xz-plane is the area of the disc, which is π, multiplied with the distance through which the center of mass is rotated (or 3 times the circumference, namely 6π). Hence the volume is $\boxed{6\pi^2}$.*

Extension Exercise 139. *(2015 USAMO P2) Quadrilateral $APBQ$ is inscribed in circle ω with $\angle P = \angle Q = 90°$ and $AP = AQ < BP$. Let X be a variable point on segment $\overline{PQ}$. Line AX meets ω again at S (other than A). Point T lies on arc AQB of ω such that $\overline{XT}$ is perpendicular to $\overline{AX}$. Let M denote the midpoint of chord $\overline{ST}$. As X varies on segment $\overline{PQ}$, show that M moves along a circle.*

Extension Solution 139. *Let K be the midpoint of $\overline{AO}$. We show that $KM = KP$, since KP is constant.*

Note that $KM^2 = \dfrac{AM^2 + OM^2}{2} - \dfrac{AO^2}{4}$ by Stewarts on $\triangle AMO$, and $KP^2 = \dfrac{AP^2 + OP^2}{2} - \dfrac{AO^2}{4}$. Thus, we show that $AM^2 + OM^2 = AP^2 + OP^2$. As $OP = OT$, we have $OP^2 - OM^2 = MT^2$.

Furthermore, as $\triangle XTS$ has circumcenter M, with circumradius $\overline{MT}$ and corresponding circumcircle ω, $AM^2 - MT^2$ is indeed the power of A with respect to ω. Since $AX \cdot AS$ is also the power of A, $AM^2 - MT^2 = AX \cdot AS$. Finally, $\triangle APX \sim \triangle ASP$, so that $AX \cdot AS = AP^2$; the desired conclusion follows.

Section 9.4: Inversion

Exercise 317.

(a) *Invert P about circle ω to get P'. If P' is outside of ω and the tangent points from P' to ω are X, Y, prove P, X', Y' are collinear.*

(b) *Consider circle ω with diameter AB. What do you get when you invert line AB about ω?*

(c) *What about inverting segment AB?*

(d) *Prove that if you invert about a unit circle centered at the origin in the Cartesian plane, $P = (x, y)$ goes to $P' = (\frac{x}{x^2+y^2} + \frac{y}{x^2+y^2})$.*

(e) *Prove that if you invert about a unit circle centered at the origin in the Complex plane, z goes to $\overline{z}^{-1}$.*

Solution 317.

(a) *By similarity, $XP \perp OQ$ and $YP \perp OQ$. Since this implies $XP \parallel YP$, X, P, Y are collinear.*

(b) *You get line AB. Letting the center be O, you can pair up any point X on ray OA with another point Y such that $OX \cdot OY = OA^2$. You can do the same thing with OB.*

(c) *You get all of line AB except for segment AB. This is because the aforementioned pairs have a group of points inside and a group of points outside the circle. You take the points inside the circle and make them the points outside the circle. (Notice that A and B remain.)*

(d) *Let $O = (0, 0)$. Since P' is the result of a dilation about P, the collinearity condition is satisfied. The magnitude of P is $\sqrt{x^2 + y^2}$, and the magnitude of P' is $\frac{1}{\sqrt{x^2+y^2}}$. Thus, $OP \cdot OP' = 1$, as desired.*

(e) *Notice that z and $\overline{z}^{-1}$ are collinear and that $z \cdot \overline{z}^{-1} = 1$, as desired.*

Exercise 318. *Given the normal construction tools and the ability to dilate by any factor, construct the inversion of a circle not passing through the center of the circle of inversion.*

Solution 318. *Let our circle of inversion be ω and let the circle we want to invert be Γ. Let the center of ω be O and let the center of Γ be O. Then let OO intersect Γ at X, Y. Then dilate Γ by a factor of $\frac{r^2}{OX \cdot OY}$, where r is the radius of ω.*

Exercise 319. *With the same tools, construct the inversion of a circle passing through the center of the circle of inversion.*

Solution 319. *Let our circle of inversion be ω and let the circle we want to invert be Γ. Let P be on Γ such that OP is a diameter of Γ. Then invert P about ω to get Q. Draw the line passing Q perpendicular to OQ to get your inversion.*

Exercise 320. *With the same tools, construct the inversion of a line.*

Solution 320. *Let the circle of inversion ω have center O. Then choose the point Q on the line such that OQ is perpendicular to the line. Let P be the result of an inversion of Q about ω. Then draw the circle with diameter OP.*

Exercise 321. *(2013 Berkeley Math Tournament) From a point A construct tangents to a circle centered at point O, intersecting the circle at P and Q respectively. Let M be the midpoint of PQ. If K and L are points on circle O such that $K, L,$ and A are collinear, prove $\angle MKO = \angle MLO$.*

Solution 321. *As $AP = AQ$, median $\overline{AM}$ is a perpendicular bisector of $\overline{PQ}$, so A, M, O must be collinear. Then observe the similarity of $\triangle OPA, \triangle PMA$, which implies that $AM \cdot AO = AP^2$. Then by Power of a Point, $OP^2 = AL \cdot AK$, so $AM \cdot AO = AL \cdot AK$ and the converse of power of a point implies the concyclicity of O, M, L, K.*

(Note: We can also observe that A and M are polars with respect to O. The desired conclusion is, in fact, equivalent to showing that O, M, L, K are collinear. Note that $O = P_\infty, M = A, L = L, K = K$, which reduces to showing that A, L, K are collinear.)

Exercise 322. *Circles C_1, C_2, C_3, and C_4 lie in $\mathbb{R}^2$ such that C_2 and C_4 are both tangent to C_1 and C_3. Prove that the points of tangency are collinear or concyclic.*

Solution 322. *Make an inversion with one point of tangency as the center. Then two circles tangent at that center of inversion will map on two parallel lines, with the other two circles mapping onto tangent circles each tangent to one of the parallel lines (by the angle preservation property of an inversion map). Observe that there is a homothety between any two parallel lines with some point in the plane, as well as between any two tangent circles with the common point of tangency as the center of homothety.*

Hence, the whole configuration is homothetic with respect to the point of tangency of the two inverted circles. Thus, by inversion, the line containing the three points of tangency of the circles with the parallel lines is mapped onto a circle that contains the center of inversion.

Hitherto, we have considered one of the points of tangency as the center of inversion. The other three points are concyclic, and say they lie on some circle ω. Under inversion, the images of those points are collinear, so ω must pass through the center of inversion, i.e. the remaining point of tangency. The conclusion follows.

Exercise 323. *Consider scalene $\triangle ABC$ with incenter I. Let the A excircle of $\triangle ABC$ intersect the circumcircle of $\triangle ABC$ at X, Y. Let XY intersect BC at Z. Then choose M, N on the A excircle of $\triangle ABC$ such that ZM, ZN are tangent to the A excircle of $\triangle ABC$. Prove I, M, N are collinear.*

Solution 323. *For reference, let the circumcircle of $\triangle ABC$ be ω_1, the A excircle of $\triangle ABC$ be ω_2, and the circumcircle of $\triangle BIC$ be ω_3.*

The main idea is to notice that we want to prove I lies on the polar of Z with respect to the A excircle, which means we can prove that Z lies on the polar of I. Let the A excenter be I_A, and let the tangents from I to ω_2 be P, Q. Then notice $\angle IPI_A = \angle IQI_A = 90° = \angle IBI_A = \angle ICI_A$, implying that $BCPQ$ are concyclic and that PQ is the polar of I with respect to ω_2.

But notice that Z is the intersection of BC and XY, the radical axes of (ω_1, ω_3) and (ω_1, ω_2), respectively. Then Z lies on the radical axis of (ω_2, ω_3), otherwise known as PQ. As PQ is the polar of I and Z lies on the PQ, Z lies on the polar of I, as desired.

Exercise 324. *Consider $\triangle ABC$ with point D on BC. Let M, N be the circumcenters of $\triangle ABD$ and $\triangle ACD$, respectively. Let the circumcircles of $\triangle ACD$ and $\triangle MND$ intersect at $H \neq D$. Prove A, H, M are collinear.*

Solution 324. *Invert about the circle with center D and radius DA. Let this send B, C, M, N, H to B', C', M', N', H', respectively.*

Notice that M' is the reflection of D about AB' and N' is the reflection of D about AB'. Then notice that H' is the intersection of $M'N'$ and AC'.

Let P be the midpoint of DM' and Q be the midpoint of DN'. Notice that by a lemma, $\angle APD = \angle AQD = 90°$, implying $APDQ$ is cyclic. So $\angle QAD = \angle QPD = \angle N'M'D = \angle H'M'D$. Since $\angle H'AD = 180° - \angle QAD = 180° - \angle H'M'D$, we notice that $H'M'DA$ is cyclic, as desired.

Exercise 325. *Consider $\triangle ABC$ with orthocenter H and circumcenter O. Let X, Y, Z be the midpoints of AH, BH, CH and let D, E, F be the midpoints of BC, CA, AB. Prove that XD, YE, ZF are concurrent.*

Solution 325. *We claim that XD, YE, ZF all pass the center of the nine-point circle. We only need to prove this for XD by symmetry. If we dilate about H with a factor of 2, we send $M \to O, X \to A, D \to A_M$, for some point A_M. But looking at the proof of the Nine-Point Circle, we notice AA_M is a diameter of the circumcircle, as desired.*

Exercise 326. *(1993 USAMO P2) Let $ABCD$ be a convex quadrilateral such that diagonals AC and BD intersect at right angles, and let E be their intersection. Prove that the reflections of E across AB, BC, CD, DA are concyclic.*

Solution 326. *Let X, Y, Z, W be the respective feet of the E-altitudes in triangles $\triangle AEB, \triangle BEC, \triangle CED, \triangle DEA$. Note that, if we reflect E over the 4 lines, we get $XYZW$, except dilated by a factor of $\frac{1}{2}$ (i.e. a homothety about E). Hence, if $XYZW$ is cyclic, then so are the reflections.*

$\angle EWA = 90°$, so $\angle EXA = 90°$ also. Thus, $EXAW$ is cyclic, with $\overline{EA}$ being the circum-diameter. It follows that $\angle EWX = \angle EAX = \angle EAB$ by virtue of inscribing the same angle. Similarly, $\angle EWZ = \angle EDC, \angle EYX = \angle EBA, \angle EYZ = \angle ECD$. Also, $\angle XYZ + \angle XWZ = \angle EWX + \angle EYX + \angle EYZ + \angle EWZ = 360° - \angle CED - \angle AEB = 180°$. Hence $\angle XYZ + \angle XWZ = 180°$, from which we conclude that $XYZW$ is cyclic.

Exercise 327. *(2003 IMO Shortlist G3) Let $\triangle ABC$ be a triangle and let P be a point in its interior. Denote by D, E, F the feet of the perpendiculars from P to the lines $\overline{BC}, \overline{CA}, \overline{AB}$, respectively. Suppose that $AP^2 + PD^2 = BP^2 + PE^2 = CP^2 + PF^2$. Denote by I_A, I_B, I_C the excenters of the triangle $\triangle ABC$. Prove that P is the circumcenter of the triangle $I_A I_B I_C$.*

Solution 327. *Let $X = DF \cap EG$. Angle chasing yields $\angle A = \angle FXG$. Since $AFPE, BGPD$ are cyclic (because they are isosceles trapezoids), $FAXE, FXBG$ are cyclic as well, and so $\overline{PX}$ is the radical axis of the circumcircles of $AFPE$ and $DPGB$. We have $CA \cdot CF = CB \cdot CG$, so C lies on the radical axis of the two circles. Then $\angle AXB = \angle AXP + \angle BXP = \angle CFP + \angle CGP = 2\angle = 180° - \angle C$, as desired.*

Exercise 328. *A hexahedron is any polyhedron with six faces. Prove that if seven of the vertices of a hexahedron lie on a sphere, then so does the eighth vertex.*

Solution 328. *Let A through H be the vertices of a cube (a prototypical hexahedron), and let U through Z be the faces of the cube. It suffices to prove this for a point A at infinity. Project the 6-8 configuration from A to obtain a configuration with U corresponding to $\overline{BC}$, W corresponding to $\overline{CE}$, and Y corresponding to $\overline{BE}$. The circles then form $\triangle BCE$.*

As for points D, F, G, we can choose them to be anywhere on the sides of the newly-formed triangle $\triangle BCE$, as well as beyond the vertices themselves. The circles V, X, Z pass through three of the points, similarly to the previous construction. We claim that all three of these circles intersect at point H.

Remove the circumcircle of $\triangle EFG$, and let $H \neq D$ be the intersection of the circumcircles of $\triangle BDF, \triangle CDG$. Draw $\overline{DH}, \overline{FH}, \overline{GH}$, thus forming quadrilaterals $BDHF, CDHG, EFHG$. $BDHF$ and $CDHG$ are cyclic, so $\angle BDH + \angle BFH = 180°$, and $\angle CDH + \angle CGH = 180°$ as well. Furthermore, the pairs $\angle EFH, \angle BFH$, $\angle BDH, \angle CDH$, and $\angle CGH, \angle EGH$ are also supplementary. So it follows that $\angle EFH + \angle EGH = 180°$. Hence $EFHG$ is cyclic, and so the circumcircles of the triangles $\triangle BDF, \triangle CDG, \triangle EFG$ concur at H, as desired.

Exercise 329. *Prove Feuerbach's Theorem:*

Theorem 9.22: Feuerbach's Theorem

The nine-point circle of triangle $\triangle ABC$ is tangent to the incircle, and all three excircles, of $\triangle ABC$.

Solution 329. *Let $\overline{BC}, \overline{C_bC_c}$ intersect at S. Then $\dfrac{AC_b}{AC_c} = \dfrac{r_b}{r_c} = \dfrac{SC_b}{SC_c}$, so it also follows that $\dfrac{PB}{PC} = \dfrac{SB}{SC}$, since A, S divide $\overline{C_bC_c}$ harmonically. This is, in turn, equivalent to stating that $LP \cdot LS = LB \cdot LC = LB^2 = LC^2$. Hence there exists an inversion mapping the circle ω with center L and radius $LB = LC$, or the circle which has BC as its diameter. Then since $2BB = AB + BC + AC = 2CC$, L is the midpoint of $\overline{BC}$.*

Furthermore, let K be the midpoint of $\overline{AB}$, and M the midpoint of $\overline{AC}$. $\triangle ABCs$ nine-point circle is thus the circumcircle of $\triangle KLM$. Since it contains L, the inversion of ω maps the nine-point circle such that it passes through the feet of the altitudes. Thus, the inversion maps it to pass through S, the image of P.

By properties of the inversion, where XX is tangent to the nine-point circle at L, we have $\angle XLK = \angle KML = \angle ABC$. By symmetry, XX is parallel to the tangent. Then the inversion maps $\overline{KL}$ onto itself, with $\overline{KL} \parallel \overline{AC}$. The image of the nine-point circle must then form the same angle with $\overline{KL}$ as it does with the circle. This is satisfied in particular by the external tangent to C_b, C_c, which also passes through S. Hence, it is precisely the image of the nine-point circle under the inversion map. The circles C_b, C_c are fixed under this map, which implies that the image of the nine-point circle is just the nine-point circle itself.

Extension Exercise 140. *(1996 IMO P2) Let P be a point inside $\triangle ABC$ such that $\angle APB - \angle C = \angle APC - \angle B$. Let D and E be the incenters of $\triangle APB$ and $\triangle APC$ respectively. Show that AP, BD, and CE meet at a point.*

Extension Solution 140. *Let $P \in \triangle ABC$ with X, Y, Z being the feet of the perpendiculars to $\overline{BC}, \overline{CA}, \overline{AB}$ respectively. Then $PA = \dfrac{YZ}{\sin A}$ by Law of Sines, and $\angle APB - \angle C = \angle XZY$. (More specifically: $\angle XZY = \angle XZP + \angle YZP = \angle XBP + \angle YAP = 90° - \angle XPB + 90° = \angle APB - \angle C$).*

Hence, XYZ is isosceles with $XY = XZ$, and it follows that $PC \cdot \sin C = PB \cdot \sin B$. But we also have $AC \cdot \sin C = AB \cdot \sin B$, so $\dfrac{AB}{PB} = \dfrac{AC}{PC}$. Let $\overline{BD}$ (an angle bisector) hit $\overline{AP}$ at Q. Then the Angle Bisector Theorem says that $\dfrac{AB}{PB} = \dfrac{QA}{QP}$; i.e. $\dfrac{AW}{PW} = \dfrac{AC}{PC}$, so $\overline{CW}$ bisects $\angle ACP$.

Extension Exercise 141. *(2002 IMO Shortlist G1) Let B be a point on a circle S_1, and let A be a point distinct from B on the tangent at B to S_1. Let C be a point not on S_1 such that the line segment $\overline{AC}$ meets S_1 at two distinct points. Let S_2 be the circle touching $\overline{AC}$ at C and touching S_1 at a point D on the opposite side of $\overline{AC}$ from B. Prove that the circumcenter of triangle $\triangle BCD$ lies on the circumcircle of triangle $\triangle ABC$.*

Extension Solution 141. *Let E be the midpoint of $\overline{BD}$, and let F be the midpoint of $\overline{CD}$. Then let K be the circumcenter of $\triangle BCD$, with T being a point such that TT is the tangent common to both circles. Then $\overline{EK} \perp \overline{BD}$, and furthermore, bisects the angles between the tangents $\overline{BA}$ and $\overline{DT}$ to S_1 at B and D. Thus, K is equidistant from both $\overline{BA}, \overline{DT}$. Hence, $\overline{KF} \perp \overline{CD}$, with K equidistant from $\overline{AC}, \overline{DT}$. K is then the center of the circle tangent to $\overline{BA}, \overline{AC}, \overline{DT}$, and $\overline{AK}$ bisects $\angle BAC$. K is also on the perpendicular bisector of $\angle BAC$, which implies the line meets the bisectors of that angle on the circumcircle of $\triangle ABC$ - QED.*

Extension Exercise 142. *(2003 IMO Shortlist G4) Let Γ_1, Γ_2, Γ_3, Γ_4 be distinct circles such that Γ_1, Γ_3 are externally tangent at P, and Γ_2, Γ_4 are also externally tangent at P. Suppose that, for $1 \leq i \leq 4$, C_i and C_{i+1} meet at A, B, C, D, respectively, and that none of these points coincides with P. Prove that*
$$\frac{AB \cdot CD}{AD \cdot CD} = \left(\frac{PB}{PD}\right)^2$$

Extension Solution 142. *Let the common tangent of Γ_1, Γ_3 and line $\overline{AB}$ intersect at Q. Then observe that $\angle APB = \angle APQ + \angle BPQ = \angle PDA + \angle PCB$.*

For $1 \leq i \leq 8$, define θ_i to be the angle adjacent to A, B, C, D (A for $i = 1, 2$, B for $i = 3, 4$, C for $i = 5, 6$, and D for $i = 7, 8$) in clockwise order. Then $\theta_2 + \theta_3 + \angle APB = \theta_2 + \theta_3 + \theta_5 + \theta_8 = 180°$. Similarly, $\theta_4 + \theta_5 + \theta_2 + \theta_7 = 180°$, as $\angle BPC = \angle PAB + \angle PDC$. If we scale the side lengths of $\triangle PAB, \triangle PBC, \triangle PCD, \triangle PDA$ by $PC \cdot PD$, $PD \cdot PA$, $PA \cdot PB$, $PB \cdot PC$ respectively, then we get a quadrilateral $ABCD$ with $AD \parallel BC$, $AB \parallel CD$; i.e. $ABCD$ is a parallelogram. Hence, $AB = CD, AD = CB$; i.e. $AB \cdot PC \cdot PD = CD \cdot PA \cdot PB$ and $AD \cdot PB \cdot PC = BC \cdot PA \cdot PD$, from which the conclusion immediately follows.

Extension Exercise 143. *(1985 IMO P5) A circle with center O passes through the vertices A and C of the triangle ABC and intersects the segments AB and BC again at distinct points K and N respectively. Let M be the point of intersection of the circumcircles of triangles ABC and KBN (apart from B). Prove that $\angle OMB = 90°$.*

Extension Solution 143. *Note that M is the Miquel point of $ACNK$, so there is some spiral similarity about M taking $\overline{KN}$ to $\overline{AC}$. Let M_1 be the midpoint of $\overline{KA}$ and K_2 the midpoint of $\overline{NC}$. Thus, $M_1 \mapsto M_2$ with BMM_1M_2 a cyclic quadrilateral. Then OM_1BM_2 is cyclic as well, with diameter $\overline{BO}$. Hence, M is concyclic with B, M_1, M_2, and so $\angle OMB = 90°$.*

Extension Exercise 144. *(2007 USAMO P6) Let ABC be an acute triangle with ω, Ω, and R being its incircle, circumcircle, and circumradius, respectively. Circle ω_A is tangent internally to Ω at A and tangent externally to ω. Circle Ω_A is tangent internally to Ω at A and tangent internally to ω. Let P_A and Q_A denote the centers of ω_A and Ω_A, respectively. Define points P_B, Q_B, P_C, Q_C analogously. Prove that*

$$8 P_A Q_A \cdot P_B Q_B \cdot P_C Q_C \leq R^3,$$

with equality if and only if triangle ABC is equilateral.

Extension Solution 144. Lemma: *We have $P_A Q_A = \dfrac{4R^2(s-a)^2(s-b)(s-c)}{rb^2c^2}$.*

Proof. P_A, Q_A lie on $\overline{AO}$, since for two tangent circles, the centers and the point of tangency are collinear (exercise to the reader to verify this).

Let ω hit $\overline{BC}, \overline{CA}, \overline{AB}$ at D, E, F respectively, so that $AE = AF = s - a$. Now invert about A with an inversion I to derive $I(\omega) = \omega$ from $\overline{IE} \perp \overline{AE}$. Then let there be ω_A such that $I(\omega_A) = \omega_A$ as well. Note that ω_A passes through A, and so ω_A is tangent to ω. Thus, $\omega_A \perp \overline{AO}$. Similarly, $I(\Omega_A) = \Omega_A$ is perpendicular to $\overline{AO}$.

Let $\omega_A \cap \overline{AO} = X$, with $\omega_A \cap \overline{AO} = X$. Define Y, Y similarly for Ω_A and Ω_A. We have $AX = 2AP_A$, $AY = 2AQ_A$, so we want to compute $AQ_A - AP_A = \dfrac{AY - AX}{2} = \dfrac{(s-a)^2}{2} \cdot \left(\dfrac{1}{AY} - \dfrac{1}{AX}\right)$ by inversion.

Observe that $\omega_A \parallel \Omega_A$. Then the distance between those lines is $2r = AX - AY$. Drop the perpendicular from I to $\overline{AO}$ so that they intersect at H; consequently, $AH = AI \cdot \cos \angle OAI = AI \cdot \cos \dfrac{\angle B - \angle C}{2}$. It follows that $AX \cdot AY = AI^2 \cdot \cos^2 \dfrac{\angle B - \angle C}{2} = r^2$.

Through some further computation on $AQ_A - AP_A$, we get $AQ_A - AP_A = \dfrac{(s-a)^2}{\frac{AI^2}{r} \cdot \cos^2 \frac{\angle B - \angle C}{2} - 1}$. Note

that $\dfrac{AI}{r} = \dfrac{1}{\sin \frac{A}{2}}$, so $AQ_A - AP_A = \dfrac{(s-a)^2(s-b)(s-c)}{rbc \sin \angle B \sin \angle C}$ from which the desired result follows. $\qquad \square$

Using the lemma, we have

$$8 \prod \frac{4R^2(s-a)^2(s-b)(s-c)}{rb^2c^2} \leq R^3$$

which eventually simplifies to $2r \leq R$ using the combined formulae $\text{area} = \dfrac{abc}{4R} = rs$. This holds true, since $OI^2 = R(R-2r)$, with equality when the circumcenter is the incenter. Hence $\angle OAI = \dfrac{1}{2}(\angle B - \angle C) = 0$, and so $\angle A = \angle B = \angle C \implies \triangle ABC$ equilateral; QED.

Chapter 9 Review Exercises

Exercise 330. *Triangle $\triangle ABC$ has $AB = 10$, $BC = 12$, $CA = 14$. The angle bisector of $\angle BAC$ hits $\overline{BC}$ at D. Non-concentric circles C_1 and C_2 respectively have $\overline{BD}$ and $\overline{CD}$ as radii, with D being the center of C_1. What is the distance between the points of intersection of C_1 and C_2?*

Solution 330. *WLOG let $A = (2, 4\sqrt{6}), B = (0,0), C = (12,0)$. Then $D = (5,0)$ by Angle Bisector Theorem. Let ω_1 be the circumcircle with radius $\overline{BD}$, and let ω_2 be the circumcircle with radius $\overline{DC}$ (since ω_2 should not contain ω_1 completely). Then $\omega_1 := (x-5)^2 + y^5 = 25$, $\omega_2 := (x-12)^2 + y^2 = 49$. These intersect at $(x,y) = \left(\dfrac{95}{14}, \pm \dfrac{15\sqrt{19}}{14} \right)$, so the desired distance is $\dfrac{15\sqrt{19}}{7} \implies \boxed{41}$.*

Exercise 331. *(2003 AIME I) Point B is on $\overline{AC}$ with $AB = 9$ and $BC = 21$. Point D is not on $\overline{AC}$ so that $AD = CD$, and AD and BD are integers. Let s be the sum of all possible perimeters of $\triangle ACD$. Find s.*

Solution 331. *This is somewhat trivialized using Stewart's Theorem, with the side being c and the cevian being d. Then $30d^2 + 21 \cdot 9 \cdot 30 = 9c^2 + 21c^2$; solving yields $d^2 + 189 = c^2$ and we can solve for the possible integral values of (c,d). Some computation from here yields $\boxed{380}$.*

Exercise 332. *$\triangle ABC$ has $AB = 4$, $BC = 5$, $CA = 6$. As point D on the incircle of $\triangle ABC$ varies, what is the maximum possible area of $\triangle BAD$?*

Solution 332. *By Herons, $[\triangle ABC] = \dfrac{15\sqrt{7}}{4}$. Then using $A = rs$, $r = \dfrac{\sqrt{7}}{2}$. The area is simply maximized when D is furthest from $\overline{AB}$, i.e. $d = 2r$. Hence, $\dfrac{AB}{2} = \boxed{2\sqrt{7}}$.*

Exercise 333. *Prove that two circles cannot intersect at more than two points.*

Solution 333. *Assume that the circles C_1, C_2 have different centers. Hence, it suffices to prove that three intersection points cannot co-exist. (For more than three intersection points, just pick any three to obtain an immediate contradiction). There are two cases: the intersection points i_1, i_2, i_3 are either collinear or non-collinear (i.e. form a non-degenerate triangle).*

Case 1: WLOG assume that i_2 is the middle point. Since $c_1 i_1 = c_1 i_2$, the perpendicular bisector of $\overline{i_1 i_2}$ must intersect c_1. Similarly the perpendicular bisector of $\overline{i_2 i_3}$ passes through c_1 as well. But then these lines formed from the pairs of intersection points would be parallel, which yields a contradiction.

Case 2: Since $c_1 i_1 = c_1 i_2$ as before, the perpendicular bisector of $\overline{i_1 i_2}$ must pass through c_1 (similarly, through c_2). Then the perpendicular bisector of $\overline{i_2 i_3}$ passes through c_1 and c_2. But two non-parallel lines can intersect only at one point - contradiction!

Exercise 334. *In $\triangle ABC$, point D lies on $\overline{BC}$ such that $CD = 6$ and $BD = 3$. Given that $AC = 11$ and $AB = 7$, compute AD.*

Solution 334. *Using Stewarts Theorem, let $a = 9 = m + n$ with $m = 3$ and $n = 6$, $b = 11, c = 7$. Then $man + dad = bmb + cnc \implies d = AD = \boxed{\sqrt{55}}$.*

Exercise 335. *Prove Apollonius' Theorem:*

Theorem 9.23: Apollonius' Theorem

In $\triangle ABC$, if $\overline{AD}$ is a median, then $|AB|^2 + |AC|^2 = 2(|AD|^2 + |BD|^2)$.

Solution 335. *This is actually an elementary proof using the Pythagorean Theorem. With M the midpoint of $\overline{BC}$ and H the foot of the A-altitude, we have $BM = CM = \frac{1}{2}BC$, so $BH + CH = BC$. It is clear that $AB^2 = AH^2 + BH^2$, $AC^2 = AH^2 + HC^2$, and $AM^2 = MH^2 + AH^2$. It then follows that*

$$AB^2 + AC^2 = 2AH^2 + BH^2 + CH^2 = 2\left(AM^2 + \left(\frac{BC}{2}\right)^2\right)$$ *using some manipulation.*

Exercise 336. *(1991 USAMO P1) In triangle ABC, angle A is twice angle B, angle C is obtuse, and the three side lengths a, b, c are integers. Determine, with proof, the minimum possible perimeter.*

Solution 336. *Let the bisector of $\angle A$ hit $\overline{BC}$ at point D. Note that $\triangle ADC \sim \triangle BAC$, and let $x = AD$. Then $x = \dfrac{bc}{a}$ by similarity, and furthermore, $BD = \dfrac{ac}{b+c}$ by Angle Bisector Theorem. Since $\triangle ABD$ is isosceles, $a^2 = b(b + c)$.*

Note that the problem implicitly imposes the constraint $\gcd(a, b, c) = 1$. b must then be a perfect square, namely $b = x^2$, with $b + c = y^2$, so $a = xy$. We want to minimize x, y. Law of Cosines implies $b^2 = a^2 + c^2 - 2ac\cos B$, so $\cos B = \dfrac{y}{2x}$. Since $\angle C$ is obtuse, $\angle B < 30° \implies \dfrac{y}{x} \in (\sqrt{3}, 2)$. For $x = 4$, we have the smallest solution, namely $y = 7$, so the side lengths are 28, 16, 33 and the minimal perimeter is $\boxed{77}$.

Exercise 337. *(1984 IMO P4) Let $ABCD$ be a convex quadrilateral with the line CD being tangent to the circle on diameter AB. Prove that the line AB is tangent to the circle on diameter CD if and only if the lines BC and AD are parallel.*

Solution 337. *Let M, N respectively be the midpoints of $\overline{AB}, \overline{CD}$. Then suppose that $AD \parallel BC$ with the circle centered at M tangent to $\overline{CD}$ at T. Then $\overline{MN} \parallel \overline{AD}$. Thus, $[\triangle AMN] = [\triangle DMN]$, so $\dfrac{d}{2} \cdot AM = \dfrac{MT \cdot DN}{2}$, where d is the distance from $\overline{AB}$ to N. $MA = MT$, so in turn, $DN = d$. It follows that the circle with center N and diameter $\overline{CD}$ is tangent to $\overline{AB}$.*

Let the point of tangency be U; then $\dfrac{MT \cdot DN}{2} = \dfrac{AM \cdot UN}{2}$. Thus, $[\triangle ANM] = [\triangle DMN]$, so A, D are equidistant from $\overline{MN}$. Hence, $\overline{AB} \parallel \overline{MN}$, and similarly, $\overline{CD} \parallel \overline{MN}$. It follows that $\overline{AB} \parallel \overline{CD}$.

Exercise 338. *(1989 USAMO P4) Let ABC be an acute-angled triangle whose side lengths satisfy the inequalities $AB < AC < BC$. If point I is the center of the inscribed circle of triangle ABC and point O is the center of the circumscribed circle, prove that line IO intersects segments AB and BC.*

Solution 338. *Extend lines $\overline{AO}, \overline{BO}, \overline{CO}$ to D, E, F respectively. Notice that $\overline{IO}$ passes through sides a, c iff $I \in AOF, COD$. Since $AO = BO = CO = R$, let $\alpha = \angle OAC = \angle OCA, \beta = \angle ABO = \angle BAO, \gamma = \angle BCO = \angle CBO$.*

We have $a > b > c$, or $\angle A > \angle B > \angle C \implies \gamma < \alpha < \beta$ by Law of Sines (in detail: $\gamma + \alpha < \beta + \gamma < \alpha + \beta$). As $\overline{AI}$ bisects $\angle A$, $I \in ABD$. Similarly, $I \in ACF, ABE$ as well. Thus, $I \in ACF \cap ABE = AOF$. It follows that $\overline{IO}$ intersects sides $a, c = \overline{AB}, \overline{BC}$.

Exercise 339. *(1968 IMO P1) Prove that there is one and only one triangle whose side lengths are consecutive integers, and one of whose angles is twice as large as another. Find this triangle.*

Solution 339. *In $\triangle ABC$, let $\angle ABC = \alpha, \angle BAC = 2\angle ABC = 2\alpha$. Law of Sines suggests that $\cos\alpha = \dfrac{a}{2b}$. Furthermore, applying Law of Cosines yields $\dfrac{a^2 + c^2 - b^2}{2ac} = \dfrac{a}{2b}$ through manipulation. Hence, $a^2 c = b(a^2 + c^2 - b^2)$. As $a, b, c \in \mathbb{N}$, $b \mid a^2 c$. Logically, b is either the smallest or the largest among a, b, c (else, b would be relatively prime with both a, c, but then $b \nmid a^2 c$, a contradiction).*

WLOG assume that $\min(a, b, c) = b$. Then either $b \mid (b+2)$ (impossible, as $b \neq 1, 2$ by the triangle inequality) or $b \mid (b+2)^2$. Then since $b \nmid (b+2)$, $b = 4$, which implies that $a^2 + c^2 - b^2 = 45 = \dfrac{1}{4}a^2 c$. Then $a = 6, c = 5$, so the triangle is a 4-5-6 triangle.

Alternately, assume that b is the maximum of the three integers. Thusly it would follow that $b \mid (b-2)$ or $b \mid (b-2)^2$, and we can apply similar logic as in the above paragraph to finish this case off. (As it turns out, both sub-cases produce no viable result.) Hence, the sole triangle is 4-5-6, and we are done.

Exercise 340. *Does there exist a triangle whose side lengths and angle measures are both in arithmetic progression? What about geometric progression?*

Solution 340. *Assume the existence of a triangle whose side lengths and angle measures were both in AP. Then one of the angle measures would have to be $60°$, and Law of Sines would imply that $\dfrac{2\sqrt{3}}{3}b = \dfrac{a}{\sin A} = \dfrac{c}{\sin C}$ with $A + C = 120°$. (WLOG we are assuming $a < b < c$.) In addition, $a + b < c$ by Triangle Inequality. But in combination with Law of Cosines, we can show that no such triangle exists (creating a contradiction).*

For the side lengths/angles in GP, if the smallest angle is $d°$, and the other two are $dr°$ and $(dr^2)°$ for $r > 1$, then $d + dr + dr^2 = 180$ and the Law of Sines again yields a contradiction. Hence no such triangles exist in either scenario.

Exercise 341. *(2012 Stanford Math Tournament) In quadrilateral $ABCD$, $m\angle ABD \cong m\angle BCD$ and $\angle ADB = \angle ABD + \angle BDC$. If $AB = 8$ and $AD = 5$, find BC.*

Solution 341. *$\angle ADB + \angle CBD = 180°$, and so we can extend $\overline{AD}$ beyond D to C such that $\triangle DBC = \triangle BDC$. Since $\angle ABD = \angle ACB$, it follows that $\triangle ABD \sim \triangle ACB$, so $\dfrac{AC}{AB} = \dfrac{AB}{AD} \implies AC = \dfrac{64}{5}$. Thus, from $AC = AD + DC$, we get $DC = BC = \dfrac{64}{5} - 5 = \boxed{\dfrac{39}{5}}$.*

Exercise 342. *(2006 Romanian NMO Grade 7 P1) Let ABC be a triangle and the points M and N on the sides AB respectively BC, such that $2 \cdot \dfrac{CN}{BC} = \dfrac{AM}{AB}$. Let P be a point on the line AC. Prove that the lines MN and NP are perpendicular if and only if PN is the interior angle bisector of $\angle MPC$.*

Solution 342. *Let L be on $\overline{BC}$ with N the midpoint of $\overline{LC}$. Then $LC = 2CN$ and $\dfrac{LC}{BC} = \dfrac{AM}{AB}$. By Thales Theorem, $AC \parallel ML$.*

Subsequently, let R be on $\overline{MN}$ such that N is the midpoint of $\overline{MR}$. Combined with $LN = NC$, it follows that $RLMC$ is a parallelogram, so $\overline{CR} \parallel \overline{ML}$. $\overline{ML}$ is itself parallel to $\overline{AC}$, hence A, C, R are collinear.

Finally, $\overline{MN} \perp \overline{NP} \iff \overline{NP}$ is the perpendicular bisector of $\overline{MC} \iff \overline{NP}$ bisects $\angle MPR \iff \overline{PN}$ bisects $\overline{MPC}$.

Exercise 343. *(2008 iTest) Points C and D lie on opposite sides of line $\overline{AB}$. Let M and N be the centroids of $\triangle ABC$ and $\triangle ABD$ respectively. If $AB = 841, BC = 840, AC = 41, AD = 609$, and $BD = 580$, find the sum of the numerator and denominator of the value of MN when expressed as a fraction in lowest terms.*

Solution 343. *Observe that $41^2 + 840^2 = 841^2$ (by difference of squares) and $580^2 + 609^2 = 841^2$, so $\triangle ABC$ and $\triangle ABD$ are both right. $ACBD$ is then cyclic. Let O be the midpoint of $\overline{AB}$. As M, N are centroids, we have $2MO = CM, 2DO = DN$. It follows that $\triangle COD \sim \triangle MON$, and so $3MN = CD$.*

Applying Ptolemys Theorem to $ACBD$ yields $CD = \dfrac{18460}{29}$, so $MN = \dfrac{18460}{87} \implies \boxed{18,547}$.

Exercise 344. *(1978 USAMO P4)*

(a) *Prove that if the six dihedral (i.e. angles between pairs of faces) of a given tetrahedron are congruent, then the tetrahedron is regular.*

(b) *Is a tetrahedron necessarily regular if five dihedral angles are congruent?*

Solution 344.

(a) *Let $ABCD$ be the tetrahedron with insphere tangent to the faces of $ABCD$ at W, X, Y, Z. Then $OW = OX = OY = OZ$ (as they are all perpendiculars to their respective faces), so $WX = 2OW \cdot \sin \dfrac{\angle WOX}{2}$. Then $WXYZ$ is itself a regular tetrahedron. Consider a rotation about $\overline{OW}$ by 120 degrees. Then $X \mapsto Y, Y \mapsto Z, Z \mapsto X$ (as this is a cyclic group under rotation). Thus $AB = AC = AD = AB$ and $BC = CD = DB$. Rotations about the other axes show that $ABCD$ is indeed regular.*

(b) *Shift X, Y, Z closer so that $\triangle XYZ$ remains equilateral, with all angles between each pair of perpendiculars less than $180°$. Then shift W such that $WX = WY = WZ$; then five distances are equal, but the sixth is not.*

Exercise 345. *(1994 AIME) Given a point P on a triangular piece of paper ABC, consider the creases that are formed in the paper when $A, B,$ and C are folded onto P. Let us call P a fold point of $\triangle ABC$ if these creases, which number three unless P is one of the vertices, do not intersect. Suppose that $AB = 36$, $AC = 72$, and $\angle B = 90°$. Then the area of the set of all fold points of $\triangle ABC$ can be written in the form $q\pi - r\sqrt{s}$, where $q, r,$ and s are positive integers and s is not divisible by the square of any prime. What is $q + r + s$?*

Solution 345. *Let O_{AB} be the intersection of perpendicular bisectors of $\overline{PA}$ and $\overline{PB}$, and define O_{AC}, O_{BC} analogously. Then O_{AB}, O_{AC}, O_{BC} are the circumcenters of $\triangle PAB, \triangle PBC, \triangle PCA$. By the problem statement, $\angle APB, \angle BPC, \angle CPA$ are all obtuse, and the region inside the semicircles with diameters $\overline{AB}, \overline{BC}, \overline{CA}$ is our region of consideration.*

Note that the circle whose diameter is $\overline{AC}$ completely contains the triangle, as it is also the circumcircle of $\triangle ABC$. Thus it suffices to take the circles intersection with respect to $\overline{AB}$ and $\overline{BC}$. They intersect within $\triangle ABC$, so the area of the region of consideration is the sum of two segments of the circles. These segments subtend a 120 degree arc with respect to the radius-18 circle, and a 60 degree arc with respect to the radius-$18\sqrt{3}$ circle. The requested area is then $\left(\dfrac{\pi}{3} \cdot 18^2 - \dfrac{18^2 \cdot \sin \frac{2\pi}{3}}{2} \right) + \left(\dfrac{\pi}{6} \cdot (18\sqrt{3})^2 - \dfrac{(18\sqrt{3})^2 \cdot \sin \frac{\pi}{3}}{2} \right) = 270\pi - 324\sqrt{3} \implies \boxed{597}$.

Extension Exercise 145. *(2008 Putnam B1) What is the maximum number of rational points that can be on a circle in $\mathbb{R}^2$ whose center is not a rational point? (A rational point is a point both of whose coordinates are rational numbers.)*

Extension Solution 145. *We claim that at most two such points exist. For contradiction, let $P = (a, b), Q = (c, d), R = (e, f)$ be three rational points on a circle. Let M be the midpoint of $\overline{PQ}$; then $M = \left(\dfrac{a+c}{2}, \dfrac{b+d}{2} \right) \in \mathbb{Q}^2$. The slope of $\overline{PQ}$ is again rational (or undefined, but certainly not irrational).*

Similarly, if we let N be the midpoint of $\overline{QR}$, then N is a rational point and the normal to $\overline{QR}$ through N has rational slope. The circles center lies on both lines, so its coordinates satisfy two linear equations with rational coefficients that have a unique solution. The center of the circle is then a rational point, which proves that the maximum number of rational points is indeed 2.

Extension Exercise 146. *(1973 IMO Shortlist Bulgaria 1) A tetrahedron $ABCD$ is inscribed in the sphere S. Find the locus of points P, situated in S, such that*

$$\frac{AP}{PA_1} + \frac{BP}{PB_1} + \frac{CP}{PC_1} + \frac{DP}{PD_1} = 4,$$

where A_1, B_1, C_1, D_1 are the other intersection points of AP, BP, CP, DP with S.

Extension Solution 146. *Let $S \in \mathbb{R}^3$ have center at the origin and radius r. The power of point P with respect to S is invariant, so we can multiply through by the power to get $AP^2 + BP^2 + CP^2 + DP^2 = 4(r^2 - OP^2)$. Law of Cosines allows us to rewrite this as $4OP^2 = \sum_{cyc}(AP \cdot OP \cdot \cos(\angle AOP))$.*

Write $P = (x, y, z), A = (a_1, a_2, a_3), B = (b_1, b_2, b_3), C = (c_1, c_2, c_3), D = (d_1, d_2, d_3)$. Then using the dot product, we obtain $4(x^2 + y^2 + z^2) = x(a_1 + b_1 + c_1 + d_1) + y(a_2 + b_2 + c_2 + d_2) + z(a_3 + b_3 + c_3 + d_3)$. Completing the square reveals that the form of this equation corresponds to a sphere with center at the midpoint of the segment from O to the centroid of $ABCD$, and with diameter equal to the distance between O and the centroid of $ABCD$.

Extension Exercise 147. *(2015 Putnam B4) Let T be the set of all triples (a, b, c) of positive integers for which there exist triangles with side lengths a, b, c. Express*

$$\sum_{(a,b,c)\in T} \frac{2^a}{3^b \cdot 5^c}$$

as a rational number in lowest terms.

Extension Solution 147. *(Adapted from the official Putnam solution literature.)*

Fix b, c. By Triangle Inequality,

$$S = \sum_{b,c} \frac{1}{3^b 5^c}\left(\sum_{a=|b-c|+1}^{b+c-1} 2^a\right) = \sum_{b,c} \frac{2^{b+c} - 2^{|b-c|+1}}{3^b 5^c}$$

Split this into two sums, S_1 and S_2. Then

$$S_1 = \sum_b \sum_{c=b} \frac{2^{b+c} - 2^{c-b+1}}{3^b 5^c}$$

$$= \sum_b \left(\left(\left(\frac{2}{3}\right)^b - \frac{2}{6^b}\right)\sum_{c=b}\left(\frac{2}{5}\right)^c\right) = \frac{85}{231}$$

Similarly, we find that

$$S_2 = \sum_c \sum_{b=c} \frac{2^{b+c} - 2^{c-b+1}}{3^b 5^c}$$

$$= \sum_c \left(\left(\left(\frac{2}{5}\right)^c - \frac{2}{10^c}\right)\sum_{b=c+1}\left(\frac{2}{5}\right)^b\right) = \frac{34}{77}$$

Hence $S = S_1 + S_2 = \boxed{\dfrac{17}{21}}$.

Extension Exercise 148. *Points A, B, and C lie in the first quadrant of the Cartesian coordinate plane such that A lies on the line $x + y = 6$, B lies on the x-axis, and C is the point $(3, 4)$. Compute the minimum possible value of $AB + BC + CA$.*

Extension Solution 148. *Hint: Let $A = (t, 6 - t)$, $B = (u, 0)$. We can reflect about the line $y = -x$, so that $(a, b) \mapsto (-a, -b)$. The answer is $\boxed{8}$, achieved for the degenerate triangle with $A = (3, 3), B = (3, 0)$.*

Extension Exercise 149. *(2015 IMO P4) Triangle ABC has circumcircle Ω and circumcenter O. A circle Γ with center A intersects the segment BC at points D and E, such that B, D, E, and C are all different and lie on line BC in this order. Let F and G be the points of intersection of Γ and Ω, such that A, F, B, C, and G lie on Ω in this order. Let K be the second point of intersection of the circumcircle of triangle BDF and the segment AB. Let L be the second point of intersection of the circumcircle of triangle CGE and the segment CA.*

Suppose that the lines FK and GL are different and intersect at the point X. Prove that X lies on the line AO.

Extension Solution 149. *Let Ω meet lines $\overline{FD}, \overline{GE}, \overline{FK}, \overline{GL}$ at H, I, M, N. Apply the lemma to FDH and GEI, FKM and BDC, FDH and BKA, GLN and CEB, and finally GEI and CLA. We get the following pairs of parallel lines: $(DE, IH), (KD, MC), (KD, AH), (LE, NB), (LE, AI) \implies (BC, IH), (MC, AH), (NB, AI)$. Thus $AN = IB = HC = AM$, with M and N symmetric with respect to $\overline{AO}$, the diameter of Ω passing through A. Hence $AF = AG$, and $\overline{FM}, \overline{GN}$ are also symmetric, meeting on $\overline{AO}$ at X.*

Extension Exercise 150. *(2016 HMIC P2) Let $\triangle ABC$ be an acute triangle with circumcenter O, orthocenter H, and circumcircle Ω. Let M be the midpoint of $\overline{AH}$ and N the midpoint of $\overline{BH}$. Assume the points M, N, O, H are distinct and lie on a circle ω. Prove that the circles ω and Ω are internally tangent to each other.*

Extension Solution 150. *$\triangle ABC$ has circumradius R, which is also the circumradius of $\triangle HBC$. A homothety with factor 2 centered at H implies that $\triangle HMN$ has circumradius $\dfrac{R}{2}$, as does ω. O is on ω; QED.*

Extension Exercise 151. *(1973 IMO P1) Point O lies on line g; $\overrightarrow{OP_1}, \overrightarrow{OP_2}, \cdots, \overrightarrow{OP_n}$ are unit vectors such that points $P_1, P_2, \cdots, P_n$ all lie in a plane containing g and on one side of g. Prove that if n is odd,*

$$\left| \overrightarrow{OP_1} + \overrightarrow{OP_2} + \cdots + \overrightarrow{OP_n} \right| \geq 1.$$

Here $\left| \overrightarrow{OM} \right|$ denotes the length of vector $\overrightarrow{OM}$.

Extension Solution 151. *Inducting upon n, we find that for $n = 1$ (the base case), the result is trivial. Assuming it holds for some $2n - 1 \geq 3$, we prove it for $2n + 1$. Given OP_i, order them such that all the OP_i are between OP_{2n}, OP_{2n+1}. Then $u = OP_{2n} + OP_{2n+1}$ lies along the bisector of $\angle P_{2n}OP_{2n+1}$, which forms an acute angle with $v = \sum_{i=1}^{2n-1} OP_i$ (which itself lies between OP_{2n} and OP_{2n+1}). By induction, $|v| \geq 1$ (left to the reader). By the triangle inequality of vectors, $|u + v| \geq |v|$, so the result holds for $2n + 1$, completing the proof. (Equality holds for all $|OP_i|$ equal for $1 \leq i \leq 2n$, with OP_{2n+1} arbitrary.)*

Extension Exercise 152. *(2004 USAMO P1) Let $ABCD$ be a quadrilateral circumscribed about a circle, whose interior and exterior angles are at least 60 degrees. Prove that*

$$\frac{1}{3}|AB^3 - AD^3| \leq |BC^3 - CD^3| \leq 3|AB^3 - AD^3|.$$

When does equality hold?

Extension Solution 152. *The sums of the pairs of opposite sides are equal, namely $a + c = b + d \implies |a - b| = |d - c|$. Thus $\dfrac{|d - c|\,(c^2 + cd + d^2)}{3} \leq |a - b|\,(a^2 + ab + b^2) \leq 3\,|d - c|\,(c^2 + cd + d^2)$. We divide through by $|a - b|$ (or $|d - c|$; it makes no difference which one we choose) and obtain $\dfrac{c^2 + cd + d^2}{3} \leq a^2 + ab + b^2 \leq 3(c^2 + cd + d^2)$.*

Draw diagonal $\overline{BD}$; then LoC allows us to conclude that $c^2 + d^2 + 2cd \cos A = BD$. Each of the angles of the quadrilateral (both interior and exterior) measures at least $60°$, so $c^2 + d^2 - cd \leq BD^2 \leq c^2 + d^2 + cd$ (upon setting $A = 60°, 120°$). Similarly, $a^2 + b^2 - ab \leq BD^2 \leq ab \implies a^2 + b^2 - ab \leq BD^2 \leq c^2 + d^2 + cd$

and $c^2 + d^2 - cd \leq BD^2 \leq a^2 + b^2 + ab$.

Then consider the latter of those inequalities. Since $c^2 + d^2 - cd \leq BD^2 \leq a^2 + b^2 + ab$, $a^2 + b^2 + ab$, the lower bound on the RHS is $3(a^2 + b^2 - ab) \geq a^2 + b^2 + ab$. Multiplying the first inequality by 3 yields a slightly modified form of the second inequality; by symmetry, both inequalities hold true. (Equality holds when $a = b$ and $c = d$.)

Index

Modus Ponens, 10
Modus Tollens, 10
monic, 68, 331
monomial, 36
monotonically increasing, 157
monotonicity, 33
Muirhead Inequality, 160
multinomial coefficient, 139
Multinomial Theorem, xvii, 138
multiplicity, 42
MVT, *see also* Mean Value Theorem

NAND, 3
negation, 2
Nesbitt's Inequality, 161
Newton's Inequality, 172
Newton's Sums, 172
Newton's sums, 171
Nine point circle, 304
non-constructive proof, 16
non-degenerate, 213
NOR, 3
norm, 213
normal vector, 276
not, 2
null space, 131, 190
nullity, 130
number theory, 23
numerical analysis, 41

obtuse, 203
OGF, *see also* ordinary generating function
Olympiad, vii
one-seventh triangle, 259
operations, 23
or, 2
order, 104
order relation, 27
ordered pair, 28
ordinary generating function, 144
orientation-preserving, 129
origin, 207
orthocenter, 293
orthogonal, 209
orthogonalization, 193
orthonormalization, 193
output, 29

parallel, 209
parallelepiped, 129, 275
parallelogram, 269
Parallelogram Law, 193
parallelogram law, 102
parameter, 115
partial sum, 71

Pascal, 36
Pascal's Identity, 104
Pascal's Theorem, 281
Pascal's Triangle, 36, 141, 356
perimeter, 204
period, x
periodicity, *see also* periodic
permutation, x, 155
perpendicular, 209, 217
perpendicular bisector, 273
PGF, *see also* Poisson generating function
Pick's function, 223
Pick's Theorem, 223
PID, *see also* principal ideal domain
planar graph, 276
plane, 275
point, 203
point at infinity, 298
point-to-line, 275
point-to-line distance, 209
point-to-plane, 275
Poisson generating function, 144
polar, 300
polar coordinates, 292
pole, 300
polygon, 221
polyhedron, 276
polynomial, xvii, 65
polynomial equation, 65
positive, 213
postulate, 14
power, 262
Power of a Point, 262, 300
power series, xvii, 72, 96, 142, 146
premise, 9
primitive, 65
primitive polynomial, 65
principal ideal domain, 116
prism, 275
probability theory, 159
problem solving, 23
proof, vii, 2, 8, 23
proof by contradiction, 16
proof by exhaustion, 17, 27
proposition, 2
Ptolemy, 268
Ptolemy's Inequality, 268
Ptolemy's Theorem, 268

QM, 55
QM-AM-GM-HM, 55
QR decomposition, 194
quadrant, 207
quadratic, 24, 55

Made in the USA
Columbia, SC
08 January 2020